INDUSTRIAL WASTE: ENVIRONMENTAL IMPACT, DISPOSAL AND TREATMENT

INDUSTRIAL WASTE: ENVIRONMENTAL IMPACT, DISPOSAL AND TREATMENT

JOHN P. SAMUELSON
EDITOR

Nova Science Publishers, Inc.
New York

For permission to use material from this book please contact us:
Telephone 631-231-7269; Fax 631-231-8175
Web Site: http://www.novapublishers.com

LIBRARY OF CONGRESS CATALOGING-IN-PUBLICATION DATA
Available upon request.

ISBN 978-1-60692-720-5

Published by Nova Science Publishers, Inc. ✢ New York

CONTENTS

PREFACE

Industrial waste is a type of waste produced by industrial activity, such as that of factories, mills and mines. It has existed since the outset of the industrial revolution. Much industrial waste is neither hazardous nor toxic, such as waste fiber produced by agriculture and logging. Toxic waste and chemical waste are two designations of industrial waste. Sewage treatment can be used to clean water tainted with industrial waste. This volume presents new research in this growing field.

As explained in Chapter 1, hexavalent chromium [Cr(VI)], an environmental toxin produced as a by-product of industrial chromium processing, is a known human carcinogen which may induce lung cancer through inhalation. Cr(VI) found in certain consumer products may induce allergic contact dermatitis. Also, Cr(VI) may possibly contribute to carcinogenesis when ingested. Upon exposure Cr(VI) enters cells and is metabolized into trivalent chromium Cr(III). This process generates harmful free radicals and the reactive intermediate chromium species Cr(V) and Cr(IV). These unstable forms of chromium interact with cellular macromolecules and damage DNA and proteins. Many different types of genetic lesions can be induced by Cr(VI) exposure, which may be mediated directly through Cr-DNA interaction or indirectly from oxidative stress. These genetic lesions may induce mutations, prevent DNA replication, and/or create chromosomal abnormalities. In response to Cr(VI)-induced DNA damage, cells respond by activating cell cycle checkpoints to arrest cell growth. Further, higher concentrations of Cr(VI) can directly induce programmed cell death (apoptosis). The cellular mechanisms which detect DNA damage and activate the response pathways determine cellular reaction to Cr(VI) exposure. Much is known about proteins involved in activating cell cycle checkpoints and DNA repair processes. However, many questions remain regarding the exact means by which Cr(VI) is able to induce carcinogenesis. Further research of thesc signaling pathways will develop our understanding of the cellular response to Cr(VI) which may be applicable to other toxic metals and potentially lead to the development of treatment and prevention strategies.

As presented in Chapter 2, adsorption, known for over a century, has remained to the present day an efficient technique for purifying water. For textile effluents containing the unfixed dyes in fibres, adsorption has shown good decolourization results. Activated carbon is a widely-used adsorbent in commercial systems due to its extended surface area, microporous structure, great adsorption capacity and high degree of surface reactivity. Nevertheless, the high prices associated with activated carbon replacement and regeneration are considerable. Several research studies have been carried out in order to find low-cost materials as possible alternatives and, in particular, industrial wastes have been tested for this purpose. The generation of large amounts of wastes by industrialized countries requires significant costs for their disposal or treatment. Valorisation by recycling or reuse is an environmentally sound option. Residues and by-products from water and wastewater treatment plants, steel-making process, power generation, metal plating, timber and tanning industries, agricultural, and food and drink companies are examples of wastes tested as adsorbents. Carbonaceous waste materials have been especially studied as precursors for the preparation of activated carbons or coal-based adsorbents. Studies reported in the literature comprise the waste characterization, possible treatments to enhance the material adsorption properties, determination of equilibrium adsorption isotherms and kinetics for different operating conditions, and experiments in packed bed columns. Attempts to identify the adsorption underlying mechanism have also been reported. Assessment of metals' leachability is also very important for some industrial wastes. In spite of the numerous papers published, the true challenge of finding good low-cost adsorbents for anionic dyes—especially for the reactive ones—using low adsorbent dosages and minimizing waste problems, has still not been met. According to the authors' knowledge, a considerable lack of information also exists in the application of these wastes to real effluents, since some studies have shown a substantial decrease in the adsorbent performance due to the presence of other species in solution (salts, other dyes, and dyeing chemicals). Materials processed from chitosan and carbonaceous wastes were identified as promising adsorbents for anionic dyes. Metal hydroxide sludges, fly ashes and leather wastes also proved to be good materials for dye adsorption.

Developing countries suffer from water pollution, the high cost of contents and treatments make the major problems in these countries. Search for a cheaper alternative, reveals that utilizing natural processes and materials significantly reduce high cost. Adsorption as one of the effective technique widely used for treatment of industrial wastewater from organic and inorganic pollutants meet the great attention from researchers. In solving the problem of industrial solid waste disposal, adsorption technique using the treated solid waste also, meet the attention of researchers. Natural adsorbents such as ores, rocks, plant straw, soil, carbonaceous material developed from fertilizer waste slurry, activated carbons from date pits and the dried aquatic plants are of the effective adsorbent materials and environmentally sound low cost materials for treatment of industrial wastewater from organic pollutants and heavy metals. Also, ores (kaolin, talc, barite and chalcopyrite) and solid wastes from food industries and mining can be used successfully as low-cost materials for treatment of polluted water. Chapter 3 presents recent researches on:-1-fundamental of adsorption. 2-recycle and reuse of industrial solid wastes for industrial wastewater treatment. 3- Naturally materials as adsorbents for the removal of pollutants from industrial wastewater.4- Case studies using adsorption technique

Agro-industry generates a wide number of waste products. As their disposal and landfill present serious environmental concerns, research studies are performed proposing value-adding technologies. Chapter 4 describes a number of biotechnological processes aimed at treatment and valorisation of wastes with high organic load derived from the industrial production of olive oil, sugar, fruit juices, etc. Employing solid-state fermentation technology, microorganisms are able to degrade solid wastes simultaneously, producing metabolites that can be further used. A special emphasis is given to the recent achievements leading to formulation of bio-control products, production of some important microbial metabolites, and biological organic-matter–based fertilizers. Concomitant solubilization of low-grade rock phosphates (introduced into solid waste microorganism systems), which are considered waste materials from the mining industry, is also discussed. The review deals with the possibility of creating a sustainable practice in industries processing agricultural products.

With ever increasing focus on industrial pollution and environment, traditional industrial design framework needs to be extended from green process design, to industrial ecology, to sustainability. The concept of overall sustainability goes beyond industrial ecology and brings in time dependent nature of the ecosystem. Decisions regarding regulations and human interactions with ecosystem come in picture. It involves dealing with various time scales and time dependent uncertainties. The available management options and decision making change from the single industrial production plant to the larger scales of an integrated industrial park, community, and natural systems. Further, green engineering concepts change the goal of engineering design from profitability to include a number of different and conflicting objectives that can define sustainability in the end. This integration poses challenging problem of discrete and continuous decisions, nonlinearities and uncertainties. Systems theory approach deals with similar ideas of disparate temporal and spatial systems, their interactions and interdependencies. This prompts one to extend systems theory approach for decision making in sustainability. This chapter presents such an idea through a case-study problem of mercury waste management in a watershed. The harmful effects of mercury pollution are well documented. Mercury in water particularly presents a grave danger because of its bioaccumulative nature. It is further complicated by the complex mercury cycling in the environment. As a result, successful control of mercury pollution in water depends on actions at various points in the mercury cycle, creating an appropriate case study for the sustainability initiative. Chapter 5 looks at three levels of pollution control: individual industry level, industrial sector level and ecosystem level. At the individual industry level, the goal is to optimize mercury treatment technology selection and design. At the industrial sector level, the option of pollutant trading to balance ecological and economic objectives is analyzed, and an optimization based decision making framework is proposed. Once mercury enters the water bodies, controlling the bioaccumulation of mercury can further reduce the harmful effects and affect the regulatory decisions. This work investigate the idea of lake pH control through time dependent liming, constituting work at the ecosystem level. Uncertainties increase as we move from the individual industry level, where model and tools are well developed, to ecosystem level. Although the application is specific to mercury pollution, the goal is to understand the generalized integrated approach for sustainable waste management from industrial perspective. To achieve these goals, appropriate modeling of the objectives, risks and uncertainties is necessary, followed by the use of appropriate tools to optimize various decision. This work takes a systems theory approach, employing techniques such as optimization, optimal control and real options theory, to come up with efficient management strategies.

Traditional systems of analysis are no longer adequate, owing to the increasing demand for real-time and in situ control of pollution (industrial wastes, heavy metals, quality of water, etc.). The appearance of electrochemical biosensors is probably one of the most promising ways to solve some problems concerning sensitive, fast and cheap analytical techniques. In Chapter 6, we focus on conductometric biosensors because of the advantages they offer over other types of transducers: they can be produced through inexpensive thin-film standard technology, no reference electrode is needed, and differential mode measurements allow cancellation of a lot of interferences. In the first part, the principles of conductometric measurements in ionic media are presented. These types of measurements are first applied to the monitoring of biocatalytic reactions. The use of conductometric microtransducers is then presented and detailed in the case of pollutant detection for environmental monitoring. The performances obtained for the detection of different pesticides, herbicides and heavy metal ions, based on enzyme inhibition, are presented as well as those obtained for the detection of formaldehyde, 4-chlorophenol, nitrate and proteins as markers of dissolved organic carbon based on enzymatic microbiosensors. Finally, whole-cell conductometric sensors using microalgea and bacteria as bioreceptors for heavy metals, selenite and pesticides detection are described.

Multilayer perceptron (MLP) and support vector machine (SVM), as two popular learning machines, are used increasingly as alternatives to classical statistical models for the ground-level ozone prediction. However, without the sufficient awareness about their limitations employing learning machines can still lead to unsatisfactory results on modeling ozone evolving mechanism, especially during ozone episodes. With some spirit of review, Chapter 7 discusses, in the ozone prediction context, the recently developed algorithms/technologies on treating the most prominent model-performance-degraded limitations which include, for MLP, the "black-box" property, i.e., hardly proving physical explanation for the trained model, overfitting and local minima problems, and for SVM, parameters identification and class imbalance problems. This commentary will impress the "would-be" environmental modelers that the underlying philosophy of using learning machines is by no means as trivial as simply fitting models to the data, because it has reflected on the difficulties, controversies or unresolved problems; meanwhile this commentary also serves as a reference point for further technical reading for experts in this field.

Chapter 8 had the fundamental objective of assessing product stewardship initiatives designed by a plethora of chemical industries in Germany. The paper justifies the need for product stewardship in Germany's chemical industry's and presents the results of a survey of some chemical industries in Germany in June 2003. A total of 122 questionnaires were administered to environmental managers in the selected industries using systematic random sampling. The return rate was 51% corresponding to 66 industries. Survey results revealed that despite the inadequacy in the implementation of some product stewardship tools like emission reduction (particularly solid waste emission), implementation of product stewardship tools by the sampled industries was advanced. It was concluded that most of the attempts made thus far by industries to address product stewardship have been major successes. However, it was emphasized that further improvement could be achieved if incentives for product stewardship implementation are promoted and the merits of strategic research in the area of reduction in solid waste emission is further developed and encouraged.

Approximately one billion tones of fish and shellfish are produced worldwide every year, but only 50–60% is used for human consumption. Fish processing wastes are defined as fish processing by-products, such as viscera, heads, bones and skins, that are not utilized in fishery products. These by-products create multiple environmental problems. Recently, great efforts have been made to utilize fish processing wastes by producing high-value compounds such as pharmaceuticals, nutraceuticals and cosmetics, which has resulted in isolation of some biologically active compounds such as functional enzymes, proteins, lipids, chitins and minerals. Bioactive substances isolated from fish processing by-products are commercially important and are used for development in biotechnological applications. In Chapter 9 we discuss recent progress that has focused on producing bioactive compounds from fish processing wastes and also review their biotechnological applications.

Advanced oxidation with Fenton's reagent is a promising technology for removal of toxic and non biodegradable organic compounds present in wastewaters produced in semiconductor and wood production industries.

The objective of Chapter 10 was to optimize COD removal from the industrial wastewaters, by applying advanced oxidation processes with Fenton's reagent.

In this study, the effect of three variables (reaction time, initial hydrogen peroxide and ferrous ion concentration) on COD removal from two different industrial wastewaters was evaluated.

For this purpose, an experimental design with two levels and three variables was used leading to the conclusion that reaction time is the only variable that has a statistical meaning for semiconductor industry wastewater treatment. The best %COD removal that could be attained for Fenton oxidation of this wastewater (with an initial COD of 21-32 g O_2/L) was 75%, when reaction started at pH = 3, adding hydrogen peroxide at 4.2M concentration and ferrous ions at a $FeSO_4$: H_2O_2 mass ratio of 1:5 for a 4h reaction time.

When optimizing the treatment of wastewater from wood production (with an initial COD of 25 g O_2/L) with Fenton's reagent, also an experimental design with two levels and three variables (reaction time, initial hydrogen peroxide and ferrous ion concentration) was used. For this wastewater, all of the three variables have shown statistical meaning, as well as the interaction between each other. The conditions that conducted to the best results (%COD removal = 82%) were: initial pH = 3, [$FeSO_4$] = 50 mM, [H_2O_2] = 4.2 M and reaction time = 60 min.

The coagulation process is essential in a number of disciplines including biochemistry and most notably in water and wastewater treatment. The treatment of water using coagulation agents has been noted from ancient times, as early as 2000 BC in ancient Egypt. In modern water and wastewater treatment, coagulation process is an integrated step of the overall suite of treatment process. The continuous need for a thorough understanding of coagulation – flocculation phenomena is as important today as it has been in the past, taking into account the water scarcity as well as the fact that the requirements for water and wastewater quality has become increasingly strict. The aim of Chapter 11 is to provide a general description of coagulation process, focusing on the theoretical aspects and mechanisms via which coagulation proceeds. Moreover, the most commonly coagulation reagents, including metal coagulants and polymers, are presented, while the recent research trend and latest developments regarding the evolution of coagulation reagents are also discussed. Finally, the significance of coagulation – flocculation process in the area of water and wastewater treatment is reviewed and evaluated, emphasizing on the series of applications employed, including destabilization of colloids, removal of inorganic and organic matter (particulate and/or dissolved), removal of metals and anions (arsenic, phosphate etc) as well as removal of pathogen microorganisms.

Red mud and fly ash are important solid wastes from alumina refinery industry and coal-firing power plants. Disposal of red mud and fly ash has been a significant issue in environmental protection. Meanwhile, chitin and chitosan, as the largest wastes from food processing industry, also produce a problem in their disposal. Recycling and utilisation of these wastes will help to reduce solid wastes and save resources. In Chapter 12, red mud and fly ash have been tested for soil amendments for humic acid adsorption,. Their characteristics were analysed using N_2 adsorption, XRD, SEM and FT-IR. It is found that red mud exhibits higher adsorption capacity than fly ash due to higher surface area and positive surface charge. The adsorption capacities of red mud and fly ash are 31.0 and 16.6 mg/g, respectively at 30 °C and pH 7. However, treatment of red mud using physical and chemical methods including thermal treatment, acid or base treatment, and H_2O_2 oxidation will decrease the adsorption capacity. For red mud, the presence of NaCl in solution will produce a competitive adsorption and thus reducing humic acid adsorption. Humic acid adsorption on red mud is also decreased with increasing solution pH.

For chitin and chitosan, they were also investigated for humic acid adsorption in aqueous solution. It is found that chitosan exhibits much higher adsorption capacity than chitin. Humic acid adsorption on chitin and chitosan follows the parallel first-order kinetics and the adsorption isotherm can be fitted by the Langmuir and Freundlich isotherms. Further investigations also indicate that humic acid adsorption on chitin was influenced by solution pH and ionic strength. Higher pH will reduce HA adsorption on chitin. The presence of cations and anions in solution will increase HA adsorption. For cation, Mg^{2+} will produce higher adsorption than Na^+ and K^+ while for anions, Cl^- will have stronger effect than NO_3^-. Isothermal calculation indicates that the maximum adsorption of HA on chitin and chitosan is 28 and 202 mg/g, respectively, at pH 7, 30 °C. These investigations indicate that red mud, fly ash, chitin and chitosan can be used as good materials for water and soil remediation to retain humic acid.

Industrialization has contributed to the increase in the generation of potentially toxic solid residues. Textile and tannery industries produce residues, called sludge, with high organic and inorganic content, including heavy metals. The most widely-used method for disposal of tannery, textile and sewage sludge is sanitary sites, presenting a high risk for environmental contamination. Concerns about environmental quality have led to the introduction of alternative disposal methods, such as the use of sludge as a nutrient source for plants and as soil conditioners. However, the use of sludge in agricultural soils needs defined action, in order to avoid damage to the environment, mainly to the soil. The knowledge of short- and long-term effects of soil amendments becomes essential when considering, for example, the conservation of soil quality. In recent years, soil microbiological parameters, as such microbial activity and biomass, along with soil enzyme activity, have been seen to be early and sensitive indicators of soil stress and can be used to predict long-term trends in soil quality. Several studies have been done aiming to evaluate the effects of use of these sludges in agricultural soil. In this way, Chapter 13 describes the state of the art about the effect of textile, tannery and sewage sludge on soil microorganisms and soil enzyme activity.

In the first Short Communication, biosorption of Cr(III), Cu(II) and Pb(II) from binary metal solutions onto *Gelidium* algae, algal waste and a composite material was investigated in batch systems, at pH 4 for Cu(II)/Cr(III) and pH 5.3 for Cu(II)/Pb(II). Three biosorbents have been tested, namely, algae *Gelidium*, algal waste from agar extraction industry and a composite material. The maximum uptake capacity followed the order: Pb > Cu > Cr. The interference of lead on copper uptake is stronger than on chromium. The experimental equilibrium data were well described by a discrete and a continuous equilibrium model. Three dimensional sorption surfaces were generated, describing the metal uptake as a function of the equilibrium concentration of both metals. From the discrete model, it has been possible to obtain the Langmuir affinity constant. The presence of co-cations decreases the apparent affinity of copper ions for the biosorbent binding sites.

Phosphorus is a vital nutrient for all life forms, but is a finite non-renewable element with a progressively declining availability and non-sustainable exploitation rate. The second Short Communication deals with novel trends in exploring alternative phosphate-bearing sources and corresponding treatment techniques aimed at substituting for the existing resources and environmentally damaging chemical processing. Approaches are presented that include biological treatment of low-grade (waste) phosphate rocks deriving from the mining industry, as well as biosolubilization of industrial ashes of animal bone meal. The novel biotechnology-based processing may employ a wide range of agro-industrial residues as substrates in fermentation operations.

In: Industrial Waste...
Editor: John P. Samuelson, pp. 1-40
ISBN: 978-1-60692-720-5

Chapter 1

MOLECULAR MECHANISMS OF CHROMIUM-INDUCED CARCINOGENESIS

***Evan Comeaux*[1,*] *and Bo Xu*[2]**

[1]Department of Pharmacology & Toxicology, University of Alabama at Birmingham,
[2]Department of Biochemistry and Molecular Biology, Southern Research Institute, Birmingham AL, 35205.

Abstract

Hexavalent chromium [Cr(VI)], an environmental toxin produced as a by-product of industrial chromium processing, is a known human carcinogen which may induce lung cancer through inhalation. Cr(VI) found in certain consumer products may induce allergic contact dermatitis. Also, Cr(VI) may possibly contribute to carcinogenesis when ingested. Upon exposure Cr(VI) enters cells and is metabolized into trivalent chromium Cr(III). This process generates harmful free radicals and the reactive intermediate chromium species Cr(V) and Cr(IV). These unstable forms of chromium interact with cellular macromolecules and damage DNA and proteins. Many different types of genetic lesions can be induced by Cr(VI) exposure, which may be mediated directly through Cr-DNA interaction or indirectly from oxidative stress. These genetic lesions may induce mutations, prevent DNA replication, and/or create chromosomal abnormalities. In response to Cr(VI)-induced DNA damage, cells respond by activating cell cycle checkpoints to arrest cell growth. Further, higher concentrations of Cr(VI) can directly induce programmed cell death (apoptosis). The cellular mechanisms which detect DNA damage and activate the response pathways determine cellular reaction to Cr(VI) exposure. Much is known about proteins involved in activating cell cycle checkpoints and DNA repair processes. However, many questions remain regarding the exact means by which Cr(VI) is able to induce carcinogenesis. Further research of these signaling pathways will develop our understanding of the cellular response to Cr(VI) which may be applicable to other toxic metals and potentially lead to the development of treatment and prevention strategies.

Key words: chromium, carcinogenesis, contact dermatitis, free radicals, cell cycle checkpoints, apoptosis

---

[*] E-mail address: xu@sri.org. Phone: 205-581-2845. Fax: 205-581-2097. Corresponding Author: Bo Xu, MD, PhD, Southern Research Institute, Laboratory of Molecular Radiobiology, 2000 9th Avenue South, Birmingham, AL 35205.

Introduction

Elemental chromium exists as one of two stable forms; trivalent [Cr(III)] and hexavalent [Cr(VI)]. Cr(III) is what most people refer to when mentioning chromium. Cr(III) is not toxic since it is the most stable form of chromium and it may be taken as a nutritional supplement. Cr(VI) is less stable than Cr(III) and is toxic when exposed to cells. Cr(VI) is a known human carcinogen, which has been well documented in studies by many independent researchers (1-6) and government agencies (7-10). This paper reviews primary literature regarding the effects of Cr(VI) exposure in living organisms. We first summarize the findings of epidemiological studies which investigate the effects observed in humans exposed to Cr(VI). Next, we review studies of Cr(VI) metabolism and its effects on cellular processes and functions. Finally, we look at specific macromolecules and signaling pathways within the cell which regulate the cellular response to Cr(VI) exposure and their role in Cr(VI)-induced carcinogenesis.

Part 1. Epidemiology

Hexavalent chromium is an environmental toxin which is mainly produced as industrial waste. Though many people are exposed to airborne Cr(VI) in the environment, the highest rates of exposure occur mainly in occupations which utilize industrial scale chromium processing procedures. Epidemiological studies point to chrome plating and pigmentation, leather tanning, and production of stainless steel as likely sources for workers occupationally exposed to chromium (11).

Epidemiology serves as a valuable tool in quantifying the risks posed by Cr(VI). The human data obtained from epidemiological studies provides a useful starting point that overcomes the limitations of data obtained in animal studies which must be extrapolated into expected effects in humans. In addition, epidemiological data includes exposure conditions which may be more applicable to human populations at risk than conditions simulated in animal experiments (12)

Many epidemiological studies which established quantitative risk assessments on exposure to chromium and lung cancer obtained different risk estimates due to differences in the data sets used and methods employed (13-17). This makes comparison and evaluation of the risk assessments difficult. The US EPA determined an excess lifetime risk (ELR) of 11.6×10^{-3} per unit of cumulative exposure to CrVI (1 $\mu g/m^3 \cdot$year) for the general population for continuous lifetime exposure based on the study by Mancuso (18). A recalculation for those not continuously exposed produced an ELR of 1.3×10^{-3}. Sorahan et al. estimated an ELR range of 0.3 and 8.1×10^{-3} by applying a linear relative risk model. The Scientific Committee on Occupational Exposure Limits of the European Union (15) calculated an even lower ELR range of $0.1\text{-}0.6\times10^{-3}$ by using a meta-analysis of 10 separate cohort studies (19). This risk assessment however was based on average individual exposure from 15 years since no data on the duration of exposure was available. The World Health Organization (20) combined several older studies (21-23) to determine conditional ELRs from 1.3×10^{-1} and 4.3×10^{-3} using a linear relative risk model for continuous lifetime exposure in the general population.

More recent studies of existing cohorts provide very useful information for quantitative risk estimation (24-26). Mancuso's study utilized one industrial hygiene survey involving high levels of Cr(VI) exposure, however the confounding variable of smoking history for the cohort was not provided. The largest study was performed by Gibb et al. using several industrial hygiene surveys along with smoking information for 93% of the cohort. In this study however the levels of exposure and contrast between groups was relatively low. Crump's study involved a much smaller study group, but with extensive information on the high levels of exposure and smoking status for part of the cohort. The risk estimation based on these three studies using a linear relative risk model produced point estimates within a small range of $3–16x10^{-3}$ excess lung cancer mortalities continuing to age 89, with the Gibb study demonstrating the highest estimates.

The EPA used data from Mancuso et al. (16) to perform a risk assessment for Cr(VI). Though the meta-analysis by Steenland et al. (19) did not provide any quantitative data of exposure, it was used by the EU Scientific Committee for Occupational Exposure Limits in risk assessment after assuming exposure levels (15). Sorahan et al. (13) also did not provide quantitative measurement of chromium exposure, but still performed quantitative risk assessment by estimating exposure level ranges. The following sections review epidemiological studies regarding the three main routes of Cr(VI) exposure.

I. Inhalation

Hexavalent chromium inhalation is the route of exposure that produces the most severe toxicity since it is able to directly interact with lung cells and enter the bloodstream. Inhaled Cr(VI) is known to cause lung cancer in humans, especially in those who are exposed to high concentrations occupationally (3). Some industrial processes produce hexavalent chromium as a waste product which may be released into the atmosphere and be exposed to entire populations. The environmental concentrations of Cr(VI) are many orders of magnitude lower than those experienced by workers in the chrome industry. For this reason, most epidemiology studies focus on populations of workers who are at the highest risk for Cr(VI)-induced lung cancer.

Hexavalent chromium continues to be a very common environmental pollutant. Millions of people are exposed to Cr(VI) regularly, however the concentration is so small that it has relatively minimal effect on Cr(VI)-induced carcinogenesis. Cigarette smoke and automobile exhaust are the major sources of non-occupational Cr inhalation. Environmental contamination occurs from landfills, industrial chromate production and use, highway construction, hazardous waste disposal sites (7), and from airborne particulate chromate (27). The atmospheric Cr concentration in the US has been estimated to range from 0.2-0.9 ng/m^3 (7). Concentrations 10-100-fold higher have been detected in urban areas, especially near ferrochrome industries. The Cr particles inhaled are not evenly spread throughout the lung tissue, but tend to accumulate at the major bifurcations (28). More detailed study of the public health impact of environmental Cr(VI) exposure is necessary to determine the populations at the highest health risk.

Workers occupationally exposed to Cr experience chronic inhalation of small particles which are able to accumulate to relatively high concentration in the lungs (28). The particles have been shown to accumulate at bronchial bifurcations and are persistent even after years of no exposure (29). The local concentration in these "hot spots" can exceed 50 mM. The amount of occupational Cr exposure via inhalation varies with industry but can reach over 100 $\mu g/cm^3$. Industrial hygiene and worker regulations have been utilized to limit the amount of Cr exposure in the past few decades.

Several carcinogenesis experiments were unable to induce tumors via Cr(VI) inhalation in animal models which has been accredited to Cr(VI) being reduced to Cr(III) outside of the cells. However it is known that Cr(VI) inhalation can induce respiratory tract carcinogenesis when exposure occurs over long periods of time, which is thought to be due to accumulation of relatively insoluble particles of Cr on the airway epithelia which greatly increases the local Cr(VI) concentration at the cell surface.

The OSHA Permissible Exposure Limit (PEL) of Cr(VI) as CrO_3 occupationally is 5 $\mu g/m^3$ as of February 2006. The previous limit was suggested to be too high since many incidences of lung cancer in workers exposed to 100 $\mu g/m^3$ have been reported (24, 30). The NIOSH Recommended Exposure Limit (REL) is even more stringent at 1 $\mu g/m^3$.

Epidemiological studies have determined the excess lifetime risk of dying of lung cancer when exposed to the previous OSHA PEL of 100 $\mu g/m^3$ to be 255 per 1000 (95% CI 109–416) (31). This data was obtained from study of 2357 chromate production workers. Extrapolation of this data estimates 255,000 lung cancer deaths per million which far exceeds the *de minimus* level of 1 death per million. This suggests that Cr(VI) is much more carcinogenic that previously thought. The US EPA has already set the 1 in 1 million cancer risk estimate for human lifetime exposure to Cr(VI) at 80 pg/m^3 (32) using data from occupational lung cancer incidence. Exposure to Cr(VI) in urban areas has been reported over 1000 times this exposure standard. However, the amount of Cr(VI) exposure in the general environment is still very small and many orders of magnitude less that the OSHA PEL in used for occupational exposure.

The largest epidemiological study on Cr(VI) inhalation by Gibb et al. (24) demonstrated the dose-dependent response between Cr exposure and lung cancer incidence. This study included controls for smoking which illustrated the synergy between Cr(VI) and other carcinogens in amplifying the risk for lung cancer. Cr(VI) exposure still showed a statistically significant effect on lung cancer rate despite the smoking covariate inclusion.

Since inhaled Cr(VI) is known persist in the lungs and lead to significantly increased lung cancer incidence and mortality, much research is being focused on epidemiology to assist in determining safe exposure limits. Refining the chromium processing procedures, enforcing strict worker hygiene, and setting lower PEL has led to significant reduction in occupational Cr-induced lung cancer incidence. However, the environmental airborne Cr(VI) concentration often exceeds the EPA limit, especially in industrialized urban areas. This continues to pose a health risk to many human populations and may be added to the growing list of airborne pollutants which can increase risk of lung cancer.

II. Ingestion

Though Cr(VI) is a known carcinogen when inhaled there is vigorous debate over its carcinogenic potential when ingested. Despite large populations having detectable levels of Cr(VI) in the drinking water, with California having detected Cr(VI) in almost 2000 drinking water sources (33), there is still little evidence of Cr(VI) carcinogenicity when exposed through the oral route (34). Sources of Cr(VI) groundwater contamination include natural deposits of chromium, industrial discharge from chromium processing and plating facilities, and Cr(VI) application to prevent corrosion in cooling water (35).

There is a wide range of beliefs concerning the potential risks of Cr(VI) exposure in drinking water. Several investigators have claimed that Cr(VI) will be completely reduced by biological fluids even if the chromium concentration exceeds the US EPA standard of 50 ppb (36-38). However, Cr(VI) is absorbed into the blood when ingested with pure drinking water which points out that not all Cr(VI) is reduced by saliva, gastric juices, or the intestinal flora (36,43).

Three separate studies have been published on the cancer rates in human populations drinking Cr(VI)-contaminated drinking water. One study published in 2001 reported no increase in total cancer incidence in separate regions of California where drinking water contained Cr(VI) released from natural gas compression facilities (35). Another study published in 1991 reported no correlation between the concentrations of Cr(VI) in the water and total cancer mortality rates in 453 communities in Nebraska (39). A third study published in 1987 reported increased cancer mortality rates in areas with Cr(VI)-contaminated water in Liaoning Province, China (40). Cr(VI) concentrations in Liaoning Province ranged up to 20 mg/L which is among the highest Cr(VI) concentration observed in drinking water. Though this study reported rates of cancer mortality, it failed to report the statistics of association and precision.

Further statistical analysis of the data obtained from the population of Liaoning Province was performed to see if such very high Cr(VI) concentrations had any effect on cancer incidence and mortality (41). Investigation of cancer mortality rates in Chinese provinces with hexavalent chromium contaminated drinking water and uncontaminated drinking water reported higher incidence of cancer in those provinces with Cr(VI)-contaminated drinking water, particularly stomach cancer. Regions with contaminated water showed markedly increased stomach cancer mortality compared to those provinces without contaminated water (rate ratio = 1.82; 95% CI = 1.11–2.91), as well as compared to the whole province (1.69; 1.12-2.44). However, the mortality rate from other types of cancer combined was not increased when compared with unexposed regions (0.86; 0.53-1.36) or the whole province (0.92; 0.58-1.38).

Three sources of Cr(VI) were located within Liaoning Province: Cr(VI)-contaminated wastewater released from a factory that flowed into a dry riverbed, equipment leaking Cr(VI)-contaminated waste onto the ground at the worksite, and the chromium ore stored on open ground all led to water soluble Cr(VI) percolating into the groundwater (41, 42). Measures implemented in 1965 to regulate pollution, such as halting production, improving the production process, repair of leaking equipment, waterproofing of the workshop to prevent leaking, and adding ferrous sulfate to the chromium ore residue and wastewater to reduce Cr(VI) to Cr(III), led to a sharp decline in Cr(VI) concentrations in the groundwater after 1967 (40-42).

Experiments in animals suggest that Cr(VI) is not distributed throughout the body in the same manner as Cr(III) since animals ingesting Cr(VI) have higher levels of chromium retained in their tissues compared to those animals fed Cr(III) (43, 44). Two lifetime animal studies have been conducted on the carcinogenicity of Cr(VI) when ingested. The first study published in 1968 demonstrated increased formation of benign and malignant neoplasms in the stomach of NMRI mice after exposure to potassium chromate in the drinking water. This study examined 3 generations of NMRI mice, but has been criticized for including mortality caused by viral exposure in the first generation (45, 46). The U.S. National Toxicology Program finished a 2-year study in 2007 with F344/N rats and B6C3F1 mice fed sodium dichromate dehydrate-contaminated water (47). Rats developed squamous cell neoplasms of the oral cavity while the mice developed neoplasms of the small intestine, thus demonstrating carcinogenic activity of Cr(VI) after exposure through the oral route in mice.

A recent study reports an increase in UVR-induced skin cancer in Skh1 hairless mice after chromium ingestion in a gender-independent manner (48). Using hairless mice to study the effect of Cr(VI) ingestion on UVR-induced skin cancer confirmed previous results (49) that Cr(VI) ingestion can enhance skin cancer formation in a dose-dependent manner. Co-treatment with vitamin E and selenomethionine did not reduce the increase in skin cancer formation after Cr(VI) ingestion, suggesting that the enhancement of UVR-induced skin cancer by Cr(VI) is not mediated by oxidative stress. The mice drinking chromium-contaminated water also contained a significantly higher skin concentration of chromium. The chromium concentration in the skin correlated with the incidence and yield of UVR-induced cancer. These data point to a potential co-carcinogenic threat posed by chromium exposure when combined with UVR exposure. Though ingestion of Cr(VI) did not induce cancer by itself, the potassium chromate served to potentiate UVR carcinogenesis, implicating a role of environmental or occupational chromium exposure in carcinogenesis in humans regularly exposed to UVR from sunlight. Since the chromium itself did not cause cancer, but only enhanced UVR carcinogenesis it is hypothesized that the chromium enhances an essential aspect of UVR carcinogenesis in the epidermal cells.

While it appears that only extremely high concentrations of Cr(VI) may be able to directly induce cancer upon ingestion, there is evidence suggesting that Cr(VI) is metabolized and distributed by the body in a different manner than Cr(III) and may accumulate in the skin and act to enhance susceptibility to UVR-induced squamous cell carcinoma formation in mice. Thus, the debate of the heath implications of ingested Cr(VI) is still ongoing and there is evidence supporting both sides. Whether or not animal studies accurately reflect the results of human exposure is a major issue in many of these studies. Meanwhile, many sources of drinking water contain Cr(VI), whether naturally-occurring or as industrial waste.

III. Skin Contact

Cr(VI) is a very common occupational dermal sensitizer which results in almost 5% of allergic dermatitis cases reported clinically in the US (50). Cr(VI)-induced allergic contact dermatitis can cause erythema, oedema, and small vesicles (50, 51). This is typical of type IV cell-mediated allergic reactions (50-52) which are generally limited to the area of skin exposed. The ability of chromium salts to penetrate the skin is proportional to their water solubility. Potassium dichromate has high water solubility and is therefore better able to pass through the skin than most Cr(III) compounds that are rather insoluble.

Occupational exposures and reported cases of Cr(VI)-induced contact dermatitis occur most often in chromium plating workers (52, 53), diesel repair shop workers (54), and leather workers (55). Additional exposure from household products that contain chromium, such as detergents, cosmetics, and shoe polish can also lead to contact dermatitis (56). Cement workers are often exposed to Cr(VI) and cement containing chromium is a major cause of occupational chromium dermatitis. 17% of the grouters constructing the Channel Tunnel for England tested positive in a patch test for potassium dichromate ($K_2Cr_2O_7$) (57). Denmark has added ferrous sulfate to cement to reduce water-soluble Cr(VI) to insoluble Cr(III) hydroxide which is then precipitated. This led to a substantial decrease in chromium allergy reports from the construction workers (58).

Leather products also serve as an important source of chromium exposure (59) since the Cr(III) sulfate used during the tanning process is oxidized to Cr(VI) which is detectable in the final leather product (60). Chromium tanning of leather is currently used in about 80% of all leather produced (61). Despite only using Cr(III) in the tanning process, several studies have reported significant amounts of both Cr(VI) and Cr(III) leaching out of Cr-tanned leather (62-64).

Epidemiological studies have been performed in many different countries to determine the frequency of Cr(VI) sensitivity amongst the general population. In Finland, a study of 822 volunteers found 1.7% (65) to be sensitive to Cr(VI). The prevalence of Cr(VI) sensitivity in a Danish population of 469 participants was calculated to be ~0.5% (66). Chromium allergy prevalence in Germany in an unselected population of 1141 people was 1.1% (67). Chromium allergy prevalence of consecutive patients from the United Arab Emirates has been calculated to be 7.5% (68). North Americans tested found a prevalence of 2.8% using a test concentration of 0.25% potassium dichromate (69).

One study determined the Cr(VI) Minimum Elicitation Threshold (MET) to range from 9-1770 ppm (70). In a separate study the Cr(VI) MET ranged from 28-885 ppm (71). These dose-response relationships are very similar to previous studies performed (72, 73, 73). These studies determined the Cr(VI) MET10% to range from 7 ppm (70) to 44 ppm (71) with a median value of 13 ppm. Data from other studies however differ from this dose-response relationship by reporting much higher eliciting thresholds (74, 75).

The data mentioned so far has reported the MET threshold in terms of the applied concentration (ppm), but in order to be of use as a risk assessment the threshold data must be reported in terms of mass of allergen per surface area of skin per unit time. One study that utilized the TRUE Test system (76) found the MET ranged from 4.4-0.018 mgCr(VI)/cm^2/48hr. They determined the MET10% dose to be 0.09 mgCr(VI)/cm^2/48hr. This data is much more useful for determining risk posed by a specific product given the amount of chromium leaching out and the duration of exposure.

Chromium allergic dermatitis represents the least harmful source of Cr(VI) exposure. This route of exposure has minimal health implications due to the protection afforded by the skin and only a small percentage of the population experience chromium dermatitis. Those who are particularly sensitive to chromium may find leather products which are not tanned using chromium and thus avoid unnecessary exposure. In addition, changes in cement production have reduced incidence of dermatitis in construction workers. Thus, industrial regulations and consumer education have been able to reduce the health impact of Cr(VI) exposure to the skin.

Part 2. Cellular Metabolism and Cr(VI)-induced DNA Damage

I. Toxicity of Different Valence States

Soluble Cr(VI) exhibits toxicity not observed by Cr(III) salts because Cr(VI) is absorbed more easily by the body and is able to cross cell membranes. At physiological pH Cr(VI) exists in the form of a chromate oxyanion which may enter into cells via non-specific phosphate/sulfate anionic transporters (77, 78). Cr(VI) gains electrons from oxygen to form an oxyanion at physiological pH. While in this state Cr(VI) may be taken into cells by a non-specific anion transporter systems which may mistake the Cr(VI) for sulphate or phosphate due to the close similarity in oxyanionic structure. Most Cr(III) is not able to enter cells and therefore does not damage intracellular macromolecules.

Cr(VI) is metabolized to ultimately produce Cr(III) inside the cell. The reduction of Cr(VI) into Cr(III) can mediate toxicity by forming stable bonds with cellular macromolecules (11, 79, 80). Though the trivalent Cr species can directly induce DNA damage and is able to the DNA inside of cells, it is not carcinogenic since it cannot readily enter cells. However, the Cr(VI) species easily enters cells where it is reduced to Cr(III) (81). Once inside of the cell Cr(III) forms stable bonds with nucleic acids and proteins which are very resistant to substitution reactions (82).

Extracellular reduction of Cr(VI) to Cr(III) can provide defense against chromium toxicity by preventing cellular uptake. Non-specific transport of Cr(VI) into cells followed by intracellular reduction allows chromium to accumulate in cells at relatively high concentrations compared to the extracellular environment (83, 84). Chromium in the hexavalent state does not readily interact with cellular molecules and displays very little biological activity. Once chromium enters the cell the toxic effects are mediated by Cr(VI), through reductive metabolism of Cr(VI) to Cr(III) via unstable intermediate species (11, 85, 86).

II. Cr(VI) Reductive Metabolism

There are two main mechanisms which mediate Cr(VI) reduction. One mechanism utilizes a series of one-electron reductions which produces the highly reactive intermediate species Cr(V) and Cr(IV) and eventually Cr(III) as the stable end form of Cr. This method of reduction generally occurs at relatively low levels of reductants (86). If there is an surplus of reductants over the total Cr(VI), then a two electron reduction may occur first, thus creating Cr(IV), which is followed by a one electron reduction to Cr(III).

Measurement of Cr(V) ions produced by one-electron reduction in animals treated with Cr(VI) provides *in vivo* evidence of Cr reductive metabolism (87). The Cr(V) species found was determined to be a Cr(V)-NADPH complex via an oxygen bond with the Cr (88, 89). When given pretreatment with ascorbate (Asc) and glutathione (GSH) the formation of Cr(V) decreased, however Cr(V) formation was increased with pretreatment of NADPH. These data suggest that GSH and Asc do not mediate one electron reduction of Cr(VI) *in vivo*, rather one-electron reduction is mainly via NADPH and flavoenzymes.

The oxidation of the reductant species during Cr(VI) metabolism produces reactive oxygen species (ROS) and carbon-based radical species, such as hydroxyl radicals, hydrogen peroxide, and ascorbate ions that can induce DNA strand breaks and form DNA adducts that cause the secondary toxicological effects of Cr(VI) exposure (86, 90, 91). Since Cr(VI) is not able to directly interact with DNA (92), reduction to the lower oxidation states is necessary to generate Cr-DNA adducts and lesions (93). The highly reactive intermediate species Cr(V) has been shown to induce DNA strand breaks and mutations *in vitro* (94-96). The ROS generated during the reductive metabolism are also likely to play an important role in Cr(VI)-induced DNA damage (97).

The Cr(VI) reduction process reduces molecular oxygen to O_2^-, which then reduces to hydrogen peroxide via dismutation. Hydrogen peroxide then reacts with the intermediate Cr-ion species Cr(V) and Cr(IV) to create hydroxyl radicals. These ROS may be neutralized by cytoprotective antioxidants which are part of the cellular defense against free radicals. The cell's protective mechanisms against oxidative injury include both enzymatic and non-enzymatic antioxidants. These defenses are able to prevent oxidation of important cellular components. However, once the antioxidant defenses are overwhelmed by the pro-oxidant Cr species, then oxidative injury occurs.

Intracellular Cr(VI) reduction generates reactive metal and oxygen species which can oxidize DNA bases (4). However, the significance of Cr(VI)-induced DNA oxidative damage and ROS in Cr(VI) toxicity is not yet apparent. Though the loss of endonuclease Apn1 function in Cr(VI)-treated yeast is suggested to be pro-mutagenic (98), Apn1, Rad1, Ntg1, and Ntg2-mutant yeast, which have impaired oxidative DNA damage repair capacity, demonstrate an insensitivity towards Cr(VI) cytotoxicity (99). This suggests that oxidative DNA damage mechanisms are separate from cytotoxic Cr-induced lesions since the same yeast strains are exceptionally sensitive to hydrogen peroxide exposure (100). The lack of increased sensitivity to Cr(VI)-induced oxidative DNA damage in NER/BER mutant yeast strains suggests that the DNA oxidation produced by Cr(VI) reduction is not robust enough to induce a lethal effect or that it is a small component of the full range of lesions caused by Cr(VI) exposure. In addition, vitamin E and selenomethionine were able to attenuate arsenite-induced carcinogenesis, but they did not protect mice from chromium-enhanced UVR carcinogenesis. These data suggest that the oxidative stress mechanism of arsenite exposure is not present in chromium enhancement of UVR-induced skin cancer.

Though the Cr(VI)-induced oxidative DNA damage may not be lethal it may still be an important part of Cr-induced carcinogenesis. Each Cr oxidation state is capable of generating hydroxyl radicals from hydrogen peroxide, especially Cr(V) and Cr(IV). Hydroxyl radicals may then react with guanine bases to create oxidized forms such as 8-hydroxy-deoxyguanosine (8-OHdG) (101). Formation of this adduct is used as a marker of ROS-mediated DNA damage and plays a role in toxicity of numerous toxicants. Cr(V)- and Cr(VI)-generated hydroxyl radicals have been shown to generate 8-OHdG (102, 103).

Cr(VI)-mediated depletion of cellular antioxidants leads to up-regulation of antioxidant genes such as HO-1 (104). The ROS-mediated reactions initiated by Cr(VI) lead to DNA damage and activation of transcription factors such as NF- κB, AP-1, and p53. Apoptosis may be induced in response to cellular oxidative damage and/or Cr-DNA lesions. If apoptosis is impaired, cell cycle arrest may occur through the activation of cell cycle control enzymes and the checkpoint activation mechanisms. Activation of MAPK signaling pathway enzymes has also been reported (105-107) in response to Cr(VI) exposure. If cellular antioxidant defense mechanisms are inadequate and the apoptosis and cell cycle arrest mechanisms are damaged and not able to function properly, then Cr(VI)-induced DNA damage can lead to up-regulation of oncogenes and carcinogenesis.

NF-κB is a transcription factor which enhances transcription of many genes involved in the oxidative stress response (108-110). Cr(VI) exposure in Jurkat cells was shown to lead to NF-κB activation (111). This was dependent on reduction of Cr(VI) since the hydroxyl radicals generated through the reductive metabolism are a key part of Cr(VI)-induced NF-κB activation. It is suggested that NF-κB activation may lead to expression of proto-oncogenes (c-myc) which may be important for initiation of neoplastic transformation by Cr(VI).

Cr(VI) is also able to induce AP-1 activation (112) which leads to the expression of proto-oncogenes such as c-jun. AP-1 activation occurs via mitogen activated protein kinase (MAPK) signaling cascades. Cr(VI)-mediated induction of AP-1 has been linked to phosphorylation of p38 and JNK. Cr(VI)-mediated induction NF-κB and AP-1 activation was diminished upon p38 and IκB kinase (IKK) inhibition as well as treatment with aspirin. Therefore it is proposed that the ROS generated by Cr(VI) reduction signal to activate both AP-1 and NF-κB, though p38 and JNK function as downstream kinases of AP-1 and NF-κB activation, respectively.

Many studies have looked at the formation of reactive intermediate Cr species produced during the reduction of Cr(VI) inside of cells in an attempt to link Cr(VI)-induced damage with these highly reactive intermediate species. The Cr(V) form has been shown to cause DNA strand breaks *in vitro* and create mutations in bacteria (94-96, 113).The reactive intermediates have also been reported to generate ROS (103, 114-120). These free radicals can then induce DNA strand breaks (102, 103), lead to base modifications (102, 103, 121), and activate NF- kB (83, 111). Reduction of Cr(VI) in Fenton and Haber-Weiss type reactions can produce persistent oxidative stress which may be a key mechanism of Cr(VI)-induced carcinogenesis.

III. Effects of Antioxidants on Cr(VI) Toxicity

Once inside the cell Cr(VI) is rapidly reduced by intracellular antioxidants such as ascorbic acid (Asc) and thiol groups of proteins, such as cysteine residues and reduced glutathione (GSH). Many other cellular components such as cytochrome P450 reductases (122, 122-124), mitochondrial inner-membrane electron transport chain protein complexes (125, 126), aldehyde oxidase (127), and hydrogen peroxide (128) may potentially reduce Cr(VI). Ascorbate and thiols are thought to be the most significant reductants of Cr(VI) since ascorbate is the most kinetically favored (129) and thiol groups are present in relatively high concentrations inside the cell.

Reductive metabolism of Cr(VI) at physiological pH is kinetically controlled. Studies of the reductive metabolism of Cr(VI) by Wetterhahn et al. determined Asc, GSH, and cysteine are the foremost non-enzymatic Cr(VI) reductants (130, 131). Microsomal P450 reductase may reduce Cr(VI) only in the near complete absence of oxygen (124), thus this is not a likely source of much Cr(VI) reduction in aerobic cells. Many cellular constituents such as glutathione (GSH) (132), cysteine (133), lipoic acid (83), NADPH and other thiol-containing molecules (116, 117) have been demonstrated to reduce Cr(VI) at physiological pH *in vitro*. The electron transport chain complexes within the mitochondrial inner membrane have been shown to reduce Cr(VI) *in vitro* (125, 126, 134, 135) where the reactive intermediates interfere with mitochondrial function, but this is not considered a major source of Cr(VI) reduction. *In vitro* studies have shown millimolar concentrations of hydrogen peroxide to reduce Cr(VI), but since this is much higher than in most living cells it is uncertain if it occurs *in vivo*. However the molecules that react with Cr(VI) at the highest rate will make the most significant contribution to its reduction.

In vitro experiments have demonstrated reduction of Cr(VI) intracellularly by glutathione (GSH) (115, 130), ascorbate, cysteine (103), lipoic acid (83), and other molecules, such as NADPH, fructose, ribose, and arabinose which contain thiol groups that are exposed for oxidation (116, 117). Ascorbate and GSH demonstrated the highest reaction rate with Cr(VI) in the cellular system (136). These molecules likely play the most significant role in reducing Cr(VI) as non-enzymatic reductants due to their ubiquitous presence in most cell types. *In vitro* experiments revealed significant reduction of Cr(VI) in the presence of high concentrations of Asc (137). Cr(VI) reduction in the presence of Asc is much faster than equal concentrations of GSH and cysteine (129).

One study injected Cr(VI) intratracheally in rat lung and reported depletion of Asc but not GSH in the exposed cells (138). Studies of Asc-mediated reduction of Cr(VI) in non-cellular systems detected Cr(V), Cr(IV) and Asc-derived free radicals. The reactive intermediate Cr species were shown to react with hydrogen peroxide to generate hydroxyl radicals which induced DNA strand breaks (103). A relatively high intracellular pH (pH ~10) increases the oxidation of Cr(III) with hydrogen peroxide. This correlates with the reported decrease in Cr(III) concentration and subsequent increase in hydroxyl radical generation after addition of hydrogen peroxide to cells.

Cells exposed to Cr(VI) and Asc produced numerous DNA strand breaks, with the extent of DNA damage depending on the relative concentrations of Cr(VI) and Asc. Exposing cells to hydrogen peroxide in addition to Cr(VI) and Asc significantly increased the observed DNA damage. When Mn(II) was added, DNA damage was inhibited. This is believed to be due to Mn(II)-mediated removal of Cr(IV) and the inhibition of Cr(IV) Fenton reactions. The extent of DNA damage was observed to correlate with the production of ROS. Asc-mediated reduction of Cr(VI) has been shown to create free radicals, Cr(V), and Cr(IV) (103). The relative concentration of each chromium species is dependent on the relative concentrations of Asc and Cr(VI). These ascorbate-derived free radicals and the hydroxyl radicals produced by the reaction of Cr(V) and Cr(IV) with hydrogen peroxide can induce DNA strand breaks (103).

Studies in rats reported reduction of Cr(VI) to Cr(III) by Asc in the lung, liver, and kidney. This has led some to suggest treatment with Asc for systemic chromium poisoning as well as direct application to the skin for chromium dermatitis to rapidly reduce Cr(VI) (10, 139). Though some *in vitro* data shows Cr(VI) reduction to Cr(III) in a dose-dependent manner upon addition of Asc to plasma, there is no clinical evidence supporting a reduction in morbidity or mortality from Asc administration in systemic chromium poisoning. This may be because most Cr(VI) is already taken into cells before the Asc is able to reduce it in the extracellular space.

Intracellular glutathione that exists in the reduced state (GSH) is a tripeptide composed of glutamine, cysteine, and glycine and plays a crucial role as intracellular antioxidant. The concentration of GSH is less than that of ascorbate, usually estimated in the millimolar range. Though the rate of Cr(VI) reduction for GSH is kinetically slower than for Asc (129), GSH has been shown to reduce Cr(VI). The high intracellular concentration of GSH likely makes it a major part of the reductive metabolism of Cr(VI). GSH can react with Cr(VI) to produce Cr(V), Cr(IV), thiol radicals, and Cr(III)-GSH complexes (140-142).

The reductive metabolism of Cr(VI) with GSH requires the reduction NADPH and molecular oxygen which produces $NADP^-$ and O_2^- and Cr(V). Reduced O_2^- then reacts with hydrogen ions to produce hydrogen peroxides (H_2O_2). The Cr(V)-NADPH complex can then react with H_2O_2 to generate hydroxyl radicals and Cr(VI) (115, 118, 143).

ESR experiments have demonstrated Cr(VI) reduction and generation of Cr(V) *in vivo* via one-electron reduction (87). Cr(V) was found to be in complex with NADPH (88, 89). Cr(V) formation was attenuated with ascorbate or GSH pretreatment, however Cr(V) formation was enhanced with NADPH pretreatment. The authors conclude that NADPH and flavoenzymes rather than GSH or ascorbate are responsible for the one-electron reduction of Cr(VI) *in vivo.*

IV. Cr(VI)-induced Genetic Lesions

The complexity of Cr-induced DNA damage is illustrated by the diversity of lesion sites; be it the bases themselves (92) or the phosphate backbone (80, 144-147). DNA damage may be directly induced by Cr-DNA interactions or indirectly through ROS and carbon radicals produced through reductive metabolism (4). Intracellular reductive metabolism of Cr(VI) generates many types of genetic lesions such as Cr-DNA binary and ternary adducts, DNA-protein crosslinks (DPCs), DNA interstrand crosslinks (ICLs), single-strand breaks (SSBs),double-strand breaks (DSBs), and oxidized bases. Cr(VI) exposure can also produce gene mutations (148-150), sister chromatid exchanges (151-153), and chromosomal aberrations (154-158).

Since Cr(VI) itself is unable to interact with DNA, reduction to the more reactive valence states is required for interaction with macromolecules. Cr(III) is the most stable metabolite of Cr(VI) and all Cr(VI) is eventually converted to Cr(III) intracellularly. Each intermediate metabolite of Cr (Cr(V), Cr(IV), and Cr(III)) is able to interact with DNA (90-92, 159). Cr(III) is able to complex with a variety of molecules including amino acids (82, 144, 160), whole proteins (79, 161, 162), and nucleic acids (92, 163-165). The reduction of Cr(VI) by macromolecules into Cr(III) causes the formation of the Cr-DNA-protein complexes which is how Cr directly damages cellular components. Intracellular reduction of Cr(VI) to Cr(III) in the presence of the Cr(III) chelator EDTA attenuated Cr-DNA binding, suggesting that Cr(III) is the Cr species that binds to DNA (166, 167). Other ligands bound to Cr(III), such as Asc, GSH, or cysteine, will affect the Cr-DNA interactions (81, 129, 163, 164, 167, 168).

Cr(VI) exposure can cause mutations, chromosomal aberrations, and lead to transformation in cell culture (2, 157). In addition many different DNA lesions can be formed including DNA protein crosslinks, interstrand crosslinks, and strand breaks (94, 102, 121). Though much is known about Cr genotoxicity the exact toxicological significance of each type of lesion is still unclear.

i. Cr-DNA Adduct Formation

Cr has been shown to bind to DNA bases and phosphate groups through both covalent and ionic interactions. Analysis of the Cr(III)-DNA adducts reveals that they are primarily in the form of ternary complexes of Cr-DNA interaction with an intracellular reducer such as Asc or GSH (80, 82, 144). The Cr(III) will bind to the phosphate backbone covalently and an intracellular reducer bound to Cr will form the ternary complex (79, 82). It appears that <1% of the Cr(III) bound to DNA forms a complex with protein while the majority forms DNA-DNA cross-links (79, 169). Cr-DNA binding has shown a preference for guanine bases which favors the coordination of two adjacent guanine residues (148, 170). Some evidence suggests Cr binding occurs preferentially at the N7 position of guanine (170, 171). These Cr-DNA adducts can induce mutations and most likely alter the DNA code during cell transformation (167, 169, 172-174).

NaCl extraction experiments were only able to remove 25% of the Cr bound to DNA (145). A similar experiments using only single-stranded DNA was able to remove 40% of Cr associated with DNA while 20% was chelatable with EDTA (175). This demonstrates the high affinity of covalent Cr(III)-DNA interactions which are formed in large part between Cr(III) and the phosphate backbone. Cr-DNA adducts could potentially distort the helical structure of the DNA by neutralizing electronegative phosphate moieties and impact DNA excision repair and replication fidelity (176).

ii. DNA Interstrand Crosslinks (ICLs)

Cr is able to form DNA interstrand crosslinks (ICLs) through six coordination sites which may form covalent bonds with the DNA molecule (177, 178). Cr-induced ICLs are thought to be largely responsible for the physical obstruction of DNA replication at guanine residues on the template DNA strand (170, 171, 179). Most types of Cr-DNA interactions involve coordination with the phosphate backbone. Two types of Cr-ICLs have been characterized; those that are heat sensitive and those that are heat stable (6). The heat stable variety make up only ~10% of ICLs observed, but they are known to lead to guanine-specific polymerase arrest (180).

iii. DNA-Protein Crosslinks (DPCs)

Intracellular reduction of Cr(VI) into the stable Cr(III) species may be mediated by the thiol groups of proteins, in addition to binding to DNA. Chromium may thus form stable bonds between both proteins and DNA to form stable DNA-protein crosslinks (DPCs) and ternary DNA adducts. Early investigation into Cr-DNA interactions determined that Cr(III) was shown to directly link proteins to DNA (181). DPCs are most persistent in double-stranded DNA (182).

A large percentage of Cr-DNA-protein crosslinks (DPCs) are reported to either be the interaction of DNA and non-histone proteins or RNA coupled with cytoplasmic proteins in studies conducted using liver and kidney cells of treated rats (183). Protein crosslinks of nuclear lamins, actin, and matrix proteins have also been reported (161, 162).Some have suggested that the majority of Cr-DNA interactions form DPCs (79) and that most DPCs are formed with single amino acids such as cysteine and histidine and small peptides like GSH (82, 144).

The majority of Cr-protein complexes with DNA appear to be primarily occurring through interactions with the phosphodiester backbone (82, 169). The observation that pre-treatment with Mg^{2+} attenuates Cr-DNA binding in the formation of Cys-Cr-DNA adducts supports this hypothesis (167). Moreover, the Cr-DPCs which are phosphate-based are proposed to correspond with Cr-induced pre-mutagenic lesions (169).

DPC formation was attenuated by antioxidant pretreatment, suggesting that oxidative mechanisms may be responsible for some Cr(VI)-induced DPCs (184). Thus Cr-DPCs may be present as a Cr-DNA ternary adduct or as a physical crosslink between DNA and protein due to Cr-mediated oxidative mechanisms.

iv. DNA Strand Breaks and Chromosomal Abnormalities

Single DNA strand breaks (SSBs) are one of the most common genetic lesions observed after Cr(VI) exposure. Cr(VI) exposure leading to increased single strand breaks has been observed in cell culture (178, 182, 185, 186), mice (187), and humans (188). SSBs may be produced through direct Cr-DNA interaction or through free radical-mediated mechanisms. It has also been shown that Cr(VI) treatment may cause DNA double-strand breaks (DSBs) (113). DSBs cause increased chromosomal instability which is characteristic of Cr(VI)-induced tumors (97).

Numerous studies have reported Cr(VI) to have clastogenic activity in addition to creating DNA strand breaks (158, 159, 189-195). Both soluble and insoluble chromates have been shown to generate damaged metaphases (196). These data suggest that exposure to Cr(VI)-containing compounds, whether water-soluble or insoluble, can lead to loss of DNA integrity and potentially carcinogenesis. Micronuclei and microsatellite instability have also been reported after Cr(VI) exposure (197). Such chromosomal abnormalities increase genomic instability and may play a major role in the initiation of cancer in exposed humans.

v. Cr-induced Mutagenesis

Cr-induced mutagenesis is well studied using *in vitro* (198, 199) and *in vivo* (148-150) models. The mutagenicity of ternary Cr-DNA adducts has been investigated using shuttle vector experiments. The rate of mutation induction by Cr(III)-amino acid ternary DNA adducts is significantly higher than through the formation of Cr(III)-DNA binary adducts (169). Pre-treatment with $MgCl_2$ reduced the formation of Cr(III)-GSH complexing with DNA by 20-fold, thus illustrating the significant role of the phosphate backbone in the formation of Cr-DNA ternary adducts. The shuttle vector studies provide valuable data on Cr(VI)-induced mutagenesis, however they are limited in the types of mutations which can be detected. Thus, the full spectrum of Cr-induced mutagenesis has likely not yet been elucidated.

vi. Effects on DNA and RNA Synthesis

Genetic lesions that prevent DNA replication will lead to cell cycle arrest or apoptosis. Cell cycle arrest allows time for DNA repair mechanisms to fix genetic insults while apoptosis is a means of cell suicide as a result of irreparable DNA damage. Cr(VI) exposure is known to impair DNA synthesis in cell culture (200). Cr-damaged DNA isolated from *in vitro* systems was unable to be replicated with bacterial enzymes (171, 179, 201-203) and human recombinant enzymes (170) in PCR-based assays.

Cr(VI) obstructs DNA replication and also hinders both RNA and protein synthesis (204). This may be due to widespread oxidation induced by the Cr species or dNTP depletion by Cr(III) complexing with dNTP precursors. Both mechanisms may work in tandem to produce the observed DNA synthesis inhibition. Cr may also bind transcription factors and thus interfere with gene expression (205, 206). A recent study reports Cr interferes with RNA chain elongation which inhibits transcription (207). RNA polymerase arrest was observed mostly in GC-rich DNA regions. This is similar to observed Cr-induced DNA polymerase arrest that occurs specifically at guanine residues (170, 171, 202, 208). Therefore the same types of lesions may be involved in each of mechanism leading to polymerase arrest.

It has been postulated that RNA synthesis inhibition may lead to p53 activation (209), which could be relevant to Cr(VI)-mediated p53 activation. Some have suggested that apoptosis may be stimulated by RNA synthesis inhibition after genotoxic exposure (209-211) which may be responsible for apoptosis observed in Cr(VI)-treated cells (212-216).

vii. Role of ROS in Cr-DNA Damage

It has been suggested that ROS produced through Cr(VI) reduction is the major mechanism by which Cr(VI) generates DNA damage. However, the importance of ROS species in mediating Cr(VI) toxicity is still not completely understood. Several studies have investigated the role of ROS-mediated DNA damage after Cr(VI) exposure using buffers to prevent ROS formation which led to a significant decrease in SSBs (217-219) Pre-treatment with Vitamin E and Asc in Chinese hamster V79 cells was able to attenuate the formation of Cr(V) and reduce the number of Cr(VI)-induced DNA strand breaks (85, 194, 220). *In vivo* studies using chick embryo red blood cells reported persistent DNA strand breaks after Cr(VI) treatment (186). Superoxide dismutase and catalase were able to reduce DNA strand breaks caused by Cr(VI) treatment in human diploid fibroblasts.

Cr(VI) and the highly reactive intermediate species, Cr(V) and Cr(IV), react with hydrogen peroxide intracellularly to generate hydroxyl radicals which are genotoxic and can oxidize DNA bases (119, 221). Hydroxyl radicals are extremely reactive and easily react with guanine residues to produce acute oxidative stress. The hydroxyl radicals thus may generate oxidized forms of guanine, such as 8-OH-dG (8-hydroxy-deoxyguanosine) (101). HPLC has been used to confirm 8-OH-dG formation from 2'-deoxyguanosine (dG) by hydroxylation from hydroxyl radicals produced by Cr(V) and Cr(VI) (102, 103). Formation of 8-OH-dG is thought to implicate ROS in the mechanism of Cr(VI)-induced genotoxicity and carcinogenicity. However, other studies report that Asc pretreatment does not prevent base oxidation or SSB formation and that DNA base oxidation by Cr is not mediated by Fenton-type reactions (222, 223). However, other studies claim that the majority of DNA lesions caused by Cr(VI) exposure *in vivo* to be DNA protein cross-links (DPCs), DNA interstrand cross-links, and Cr-DNA adducts (177, 183, 186). Thus there is evidence to support both sides of the argument of the significance of ROS in Cr(VI) genotoxicity.

The wide range of genetic lesions which are observed after Cr(VI) exposure demonstrate the complexity of Cr(VI) toxicity. The intracellular metabolism of Cr(VI) produces both highly unstable intermediate species which may react with macromolecules and form cross-links between them. In addition, free radicals such as hydrogen peroxide, hydroxyl radicals, and carbon-based radicals may be generated by reduction of Cr(VI) to Cr(III). This allows Cr(VI) to mediate direct damage to the cell and the DNA, in addition to indirectly damaging the cell through oxidation. Thus, Cr(VI) induces a range of effects which all combined generate the observed toxicity.

Part 3. Molecular Mechanisms of Cellular Response to Cr(VI)

I. Regulation of Cell Cycle Checkpoint Arrest

Cell cycle checkpoints refer to those points in the cell growth cycle where certain criteria must be met before the cell will proceed on to the next phase of growth. Cell cycle checkpoints are typically activated in response to DNA damage in order to arrest cell growth (224, 225). This is a mechanism to maintain genomic fidelity and integrity after genotoxic insult. If cell cycle arrest does not occur, then mutations may be passed on to daughter cells which can potentially lead to carcinogenesis. The cell cycle checkpoints represent one of the primary mechanisms by which a cell responds to loss of genomic integrity and may be used as a marker for DNA damage.

A recent study reported several dose-dependent responses in cells which is suggested to correlate with the extent of DNA damage (226). Transient cell cycle arrest was observed with Cr(VI) doses of < 2 μM for 24 hr exposure. The cell growth arrest allows for repair of low to moderate levels of Cr-induced DNA damage. The higher doses of > 5 μM for 24 hr exposure led to terminal growth arrest or apoptosis due to cell cycle blockade, suggesting extensive cellular damage and/or irreparable genotoxic insult.

i. S-phase Checkpoint

S-phase refers to the part of the cell cycle where the DNA is replicated in preparation for cell division. Genotoxic insult which affects genomic integrity and results in S-phase checkpoint activation represents inhibition of DNA replication. S-phase arrest induced by DNA damage is a cellular mechanism which increases cell survival and diminishes genetic abnormalities (225). Inability to initiate S-phase arrest could prevent repair of genetic mutations which may then be passed on to the daughter cells. The cellular mechanisms which respond to DNA damage to activate the S-phase checkpoint are crucial for maintaining genomic integrity and preventing carcinogenesis.

ATM and ATR are kinases which play an important role in responding to DNA damage and transduce signals to activate pathways involved in cell cycle arrest and apoptosis (227) by simultaneously phosphorylating proteins which ultimately activate cell cycle arrest. ATR (ATM-Rad3-related) is structurally similar to ATM and functions to signal through some of the same pathways as ATM. While ataxia-telangiectasia-mutated (ATM) has been shown to respond mainly to DNA DSBs which are induced by ionizing radiation, ATR is more responsive to DNA damage due to UV radiation (228). Both ATM and ATR have been reported to mediate Cr(VI)-induced S-phase arrest (229, 230). Cr(VI)-induced S-phase checkpoint activation may occur by an ATM-dependent an SMC1-dependent mechanism (230). Studies have shown ATM autophosphorylates itself at Ser1981 in response DNA damage induced by ionizing radiation (IR) (231). Once activated, ATM then phosphorylates numerous downstream target proteins such as p53, BRCA1, Chk2, MDM2, NBS1, Rad17, and SMC1 which can lead to cell cycle checkpoint arrest (232). ATM initiates S-phase activation by phosphorylating NBS1 and structural maintenance of chromosomal protein 1 (SMC1) (233-236).

However, in response to Cr(VI) doses higher than 20 μM an ATM-independent pathway is able to induce S-phase arrest. SMC1 is phosphorylated at Ser966 upon exposure to Cr(VI) concentrations greater than 20 μM in ATM-deficient cells, suggesting another kinase is responding to DNA damage and activating SMC1 (230). ATR was reported to be necessary for SMC1 phosphorylation in response to 40 μM Cr(VI) treatment since ATR-deficient cells treated with high concentrations of Cr(VI) displayed inhibited ability to initiate S-phase arrest.

ATM and ATR do not directly detect DNA damage. Rather they serve to transduce the signals from those proteins that sense the damage in order to activate downstream pathways. Four proteins identified in yeast are thought to have the ability to bind to damaged DNA and may be able to interact with PIK-related kinases ATM and ATR. These proteins are Rad9, Rad1, Rad17, and Hus1 (237-239). Rad17 has been shown to be phosphorylated by ATR and localizes ATR to the site of DNA damage (238, 240). It is suggested that Rad17 functions to connect the Rad9-Rad1-Hus1 complex to the site of DNA damage until DNA repair is complete (241). Cells transfected with mutant Rad17 and treated with 40 μM Cr(VI) displayed an impaired ability to initiate S-phase arrest and reduced SMC1 phosphorylation. Overall these data implicate Rad17 DNA binding as critical for ATR-mediated SMC1 phosphorylation which leads to S-phase arrest.

ii. G2/M Checkpoint

G2/M-phase refers to the final stage of cell growth and the mitosis phase where the mature cell divides into two identical daughter cells. At this phase the cell's genome has already been replicated. Less research has been done to investigate the effects of Cr(VI) at the G2/M-phase and the process of cell division. However, this is an important checkpoint since any loss of chromosomal integrity during mitosis may be directly passed onto daughter cells which then may continue to be exposed to the original genotoxin.

Cell cycle studies in S. *cerevisiae* treated with hexavalent chromium reported Cr(VI)-induced G2 arrest using low doses that caused very little cytotoxicity (99). G2 phase arrest may be more sensitive to Cr(VI) exposure since cells must undergo homologous recombination at this point in the cell cycle in order to complete mitosis. Further investigation observed mutant recombination-deficient yeast (rad52 and rad53 rev3) experience terminal G2/M-phase arrest upon exposure to Cr(VI). This led to increased sensitivity to Cr(VI) toxicity in the recombination-deficient yeast strains. This demonstrates that the rad52-mediated recombinational repair mechanism is impaired by Cr(VI) exposure and must be functional in order for yeast to exit G2/M-phase arrest. Overall these data demonstrate G2/M phase arrest with lower Cr(VI) concentrations, while higher Cr(VI) doses induced an S phase delay.

Cr(VI) treatment has been shown to generate both S-phase and G2 checkpoint arrest in a dose-dependent manner and inhibition of p38 prevents activation of these cell cycle checkpoints (242). This suggests that p38 functions to enhance cell survival and prevent apoptosis in response to Cr(VI)-induced stress. Upstream mechanisms mediated Cr(VI)-induced p38 activation are still not fully known. It has been shown that Cr(VI)-induced oxidative stress is necessary for p38 activation in response to Cr(VI) exposure (107). Also, ATM is known to regulate p38 activity after IR, which is required for IR-induced G2 checkpoint activation. Since Cr(VI) exposure leads to activated ATM and ATR, it is possible that these kinases may induce p38 activation.

II. Apoptosis

Apoptosis is a means of eliminating damaged cells with abnormal DNA. Cell cycle arrest transiently terminates cell growth and division to prevent the expression of any mutations induced by a toxicant. These DNA damage responses are critical for maintaining genomic stability and prevention of tumorigenesis (225). Apoptosis is regulated through activation and/or synthesis of proteins which initiate the systematic destruction of the cell. Apoptosis is initiated in response to stresses which disturb the equilibrium between cell division and cell elimination. Such an imbalance between cell production and cell destruction can lead to diseases such as cancer where neoplastic cells gather to form tumors.

Cr(VI) causes apoptosis (243) via signaling through activated p53 as well as through ROS generated from Cr(VI) reductive metabolism. ATM is a kinase which responds to DNA damage by phosphorylating Chk2 which in turn phosphorylates CDC25A (244). Chk2 phosphorylation mediates the apoptotic response to IR through p53 activation. In addition, free radicals produced by the metabolism of Cr(VI) into Cr(III) may activate NF-κB which is involved in regulating the apoptotic response (245). The ROS created by Cr(VI) reduction can initiate apoptosis independent of p53 activation. Cr(VI)-induced p53 activation is not required for Cr(VI)-induced apoptosis but can serve to enhance apoptosis. Thus, Cr(VI) can induce apoptosis in both p53-independent and p53-dependent mechanisms. In addition, ROS also damage DNA which leads to other cellular responses. The balance of both apoptotic pathways may play an important role in Cr(VI)-induced carcinogenesis.

In vitro studies of Cr(VI)-induced apoptosis in both CHO cells (246) and telomerase-transfected fibroblasts (226) has been demonstrated to disturb mitochondrial stability and lead to the release of cytochrome c. Cytochrome c is released from the mitochondria via the mitochondrial transition pore which is regulated by the pro- and anti-apoptotic Bcl family proteins (247). Cr(VI) interactions with DNA may affect the transcription of Bcl family proteins and alter the ratio of pro- and anti-apoptotic proteins which can facilitate the opening of the transition pore. A study investigating the transcriptional activity of Bcl family proteins in HLF cells upon Cr(VI) exposure found upregulated bclXS which is pro-apoptotic. Also, down-regulation of certain anti-apoptotic genes bcl-W and bcl-XL was observed after Cr(VI) exposure (248). The exact function of bclXS, which mediates apoptosis in a p53-independent manner, is not well understood, though there is evidence that p53 activation occurs prior to cytochrome c release (215, 249). These data suggest a Bcl-mediated mechanism may be involved in Cr(VI)-induced p53-dependent apoptosis pathways (214, 215, 249).

III. Molecular Determinants of Cellular Response to Cr(VI)

ATM

Ataxia telangiectasia mutated (ATM) is a 370 kDa serine/threonine kinase (250, 251) which has a catalytic domain at its C-terminus which is very similar to that of phosphatidylinositol

3-kinases. ATM is crucial for detection of DNA double strand breaks (DSBs) in mammalian cells. ATM responds to DNA DSBs by activating multiple substrates which mediate cell cycle checkpoint, apoptosis, and DNA repair mechanisms (232, 252). Thus, ATM serves as a critical regulator of the cell cycle response to DNA damage (253, 254). ATM-deficient mouse models demonstrate abnormal cell cycle checkpoints with increased susceptibility to DNA damage and tumor formation (255-257). DNA DSBs were observed in HeLa cells treated with Cr(VI) for 4 hr (113, 229). In response to DNA damage ATM autophosphorylates at Ser1981 which mediates the dissociation of the ATM dimers and releases the active monomers. Phosphorylated ATM was subsequently detected in Cr(VI) treated HeLa cells.

ATM is necessary for Cr(VI)-induced apoptosis and promoting recovery of replicative potential in order to escape terminal growth arrest. ATM functions as a signal transducer upstream of p53 (258). Studies using ionizing radiation have reported p53 phosphorylation at Ser15 by ATM (259, 260) and Chk2 phosphorylation at Ser20 (261-263). ATM can regulate p53 stability by directly phosphorylating p53 or indirectly through Chk2-mediated p53 phosphorylation. Chk2 and p53 phosphorylation were also observed to occur in an ATM-dependent manner after Cr(VI) treatment. Inhibition of ATM function reduced Cr(VI)-induced apoptosis but increased sensitivity to Cr(VI)-induced clonogenic lethality. ATM -/- cells also displayed impaired ability to recover from Cr(VI)-induced cell growth arrest.

p53

p53 plays a crucial role in the DNA damage response and signals to the cell to cause cell cycle arrest or undergo apoptosis (264). p53 mediates the DNA damage response by transactivating many proteins involved in regulated the cell cycle and apoptosis. Cr(VI)-induced apoptosis has been shown to require functional p53 in HLF cells (215). Cr(VI) exposure led to increased p53 levels with increased p53 translocation to the nucleus. Also, Studies in human lung cancer cells confirmed an increase in p53 levels and transactivation activity after Cr(VI) treatment (243, 265).

p53 is phosphorylated in response to DNA damage and undergoes conformational change which increases the activity and stability of p53 (266). It has been shown that phosphorylation of Ser20 is necessary for p53 stability in response to DNA damage (6, 261). Activated p53 enhances the rate of transcription of genes such as *p21*, *Mdmd2*, *Gadd45*, and *Bax* (267, 268). These genes mediate p53-dependent cell functions such as initiation of apoptosis and cell cycle arrest.

Cr(VI) reduction into Cr(III) involves interactions with proteins which is likely important for mediating Cr(VI) cytotoxicity. Cr-protein-DNA adducts may activate DNA damage response pathways to activate p53 and cause cell cycle arrest. If the DNA damage is irreparable then the cell will signal to undergo apoptosis (107, 269, 269). The production of free radicals during Cr(VI) reductive metabolism can work in tandem to activate signaling pathways through PI-3 kinases and Akt to inhibit apoptosis (102, 103). In addition, catalase treatment was shown to inhibit p53 activation. Sodium formate and aspirin were also able to attenuate p53 activation. Deferoxamine was used to chelate Cr(V) to prevent free radical production from H_2O_2 which led to reduced p53 activation. NAPDH facilitates one-electron reduction of Cr(VI) to Cr(V) and increased NAPDH in cells led to enhanced p53 activation. Overall these data suggest that hydroxyl radicals produced by Cr(VI) reduction may play a prominent role in Cr(VI)-induced p53 activation. Activation was shown to occur at the protein level rather than through transcriptional activity (265). Thus, Cr(VI) treatment increases p53 stability and reduces p53 degradation.

Chk2

ATM phosphorylates Chk2 (270-273) thus activating Chk2 which phosphorylates several Cdc25 proteins which may inhibit cyclin-dependent kinase activation (274-276). Chk2 phosphorylation occurs via an ATM-independent pathway in response to UV-irradiation which may be mediated by ATR (271, 277). ATM, ATR, and Chk2 have each been shown to phosphorylate p53 (259-261, 263, 278, 279).

p53 phosphorylation may be mediated by Chk2 without functional ATM, suggesting that a separate kinase is involved in sensing DNA damage and phosphorylating p53 to initiate apoptosis. Activated p53 also functions differently depending on whether it is phosphorylated by Chk2 or ATM. ATM appears to be crucial for regulating G1/S phase checkpoint arrest through p53, while Chk2 is necessary for proper regulation of IR-induced apoptosis through p53.

In a study comparing ATM -/- and Chk2-/- mice using IR to induce apoptosis and cell cycle arrest it is suggested that Chk2 may induce p53-dependent apoptosis without functional ATM (280). Chk2 -/- mice were not able to induce apoptosis or G1/S phase arrest in response to IR, however IR-induced apoptosis was re-established after adding the wild-type Chk2 gene. Mutant Chk2 with amino acid substitutions at the residues phosphorylated by ATM and rad3^{+} related (ATR) did not restore IR-induced apoptosis. This illustrates a role for ATR in p53-mediated apoptosis and pathways which regulate p53 activation which initiates cell cycle arrest and apoptosis.

This study demonstrates differential regulation of p53 activation leading to either cell cycle arrest or apoptosis. ATM is not required for Chk2 phosphorylation during p53-mediated apoptosis.

MAP Kinases

Mitogen-activated protein kinases (MAPKs) are signaling enzymes which regulate many cellular processes, including response to genotoxic compounds. MAPK signaling leads to transcription factor activation and gene expression (281). MAPKs are divided into 3 subclasses in mammalian cells: the Jun-NH2- terminal kinases (JNK), extracellular signal-regulated kinases (ERK1/2), and p38 kinases.

Many studies have looked at Cr(VI) exposure on MAPK activation and signaling. Often very high Cr(VI) concentrations are used to observe a robust response. For example, in BEAS-2B cells 500 μM Cr(VI) induced ERK, JNK, and p38 activation (105). JNK, ERK, and p38 activation was reported in CL3 human lung adenocarcinoma cells after 10-80 μM Cr(VI) exposure. The degree of activation varied between kinases and was observed to be both time- and dose-dependent (106, 107). MAPK activation was maintained until Cr(VI) was removed.

It is hypothesized that the p38/JNK pathway mediates Cr(VI)-induced cell growth arrest. p38 is activated in response to Cr(VI) exposure (107). p38 has been reported to play a role in regulating apoptosis, however it may function to promote or repress apoptosis depending on cell type and toxic agent (282). While p38 activity is necessary for IR-induced G2 arrest, it has also been suggested to be involved in initiation of UV-induced G1, S-phase, and G2 arrest

(283-285). Cr(VI) exposure activates p38 and inhibition of p38 during Cr(VI) exposure prevents S-phase and G2 cell cycle checkpoint activation. p38 inhibition decreases cell survival and increases apoptosis following Cr(VI) exposure. Therefore p38 function is necessary for proper cellular stress response induced by Cr(VI) (242). Though Cr(VI) exposure activates JNK and p38 pathways, this was shown not to be involved in Cr(VI) cytotoxicity (106, 107).

ERK activation enhances cell proliferation in normal fibroblast and epithelial cells (286). Some studies have reported ERK activation in response to DNA damage induced by different stimuli. ERK activation in response to DNA damage was shown to occur by p53-dependent and independent mechanisms and led to either cytoprotective or cytotoxic outcomes (265, 287-290).

Studies in normal fibroblast and epithelial cells have shown ERK to be necessary for S phase entry (291, 292). ERK inhibition did not effect Cr(VI)-induced clonogenic lethality, suggesting that ERK signaling is not required for Cr(VI)-induced apoptosis or terminal growth arrest (293). A more recent study using CL3 cells demonstrated Cr(VI)-induced ERK activation, however no change in growth rate or clonogenic survival was observed with co-treatment with MKK1 inhibitor PD98059 (107).

Data from several studies points towards dose concentration and duration as determinants of the outcome of ERK activation, possibly by affecting MAPK crosstalk (294, 295). It is suggested that transient ERK activation may induce a response which opposes that of sustained ERK activation in regards to cell proliferation (295). The duration of activation and the observed cellular response can differ considerably between cell types. Cr(VI)-induced ERK phosphorylation has been reported in A549 cells after 50 μM exposure for 2 hr (265) while 30 μM Cr(VI) exposure in CL3 cells increased ERK phosphorylation 3.5-fold in 12 hr (107). HLF cells treated with a much lower concentration of 6 μM observed transient ERK phosphorylation which was absent after 7 hr of exposure (248). Therefore, understanding the kinetics of ERK activation may clear up our understanding of the downstream effects of ERK activation.

AP-1

CR(VI) has been shown to induce AP-1 activation (111). MAPK signaling cascades are involved in mediating Cr(VI)-induced AP-1 activation. JNK and p38 phosphorylation, but not ERK phosphorylation, has been associated with AP-1 activation by Cr(VI). Treatment with aspirin inhibited both AP-1 and NF-κB activation after Cr(VI) exposure. In addition, p38 inhibition reduced AP-1 activation, while IκB kinase (IKK) inhibition reduced NF-κB activation. Therefore, it may be that ROS created by Cr(VI) reduction signal to initiate the MAPK cascade which activates AP-1 and NF-κB.

HIF-1 and VEGF

Hypoxia-inducible factor 1 (HIF-1) is a transcription factor which is induced by decreased cellular O_2 concentration. HIF-1 regulates expression of genes such as VEGF, HO-1, aldolase, enolase, and lactate dehydrogenase. Vascular endothelial growth factor (VEGF) has been demonstrated to be required for tumor angiogenesis. HIF-1 activity has been correlated with tumorigenicity and angiogenesis in a nude mouse model. HIF-1 is also induced by oncogenes like v-Src and Ras which are often over-expressed in human cancers. Cr(VI) treatment has been shown to induce HIF-1 and lead to increased VEGF expression in DU145 human prostate carcinoma cells. p38 MAPK signaling was found to be necessary for Cr(VI)-induced HIF-1α expression while PI3-K and ERK activity were not necessary. It is suggested that H_2O_2 produced by Cr(VI) in DU145 cells induces HIF-1 and VEGF expression and that expression of HIF-1 and VEGF may serve as an important mechanism of Cr(VI)-induced carcinogenesis.

NF-kB

NF-κB is a transcription factor which functions in many cellular processes including cell proliferation and apoptosis (296). DNA damage induced by Cr can inhibit transcription through RNA polymerase arrest (207). Cr-protein interactions may also function to inhibit transcription without necessarily damaging the DNA template (297). This has been shown in A549 cells where Cr(VI) was able to block interaction of NF-κB p65 subunit and CREB (298). In BEAS-2B cells however, Cr(VI) activated NF-κB which was thought to prevent cytotoxicity (299). The differences observed in Cr(VI) exposure's effects on NF-κB activity may be due to differences in cell type and exposure concentration and duration (300).

NF-κB is a thought to be involved in the oxidative stress response (109, 110, 301). Cr(VI) exposure has been shown to cause NF-κB activation in Jurkat cells (111). Hydroxyl radicals produced Fenton-like reactions of Cr(V) and Cr(IV) may mediate Cr(VI)-induced NF-κB activation. In addition, activated NF-κB may lead to expression of proto-oncogenes which may be involved in Cr(VI)-induced neoplastic transformation.

Conclusion

In this paper we have reviewed the literature of Cr(VI) toxicity. Specifically we discussed Cr(VI) epidemiological studies and the different effects of the main routes of exposure. Then, we reviewed the current knowledge of the reductive metabolism in which Cr(VI) is reduced intracellularly to ultimately generate the stable Cr(III) form. This way it is not Cr(VI) which actually harms the cell, but rather the unstable intermediate species Cr(V) and Cr(IV) which are able to bind macromolecules and generate free radicals. Then, we covered many different types of genetic lesions that may be induced by Cr(VI). Finally, we discussed the cellular response mechanisms observed after Cr(VI)-induced DNA damage and specific proteins which regulate the pathways that allow for cellular response and repair. These mechanisms are ultimately what determine the fate of cells exposed to Cr(VI).

The enzymes discussed in the final section are far from a complete list of all molecular determinants of the cellular response to Cr(VI) exposure. The exact mechanisms of Cr(VI)-induced carcinogenesis are still poorly understood. In addition, the relative importance of the different types of genetic lesions caused either directly or indirectly by Cr(VI) metabolism is still being investigated. Here we try to highlight some of the proteins which are well-studied and illustrate how their functions are important for the cell cycle activation and apoptotic response observed in cells after Cr(VI) treatment.

Investigation of Cr(VI) and its effects on living organisms, whether through epidemiology, animal studies, or *in vitro* cultures, has become a fertile area of research. Studies of Cr(VI) have led to our current understanding of the complexities of Cr(VI)-induced carcinogenesis, as well as the processes of cellular response to DNA damage and repair mechanisms. Understanding of Cr(VI)-mediated damage to cellular macromolecules and the cellular response mechanisms has become a model which is being applied to other toxic metals such as nickel and mercury. Thus, the knowledge gained through studies of hexavalent chromium may provide useful for studies of other reactive heavy metals.

Considering that millions of people are exposed to airborne hexavalent chromium in the environment which may persist in the lungs for years, our understanding of Cr(VI) toxicity and the mechanisms by which it overcomes cellular defenses to lead to carcinogenesis is not relevant only to those exposed occupationally. However, many questions remain unanswered. A complete understanding of the mechanisms of Cr(VI) toxicity in different cell types and animal models may provide the knowledge for the development of treatments for the cellular injuries caused by Cr(VI) exposure. Further analysis of the genes involved in protecting cells from Cr(VI)-induced carcinogenesis may also provide evidence to support preventative strategies for protecting those who may be most susceptible to Cr(VI)-induced carcinogenesis.

Acknowledgement

Additional credit for this work goes to Xi Tang and Xiao-li Cui. This work was made possible by funding from NIH ES013301.

Reference

[1] Bianchi V, Levis AG. Review of genetic effects and mechanisms of action of chromium compounds. *Sci Total Environ.* 1988;71:351-355.

[2] De FS, Bagnasco M, Serra D, Zanacchi P. Genotoxicity of chromium compounds. A review. *Mutat Res.* 1990;238:99-172.

[3] De FS. Threshold mechanisms and site specificity in chromium(VI) carcinogenesis. *Carcinogenesis.* 2000;21:533-541.

[4] Sugden KD, Stearns DM. The role of chromium(V) in the mechanism of chromate-induced oxidative DNA damage and cancer. *J Environ Pathol Toxicol Oncol.* 2000;19:215-230.

[5] Snow ET. Metal carcinogenesis: mechanistic implications. *Pharmacol Ther.* 1992;53:31-65.

[6] Singh J, Carlisle DL, Pritchard DE, Patierno SR. Chromium-induced genotoxicity and apoptosis: relationship to chromium carcinogenesis (review). *Oncol Rep.* 1998;5:1307-1318.

[7] Agency for Toxic Substances and Disease Registry. Toxicological Profile for Chromium. P.H.S.US Department of Health and Human Services; 2000.

[8] Environmental Assessment and Criteria Office. EPA Health Assessment Document for Chromium. US Environmental Protection Agency, Research Triangle Park, NC; 1984.

[9] Office of Air Quality Planning and Standards. EPA Locating and Estimating Air Emissions from Sources of Chromium. *US Environmental Protection Agency, Research Triangle Park, NC.* 1984.
[10] Office of Emergency and Remedial Response. EPA Health Effects Assessment for Hexavalent Chromium. US Environmental Protection Agency, Cincinnati, OH; 1984.
[11] Cohen MD, Kargacin B, Klein CB, Costa M. Mechanisms of chromium carcinogenicity and toxicity. *Crit Rev Toxicol.* 1993;23:255-281.
[12] Vermeire T, Stevenson H, Peiters MN, Rennen M, Slob W, Hakkert BC. Assessment factors for human health risk assessment: a discussion paper. *Crit Rev Toxicol.* 1999;29:439-490.
[13] Sorahan T, Hamilton L, Gompertz D, Levy LS, Harrington JM. Quantitative risk assessments derived from occupational cancer epidemiology: a worked example. *Ann Occup Hyg.* 1998;42:347-352.
[14] Canadian Environmental Protection Act. Chromium and its compounds. 1994.
[15] ScientiWc Committee on Occupational Exposure Limits. Risk Assessment for Hexavalent Chromium. 2003.
[16] U.S. Environmental Protection Agency. Toxicological Review of Hexavalent Chromium (CAS No. 18540-29-9) in support of summary information on the Integrated Risk Information System (IRIS). 1998.
[17] WHO. Chromium. 2000.
[18] Mancuso TF. Consideration of Chromium as an industrial carcinogen. *Conference Proceedings International Conference on Heavy Metals in the Environment, Toronto, Ontario, Canada.* 1975.
[19] Steenland K, Loomis D, Shy C, Simonsen N. Review of occupational lung carcinogens. *Am J Ind Med.* 1996;29:474-490.
[20] WHO. Evaluation and use of epidemiological evidence for environmental health risk assessment: WHO guideline document. *Environ Health Perspect.* 2000;108:1002.
[21] Langard S, Andersen A, Ravnestad J. Incidence of cancer among ferrochromium and ferrosilicon workers: an extended observation period. *Br J Ind Med.* 1990;47:14-19.
[22] Braver ER, Infante P, Chu K. An analysis of lung cancer risk from exposure to hexavalent chromium. *Teratog Carcinog Mutagen.* 1985;5:365-378.
[23] Hayes RB, Lilienfeld AM, Snell LM. Mortality in chromium chemical production workers: a prospective study. *Int J Epidemiol.* 1979;8:365-374.
[24] Gibb HJ, Lees PS, Pinsky PF, Rooney BC. Lung cancer among workers in chromium chemical production. *Am J Ind Med.* 2000;38:115-126.
[25] Crump C, Crump K, Hack E, et al. Dose-response and risk assessment of airborne hexavalent chromium and lung cancer mortality. *Risk Anal.* 2003;23:1147-1163.
[26] Mancuso TF. Chromium as an industrial carcinogen: Part I. *Am J Ind Med.* 1997;31:129-139.
[27] Sheehan PJ, Meyer DM, Sauer MM, Paustenbach DJ. Assessment of the human health risks posed by exposure to chromium-contaminated soils. *J Toxicol Environ Health.* 1991;32:161-201.
[28] Ishikawa Y, Nakagawa K, Satoh Y, et al. Characteristics of chromate workers' cancers, chromium lung deposition and precancerous bronchial lesions: an autopsy study. *Br J Cancer.* 1994;70:160-166.

[29] Ishikawa Y, Nakagawa K, Satoh Y, et al. "Hot spots" of chromium accumulation at bifurcations of chromate workers' bronchi. *Cancer Res.* 1994;54:2342-2346.

[30] Langard S. One hundred years of chromium and cancer: a review of epidemiological evidence and selected case reports. *Am J Ind Med.* 1990;17:189-215.

[31] Park RM, Bena JF, Stayner LT, Smith RJ, Gibb HJ, Lees PS. Hexavalent chromium and lung cancer in the chromate industry: a quantitative risk assessment. *Risk Anal.* 2004;24:1099-1108.

[32] U.S. Environmental Protection Agency. Chromium Compounds. 1992.

[33] California Department of Health Services. Chromium-6 in Drinking Water: Regulation Update. 2007.

[34] Alexeeff GV, Satin K, Painter P, Zeise L, Popejoy C, Murchison G. Chromium carcinogenicity: California strategies. *Sci Total Environ.* 1989;86:159-168.

[35] Fryzek JP, Mumma MT, McLaughlin JK, Henderson BE, Blot WJ. Cancer mortality in relation to environmental chromium exposure. *J Occup Environ Med.* 2001;43:635-640.

[36] Finley BL, Kerger BD, Katona MW, Gargas ML, Corbett GC, Paustenbach DJ. Human ingestion of chromium (VI) in drinking water: pharmacokinetics following repeated exposure. *Toxicol Appl Pharmacol.* 1997;142:151-159.

[37] De FS, Camoirano A, Bagnasco M, Bennicelli C, Corbett GE, Kerger BD. Estimates of the chromium(VI) reducing capacity in human body compartments as a mechanism for attenuating its potential toxicity and carcinogenicity. *Carcinogenesis.* 1997;18:531-537.

[38] Paustenbach DJ, Finley BL, Mowat FS, Kerger BD. Human health risk and exposure assessment of chromium (VI) in tap water. *J Toxicol Environ Health A.* 2003;66:1295-1339.

[39] Bednar CM, Kies C. Inorganic contaminants in drinking water correlated with disease occurrence in Nebraska. *J Am Water Resour Assoc.* 1991;27:631-635.

[40] Zhang J, Li X. Chromium pollution of soil and water in Jinzhou (Chinese language with English abstract). *Zhonghua Yu Fang Yi Xue ZaZhi (Chinese Journal of Preventive Medicine).* 1987;21:262-264.

[41] Beaumont JJ, Sedman RM, Reynolds SD, et al. Cancer mortality in a Chinese population exposed to hexavalent chromium in drinking water. *Epidemiology.* 2008;19:12-23.

[42] Zhang J, Li X. *Chromium Contamination in the City of Jinzhou. Jinzhou Health and Anti-epidemic Station.* 1986.

[43] Costa M. Toxicity and carcinogenicity of Cr(VI) in animal models and humans. *Crit Rev Toxicol.* 1997;27:431-442.

[44] Sutherland JE, Zhitkovich A, Kluz T, Costa M. Rats retain chromium in tissues following chronic ingestion of drinking water containing hexavalent chromium. *Biol Trace Elem Res.* 2000;74:41-53.

[45] Borneff J, Engelhardt K, Griem W, Kunte H, Reichert J. [Carcinogens in water and soil. XXII. Experiment with 3,4-benzopyrene and potassium chromate in mice drink]. *Arch Hyg Bakteriol.* 1968;152:45-53.

[46] Sedman RM, Beaumont J, McDonald TA, Reynolds S, Krowech G, Howd R. Review of the evidence regarding the carcinogenicity of hexavalent chromium in drinking water. *J Environ Sci Health C Environ Carcinog Ecotoxicol Rev.* 2006;24:155-182.

[47] National Toxicology Program (NTP). Toxicology and Carcinogenesis Studies of Sodium Dichromate Dihydrate (CAS No. 7789-12-0) in F344/N Rats and B6C3F1 Mice (Drinking Water Studies). 2007.

[48] Uddin AN, Burns FJ, Rossman TG, Chen H, Kluz T, Costa M. Dietary chromium and nickel enhance UV-carcinogenesis in skin of hairless mice. *Toxicol Appl Pharmacol.* 2007;221:329-338.

[49] Davidson T, Kluz T, Burns F, et al. Exposure to chromium (VI) in the drinking water increases susceptibility to UV-induced skin tumors in hairless mice. *Toxicol Appl Pharmacol.* 2004;196:431-437.

[50] Marks JG, DeLeo VA. Contact and occupational dermatology. *St Louis: Mosbey Year Book.* 1992.

[51] Allergic contact dermatitis. In: Adams RM, ed. *Occupational skin disease 2nd ed.* 1990:26-31.

[52] Adverse chromate reactions on the skin. In: Burrows D, ed. *Chromium metabolism and toxicity.* 1983:138-158.

[53] Burrows D. Metals. In: Adams RM, ed. *Occupational skin disease 2nd ed.* 1990:349-386.

[54] Winston JR, WALSH EN. Chromate dermatitis in railroad employees working with diesel locomotives. *J Am Med Assoc.* 1951;147:1133-1134.

[55] Morris GE. Chrome dermatitis: a study of the chemistry of shoe leather with particular reference to basic chromic sulfate. *AMA Arch Derm.* 1958;78:612-618.

[56] Hostynek JJ, Maibach HI. Chromium in US household bleach. *Contact Dermatitis.* 1988;18:206-209.

[57] Irvine C, Pugh CE, Hansen EJ, Rycroft RJ. Cement dermatitis in underground workers during construction of the Channel Tunnel. *Occup Med (Lond).* 1994;44:17-23.

[58] Avnstorp C, Menne T. [Cement eczema]. *Ugeskr Laeger.* 1982;144:59-60.

[59] Zachariae CO, Agner T, Menne T. Chromium allergy in consecutive patients in a country where ferrous sulfate has been added to cement since 1981. *Contact Dermatitis.* 1996;35:83-85.

[60] Hansen MB, Rydin S, Menne T, Duus JJ. Quantitative aspects of contact allergy to chromium and exposure to chrome-tanned leather. *Contact Dermatitis.* 2002;47:127-134.

[61] Heidemann E. Leather. In: Ullmann's Encyclopedia of Industrial Chemistry. *Elvers B, Hawkins S, Schulz G (eds): Weinheim, VHC Verlag.* 1985;259-282.

[62] Nygren O, Wahlberg JE. Speciation of chromium in tanned leather gloves and relapse of chromium allergy from tanned leather samples 9. *Analyst.* 1998; 123:935-937.

[63] Samitz MH, GROSS S. Extraction by sweat of chromium from chrome tanned leathers. *J Occup Med.* 1960;2:12-14.

[64] Fregert S, Gruvberger B. Chromium in industrial leather gloves. *Contact Dermatitis.* 1979;5:189.

[65] Peltonen L, Fraki J. Prevalence of dichromate sensitivity. *Contact Dermatitis.* 1983;9:190-194.

[66] Nielsen NH, Linneberg A, Menne T, et al. Allergic contact sensitization in an adult Danish population: two cross-sectional surveys eight years apart (the Copenhagen Allergy Study). *Acta Derm Venereol.* 2001;81:31-34.

[67] Schafer T, Bohler E, Ruhdorfer S, et al. Epidemiology of contact allergy in adults. *Allergy.* 2001;56:1192-1196.

[68] Lestringant GG, Bener A, Sawaya M, Galadari IH, Frossard PM. Allergic contact dermatitis in the United Arab Emirates. *Int J Dermatol.* 1999;38:181-186.

[69] Marks JG, Jr., Belsito DV, Deleo VA, et al. North American Contact Dermatitis Group patch-test results, 1998 to 2000. *Am J Contact Dermat.* 2003;14:59-62.

[70] Allenby CF, Goodwin BF. Influence of detergent washing powders on minimal eliciting patch test concentrations of nickel and chromium 10. *Contact Dermatitis.* 1983;9:491-499.

[71] Kosann MK, Brancaccio RR, Shupack JL, Franks AG, Jr., Cohen DE. Six-hour versus 48-hour patch testing with varying concentrations of potassium dichromate. *Am J Contact Dermat.* 1998;9:92-95.

[72] Geiser JD. [Cement eczema]. *Praxis.* 1966;55:425-428.

[73] Rudzki E, Rebandel P, Karas Z. Patch testing with lower concentrations of chromate and nickel. *Contact Dermatitis.* 1997;37:46.

[74] Basketter D, Horev L, Slodovnik D, Merimes S, Trattner A, Ingber A. Investigation of the threshold for allergic reactivity to chromium. *Contact Dermatitis.* 2001;44:70-74.

[75] Skog E, Wahlberg JE. Patch testing with potassium dichromate in different vehicles. *G Ital Dermatol Minerva Dermatol.* 1969;44:467-469.

[76] Nethercott J, Paustenbach D, Adams R, et al. A study of chromium induced allergic contact dermatitis with 54 volunteers: implications for environmental risk assessment. *Occup Environ Med.* 1994;51:371-380.

[77] Arslan P, Beltrame M, Tomasi A. Intracellular chromium reduction. *Biochim Biophys Acta.* 1987;931:10-15.

[78] Buttner B, Beyersmann D. Modification of the erythrocyte anion carrier by chromate. *Xenobiotica.* 1985;15:735-741.

[79] Zhitkovich A, Voitkun V, Costa M. Glutathione and free amino acids form stable complexes with DNA following exposure of intact mammalian cells to chromate. *Carcinogenesis.* 1995;16:907-913.

[80] Salnikow K, Zhitkovich A, Costa M. Analysis of the binding sites of chromium to DNA and protein in vitro and in intact cells. *Carcinogenesis.* 1992;13:2341-2346.

[81] Zhitkovich A, Quievryn G, Messer J, Motylevich Z. Reductive activation with cysteine represents a chromium(III)-dependent pathway in the induction of genotoxicity by carcinogenic chromium(VI). *Environ Health Perspect.* 2002;110 Suppl 5:729-731.

[82] Zhitkovich A, Voitkun V, Costa M. Formation of the amino acid-DNA complexes by hexavalent and trivalent chromium in vitro: importance of trivalent chromium and the phosphate group. *Biochemistry.* 1996;35:7275-7282.

[83] Chen F, Ye J, Zhang X, Rojanasakul Y, Shi X. One-electron reduction of chromium(VI) by alpha-lipoic acid and related hydroxyl radical generation, dG hydroxylation and nuclear transcription factor-kappaB activation. *Arch Biochem Biophys.* 1997;338:165-172.

[84] Wetterhahn KE, Hamilton JW. Molecular basis of hexavalent chromium carcinogenicity: effect on gene expression. *Sci Total Environ.* 1989;86:113-129.

[85] Sugiyama M. Effects of vitamins on chromium(VI)-induced damage. *Environ Health Perspect.* 1991;92:63-70.

[86] Stearns DM, Kennedy LJ, Courtney KD, Giangrande PH, Phieffer LS, Wetterhahn KE. Reduction of chromium(VI) by ascorbate leads to chromium-DNA binding and DNA strand breaks in vitro. *Biochemistry.* 1995;34:910-919.

[87] Liu KJ, Jiang J, Swartz HM, Shi X. Low-frequency EPR detection of chromium(V) formation by chromium(VI) reduction in whole live mice. *Arch Biochem Biophys.* 1994;313:248-252.

[88] Liu KJ, Shi X, Jiang JJ, Goda F, Dalal N, Swartz HM. Chromate-induced chromium(V) formation in live mice and its control by cellular antioxidants: an L-band electron paramagnetic resonance study. *Arch Biochem Biophys.* 1995;323:33-39.

[89] Liu KJ, Shi X, Jiang J, Goda F, Dalal N, Swartz HM. Low frequency electron paramagnetic resonance investigation on metabolism of chromium (VI) by whole live mice. *Ann Clin Lab Sci.* 1996;26:176-184.

[90] Stearns DM, Courtney KD, Giangrande PH, Phieffer LS, Wetterhahn KE. Chromium(VI) reduction by ascorbate: role of reactive intermediates in DNA damage in vitro. *Environ Health Perspect.* 1994;102 Suppl 3:21-25.

[91] Stearns DM, Wetterhahn KE. Reaction of chromium(VI) with ascorbate produces chromium(V), chromium(IV), and carbon-based radicals. *Chem Res Toxicol.* 1994;7:219-230.

[92] Tsapakos MJ, Wetterhahn KE. The interaction of chromium with nucleic acids. *Chem Biol Interact.* 1983;46:265-277.

[93] Jennett KW. Microsomal reduction of the carcinogen chromate produces Cr(V). *J Am Chem Soc.* 198;104:874-875.

[94] Standeven AM, Wetterhahn KE. Possible role of glutathione in chromium(VI) metabolism and toxicity in rats. *Pharmacol Toxicol.* 1991;68:469-476.

[95] Kortenkamp A, Ozolins Z, Beyersmann D, O'Brien P. Generation of PM2 DNA breaks in the course of reduction of chromium(VI) by glutathione. *Mutat Res.* 1989;216:19-26.

[96] Farrell RP, Judd RJ, Lay PA, Dixon NE, Baker RS, Bonin AM. Chromium(V)-induced cleavage of DNA: are chromium(V) complexes the active carcinogens in chromium(VI)-induced cancers? *Chem Res Toxicol.* 1989;2:227-229.

[97] Shi X, Chiu A, Chen CT, Halliwell B, Castranova V, Vallyathan V. Reduction of chromium(VI) and its relationship to carcinogenesis. *J Toxicol Environ Health B Crit Rev.* 1999;2:87-104.

[98] Cheng L, Liu S, Dixon K. Analysis of repair and mutagenesis of chromium-induced DNA damage in yeast, mammalian cells, and transgenic mice. *Environ Health Perspect.* 1998;106 Suppl 4:1027-1032.

[99] O'Brien TJ, Fornsaglio JL, Ceryak S, Patierno SR. Effects of hexavalent chromium on the survival and cell cycle distribution of DNA repair-deficient S. cerevisiae. *DNA Repair (Amst).* 2002;1:617-627.

[100] Swanson RL, Morey NJ, Doetsch PW, Jinks-Robertson S. Overlapping specificities of base excision repair, nucleotide excision repair, recombination, and translesion synthesis pathways for DNA base damage in Saccharomyces cerevisiae. *Mol Cell Biol.* 1999;19:2929-2935.

[101] Dizdaroglu M. Chemical determination of free radical-induced damage to DNA. *Free Radical Biol Med.* 1991;10:225-242.

[102] Shi XG, Sun XL, Gannett PM, Dalal NS. Deferoxamine inhibition of Cr(V)-mediated radical generation and deoxyguanine hydroxylation: ESR and HPLC evidence. *Arch Biochem Biophys.* 1992;293:281-286.

[103] Shi X, Mao Y, Knapton AD, et al. Reaction of Cr(VI) with ascorbate and hydrogen peroxide generates hydroxyl radicals and causes DNA damage: role of a Cr(IV)-mediated Fenton-like reaction. *Carcinogenesis.* 1994;15:2475-2478.

[104] Joseph P, He Q, Umbright C. Heme-oxygenase 1 gene expression is a marker for hexavalent chromium-induced stress and toxicity in human dermal fibroblasts. *Toxicol Sci.* 2008;103:325-334.

[105] Samet JM, Graves LM, Quay J, et al. Activation of MAPKs in human bronchial epithelial cells exposed to metals. *Am J Physiol.* 1998;275:L551-L558.

[106] Chuang SM, Yang JL. Comparison of roles of three mitogen-activated protein kinases induced by chromium(VI) and cadmium in non-small-cell lung carcinoma cells. *Mol Cell Biochem.* 2001;222:85-95.

[107] Chuang SM, Liou GY, Yang JL. Activation of JNK, p38 and ERK mitogen-activated protein kinases by chromium(VI) is mediated through oxidative stress but does not affect cytotoxicity. *Carcinogenesis.* 2000;21:1491-1500.

[108] Chen F, Castranova V, Shi X, Demers LM. New insights into the role of nuclear factor-kappaB, a ubiquitous transcription factor in the initiation of diseases. *Clin Chem.* 1999;45:7-17.

[109] Baldwin AS, Jr. The NF-kappa B and I kappa B proteins: new discoveries and insights. *Annu Rev Immunol.* 1996;14:649-683.

[110] Baeuerle PA, Henkel T. Function and activation of NF-kappa B in the immune system. *Annu Rev Immunol.* 1994;12:141-179.

[111] Ye J, Zhang X, Young HA, Mao Y, Shi X. Chromium(VI)-induced nuclear factor-kappa B activation in intact cells via free radical reactions. *Carcinogenesis.* 1995;16:2401-2405.

[112] Chen F, Ding M, Lu Y, et al. Participation of MAP kinase p38 and IkappaB kinase in chromium (VI)-induced NF-kappaB and AP-1 activation. *J Environ Pathol Toxicol Oncol.* 2000;19:231-238.

[113] Xie H, Wise SS, Holmes AL, et al. Carcinogenic lead chromate induces DNA double-strand breaks in human lung cells. *Mutat Res.* 2005;586:160-172.

[114] Liu KJ, Shi X, Dalal NS. Synthesis of Cr(IV)-GSH, its identification and its free hydroxyl radical generation: a model compound for Cr(VI) carcinogenicity. *Biochem Biophys Res Commun.* 1997;235:54-58.

[115] Shi XL, Dalal NS. Chromium (V) and hydroxyl radical formation during the glutathione reductase-catalyzed reduction of chromium (VI). *Biochem Biophys Res Commun.* 1989;163:627-634.

[116] Shi XL, Dalal NS. ESR spin trapping detection of hydroxyl radicals in the reactions of Cr(V) complexes with hydrogen peroxide. *Free Radic Res Commun.* 1990;10:17-26.

[117] Shi XG, Dalal NS. On the hydroxyl radical formation in the reaction between hydrogen peroxide and biologically generated chromium(V) species. *Arch Biochem Biophys.* 1990;277:342-350.

[118] Shi XL, Dalal NS. NADPH-dependent flavoenzymes catalyze one electron reduction of metal ions and molecular oxygen and generate hydroxyl radicals. *FEBS Lett.* 1990;276:189-191.

[119] Shi XL, Dalal NS. Evidence for a Fenton-type mechanism for the generation of .OH radicals in the reduction of Cr(VI) in cellular media. *Arch Biochem Biophys.* 1990;281:90-95.

[120] Shi XL, Dalal NS. The role of superoxide radical in chromium (VI)-generated hydroxyl radical: the Cr(VI) Haber-Weiss cycle. *Arch Biochem Biophys.* 1992;292:323-327.

[121] Luo H, Lu Y, Shi X, Mao Y, Dalal NS. Chromium (IV)-mediated fenton-like reaction causes DNA damage: implication to genotoxicity of chromate. *Ann Clin Lab Sci.* 1996;26:185-191.

[122] Mikalsen A, Alexander J, Ryberg D. Microsomal metabolism of hexavalent chromium. Inhibitory effect of oxygen and involvement of cytochrome P-450. *Chem Biol Interact.* 1989;69:175-192.

[123] Mikalsen A, Alexander J, Andersen RA, Daae HL. Reduction of hexavalent chromium in a reconstituted system of cytochrome P-450 and cytochrome b5. *Chem Biol Interact.* 1989;71:213-221.

[124] Mikalsen A, Alexander J, Wallin H, Ingelman-Sundberg M, Andersen RA. Reductive metabolism and protein binding of chromium(VI) by P450 protein enzymes. *Carcinogenesis.* 1991;12:825-831.

[125] Ryberg D, Alexander J. Inhibitory action of hexavalent chromium (Cr(VI)) on the mitochondrial respiration and a possible coupling to the reduction of Cr(VI). *Biochem Pharmacol.* 1984;33:2461-2466.

[126] Ryberg D, Alexander J. Mechanisms of chromium toxicity in mitochondria. *Chem Biol Interact.* 1990;75:141-151.

[127] Banks RB, Cooke RT, Jr. Chromate reduction by rabbit liver aldehyde oxidase. *Biochem Biophys Res Commun.* 1986;137:8-14.

[128] Kawanishi S, Inoue S, Sano S. Mechanism of DNA cleavage induced by sodium chromate(VI) in the presence of hydrogen peroxide. *J Biol Chem.* 1986;261:5952-5958.

[129] Quievryn G, Peterson E, Messer J, Zhitkovich A. Genotoxicity and mutagenicity of chromium(VI)/ascorbate-generated DNA adducts in human and bacterial cells. *Biochemistry.* 2003;42:1062-1070.

[130] Borges KM, Wetterhahn KE. Chromium cross-links glutathione and cysteine to DNA. *Carcinogenesis.* 1989;10:2165-2168.

[131] Standeven AM, Wetterhahn KE. Is there a role for reactive oxygen species in the mechanism of chromium(VI) carcinogenesis? *Chem Res Toxicol.* 1991;4:616-625.

[132] Shi XL, Dalal NS. On the mechanism of the chromate reduction by glutathione: ESR evidence for the glutathionyl radical and an isolable Cr(V) intermediate. *Biochem Biophys Res Commun.* 1988;156:137-142.

[133] Shi X, Dong Z, Dalal NS, Gannett PM. Chromate-mediated free radical generation from cysteine, penicillamine, hydrogen peroxide, and lipid hydroperoxides. *Biochim Biophys Acta.* 1994;1226:65-72.

[134] Rossi SC, Gorman N, Wetterhahn KE. Mitochondrial reduction of the carcinogen chromate: formation of chromium(V). *Chem Res Toxicol.* 1988;1:101-107.

[135] Rossi SC, Wetterhahn KE. Chromium(V) is produced upon reduction of chromate by mitochondrial electron transport chain complexes. *Carcinogenesis.* 1989;10:913-920.

[136] Standeven AM, Wetterhahn KE. Ascorbate is the principal reductant of chromium (VI) in rat liver and kidney ultrafiltrates. *Carcinogenesis.* 1991;12:1733-1737.

[137] Standeven AM, Wetterhahn KE. Ascorbate is the principal reductant of chromium(VI) in rat lung ultrafiltrates and cytosols, and mediates chromium-DNA binding in vitro. *Carcinogenesis.* 1992;13:1319-1324.

[138] Suzuki Y, Fukuda K. Reduction of hexavalent chromium by ascorbic acid and glutathione with special reference to the rat lung. *Arch Toxicol.* 1990;64:169-176.

[139] Office of Air Quality Planning and Standards. EPA Locating and Estimating Air Emissions from Sources of Chromium. *US Environmental Protection Agency, Research Triangle Park, NC.* 1984.

[140] Aiyar J, Berkovits HJ, Floyd RA, Wetterhahn KE. Reaction of chromium (VI) with hydrogen peroxide in the presence of glutathione: reactive intermediates and resulting DNA damage. *Chem Res Toxicol.* 1990;3:595-603.

[141] Aiyar J, Berkovits HJ, Floyd RA, Wetterhahn KE. Reaction of chromium(VI) with glutathione or with hydrogen peroxide: identification of reactive intermediates and their role in chromium(VI)-induced DNA damage. *Environ Health Perspect.* 1991;92:53-62.

[142] Wetterhahn KE, Hamilton JW, Aiyar J, Borges KM, Floyd R. Mechanism of chromium(VI) carcinogenesis. Reactive intermediates and effect on gene expression. *Biol Trace Elem Res.* 1989;21:405-411.

[143] Shi XL, Dalal NS, Vallyathan V. One-electron reduction of carcinogen chromate by microsomes, mitochondria, and Escherichia coli: identification of Cr(V) and .OH radical. *Arch Biochem Biophys.* 1991;290:381-386.

[144] Voitkun V, Zhitkovich A, Costa M. Complexing of amino acids to DNA by chromate in intact cells. *Environ Health Perspect.* 1994;102 Suppl 3:251-255.

[145] Quievryn G, Messer J, Zhitkovich A. Carcinogenic chromium(VI) induces cross-linking of vitamin C to DNA in vitro and in human lung A549 cells. *Biochemistry.* 2002;41:3156-3167.

[146] Sugden KD, Wetterhahn KE. EPR Evidence for Chromium(V) Binding to Phosphate and Pyrophosphate: Implications for Chromium(V)-DNA Interactions. *Inorg Chem.* 1996;35:3727-3728.

[147] Gulanowski B, Swiatek J, Kozlowski H. Impact of chromium ions on nucleoside triphosphates and nucleic acids. *J Inorg Biochem.* 1992;48:289-298.

[148] Chen J, Thilly WG. Use of denaturing-gradient gel electrophoresis to study chromium-induced point mutations in human cells. *Environ Health Perspect.* 1994;102 Suppl 3:227-229.

[149] Chen J, Thilly WG. Mutational spectrum of chromium(VI) in human cells. *Mutat Res.* 1994;323:21-27.

[150] Yang JL, Hsieh YC, Wu CW, Lee TC. Mutational specificity of chromium(VI) compounds in the hprt locus of Chinese hamster ovary-K1 cells. *Carcinogenesis.* 1992;13:2053-2057.

[151] Elias Z, Schneider O, Aubry F, Daniere MC, Poirot O. Sister chromatid exchanges in Chinese hamster V79 cells treated with the trivalent chromium compounds chromic chloride and chromic oxide. *Carcinogenesis.* 1983;4:605-611.

[152] Sen P, Costa M. Incidence and localization of sister chromatid exchanges induced by nickel and chromium compounds. *Carcinogenesis.* 1986;7:1527-1533.

[153] Sarto F, Cominato I, Bianchi V, Levis AG. Increased incidence of chromosomal aberrations and sister chromatid exchanges in workers exposed to chromic acid (CrO3) in electroplating factories. *Carcinogenesis.* 1982;3:1011-1016.

[154] Venier P, Montaldi A, Majone F, Bianchi V, Levis AG. Cytotoxic, mutagenic and clastogenic effects of industrial chromium compounds. *Carcinogenesis.* 1982;3:1331-1338.

[155] Bianchi V, Celotti L, Lanfranchi G, et al. Genetic effects of chromium compounds. *Mutat Res.* 1983;117:279-300.

[156] Sugiyama M, Tsuzuki K, Hidaka T, Ogura R, Yamamoto M. Reduction of chromium(VI) in Chinese hamster V-79 cells. *Biol Trace Elem Res.* 1991;30:1-8.

[157] Sugiyama M. Role of physiological antioxidants in chromium(VI)-induced cellular injury. *Free Radic Biol Med.* 1992;12:397-407.

[158] Wise JP, Sr., Stearns DM, Wetterhahn KE, Patierno SR. Cell-enhanced dissolution of carcinogenic lead chromate particles: the role of individual dissolution products in clastogenesis. *Carcinogenesis.* 1994;15:2249-2254.

[159] Wise JP, Orenstein JM, Patierno SR. Inhibition of lead chromate clastogenesis by ascorbate: relationship to particle dissolution and uptake. *Carcinogenesis.* 1993;14:429-434.

[160] Lin X, Zhuang Z, Costa M. Analysis of residual amino acid--DNA crosslinks induced in intact cells by nickel and chromium compounds. *Carcinogenesis.* 1992; 13: 1763-1768.

[161] Miller CA, III, Costa M. Characterization of DNA-protein complexes induced in intact cells by the carcinogen chromate. *Mol Carcinog.* 1988;1:125-133.

[162] Miller CA, III, Cohen MD, Costa M. Complexing of actin and other nuclear proteins to DNA by cis-diamminedichloroplatinum(II) and chromium compounds. *Carcinogenesis.* 1991;12:269-276.

[163] Krupa R, Stanczak M, Walter Z. Chromium incorporated in RNA and DNA. *Z Naturforsch [C].* 2002;57:951-953.

[164] Tamino G, Peretta L, Levis AG. Effects of the trivalent and hexavalent chromium on the physicochemical properties of mammalian cell nucleic acids and synthetic polynucleotides. *Chem Biol Interact.* 1981;37:309-319.

[165] Wolf T, Kasemann R, Ottenwalder H. Molecular interaction of different chromium species with nucleotides and nucleic acids. *Carcinogenesis.* 1989;10:655-659.

[166] O'Brien T, Mandel HG, Pritchard DE, Patierno SR. Critical role of chromium (Cr)-DNA interactions in the formation of Cr-induced polymerase arresting lesions. *Biochemistry.* 2002;41:12529-12537.

[167] Zhitkovich A, Song Y, Quievryn G, Voitkun V. Non-oxidative mechanisms are responsible for the induction of mutagenesis by reduction of Cr(VI) with cysteine: role of ternary DNA adducts in Cr(III)-dependent mutagenesis. *Biochemistry.* 2001;40:549-560.

[168] Wolf T, Kasemann R, Ottenwalder H. Molecular interaction of different chromium species with nucleotides and nucleic acids. *Carcinogenesis.* 1989;10:655-659.

[169] Voitkun V, Zhitkovich A, Costa M. Cr(III)-mediated crosslinks of glutathione or amino acids to the DNA phosphate backbone are mutagenic in human cells. *Nucleic Acids Res.* 1998;26:2024-2030.

[170] Bridgewater LC, Manning FC, Patierno SR. Arrest of replication by mammalian DNA polymerases alpha and beta caused by chromium-DNA lesions. *Mol Carcinog.* 1998;23:201-206.

[171] Bridgewater LC, Manning FC, Patierno SR. Base-specific arrest of in vitro DNA replication by carcinogenic chromium: relationship to DNA interstrand crosslinking. *Carcinogenesis.* 1994;15:2421-2427.

[172] Biedermann KA, Landolph JR. Induction of anchorage independence in human diploid foreskin fibroblasts by carcinogenic metal salts. *Cancer Res.* 1987;47:3815-3823.

[173] Biedermann KA, Landolph JR. Role of valence state and solubility of chromium compounds on induction of cytotoxicity, mutagenesis, and anchorage independence in diploid human fibroblasts. *Cancer Res.* 1990;50:7835-7842.

[174] Patierno SR, Banh D, Landolph JR. Transformation of C3H/10T1/2 mouse embryo cells to focus formation and anchorage independence by insoluble lead chromate but not soluble calcium chromate: relationship to mutagenesis and internalization of lead chromate particles. *Cancer Res.* 1988;48:5280-5288.

[175] Snow ET, Xu LS. Chromium(III) bound to DNA templates promotes increased polymerase processivity and decreased fidelity during replication in vitro. *Biochemistry.* 1991;30:11238-11245.

[176] Strauss JK, Maher LJ, III. DNA bending by asymmetric phosphate neutralization. *Science.* 1994;266:1829-1834.

[177] Tsapakos MJ, Hampton TH, Wetterhahn KE. Chromium(VI)-induced DNA lesions and chromium distribution in rat kidney, liver, and lung. *Cancer Res.* 1983;43:5662-5667.

[178] Tsapakos MJ, Hampton TH, Sinclair PR, Sinclair JF, Bement WJ, Wetterhahn KE. The carcinogen chromate causes DNA damage and inhibits drug-mediated induction of porphyrin accumulation and glucuronidation in chick embryo hepatocytes. *Carcinogenesis.* 1983;4:959-966.

[179] Bridgewater LC, Manning FC, Woo ES, Patierno SR. DNA polymerase arrest by adducted trivalent chromium. *Mol Carcinog.* 1994;9:122-133.

[180] O'Brien TJ, Ceryak S, Patierno SR. Complexities of chromium carcinogenesis: role of cellular response, repair and recovery mechanisms. *Mutat Res.* 2003;533:3-36.

[181] Shi X, Mao Y, Knapton AD, et al. Reaction of Cr(VI) with ascorbate and hydrogen peroxide generates hydroxyl radicals and causes DNA damage: role of a Cr(IV)-mediated Fenton-like reaction. *Carcinogenesis.* 1994;15:2475-2478.

[182] Fornace AJ, Jr., Seres DS, Lechner JF, Harris CC. DNA-protein cross-linking by chromium salts. *Chem Biol Interact.* 1981;36:345-354.

[183] Cupo DY, Wetterhahn KE. Binding of chromium to chromatin and DNA from liver and kidney of rats treated with sodium dichromate and chromium(III) chloride in vivo. *Cancer Res.* 1985;45:1146-1151.

[184] Mattagajasingh SN, Misra HP. Mechanisms of the carcinogenic chromium(VI)-induced DNA-protein cross-linking and their characterization in cultured intact human cells. *J Biol Chem.* 1996;271:33550-33560.

[185] Christie NT, Cantoni O, Evans RM, Meyn RE, Costa M. Use of mammalian DNA repair-deficient mutants to assess the effects of toxic metal compounds on DNA. *Biochem Pharmacol.* 1984;33:1661-1670.

[186] Hamilton JW, Wetterhahn KE. Chromium (VI)-induced DNA damage in chick embryo liver and blood cells in vivo. *Carcinogenesis.* 1986;7:2085-2088.

[187] Ueno S, Kashimoto T, Susa N, et al. Detection of dichromate (VI)-induced DNA strand breaks and formation of paramagnetic chromium in multiple mouse organs. *Toxicol Appl Pharmacol.* 2001;170:56-62.

[188] Werfel U, Langen V, Eickhoff I, et al. Elevated DNA single-strand breakage frequencies in lymphocytes of welders exposed to chromium and nickel. *Carcinogenesis.* 1998;19:413-418.

[189] Bagchi D, Stohs SJ, Downs BW, Bagchi M, Preuss HG. Cytotoxicity and oxidative mechanisms of different forms of chromium. *Toxicology.* 2002;180:5-22.

[190] Balansky RM, D'Agostini F, Izzotti A, De FS. Less than additive interaction between cigarette smoke and chromium(VI) in inducing clastogenic damage in rodents. *Carcinogenesis.* 2000;21:1677-1682.

[191] Benova D, Hadjidekova V, Hristova R, et al. Cytogenetic effects of hexavalent chromium in Bulgarian chromium platers. *Mutat Res.* 2002;514:29-38.

[192] Seoane AI, Dulout FN. Genotoxic ability of cadmium, chromium and nickel salts studied by kinetochore staining in the cytokinesis-blocked micronucleus assay. *Mutat Res.* 2001;490:99-106.

[193] Stearns DM, Wise JP, Sr., Patierno SR, Wetterhahn KE. Chromium(III) picolinate produces chromosome damage in Chinese hamster ovary cells. *FASEB J.* 1995;9:1643-1648.

[194] Sugiyama M, Lin XH, Costa M. Protective effect of vitamin E against chromosomal aberrations and mutation induced by sodium chromate in Chinese hamster V79 cells. *Mutat Res.* 1991;260:19-23.

[195] Sugiyama M, Tsuzuki K, Lin X, Costa M. Potentiation of sodium chromate(VI)-induced chromosomal aberrations and mutation by vitamin B2 in Chinese hamster V79 cells. *Mutat Res.* 1992;283:211-214.

[196] Wise JP, Sr., Wise SS, Little JE. The cytotoxicity and genotoxicity of particulate and soluble hexavalent chromium in human lung cells. *Mutat Res.* 2002;517:221-229.

[197] Hirose T, Kondo K, Takahashi Y, et al. Frequent microsatellite instability in lung cancer from chromate-exposed workers. *Mol Carcinog.* 2002;33:172-180.

[198] De FS, De Renzi GP, Camoirano A, et al. Genotoxicity assay of oil dispersants in bacteria (mutation, differential lethality, SOS DNA-repair) and yeast (mitotic crossing-over). *Mutat Res.* 1985;158:19-30.

[199] Llagostera M, Garrido S, Guerrero R, Barbe J. Induction of SOS genes of Escherichia coli by chromium compounds. *Environ Mutagen.* 1986;8:571-577.

[200] Bianchi V, Dal TR, Debetto P, et al. Mechanisms of chromium toxicity in mammalian cell cultures. *Toxicology.* 1980;17:219-224.

[201] Sugden KD, Campo CK, Martin BD. Direct oxidation of guanine and 7,8-dihydro-8-oxoguanine in DNA by a high-valent chromium complex: a possible mechanism for chromate genotoxicity. *Chem Res Toxicol.* 2001;14:1315-1322.

[202] Tsou TC, Lin RJ, Yang JL. Mutational spectrum induced by chromium(III) in shuttle vectors replicated in human cells: relationship to Cr(III)-DNA interactions. *Chem Res Toxicol.* 1997;10:962-970.

[203] Xu J, Bubley GJ, Detrick B, Blankenship LJ, Patierno SR. Chromium(VI) treatment of normal human lung cells results in guanine-specific DNA polymerase arrest, DNA-DNA cross-links and S-phase blockade of cell cycle. *Carcinogenesis.* 1996;17:1511-1517.

[204] Levis AG, Bianchi V, Tamino G, Pegoraro B. Cytotoxic effects of hexavalent and trivalent chromium on mammalian cells in vitro. *Br J Cancer.* 1978;37:386-396.

[205] Wang JF, Engelsberg BN, Johnson SW, et al. DNA binding activity of the mammalian translation elongation complex: recognition of chromium- and transplatin-damaged DNA. *Arch Toxicol.* 1997;71:450-454.
[206] Wang JF, Bashir M, Engelsberg BN, Witmer C, Rozmiarek H, Billings PC. High mobility group proteins 1 and 2 recognize chromium-damaged DNA. *Carcinogenesis.* 1997;18:371-375.
[207] Xu J, Manning FC, O'Brien TJ, Ceryak S, Patierno SR. Mechanisms of chromium-induced suppression of RNA synthesis in cellular and cell-free systems: relationship to RNA polymerase arrest. *Mol Cell Biochem.* 2004;255:151-160.
[208] Tsou TC, Lai HJ, Yang JL. Effects of mannitol or catalase on the generation of reactive oxygen species leading to DNA damage by Chromium(VI) reduction with ascorbate. *Chem Res Toxicol.* 1999;12:1002-1009.
[209] Ljungman M, Zhang F, Chen F, Rainbow AJ, McKay BC. Inhibition of RNA polymerase II as a trigger for the p53 response. *Oncogene.* 1999;18:583-592.
[210] Ljungman M, Zhang F. Blockage of RNA polymerase as a possible trigger for u.v. light-induced apoptosis. *Oncogene.* 1996;13:823-831.
[211] Ljungman M, Paulsen MT. The cyclin-dependent kinase inhibitor roscovitine inhibits RNA synthesis and triggers nuclear accumulation of p53 that is unmodified at Ser15 and Lys382. *Mol Pharmacol.* 2001;60:785-789.
[212] Blankenship LJ, Carlisle DL, Wise JP, Orenstein JM, Dye LE, III, Patierno SR. Induction of apoptotic cell death by particulate lead chromate: differential effects of vitamins C and E on genotoxicity and survival. *Toxicol Appl Pharmacol.* 1997;146:270-280.
[213] Manning FC, Blankenship LJ, Wise JP, Xu J, Bridgewater LC, Patierno SR. Induction of internucleosomal DNA fragmentation by carcinogenic chromate: relationship to DNA damage, genotoxicity, and inhibition of macromolecular synthesis. *Environ Health Perspect.* 1994;102 Suppl 3:159-167.
[214] Blankenship LJ, Manning FC, Orenstein JM, Patierno SR. Apoptosis is the mode of cell death caused by carcinogenic chromium. *Toxicol Appl Pharmacol.* 1994;126:75-83.
[215] Carlisle DL, Pritchard DE, Singh J, Patierno SR. Chromium(VI) induces p53-dependent apoptosis in diploid human lung and mouse dermal fibroblasts. *Mol Carcinog.* 2000;28:111-118.
[216] Pritchard DE, Singh J, Carlisle DL, Patierno SR. Cyclosporin A inhibits chromium(VI)-induced apoptosis and mitochondrial cytochrome c release and restores clonogenic survival in CHO cells. *Carcinogenesis.* 2000;21:2027-2033.
[217] Tsou TC, Chen CL, Liu TY, Yang JL. Induction of 8-hydroxydeoxyguanosine in DNA by chromium(III) plus hydrogen peroxide and its prevention by scavengers. *Carcinogenesis.* 1996;17:103-108.
[218] Ueno S, Sugiyama M, Susa N, Furukawa Y. Effect of dimethylthiourea on chromium (VI)-induced DNA single-strand breaks in Chinese hamster V-79 cells. *Mutat Res.* 1995;346:247-253.
[219] Kortenkamp A, Oetken G, Beyersmann D. The DNA cleavage induced by a chromium(V) complex and by chromate and glutathione is mediated by activated oxygen species. *Mutat Res.* 1990;232:155-161.
[220] Sugiyama M. Effects of vitamin E and vitamin B2 on chromate-induced DNA lesions. *Biol Trace Elem Res.* 1989;21:399-404.

[221] Tsou TC, Yang JL. Formation of reactive oxygen species and DNA strand breakage during interaction of chromium (III) and hydrogen peroxide in vitro: evidence for a chromium (III)-mediated Fenton-like reaction. *Chem Biol Interact.* 1996;102:133-153.
[222] Kortenkamp A, Casadevall M, Faux SP, et al. A role for molecular oxygen in the formation of DNA damage during the reduction of the carcinogen chromium (VI) by glutathione. *Arch Biochem Biophys.* 1996;329:199-207.
[223] Casadevall M, da Cruz FP, Kortenkamp A. Chromium(VI)-mediated DNA damage: oxidative pathways resulting in the formation of DNA breaks and abasic sites. *Chem Biol Interact.* 1999;123:117-132.
[224] Hartwell LH, Kastan MB. Cell cycle control and cancer. *Science.* 1994;266:1821-1828.
[225] Elledge SJ. Cell cycle checkpoints: preventing an identity crisis. *Science.* 1996;274:1664-1672.
[226] Pritchard DE, Ceryak S, Ha L, et al. Mechanism of apoptosis and determination of cellular fate in chromium(VI)-exposed populations of telomerase-immortalized human fibroblasts. *Cell Growth Differ.* 2001;12:487-496.
[227] Shiloh Y. ATM and ATR: networking cellular responses to DNA damage. *Curr Opin Genet Dev.* 2001;11:71-77.
[228] Garg R, Callens S, Lim DS, Canman CE, Kastan MB, Xu B. Chromatin association of rad17 is required for an ataxia telangiectasia and rad-related kinase-mediated S-phase checkpoint in response to low-dose ultraviolet radiation. *Mol Cancer Res.* 2004;2: 362-369.
[229] Wakeman TP, Kim WJ, Callens S, Chiu A, Brown KD, Xu B. The ATM-SMC1 pathway is essential for activation of the chromium[VI]-induced S-phase checkpoint. *Mutat Res.* 2004;554:241-251.
[230] Wakeman TP, Xu B. ATR regulates hexavalent chromium-induced S-phase checkpoint through phosphorylation of SMC1. *Mutat Res.* 2006;610:14-20.
[231] Bakkenist CJ, Kastan MB. DNA damage activates ATM through intermolecular autophosphorylation and dimer dissociation. *Nature.* 2003;421:499-506.
[232] Shiloh Y. ATM: from phenotype to functional genomics--and back. *Ernst Schering Res Found Workshop.* 2002;51-70.
[233] Lim DS, Kim ST, Xu B, et al. ATM phosphorylates p95/nbs1 in an S-phase checkpoint pathway. *Nature.* 2000;404:613-617.
[234] Kim ST, Xu B, Kastan MB. Involvement of the cohesin protein, Smc1, in Atm-dependent and independent responses to DNA damage. *Genes Dev.* 2002;16:560-570.
[235] Yazdi PT, Wang Y, Zhao S, Patel N, Lee EY, Qin J. SMC1 is a downstream effector in the ATM/NBS1 branch of the human S-phase checkpoint. *Genes Dev.* 2002;16:571-582.
[236] Zhao S, Weng YC, Yuan SS, et al. Functional link between ataxia-telangiectasia and Nijmegen breakage syndrome gene products. *Nature.* 2000;405:473-477.
[237] Kondo T, Wakayama T, Naiki T, Matsumoto K, Sugimoto K. Recruitment of Mec1 and Ddc1 checkpoint proteins to double-strand breaks through distinct mechanisms. *Science.* 2001;294:867-870.
[238] Zou L, Cortez D, Elledge SJ. Regulation of ATR substrate selection by Rad17-dependent loading of Rad9 complexes onto chromatin. *Genes Dev.* 2002;16:198-208.
[239] Melo JA, Cohen J, Toczyski DP. Two checkpoint complexes are independently recruited to sites of DNA damage in vivo. *Genes Dev.* 2001;15:2809-2821.

[240] Bao S, Tibbetts RS, Brumbaugh KM, et al. ATR/ATM-mediated phosphorylation of human Rad17 is required for genotoxic stress responses. *Nature.* 2001;411:969-974.

[241] Lindsey-Boltz LA, Bermudez VP, Hurwitz J, Sancar A. Purification and characterization of human DNA damage checkpoint Rad complexes. *Proc Natl Acad Sci U S A.* 2001;98:11236-11241.

[242] Wakeman TP, Wyczechowska D, Xu B. Involvement of the p38 MAP kinase in Cr(VI)-induced growth arrest and apoptosis. *Mol Cell Biochem.* 2005;279:69-73.

[243] Ye J, Wang S, Leonard SS, et al. Role of reactive oxygen species and p53 in chromium(VI)-induced apoptosis. *J Biol Chem.* 1999;274:34974-34980.

[244] Cortez D, Wang Y, Qin J, Elledge SJ. Requirement of ATM-dependent phosphorylation of brca1 in the DNA damage response to double-strand breaks. *Science.* 1999;286:1162-1166.

[245] Falck J, Mailand N, Syljuasen RG, Bartek J, Lukas J. The ATM-Chk2-Cdc25A checkpoint pathway guards against radioresistant DNA synthesis. *Nature.* 2001;410:842-847.

[246] Pritchard DE, Singh J, Carlisle DL, Patierno SR. Cyclosporin A inhibits chromium(VI)-induced apoptosis and mitochondrial cytochrome c release and restores clonogenic survival in CHO cells. *Carcinogenesis.* 2000;21:2027-2033.

[247] Korsmeyer SJ, Gross A, Harada H, et al. Death and survival signals determine active/inactive conformations of pro-apoptotic BAX, BAD, and BID molecules. *Cold Spring Harb Symp Quant Biol.* 1999;64:343-350.

[248] Ceryak S, Zingariello C, O'Brien T, Patierno SR. Induction of pro-apoptotic and cell cycle-inhibiting genes in chromium (VI)-treated human lung fibroblasts: lack of effect of ERK. *Mol Cell Biochem.* 2004;255:139-149.

[249] Carlisle DL, Pritchard DE, Singh J, et al. Apoptosis and P53 induction in human lung fibroblasts exposed to chromium (VI): effect of ascorbate and tocopherol. *Toxicol Sci.* 2000;55:60-68.

[250] Abraham RT. Cell cycle checkpoint signaling through the ATM and ATR kinases. *Genes Dev.* 2001;15:2177-2196.

[251] Savitsky K, Sfez S, Tagle DA, et al. The complete sequence of the coding region of the ATM gene reveals similarity to cell cycle regulators in different species. *Hum Mol Genet.* 1995;4:2025-2032.

[252] Khanna KK, Jackson SP. DNA double-strand breaks: signaling, repair and the cancer connection. *Nat Genet.* 2001;27:247-254.

[253] Meyn MS. Ataxia-telangiectasia and cellular responses to DNA damage. *Cancer Res.* 1995;55:5991-6001.

[254] Shiloh Y. Ataxia-telangiectasia: closer to unraveling the mystery. *Eur J Hum Genet.* 1995;3:116-138.

[255] Elson A, Wang Y, Daugherty CJ, et al. Pleiotropic defects in ataxia-telangiectasia protein-deficient mice. *Proc Natl Acad Sci U S A.* 1996;93:13084-13089.

[256] Xu Y, Baltimore D. Dual roles of ATM in the cellular response to radiation and in cell growth control. *Genes Dev.* 1996;10:2401-2410.

[257] Xu Y, Ashley T, Brainerd EE, Bronson RT, Meyn MS, Baltimore D. Targeted disruption of ATM leads to growth retardation, chromosomal fragmentation during meiosis, immune defects, and thymic lymphoma. *Genes Dev.* 1996;10:2411-2422.

[258] Ha L, Ceryak S, Patierno SR. Chromium (VI) activates ataxia telangiectasia mutated (ATM) protein. Requirement of ATM for both apoptosis and recovery from terminal growth arrest. *J Biol Chem.* 2003;278:17885-17894.

[259] Banin S, Moyal L, Shieh S, et al. Enhanced phosphorylation of p53 by ATM in response to DNA damage. *Science.* 1998;281:1674-1677.

[260] Canman CE, Lim DS, Cimprich KA, et al. Activation of the ATM kinase by ionizing radiation and phosphorylation of p53. *Science.* 1998;281:1677-1679.

[261] Chehab NH, Malikzay A, Appel M, Halazonetis TD. Chk2/hCds1 functions as a DNA damage checkpoint in G(1) by stabilizing p53. *Genes Dev.* 2000;14:278-288.

[262] Hirao A, Kong YY, Matsuoka S, et al. DNA damage-induced activation of p53 by the checkpoint kinase Chk2. *Science.* 2000;287:1824-1827.

[263] Shieh SY, Ahn J, Tamai K, Taya Y, Prives C. The human homologs of checkpoint kinases Chk1 and Cds1 (Chk2) phosphorylate p53 at multiple DNA damage-inducible sites. *Genes Dev.* 2000;14:289-300.

[264] Flatt PM, Polyak K, Tang LJ, et al. p53-dependent expression of PIG3 during proliferation, genotoxic stress, and reversible growth arrest. *Cancer Lett.* 2000; 156:63-72.

[265] Wang S, Shi X. Mechanisms of Cr(VI)-induced p53 activation: the role of phosphorylation, mdm2 and ERK. *Carcinogenesis.* 2001;22:757-762.

[266] Giaccia AJ, Kastan MB. The complexity of p53 modulation: emerging patterns from divergent signals. *Genes Dev.* 1998;12:2973-2983.

[267] El-Deiry WS. Regulation of p53 downstream genes. *Semin Cancer Biol.* 1998;8: 345-357.

[268] Zhao R, Gish K, Murphy M, et al. Analysis of p53-regulated gene expression patterns using oligonucleotide arrays. *Genes Dev.* 2000;14:981-993.

[269] Zhang Q, Kluz T, Salnikow K, Costa M. Comparison of the cytotoxicity, cellular uptake, and DNA-protein crosslinks induced by potassium chromate in lymphoblast cell lines derived from three different individuals. *Biol Trace Elem Res.* 2002;86:11-22.

[270] Ahn JY, Schwarz JK, Piwnica-Worms H, Canman CE. Threonine 68 phosphorylation by ataxia telangiectasia mutated is required for efficient activation of Chk2 in response to ionizing radiation. *Cancer Res.* 2000;60:5934-5936.

[271] Matsuoka S, Rotman G, Ogawa A, Shiloh Y, Tamai K, Elledge SJ. Ataxia telangiectasia-mutated phosphorylates Chk2 in vivo and in vitro. *Proc Natl Acad Sci U S A.* 2000;97:10389-10394.

[272] Melchionna R, Chen XB, Blasina A, McGowan CH. Threonine 68 is required for radiation-induced phosphorylation and activation of Cds1. *Nat Cell Biol.* 2000;2: 762-765.

[273] Zhou BB, Chaturvedi P, Spring K, et al. Caffeine abolishes the mammalian G(2)/M DNA damage checkpoint by inhibiting ataxia-telangiectasia-mutated kinase activity. *J Biol Chem.* 2000;275:10342-10348.

[274] Blasina A, Price BD, Turenne GA, McGowan CH. Caffeine inhibits the checkpoint kinase ATM. *Curr Biol.* 1999;9:1135-1138.

[275] Chaturvedi P, Eng WK, Zhu Y, et al. Mammalian Chk2 is a downstream effector of the ATM-dependent DNA damage checkpoint pathway. *Oncogene.* 1999;18:4047-4054.

[276] Matsuoka S, Huang M, Elledge SJ. Linkage of ATM to cell cycle regulation by the Chk2 protein kinase. *Science.* 1998;282:1893-1897.

[277] Shiloh Y. ATM and ATR: networking cellular responses to DNA damage. *Curr Opin Genet Dev.* 2001;11:71-77.
[278] Hirao A, Kong YY, Matsuoka S, et al. DNA damage-induced activation of p53 by the checkpoint kinase Chk2. *Science.* 2000;287:1824-1827.
[279] Tibbetts RS, Brumbaugh KM, Williams JM, et al. A role for ATR in the DNA damage-induced phosphorylation of p53. *Genes Dev.* 1999;13:152-157.
[280] Hirao A, Cheung A, Duncan G, et al. Chk2 is a tumor suppressor that regulates apoptosis in both an ataxia telangiectasia mutated (ATM)-dependent and an ATM-independent manner. *Mol Cell Biol.* 2002;22:6521-6532.
[281] Ballif BA, Blenis J. Molecular mechanisms mediating mammalian mitogen-activated protein kinase (MAPK) kinase (MEK)-MAPK cell survival signals. *Cell Growth Differ.* 2001;12:397-408.
[282] Wada T, Penninger JM. Mitogen-activated protein kinases in apoptosis regulation. *Oncogene.* 2004;23:2838-2849.
[283] Wang X, McGowan CH, Zhao M, et al. Involvement of the MKK6-p38gamma cascade in gamma-radiation-induced cell cycle arrest. *Mol Cell Biol.* 2000;20:4543-4552.
[284] Manke IA, Nguyen A, Lim D, Stewart MQ, Elia AE, Yaffe MB. MAPKAP kinase-2 is a cell cycle checkpoint kinase that regulates the G2/M transition and S phase progression in response to UV irradiation. *Mol Cell.* 2005;17:37-48.
[285] Bulavin DV, Higashimoto Y, Popoff IJ, et al. Initiation of a G2/M checkpoint after ultraviolet radiation requires p38 kinase. *Nature.* 2001;411:102-107.
[286] Rivard N, Boucher MJ, Asselin C, L'Allemain G. MAP kinase cascade is required for p27 downregulation and S phase entry in fibroblasts and epithelial cells. *Am J Physiol.* 1999;277:C652-C664.
[287] Tang D, Wu D, Hirao A, et al. ERK activation mediates cell cycle arrest and apoptosis after DNA damage independently of p53. *J Biol Chem.* 2002;277:12710-12717.
[288] Buckley S, Driscoll B, Barsky L, Weinberg K, Anderson K, Warburton D. ERK activation protects against DNA damage and apoptosis in hyperoxic rat AEC2. *Am J Physiol.* 1999;277:L159-L166.
[289] Persons DL, Yazlovitskaya EM, Cui W, Pelling JC. Cisplatin-induced activation of mitogen-activated protein kinases in ovarian carcinoma cells: inhibition of extracellular signal-regulated kinase activity increases sensitivity to cisplatin. *Clin Cancer Res.* 1999;5:1007-1014.
[290] DeHaan RD, Yazlovitskaya EM, Persons DL. Regulation of p53 target gene expression by cisplatin-induced extracellular signal-regulated kinase. *Cancer Chemother Pharmacol.* 2001;48:383-388.
[291] Greulich H, Erikson RL. An analysis of Mek1 signaling in cell proliferation and transformation. *J Biol Chem.* 1998;273:13280-13288.
[292] Rivard N, Boucher MJ, Asselin C, L'Allemain G. MAP kinase cascade is required for p27 downregulation and S phase entry in fibroblasts and epithelial cells. *Am J Physiol.* 1999;277:C652-C664.
[293] Ceryak S, Zingariello C, O'Brien T, Patierno SR. Induction of pro-apoptotic and cell cycle-inhibiting genes in chromium (VI)-treated human lung fibroblasts: lack of effect of ERK. *Mol Cell Biochem.* 2004;255:139-149.
[294] Moelling K, Schad K, Bosse M, Zimmermann S, Schweneker M. Regulation of Raf-Akt Cross-talk. *J Biol Chem.* 2002;277:31099-31106.

[295] Murphy LO, Smith S, Chen RH, Fingar DC, Blenis J. Molecular interpretation of ERK signal duration by immediate early gene products. *Nat Cell Biol.* 2002;4:556-564.
[296] Karin M, Cao Y, Greten FR, Li ZW. NF-kappaB in cancer: from innocent bystander to major culprit. *Nat Rev Cancer.* 2002;2:301-310.
[297] Shumilla JA, Broderick RJ, Wang Y, Barchowsky A. Chromium(VI) inhibits the transcriptional activity of nuclear factor-kappaB by decreasing the interaction of p65 with cAMP-responsive element-binding protein-binding protein. *J Biol Chem.* 1999;274:36207-36212.
[298] Shumilla JA, Wetterhahn KE, Barchowsky A. Inhibition of NF-kappa B binding to DNA by chromium, cadmium, mercury, zinc, and arsenite in vitro: evidence of a thiol mechanism. *Arch Biochem Biophys.* 1998;349:356-362.
[299] Chen F, Bower J, Leonard SS, et al. Protective roles of NF-kappa B for chromium(VI)-induced cytotoxicity is revealed by expression of Ikappa B kinase-beta mutant. *J Biol Chem.* 2002;277:3342-3349.
[300] Kaltreider RC, Pesce CA, Ihnat MA, Lariviere JP, Hamilton JW. Differential effects of arsenic(III) and chromium(VI) on nuclear transcription factor binding. *Mol Carcinog.* 1999;25:219-229.
[301] Chen F, Castranova V, Shi X, Demers LM. New insights into the role of nuclear factor-kappaB, a ubiquitous transcription factor in the initiation of diseases. *Clin Chem.* 1999;45:7-17.

In: Industrial Waste...
Editor: John P. Samuelson, pp. 41-72
ISBN: 978-1-60692-720-5

Chapter 2

INDUSTRIAL WASTES AS LOW-COST ADSORBENTS FOR TEXTILE DYES: A REVIEW

Sílvia C.R. Santos and Rui A.R. Boaventura
LSRE—Laboratory of Separation and Reaction Engineering, Departamento de Engenharia Química, Faculdade de Engenharia da Universidade do Porto, Rua Dr. Roberto Frias, 4200-465 Porto, Portugal

Abstract

Adsorption, known for over a century, has remained to the present day an efficient technique for purifying water. For textile effluents containing the unfixed dyes in fibres, adsorption has shown good decolourization results. Activated carbon is a widely-used adsorbent in commercial systems due to its extended surface area, microporous structure, great adsorption capacity and high degree of surface reactivity. Nevertheless, the high prices associated with activated carbon replacement and regeneration are considerable. Several research studies have been carried out in order to find low-cost materials as possible alternatives and, in particular, industrial wastes have been tested for this purpose. The generation of large amounts of wastes by industrialized countries requires significant costs for their disposal or treatment. Valorisation by recycling or reuse is an environmentally sound option. Residues and by-products from water and wastewater treatment plants, steel-making process, power generation, metal plating, timber and tanning industries, agricultural, and food and drink companies are examples of wastes tested as adsorbents. Carbonaceous waste materials have been especially studied as precursors for the preparation of activated carbons or coal-based adsorbents. Studies reported in the literature comprise the waste characterization, possible treatments to enhance the material adsorption properties, determination of equilibrium adsorption isotherms and kinetics for different operating conditions, and experiments in packed bed columns. Attempts to identify the adsorption underlying mechanism have also been reported. Assessment of metals' leachability is also very important for some industrial wastes. In spite of the numerous papers published, the true challenge of finding good low-cost adsorbents for anionic dyes—especially for the reactive ones—using low adsorbent dosages and minimizing waste problems, has still not been met. According to the authors' knowledge, a considerable lack of information also exists in the application of these wastes to real effluents, since some studies have shown a substantial decrease in the adsorbent performance due to the presence of other species in solution (salts, other dyes, and dyeing chemicals). Materials processed from chitosan and carbonaceous wastes were identified as promising adsorbents for anionic dyes. Metal hydroxide sludges, fly ashes and leather wastes also proved to be good materials for dye adsorption.

1. Introduction

Textile, paper, cosmetics, plastics, leather and food industries use dyes to colour their products. The textile industry is, however, the most important polluting industry, considering the volume of wastewaters generated and the significant release of dyes to the environment. Dyehouse effluents contain unfixed dyes in the fibres, auxiliary dyeing chemicals, salts, acids, bases, chlorinated compounds and, occasionally, heavy metals [1, 2]. These wastewaters are characterized by a moderate organic content, low biodegradability (due to the nonbiodegradable and recalcitrant nature of many dyes), variable pH values—usually in the alkaline range—and colour. It is estimated that up to 10% of the amount of dyes produced in the world is released to the environment during their application on the fibres [3].

Modern societies have become more demanding of high standards in clothes and textiles, desiring bright and stable colours that survive washing and light. In order to meet these demands, dyes have become less amenable to degradation by aerobic biological processes [4]; as a result, colour removal is a real problem to solve in textile wastewater treatment.

The entrance of dyes into water streams results in an undesirable and unnatural aspect, easily detected, calling for the attention of the public and authorities. The presence of metals and chlorides in the dyes has been referred to as toxic to the aquatic life [5], although there are studies from ETAD (Ecological and Toxicological Association of Dyes and Organic Pigment Manufacturers) showing low evidence of toxicity in the concentrations of dyes that are likely to be found in rivers [4, 6]. No prediction of bioaccumulation in fish was reported [7]. The release of dyes into water streams, particularly with strong colours, reduces the light penetration in waters, affecting the primary productivity, the growth of plants and limiting invertebrate and other forms of wildlife [4]. An appropriate decolourization of textile dyehouse effluents is then required to avoid a significant and continuous input of dyes into the receiving waters.

Conventional wastewater treatment techniques involve biological and chemical (coagulation/flocculation) methods that remove a significant part of the organic content but are often ineffective for total colour removal. Biological methods, such as the activated sludge process, are advantageous for economical reasons but present some limitations, especially because degradation is slow [8] and an adequate colour removal, particularly for reactive and some acid dyes, is unlikely to be achieved [9]. Chemical coagulation/flocculation could be expected to decolourize waters when certain organic cationic polymers are used [9, 10], in substitution for the classic coagulants based on ferrous, ferric, aluminium and calcium salts. Nevertheless, the good performance attained in small-scale Jar-tests is sometimes difficult to achieve in full scale, due to inadequate mixing and flocculation conditions [11]. Residual concentrations of cationic polymers in the treated wastewater are also toxic to fish [9, 11], and the use of a co-flocculant salt (to enhance settling characteristics of the sludge) further increases the amount of waste sludge generated [10]. Oxidation, advanced oxidation processes and membrane separations are generally effective but highly limited by the high costs requirements in energy, chemicals and materials. Adsorption is an important method to remove contaminants from water streams, especially used for organic molecules that resist to biological degradation such as some soluble organic dyes. Activated carbon is the most widely-used adsorbent in commercial systems [8]. Its high efficiency is due to its developed porosity, large surface area, surface chemistry (functional groups, pH on the zero point of charge) and high degree of surface reactivity [12, 13]. Granular activated carbon (GAC) is generally used in packed bed columns, and powdered activated carbon (PAC) is dosed as a slurry into the effluent stream. A combination of biological treatment with adsorption on activated carbons is also a common option since no single method is able to treat textile wastewaters efficiently. According to Hazel (1995) [9], in many cases, a biological or a coagulation/flocculation treatment is required before a tertiary treatment to remove colour residues. Adsorption could be then employed in this tertiary step, but the high costs of activated carbon acquisition and regeneration have inhibited its large-scale usage as adsorbent.

In recent years, studies on the search for alternative low-cost adsorbents have proliferated, not only for adsorption of dyes but also for other organics [14-17] and heavy metals [18-22]. Literature reports an extensive investigation using a wide range of industrial solid wastes as potential adsorbents for dyes. In fact, quotidian life of modern societies involves the generation of numerous and diversified wastes that if not reused or recycled will be disposed in landfills or incinerated. Valorisation of these wastes could be a good solution, and its reutilization in wastewater treatment of other industries could be an attractive alternative. The low cost and availability of this kind of material is its great advantage, so selectivity, great adsorption capacity and high adsorption rate for different dyes are its further desirable properties. Several studies have reported the use of waste materials with no previous treatment and thus only costs in transportation have to be taken into account. Other studies reported modifications of these wastes in order to enhance their adsorption properties. In that case, savings were achieved in the start material, but a careful selection of the modification methods has to be done to avoid an offset of the savings in the precursor. Carbonaceous wastes, such as sludges, papers, and plastics, are reported as possible raw materials for activated carbon preparation. In fact, non-renewable and relatively expensive materials such as coal are unjustified in most pollution control applications [23] since almost any carbonaceous material may be used as an activated carbon precursor [24]. The performance of the prepared adsorbents depends, of course, on the precursor origin and on all of the conditions used in the activation (activating agent, temperature and time). Dias et al. [13] considered that activated carbons prepared by non-conventional starting materials could effectively compete with commercial ones.

In this chapter, a literature review on the possibilities of reusing industrial wastes as adsorbents for dye removal from textile wastewaters is presented. The readers are encouraged to consult the original references for detailed information.

2. Some Considerations

2.1. Textile Wastewaters and Dyes

Dyeing bath recipes are developed according to the desired results in the fibres and are composed of dye mixtures, salts (electrolyte function), acids, bases and chemicals such as leveling, wetting, sequestering and lubrificant agents. More than 100,000 different dyes are commercially available in the world [2]. Dyes are molecules with a high stability and not always with a known chemical structure. Nowadays, there are very few natural dyes in use. Almost all are of synthetic origin being produced using hydrocarbons like benzene, toluene, naphthalene, and anthracene [25]. Colour Index (developed by Society of Dyes and Colourists, 1924) is an international system of dyes classification by their generic name (determined by the application characteristics) followed by a constitution number based on the chemical structure when known [6]. According to its application in dyeing of different fibres, dyes can be classified into: Sulphur, Azoic, Reactive, Ingrain, Direct, Basic, Vat, Acid, Mordant, Metal complex, Solvent and Disperse [25]. Classification in chemical classes (Anthraquinone, Azo, Acridine, Methine, etc.) is also used. The present chapter refers exclusively to basic, direct, reactive and disperse dyes since they are the most commonly cited in literature. Table 1 summarizes the suitable substrates for dye with each one of these classes.

Generally, a complete fixation of the dye on the fibre does not occur during the dyeing process. The fixation degree depends on the combination dye/substrate, the application method and all the conditions used. Table 2 indicates degrees of fixation for different dye/fibre combinations estimated by Easton (1995) [6]. The problem of partial lost of dyes into the effluents is exacerbated for the reactive dye class, which is commercially very important. In reactive dyeing, dyes bind to the fibre by a covalent bond, under alkaline pH and high temperature. Under these conditions a secondary reaction, hydrolysis of the dye, occurs leading to a low degree of fixation (as could be seen in Table 2) and then considerable dye losses. The reactive class is still the less amenable to the biological treatment which make the cotton dyeing industries strongly associated with coloured effluents [26].

Table 1. Suitability of substrates for different types of dyes (adapted from [27])

	Natural fibres		Synthetic fibres		
	Protein (Wool, silk)	**Cellulosic**	**Polyamide**	**Polyester**	**Acrylic**
Basic					●
Direct	○	●	○		
Acid	●		●		
Reactive	●	●	○		
Azoic		●	○	○	○
Vat	○	●			
Sulphur		●			
Disperse			●	●	●
Metal complex	●		●		
Mordant	●				

● very suitable
○ suitable

Exhausted dye baths are mixed with waters from rinsing and soaping dyed fibres, contributing to diluting the colour in the final effluent. The complex nature of effluents, due to the wide range of dyes and chemicals involved, make difficult the prediction of an average dye concentration, at least in terms of mg/L, in the effluents. Dyeing concentrations of 10–250 mg/L have been cited as being present in dyehouse effluents [26]. In fact, seasons and fashion (influencing the colour and fibres to be dyed) contribute even better to the variability found in textile wastewaters. Table 3 presents values of parameters analysed in samples from a Portuguese textile industry effluent, during three days of operation. Table 4 indicates average values obtained in wastewaters from five Portuguese textile dyeing companies.

Table 2. Percentage of unfixed dyes on fibres for different substrates (estimated by Easton [6])

Dye Class	Fibre	Loss to effluent (%)
Acid	Polyamide	5-20
Basic	Acrylic	0-5
Direct	Cellulose	5-30
Reactive	Cellulose	10-50

Most research studies on dye adsorption have used synthetic dye effluents containing a single dye dissolved in distilled water. Methylene Blue (Basic Blue 9), Malachite Green (Basic Green 4) and Congo Red (Direct Red 28) are examples of dyes traditionally reported. Figure 1 shows the chemical structures of some textile dyes reported in literature. Few studies have also reported multi-component dye adsorption [28-31], which is a more realistic simulation of a real effluent. In this case, the determination of individual dye concentration (usually done by spectrometry) could be more complex, especially if chemical interaction occurs between dye molecules. Several studies also report the effect of ionic strength (or presence of salts) in dye solutions [32-34], which is due to the electrolytes widely used in the dyeing baths. Santos et al. (2008) [35] used a more realistic effluent, although with a single reactive dye, simulating the hydrolysis of the dye and using salts, lubrificant, wetting and sequestrant agents. Additionally, very few papers reporting adsorption tests in true real effluents are available [36-39].

Table 3. Parameters measured in three effluent samples collected on different days in a Portuguese textile industry [40]

	Sample 1	Sample 2	Sample 3
pH	10.0	10.7	10.6
Total alkalinity (mg $CaCO_3$/L)	121	829	522
Total hardness (mg $CaCO_3$/L)	36.8	23.8	26.3
Conductivity (mS/cm)	530	11000	7000
Chloride (mg/L)	101	6358	3777
Sulphates (mg/L)	23	25	37
Phosphate (mg/L)	<0.010	0.02	<0.010
Total suspended solids (mg/L)	6	8	14
Volatile suspended solids (mg/L)	6	5	13
Total dissolved solids (mg/L)	536	6573	3638
Volatile dissolved solids (mg/L)	233	333	277
BOD_5 (mg/L)	114	37	99
COD (mg/L)	320	672	520

Table 4. Average values of parameters measured in the effluents from different Portuguese textile industries [40]

	A	B	C	D	E
Total suspended solids (mg/L)	44	114	16	20	15
Volatile suspended solids (mg/L)	34	46	14	20	15
Total dissolved solids (mg/L)	1.5	4.2	3.5	2.9	3.1
Volatile dissolved solids (mg/L)	1.0	0.6	1.0	0.3	0.4
BOD_5 (mg/L)	82	459	133	105	270
COD (mg/L)	211	1020	326	260	661
Phenolic compounds (mg/L)	9.6	38.9	<1	16.5	17.5
Chromium (mg/L)	<0.05	<0.05	<0.05	0.07	<0.05

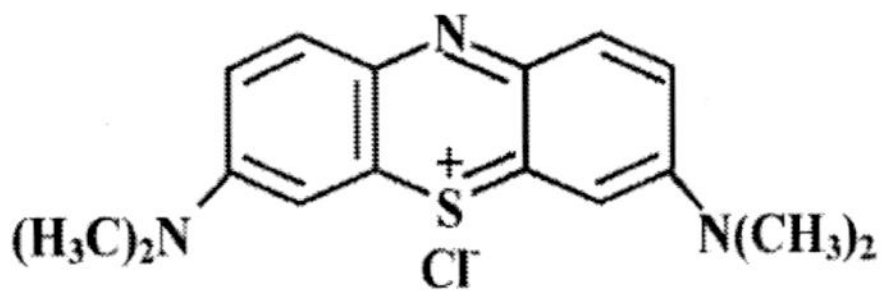

Methylene Blue
(Basic Blue 9)

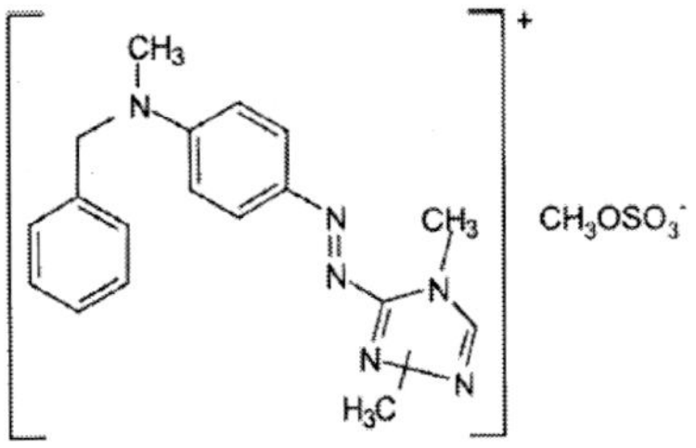

Basic Red 46

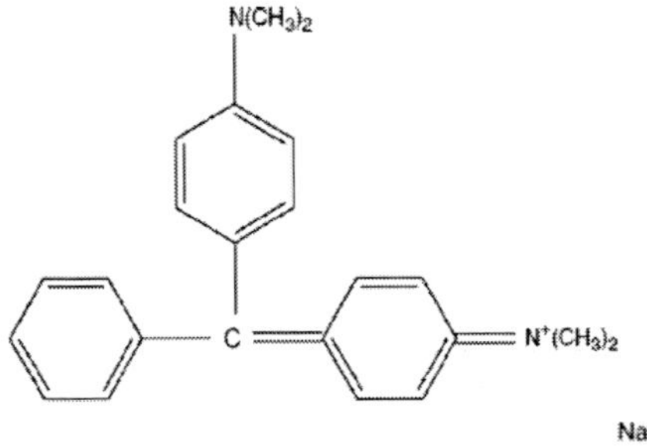

Malachite Green
(Basic Green 4)

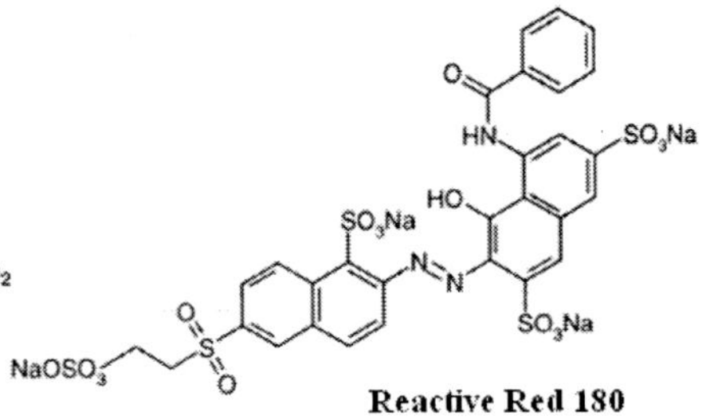

Reactive Red 180

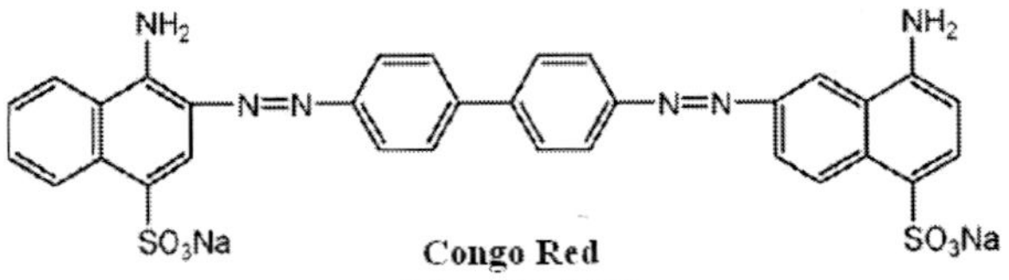

Congo Red
(Direct Red 28)

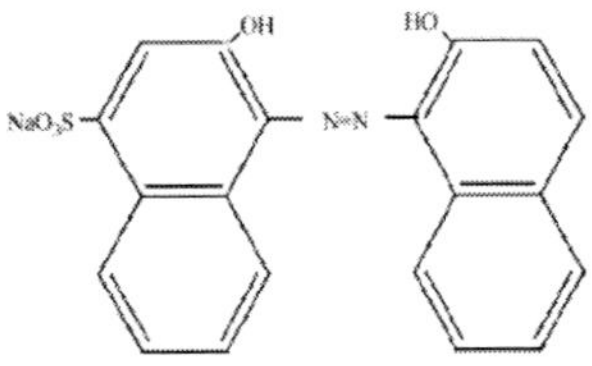

Acid Blue 193

Figure 1. Chemical structures of some textile dyes.

2.2. Adsorption

The potential of the adsorbents can be evaluated according to the adsorption capacities demonstrated for different dyes. This is, in fact, the first and the most important measure to identify how much promising could be an adsorbent. Equilibrium data is then an essential tool. Isotherms are the relation between the adsorbate concentrations in the liquid phase (or in the gas) and in the solid phase, when equilibrium is attained at a pre-established temperature. Several adsorption models such as Langmuir, Freundlich, BET, Toth, Temkin, Redlich-Peterson, Dubinin-Radushkevich, were proposed to describe equilibrium data. Langmuir and Freundlich are the most popular and have been successfully used for modelling dyes adsorption. Langmuir isotherm [41] was developed considering that (i) adsorption is limited to a monolayer coverage of adsorbed species (which corresponds to the maximum adsorption capacity, Q_m), (ii) adsorption takes place at finite and identical specific sites, (iii) energy of adsorption is constant and (iv) there is no interaction between adsorbed species. Langmuir model parameters are the monolayer adsorption capacity (Q_{max}) and a constant related to the energy of adsorption (K_L). Variables are then the adsorbate concentration in the fluid phase (C_{eq}) and in the solid phase (q_{eq}) at the thermodynamic equilibrium. Langmuir isotherm is expressed by Eq. (1).

$$q_{eq} = \frac{Q_m K_L C_{eq}}{1 + K_L C_{eq}} \tag{1}$$

Freundlich isotherm [42], represented by Eq. (2), is an empirical model used to describe multilayer adsorption in heterogeneous surfaces, considering interaction between adsorbed species and a non constant adsorption energy. K_F and 1/n are Freundlich model parameters related to the adsorption capacity and adsorption intensity, respectively.

$$q_{eq} = K_F C_{eq}^{1/n} \tag{2}$$

The adsorption capacity frequently referred in this chapter is the Langmuir parameter and has been systematically used to compare adsorption capacities for different dye/adsorbent systems.

3. Commercial Activated Carbons

Commercial activated carbons are the conventional adsorbents used for water and wastewater treatments. Around 80 % of its world production is used in liquid phase applications [13, 43]. Wood, coconut shell and coal are traditionally used as raw materials for activated carbon production, which involves chemical activation (with H_2SO_4, H_3PO_4, $ZnCl_2$, etc.) or physical activation (CO_2, H_2O, CO). The adsorption on activated carbons surface could be a result of electrostatic interactions or non electrostatic interactions (Van der Waals forces, hydrophobic interactions and hydrogen bonding) [13]. Molecular size, solubility, pK_a and nature of substituents (in the case of aromatic adsorbates) are the adsorbate properties that mainly influence the adsorption process [44]. The use of activated carbon on dye removal was already been widely investigated. Table 5 indicates some maximum reported capacities for different commercial activated carbon samples and different dyes.

Al-Degs et al. (2000) [45] studied the adsorption of three reactive dyes in single and multi-component mixtures by Filtrasorb-400 activated carbon and verified a decrease in the global adsorption capacity in multi-component dyes solution. The decrease was attributed to dyes interaction in solution, modifications in the charge and in the chemistry of the adsorbent surface due to the adsorption of the yellow dye and to the competitive adsorption. The high adsorption capacities obtained are a result of the net positive surface charge during adsorption (pH below the pH_{ZPC} of the carbon). A similar commercial activated carbon showed a low adsorption capacity for Direct Brown 1:1 dye (7.69 mg/g), which was suggested to be of physical nature [46].

Some researchers [36, 47] modified the surface chemistry of commercial activated carbons by chemical and thermal treatments, without changing significantly their textural properties. These authors proved the important role of the carbon surface chemistry on the dyes adsorption by obtaining different results for each modified activated carbon. Faria et al. (2004) [47] obtained an almost linear correlation between anionic dyes maximum adsorption capacity and activated carbon basicity. The most efficient modified activated carbon was found to be the one submitted to thermal treatment with hydrogen at 700 °C after oxidation with HNO_3. For this adsorbent, maximum adsorption capacities of 345, 246 and 714 mg/g were found respectively for Acid Blue 113, Reactive Red 241 and Basic red 14. Orfão et al. (2006) [48] prepared carbons with acidic and basic properties using a commercial activated carbon and studied adsorption of a reactive dye. pH was verified to be an important factor since it influences the activated carbon surface charge and then the electrostatic interaction between dye and adsorbent. Maximum adsorption occurred for solutions pH below the pH_{ZPC} of the carbon. Maximum reactive dye adsorption capacities followed the order acidic carbon < original carbon < basic carbon. Choy et al. (2004) [31] found adsorption capacities in the range 0.125-0.253 mmol/g for acid dyes on Filtrasorb-400 at 20 °C and identified intraparticle diffusion as the prevailing mechanism limiting adsorption kinetics.

Qada et al. (2008) [49] studied adsorption of three basic dyes on activated carbons from different origins. The adsorption extent was verified to be generally favoured by the increase in the pH as a result of the increase in the number of hydroxyl groups and then higher attraction between the negative surface of the adsorbent and the positive charge of the dye (explanation on the basis of pH_{ZPC}). Physical properties and surface chemistry of the activated carbons proved to be both important for the adsorption extent and explained the different results obtained for the three different carbons used. Al-Degs et al. (2008) [34] studied the adsorption of reactive dye on commercial activated carbons. Results obtained at different pHs indicate that electrostatic forces are not the only mechanism involved in adsorption, and hydrogen bonding and hydrophobic interactions were probably involved too. The effect of ionic strength was studied by the addition of NaCl. Adsorption was favoured up to a salt concentration of 0.1 M and for values above this, adsorption was not affected. Reactive dyes adsorption was verified to be endothermic. Mathematical modelling considering the resistances involved in the adsorption were also studied for dyes removal by activated carbons either in batch stirred conditions or in fixed bed [31, 38, 50, 51]. Walker and Weatherley (1999) [51] modelled the kinetic acid dye uptake curves by granular activated carbon considering film and pore diffusion resistances. Choy et al. (2001) [52] developed a three-resistance mass transport model based on film, pore and surface diffusion control to model the acid dye concentration decay curves.

Table 5. Reported maximum adsorption capacities of commercial activated carbons for different dyes

Activated Carbon	Dyes	Q_m (mg/g)	Conditions	Ref.
Filtrasorb-400 (*Chemviron* Carbon)	Remazol Golden Yellow	714	18 °C, pH 5.5	[45]
	Remazol Black B	278	18 °C, pH 5.0	
	Remazol Red	213	18 °C, pH 5.0	
Filtrasorb-400 (*Chemviron* Carbon)	Basic Blue 9	455	20 °C, pH 7	[49]
	Basic Red 22	556		
	Basic Yellow 21	833		
Filtrasorb-400 (*Chemviron* Carbon)	Acid Blue 80	171	20 °C	[31]
	Acid Red 114	104		
	Acid Yellow 117	186		
Filtrasorb-400 (*Calgon* Co.)	Direct Brown 1:1	7.69	27 °C, pH 7	[46]
Norit GAC 1240 PLUS	Basic Red 14	546	pH 6-7	[47]
	Acid Blue 113	310		
	Reactive Red 241	190		
Calgon Co. (USA)	Reactive Blue 2	209	25 °C, pH 7	[34]
	Reactive Yellow 2	194		
	Reactive Red 4	96		
Powdered Chemviron GW (*Chemviron Carbon*)	Basic Red 46	106	20 °C, pH 7.4	[23]
	Direct Black 168	18.7		
	Acid Brown 283	22.0		

4. Wastewater Treatment Sludges

4.1. Sewage Sludges

Sewage sludges are residual products generated in treatment processes of urban and industrial wastewaters and thus exhibit wide variations in their properties. Considering only the urban wastewater treatment plants, sludge production is estimated to be about 40-60 g of dry matter per inhabitant per day [53]. Sludge minimization is however limited by sustainment or even improvement of the treated effluent quality. Sewage sludges can be used as fertilizers or soil additives in agriculture and forestry, but this practice is restricted by their quality. At high levels of toxic substances or heavy metals contamination, sludge destination has to be landfill disposal or incineration.

Reusing these wastes is desirable and the application of this material, in fact "waste of waste"[37], in further treatment of other industrial wastewaters is appellative. Sewage sludges from different origins (activated sludge processes and coagulation-flocculation methods) have been studied as adsorbents for textile dyes in dried state and in modified forms. Activated carbon production from sewage sludge is a considerable way reported in literature. Carbonisation of sewage sludge, using different chemical agents and conditions, has produced carbon adsorbents with surface areas of 100-400 m^2/g [54-58]. Results reported in the literature showed however that the surface area development as well as the porosity do not assure better dye adsorption properties. Otero et al. (2003) [59] used two different kinds of sewage sludge respectively provided by an urban wastewater treatment plant and an agrofood industry. Methylene Blue adsorption capacities of 115 mg/g and 87 mg/g were obtained using these dried wastes. Carbon adsorbents were further prepared by pyrolysis and chemical activation with H_2SO_4 but showed lower adsorption capacities than the respective raw materials. Calvo et al. [60] presented similar results and attributed them to the high proportion of micropores developed during the preparation of the adsorbents. Relatively low adsorption values for Methylene Blue and Saphranine dyes were also found [61] for a chemically activated urban sewage sludge: 194.3 mg/L e 250.2 mg/L (recalculated for the adsorbent dose used: 19.4 and 25.0 mg/g), respectively. Martin et al. [23] prepared activated carbons by pyrolysis in the presence of H_2SO_4 using an aerobically digested sludge, from an urban wastewater treatment plant. Specific surface area, micropore and mesopore volumes obtained (253 m^2/g, 0.08 and 0.12 cm^3/g, respectively) were lower than the presented by commercial activated carbons. Direct dyes were however better removed by the sewage sludge-based carbons than by the commercial one used for comparison. Better affinity for basic dye (Basic Red 46) was on the contrary obtained for commercial activated carbon. This behaviour was explained by the wide pore size of the sludge-based carbon (2.3 nm) favouring the adsorption of big molecules, such as direct dyes, than the commercial activated carbon (pore size of 1.8 nm). The higher specific surface area of the commercial activated carbon was probably decisive for the adsorption of the lower molecular size of the basic dye. Dhaouadi and M'Henni [37] used a crude dehydrated sewage sludge (from activated sludge process) from an urban wastewater treatment plant as adsorbent for textile dyes. Langmuir adsorption capacities for Direct Red 79 and Vat Blue 4 were 19.6 mg/g (25 °C, pH 6.4) and 248.3 mg/g (25°C, pH 9.2). Adsorption experiments conducted with a real textile mill effluent showed that 8 mg/g of Indigo (dominant dye in the effluent) were adsorbed (25 °C, pH 12.3). Table 6 summarizes the literature reported adsorption capacities for sewage sludges and for the activated carbons prepared using them as starting materials.

Table 6. Reported maximum adsorption capacities of sewage sludge-based materials for different dyes

Treatment	Dyes	Q_m (mg/g)	Conditions	Ref.
Thermal treatment (700 °C), H_2SO_4	Basic Red 46	188	20 °C, pH 4.2	[23]
Thermal treatment (700 °C), H_2SO_4	Acid Brown 283	20.5	20 °C, pH 4.2	[23]
Thermal treatment (700 °C), H_2SO_4	Direct Red 89	49.2	20 °C, pH 4.2	[23]
Thermal treatment (700 °C), H_2SO_4	Direct Black 168	28.9	20 °C, pH 4.2	[23]
(no treatment)	Basic Blue 9	115	25 °C	[59]
Pyrolysis (625 °C)	Basic Blue 9	32	25 °C	[59]
Chemical activation (H_2SO_4)	Basic Blue 9	25	25 °C	[59]
(no treatment)	Basic Blue 9	87	25 °C	[59]
Pyrolysis, N_2 550 °C	Basic Blue 9	29	25 °C	[59]
Chemical activation (H_2SO_4)	Basic Blue 9	28	25 °C	[59]
(no treatment)	Direct Red 79	19.6	25 °C	[37]
(no treatment)	Vat Blue 4	248.3	25 °C	[37]

4.2. Waste Sludges of Inorganic Matrix

Sludges generated by precipitation or coagulation of metals were included in this group. Metal hydroxides waste sludges appear as promising adsorbents particularly for anionic dyes due to their positive surface charge under around neutral pH conditions. Metal hydroxides waste sludges are generated in metal finishing or plating industries, by the alkalinisation of the wastewaters with calcium hydroxide and subsequent precipitation. The result is a sludge in the form of insoluble metallic hydroxide compounds ($M(OH)_n$, n=2 or 3). Ferrous sulphate has been used to reduce the hexavalent chromium to the trivalent state and then iron is also found in this kind of waste [32, 62].

There is agreement among several authors about the probable mechanism (ion-exchange) of dye adsorption onto hydroxide metals sludges but some different behaviours about the pH effect are reported. Waste Fe(III)/Cr(III) hydroxide generated in the treatment of Cr(VI) bearing wastewaters from a fertilizer industry was used for Congo Red direct dye adsorption [63]. The low desorption observed indicated ion-exchange as the main process involved. Netpradit research team [32, 62, 64] investigated azo reactive dyes removal by electroplating industry hydroxide metals sludge. Maximum adsorption capacities in the range 48.31-62.5 mg/g were obtained. The researchers found out that the pH played an important effect on the adsorption and formation of dye-metal complexes, with the optimum pH referred to as 8-9, close to the pH_{ZPC}. The amount of sulfonic groups ($-SO_3$) in the reactive dyes was found to be an important factor for the adsorption due to the ion-exchange mechanism. For initial pH 2 (predicted final system pH above 5) the treated water with the waste sludge presented no significant concentrations of metals leachated from the sludge [62]. Enthalpy values suggest that adsorption of Red 2 (less charged dye) is mainly of physical nature and adsorption of Red 120 and Red 141 (more charged dyes) is chemical. Higher concentration and more valence of ions of electrolytes (NaCl and Na_2SO_4) caused a decrease in the adsorption efficiency [32]. Coarse size metal hydroxide sludge in a fixed bed column was verified to be feasible for the adsorption of Reactive Red 141, measured as 27-29 mg/g, lower than 45 mg/g obtained in a batch system. Santos et al. [35] studied the adsorption of Reactive Blue 19 by a waste sludge generated in a Portuguese zinc plating industry (zinc and iron were the major metal constituents). Maximum adsorption capacities at 25 °C, predicted by Langmuir isotherm fittings, were in the range 21.9- 275 mg/g (respectively for pH values of 10 and 4). Ion-exchange was also appointed as the main adsorption mechanism. The authors showed that maximum adsorption capacity of the dye decayed to 31 mg/g when a simulated effluent (containing NaHO, Na_2CO_3 and dyeing auxiliary chemicals) was used. The safe use of this sludge was assured using an initial solution pH above 5.

Hydroxide metals sludge may also be generated during the removal of heavy metals by electrocoagulation processes. In this case, the sludge is mainly composed by hydroxides of the electrode metal and heavy metals. In spite of the high toxic metal content, this kind of sludge was referred to be chemically inert, not leaching back any contaminants [65].

An electrocoagulated metal hydroxide sludge generated during the removal of Cr^{3+} using an aluminium electrode was applied to Congo Red dye adsorption [66]. Maximum adsorption capacities, given by Langmuir fittings, were in the range 271 mg/g (initial pH 10.4/final pH 7.8) – 513 mg/g (initial pH 3.0/final pH 4.41) at 30 °C. Adsorption was highly pH dependent due to the formation of various charged hydroxylated species, as $Al(OH)^{2+}$ and $Cr(OH)^{2+}$.

Chemical interaction was suggested to be involved in the dye removal. Taking in account leached Cr^{3+} concentration, safe use of this sludge requires a pH range of 3-10.

Another paper [67] refers to a metal hydroxide sediment sludge generated in wastewater treatment of the electroplating and etching processes from the manufacturing of printing circuit boards. These wastewaters contain high levels of Cu^{2+}, Zn^{2+}, Ni^{2+}, Cr^{3+} and Pb^{2+} and are treated by polyelectrolyte flocculation and hydroxide precipitation. Microwave radiation was successfully used to detoxify this sludge with a remarkable effect on the immobilization of heavy metals but still with an underlying mechanism not clearly identified. The resultant material (specific surface area of 155 m^2/g) was shown to have good adsorption properties for Reactive Red dye (maximum adsorption around 80-100 mg/g) and worse for Methylene Blue dyes (< 50 mg/g). Lee and al. [68] obtained a similar trend, better affinity of a waste chrome sludge (from electroplating industries) to adsorb anionic (acid) dyes than basic ones. Table 7 presents maximum adsorption capacities reported in literature for industrial waste sludges of inorganic matrix.

Table 7. Maximum adsorption capacities reported in literature for inorganic industrial waste sludges and different dyes

Origin	Dyes	Q_{max} (mg/g)	Conditions	Ref.
Precipitation (Electroplating industry)	Reactive Red 120	48.3	30 °C, pH 8-9	[62]
	Reactive Red 141	56.2		
	Reactive Red 2	62.5		
Precipitation (Zinc plating industry)	Reactive Blue 19	91.0	25 °C, pH 7	[35]
	Reactive Blue 19	31	25 °C, pH 7 [a]	
Electrocoagulation	Direct Dye 28	293	30 °C, pH 7	[66]

[a] in a simulated real effluent

5. Agricultural/AGROFood Industrial Wastes

A wide variety of agricultural and agrofood wastes as fruit peels, seeds and kernels and by-products from cereals processing have been extensively studied as potential adsorbents for dyes.

Orange peel, a waste discarded by the orange-juice and soft-drink industries, revealed low adsorption capacities of 22.44, 1.33 and 3.23 mg/g respectively for Congo Red, Procion Orange and Rhodamine B dyes [69]. Annadurai et al. [70] found similar values for the adsorpion of various dyes by orange and banana peels. Adsorption capacities were in the range of 6.5 - 21.0 mg/g at 30 °C and banana peel showed slightly higher adsorption capacities than the orange peel. Arami and co-authors [71] reported a sharp decrease on direct dyes removal with increasing pH but, even at the optimum pH of 2, quite low values (10.72 and 21.05 mg/g) were obtained. Strong alkaline conditions were verified to desorb more than 90 % of the adsorbed dye. Hameed and El-Khayari [72] studied the oil Palm trunk fibre, an agro-industry waste, composed essentially by cellulose and hemicellulose, on the adsorption of Malachite green. Equilibrium adsorption results at 30 °C and pH 5 were fitted to Langmuir, Freundlich and multilayer adsorption isotherms. The multilayer model described better the obtained data. Langmuir adsorption maximum capacity was calculated to be 149.35 mg/g. Studies about the pH effect on the adsorption revealed 4-8 as the optimum pH range.

Wheat husk was also tested for dye removal after being activated with hydrogen peroxide [73, 74]. Adsorption results showed lower performances than an activated charcoal for the azo dyes Reactofix Golden Yellow 3 RFN and Reactofix Navy BLue 2 GFN. Wheat husk-adsorbent showed maximum removals at very acidic pH and a strong decrease was observed when increasing pH. The activated charcoal, used for comparison, keeps its efficiency almost constant over all the pH range.

Olive-seed is an abundant waste in Mediterranean countries. Stavropoulos and Zabaniotou [75] tried the valorisation of this residue by using it as the raw material for production of activated carbons by chemical activation with KOH. The authors verified that adsorption capacities obtained for Methylene Blue dye (190-263 mg/g) are of the same order of 240 mg/g obtained for the commercial activated carbon F300 (Calgon Carbon Co., Pittsburg, USA).

Plum kernels were also used for activated carbons preparation [14, 76, 77]. Optimization of the procedure (activation temperature and time) was carried out and the predicted Langmuir monolayer capacities were 520.6 mg/g, 710.1 mg/g and 904.4 mg/g (30 °C), respectively for Methylene Blue dye, Basic Red 22 and Acid Blue 25. The authors concluded that high activation temperatures are preferable for high molecular weight adsorbates [76]. Chemical activation with NaHO also showed good adsorption properties of the prepared carbons [77], resulting in very high surface areas and adsorbed amounts of 765-828 mg/g, 549-567 mg/g and 1453-1845 mg/g (30 °C), respectively for Methylene Blue, Acid Blue 74 and Basic Brown 1.

Rice-husk, another residue from agriculture and food industry, has also been used to produce activated carbons under different conditions [78-85]. Rice husk submitted to NaHO or H_3PO_4 treatment followed by carbonization [80] presented 77-93 mg/g adsorption capacities for Malachite Green dye, comparable to the obtained values for activated carbons. Rice husk-based activated carbons prepared by carbonisation, steam digestion and acid treatment revealed an adsorption performance for Mehylene Blue (343.5 mg/g) about three times lower than commercial activated carbon but is an alternative nearly five times cheaper [84].

Sago waste is generated in the preparation process of sago, which is a major staple food for the lowland peoples of New Guinea and the Moluccas. It has been mentioned as a major waste accumulating in India sago industries. Activated carbons prepared from chemical activation of sago waste presented surface areas of 421-625 m^2/g and pH_{ZPC} of 5.7 but showed a lower adsorption capacity for Rhodamine B dye [86]. Other agricultural residues as bamboo dust, coconut shell, groundnut shell, and straw [84, 87] were converted to activated carbons showing high adsorption capacities for Methylene Blue dye.

Thinakaran et al. [88] produced carbon adsorbents using sunflower seed hull by H_2SO_4 activation at low temperatures. Specific surface areas were in the range 10.35- 21.06 m^2/g and Acid Violet 17 dye adsorption equilibrium isotherms at 30 °C predicted maximum adsorption capacities of 65.78-116.27 mg/g.

Table 8. Maximum adsorption capacities (Q_m) reported in literature for agricultural, agrofood and timber industrial wastes-based adsorbents

Raw material	Treatment	Dyes	Q_m (mg/g)	Conditions	Ref.
Orange peel	(no treatment)	Basic Blue 9	18.6	30 °C, pH 7.2	[70]
Orange peel	(no treatment)	Direct Red 28	14.0	30 °C, pH 7.9	[70]
Orange peel	(no treatment)	Methyl Violet	20.5	30 °C, pH 5.7	[70]
Banana peel	(no treatment)	Direct Red 28	18.2	30 °C, pH 7.9	[70]
Banana peel	(no treatment)	Basic Blue 9	20.8	30 °C, pH 7.2	[70]
Banana peel	(no treatment)	Methyl Violet	21.0	30 °C, pH 5.7	[70]
Coffee husks	(no treatment)	Basic Blue 9	90.1	30 °C, pH 8	[96]
Oil palm trunk fibre	(no treatment)	Malachite Green	149.35	30 °C, pH 5	[72]
Rice husk	Carbonisation (300 °C), steam digestion, acid treatment	Basic Blue 9	343.5	30 °C pH 7.2-7.4	[84]
Sago waste	Chemical activation $H_2SO_4/(NH_4)_2S_2O_8/NaHCO_3$	Basic Violet 10	16.2	30 °C pH 5.7	[86]
Straw	Carbonisation (300 °C), steam digestion, acid treatment	Basic Blue 9	472.1	30 °C pH 7.2	[84]
Coconut husk	Impregnation (KOH) and CO_2 gasification	Basic Blue 9	435	30 °C pH 6.5	[87]
Bagasse	Activation (840 °C)	Basic Red 22	942	30 °C pH 4.1	[91]
Bagasse	Activation (840 °C)	Acid Blue 25	674	30 °C pH 5.9	[91]
Mahogany sawdust	Carbonisation (500 °C), steam activation	Direct Blue 2B	518	30 °C, pH 3 [a]	[39]
Mahogany sawdust	Carbonisation (500 °C), steam activation	Direct Green B	327.9	30 °C, pH 3 [a]	[39]
Plum kernels	Carbonisation , activation (900 °C, 2h)	Basic Blue 9	520.6	30 °C	[76]
Plum kernels	Carbonisation , activation (900 °C, 2h)	Acid Blue 25	904.4	30 °C	[76]
Pinewood	Steam activation (900 °C)	Acid Blue 264	1176	30 °C, pH 6.5	[97]
Pinewood	Steam activation (900 °C)	Basic Blue 9	556	30 °C, pH 6.6	[97]

[a] in a spent textile dyeing wastewater

Sugarcane bagasse, the biomass remaining after the sugarcane stalks are crushed to extract their juice, has been reported as another raw material for activated carbons preparation. Tsai et al. [89] prepared chemical activated carbons ($ZnCl_2$) from this waste and verified a linear increase of the specific surface area and pore volume with the increase in $ZnCl_2$ impregnation ratio. Adsorption capacities of Acid Orange 10 was improved by increasing the impregnation ratio but remained poor (<4.86 mg/g at 20 °C). Another study [90] indicates H_2SO_4 as the activation agent and a preparation method including impregnation, carbonisation and CO_2 gasification. Bagasse raw material contained high ash content but even so, the prepared activated carbons presented great porosity and high values of specific surface areas (614 m2/g - 1433 m2/g). Acid Blue 80 adsorption capacities in the range 59.9-384.6 mg/g were obtained for different holding times of gasification carbons. Juang et al. [91] obtained activated carbons from bagasse with specific surface areas of 446–667 m^2/g. The increase in the activation temperature led to an increase in the surface area and pore volume and a decrease in the proportion of the microporous, which favoured the adsorption.

Table 8 indicates reported adsorption capacities using agricultural and agrofood industrial wastes adsorbents and derived materials.

6. Timber Industrial Wastes

Wood industries and furniture manufacturing generate residues as sawdusts, offcuts and chips. Sawdust, either used as a cooking fuel or as packing material, is easily available in the countryside at zero or negligible price [92]. Sawdust contains various organic compounds (lignin, cellulose and hemicellulose) with polyphenolic groups that might be useful for binding dyes through different mechanisms [8]. Sawdusts have been also tested as adsorbent for heavy metals, oils and toxic salts [93]. Table 8 shows some reported capacities for adsorption of dyes by timber industrial waste materials.

Formaldehyde and H_2SO_4 treated sawdusts were studied for adsorption of the basic dyes Malachite Green [94] and Methylene Blue [92]. Adsorption is favoured by the increase in pH from 2 to 6 (formaldehyde) or 2 to 5 (sulphuric acid) for Malachite Green dye [94] but no relevant pH effect was observed for Methylene Blue [92]. After 2 hours of contact, 43.1 % and 95.1 % of Malachite Green and 82% and 74 % of Methylene Blue (initially at 250 mg/L) were removed by formaldehyde and sulphuric acid treated sawdusts (4g/L) at neutral pH and 26 °C. The adsorbed amounts were calculated to be in the range of 27-59 mg/g.

Mansonia is a hard wood extensively used in the furniture industry and in interior applications in the automobile industry. Mansonia sawdust has its maximum adsorption ability for Methyl Violet Basic dye for pH above 6.91 (pH_{ZPC}) and for increasing temperatures [95]. At 329 K, monolayer adsorption capacity was referred to be 24.6 mg/g.

Cedar sawdust presented maximum Methylene Blue adsorption capacity of 142.36 mg/g (20 °C) [98]. Salts [99, 100] and acid-treatments (H_2SO_4) [101] showed enhancement of the sawdust adsorption properties for basic dyes, but adsorption capacities obtained did not exceed 30.5 mg/g. Studies on the pH effect on the adsorption of basic dye by original Beech sawdust (pH_{ZPC} 5.2) confirmed that capacity was favoured by the increase in the pH [102].

Bark, another waste product from the timber industry, is a polyphenol-rich material with high tannin content, which makes it an effective adsorbent [103, 104]. The polyhydroxy phenol and polyphenol groups of tannin are thought to be the active species in the adsorption process [8]. Remazol BB reactive dye adsorption onto Eucalyptus bark was shown to be affected by temperature, initial pH, NaCl concentration and initial dye concentration/adsorbent dosage [104]. Maximum adsorbed amount (90 mg/g) was obtained at pH 2.5, 50 g/L of NaCl, initial dye concentration of 500 mg/L, adsorbent dose of 2 g/L and temperature of 18 °C. This value was verified to be about a half than the obtained with commercial activated carbons, under the same conditions.

Activated carbons prepared from Mahogany sawdust (carbonisation followed by steam activation) showed considerable surface area (516.3 m^2/g) and good adsorption properties for acid [85] and direct dyes [39]. Lower pH values (optimum below 3) favoured the adsorption.

Medium density fibreboard (MDF) is a wood-based composite material used in the production of architectural mouldings for construction and in the do-it-yourself industries. In the mechanical processing of the MDF board, 18–20% of the total input mass appears as offcuts, chips, and sawdust wastes [105]. Gan et al. [105] submitted MDF sawdusts to thermal carbonisation and chemical activation. The obtained materials were verified to have surprisingly negligible surface areas probably due to the presence of urea formaldehyde resins (binding agents) in MDF sawdusts. Adsorption of the Reactive dye Levafix Brilliant Red E-4BA occurred up to the maximum amount of 14.47 mg/g for the carbonised MDF (550 °C, 30 min), which was not improved by activation.

Activated carbons [97] prepared from pinewood at 900 °C and different holding times presented good adsorption capacities for Basic blue 69, Methylene Blue and Acid Blue 264 dyes, specially for the maximum activation time (maximum surface area of 902 m^2/g and minor ratio microporous volume/total pore volume), with adsorption capacities in the range 556-1176 mg/g.

7. Tanning/Leather Industrial Wastes

Tanning operation consists in the conversion of putrescible skins into stable, durable and long-lasting leathers by chemical bound of organic or inorganic substances to the protein structure of the hide. Chromium or extracts from bark of trees, such as chestnut, are generally used as tanning agents. The leather making process generates substantial quantities of wastes such as hides and skins, fat, shavings, trimmings, buffing dusts, sludge and effluents. The most common way to manage solid wastes is by disposing them on land sites. Chromium-containing wastes require special attention due to the legislative restrictions. Buffing dust is a proteinous solid waste impregnated with chromium, synthetic fat, oil, tanning agents and dye chemicals [106].

Oliveira et al. [107] tested three types of leather: natural non-tanned leather, chromium-containing leather waste, and leather waste after being passed through a chromium extraction process. Reactive red dye adsorption intensity increased in the order of natural leather < leather waste after Cr extraction < chromium-containing leather, presenting capacities in the range 56 mg/g-163 mg/g at 25 °C and pH 7. For Methylene Blue adsorption, lower adsorbed amounts were obtained (3-83 mg/g, 25 °C, pH 7), increasing in the inverse order found for the reactive dye. Results suggested that adsorption can be due to electrostatic interaction caused by the charge of chromium cations in the solid matrix. The authors referred that no chromium leaching was detected during the adsorption experiments. Chromium and vegetable tanned shavings and buffing were also converted to chars (pyrolysis at 450 °C and 600 °C under N_2 atmosphere) and activated carbons (pyrolysis, acid demineralization and activation with CO_2) [108]. Activated carbons prepared from buffing dust and chromium tannery shavings presented adsorption capacities for Methylene Blue of 200 and 166.7 mg/g, respectively. Oil and chars from pyrolysis demonstrated further fuel properties. Sekaran [106] produced activated carbons from the buffing dust but insignificant adsorption of Acid brown dye (7.86 mg/g for 20 °C and pH3.5) was observed. Table 10 indicates some dyes adsorption capacities reported for leather industrial wastes.

8. Seafood Wastes

Crustacean shell wastes (shrimps, squid, and crabs) from seafood processing industry also present disposal problems. Chitin can be extracted and chitosan can be then prepared from these wastes and used in agriculture, biomedicine, biotechnology, water purification and food industry [2]. Recent literature has focused on the use of chitin and mainly chitosan materials as potential biosorbents [2, 103, 109-123].

Chitin is a linear long-chain polymer of N-acetylglucosamine (Figure 2). It is commercially extracted from crustaceans waste by a first demineralization step (with a dilute acidic solution to dissolve the calcium carbonate), followed by deproteinization (with dilute alkaline solution) and finally by decolourization [124-126]. Chitosan is a polysaccharide composed by polymers of glucosamine and N-acetyl glucosamine (Figure 2). It is obtained from chitin by deacetylation with concentrated alkali solution [127]. Its unique polycationic nature structure is responsible for the special good adsorption capacity for anionic dyes. Performance of chitosan on the adsorption depends however on its origin and nature since chitosan designate various stages of deacetylated and depolymerizated chitins. The high content of amino and hydroxy functional groups in chitosan can serve as coordination and reaction sites [115, 116, 128] and its chemical structure is also advantageous in allowing specific modifications without too many difficulties [2]. Chitosan derivatives could be classified into modified polymers, cross-linked chitosans, chitosan-based composites and membranes [2]. Chitosan, in its own form, is soluble in acidic media and thus several applications as adsorbent used cross-linked reagents for its stabilization under these conditions [110]. To avoid pressure drop in columns, chitin and chitosan beads have also been developed [111]. Table 9 present some reported capacities for anionic dyes adsorption by chitin and chitosan materials.

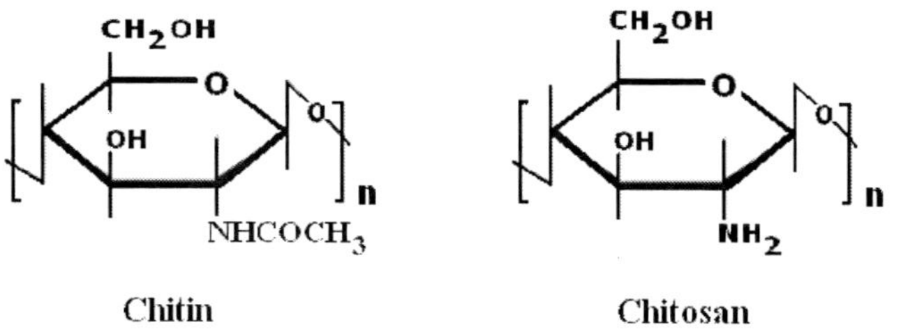

Figure 2. Chemical structure of chitin and Chitosan.

Natural waste seafood materials, Anodonta shell, Sepia and Squid pens, without any treatment, revealed relatively low adsorption capacities for anionic dyes and also high internal resistances which impairs their satisfactory use as adsorbents [124]. Demineralization and deproteinization of the materials enhanced the adsorption capacity to greater values. Prado [113] obtained a much higher adsorpion capacity for Indigo Carmine dye by the extracted chitosan (more nitrogen centers), 0.154 mmol/g than by chitin, 0.0124 mmol/g. Annadurai [114] obtained Chitosan adsorption capacities in the range 91.47-130 mg/g, respectively at 30 and 60°C, and pH 6.7, for the Reactive Black 13 dye. The amino groups of chitosan provided a fair dye removal capability. Adsorption process was verified to be endothermic, and so the temperature favours its extent. In the pH range 6.7-9.5, dye removal decreased with the increase in pH.

Table 9. Maximum adsorption capacities (Q_m) reported in literature for chitin and chitosan derived adsorbents

Adsorbent	Dyes	Q_m (mg/g)	Conditions	Ref.
Squid pen	Reactive Green 12	39.8	20 °C	[124]
Squid pen	Direct Green 26	4.83	20 °C	[124]
Sepia pen	Reactive Green 12	3.46	20 °C	[124]
Sepia pen	Direct Green 26	56.0	20 °C	[124]
Sepia pen demineralized	Reactive Green 12	119	20 °C	[124]
Squid pen deproteinized	Direct Green 26	37.3	20 °C	[124]
Squid pen deproteinized	Reactive Green 12	202	20 °C	[124]
Chitosan flakes	Reactive Blue 222	199	30 °C	[112]
Swollen Chitosan beads	Reactive Blue 222	1009	30 °C	[112]
Cross-linked Chitosan beads	Reactive Yellow 2	2436	30 °C, pH 4	[110]
Cross-linked Chitosan beads	Acid Orange 7	1940	30 °C, pH 4	[110]
Cross-linked Chitosan beads	Direct Red 81	2383	30 °C, pH 4	[110]
Chitosan	Reactive Black 5	91.47	30 °C, pH 6.7	[114]
Chitosan (crab shell)	Acid Green 25	645.1	25 °C, pH 4	[117]
Chitosan (crab shell)	Acid Orange 12	973.3	25 °C, pH 4	[117]

Chitosan derived from deacetylated crab shell chitin [117] presented high adsorption capacities for various acid dyes, in the range of 645.1 -973.3 mg/g (pH 4 and 25 °C). Wu and co-authors [112] obtained significant monolayer adsorption capacities for reactive dyes by chitosan flakes obtained from cuttlebone wastes (188-339 mg/g) and proved that these values can be greatly enhanced to 885-1653 mg/g using swollen chitosan beads. Chiou et al. [110] prepared ionically and chemically cross-linked chitosan beads and obtained maximum adsorption capacities for reactive, acid and direct dyes in the range of 1911-2498 mg/g (30 °C and pH 3-4), which are values 3-15 times higher than for activated carbon and 3-27 times greater than for chitin. Bekçi et al. [118] found adsorption capacities of synthesized chitosan beads in the range of 72.83-93.55 mg/g (303-323 K) for Malachite Green basic dye and verified an increase in the adsorption capacity when the pH increased from 2 to 8.

9. Wastes from Power Plants

The combustion of coal in power plants generates two main types of wastes: fly ashes and bottom ashes. Fly ash particles are captured from the chimneys in cyclones or electrostatic precipitators whereas bottom ash is retained at the bottom of the furnace. Fly ashes are mainly composed by silica, alumina, calcium oxide, iron oxide and residual carbon. Classification as type F and type C [129] distinguishes fly ashes according to their lime (CaO) content. Type F is generated in the harder, older anthracite and bituminous coal burning and presents a lime content lower than 10%. Type C fly ashes are generated when a younger lignite and sub-bituminous coal are used and present high content of lime (higher than 20%). Significant quantities of fly ash have been reused in civil engineering as a mineral admixture in concrete and brick production and as a filling material in road works but superfluous fly ashes still cause environmental and disposal problems in some countries.

Fly ash has demonstrated considerable adsorption properties for anionic dyes. Adsorption capacities of reactive dyes in the range 37.26-135.69 mg/g (20 °C) were obtained using a fly ash with a pH_{ZPC} of 7.0 [130]. In this study, adsorption of Reactive Blue 19 dye was found to be pH dependent but no significant pH effect was observed for both Reactive Red 198 and Reactive Yellow 84. Intraparticle diffusion was verified to be the rate limiting step. Chemisorption was proposed as the possible phenomenon involved in adsorption. Acemioglu [129] used a type C (23.66% CaO) fly ash, from a lignite-burning thermal power plant, for testing Congo Red adsorption. The amount of dye adsorbed decreased slightly with the increase in pH. Desorption studies and FTIR spectrometry show that chemisorption occurs between fly ash and Congo Red dye, probably due to dye/fly ash complexing.

Kumar et al. [131] reported poor adsorption capacity for Methylene Blue (5.574 mg/g) by a fly ash with a low content of lime (around 1%). Karagozoglu and co-authors [132] found however a reasonable removal of Astrazon Blue basic dye (128 mg/g, 30 °C) using a fly ash with 23.66 % lime content. The two different fly ash types used in these studies may be the explanation for the different adsorption behaviours observed for the cationic dyes.

Several treatments were also tried to enhance fly ash adsorption properties, either thermal [133] or chemical [133, 134]. Hsu [133] concluded that the raw coal fly ash had superior adsorbing ability for the anionic dye Acid Red 1 than the thermally (600 °C) and chemically (NaHO, 90 °C) modified samples. Adsorption was verified to increase with temperature and slightly varied with pH. The treatments carried out improved porous properties of the fly ash but decrease the pH_{ZPC}. Original fly ash was then positively charged at a solution pH of 6 (ion-exchange is operative) whereas the thermally treated sample was almost neutral and NaHO treated sample was negatively charged inhibiting ion-exchange to occur. Lin et al. [134] reported an increase in the specific surface area and Methylene Blue dye adsorption capacity by fly ash after acid treatment but adsorption capacities measured remained poor (< 1 mg/g). Wang et al. [33] used a fly ash from sub-bituminous black coal burning and carried out acid treatments under different conditions. Acid treatment enhanced basic dyes adsorption properties, mainly after a microwave treatment, but considerable low adsorption capacities were found for basic dyes. The presence of inorganic salts (KCl) was verified to favour

Methylene Blue and Crystal Violet adsorption. Results reported by Wang et al. [135] are in agreement with the effect of the acid treatment. Thermal treatment in this work was also studied and a decrease of fly ash adsorption capacity for Methylene Blue was observed.

Fly ash has then shown good adsorption properties for anionic dyes, without the need of further treatments. The studies generally agree as regards the effect of pH. As the pH of the system increases, the number of negatively charged sites increases, thus favouring the adsorption of cationic (basic) dyes [33, 134, 135]. The inverse effect is observed when decreasing the solution pH, and the adsorption of anionic dyes is enhanced [129]. Table 10 shows some adsorption capacities reported for fly ashes.

10. Steel Plant Wastes

Steel industries generate wastes as dusts and sludge and, in major amounts, slag. The gases generated in the manufacture of pig iron carry a dust load, which is removed before their release to the atmosphere. The coarse particles of the dust are settled and named blasted furnace dust and the smaller ones, removed in wet scrubbers, form a waste designated as blasted furnace sludge [136]. Slag is a by-product obtained in the separation of the molten steel from the impurities in steel-making furnaces [137]. It is a molten liquid that solidifies upon cooling, usually available in the form of granular solids [138] and mainly composed by heavy metals and trace metals, essentially in the form of oxides and silicates. Slag has useful properties which make it reusable in cementing applications during drilling and as an additive for high performance concrete [139-141]. Sorption properties of steel plants wastes are not relevant since low adsorption capacities have been reported. The effect of pH on adsorption is generally similar to the observed for other adsorbents. The metal oxides present in the slag form metal–hydroxide complexes in solution, and their subsequent acidic or basic dissociation at the solid–solution interface leads to the development of positive (favouring anionic dyes adsorption) or negative charges (favouring basic dyes adsorption) on the surface [138]. The use of strong acidic conditions for anionic dyes adsorption is however economically unattractive and environmentally undesirable due to the probability of metal leaching from the slag. Results from Ramakrishna and Viraraghavan [138], show that even at

Table 10. Maximum adsorption capacities (Q_m) reported in literature for leather, steel plants and other industrial wastes

Raw material	Treatment	Dyes	Q_m (mg/g)	Conditions	Ref.
Cr-leather waste	(no treatment)	Reactive red dye	163	25 °C, pH 7	[107]
Cr-leather waste	(no treatment)	Basic Blue 9	3	25 °C, pH 7	[107]
Fly-ash	(no treatment)	Reactive Blue 19	135.69	20 °C	[130]
Fly-ash	(no treatment)	Reactive Yellow 84	37.26	20 °C	[130]
Fly-ash	(no treatment)	Astrazon Blue	128	30 °C	[132]
Fly-ash	(no treatment)	Acid Red 1	92.59	30 °C, pH 6	[133]
Fly-ash	Thermal (600 °C)	Acid Red 1	32.79	30 °C, pH 6	[133]
Fly-ash	Chemical (NaHO)	Acid Red 1	12.66	30 °C, pH 6	[133]
Red mud	(no treatment)	Direct Red 28	4.05	30 °C, pH 7.3	[144]
Carbonaceous slurry	Thermal and Acid	Acid Blue 113	221.2	25 °C	[136]
Carbonaceous slurry	Thermal and Acid	Basic Violet 10	91.1	25 °C, pH 5.5-6.5	[145]

optimum pH, low adsorbed amounts could be expected. Disperse Red 1 dye was found to be, under certain conditions, more adsorbed by slag than by a commercial activated carbon. Column studies indicated 1.3 mg/g as the capacity of slag to adsorb this dye [138]. Jain, Bhatnagar and co-workers [136, 142] also referred adsorbents prepared from blast furnace slag, dusts and sludge as unsuitable for anionic and cationic dyes removal. Thermal treatment of slag proved to be ineffective for adsorption enhancement of the Basic Red 9 (0.0102 mmol/g, 30 °C) [143].

11. Other Industrial Wastes

In this section, wastes not matching with the previous classification were included. Old newspapers have been reused as raw materials for activated carbons preparation [146]. Different methods of activation and different agents (alkali carbonates and hydroxides) were tested. Activated carbons prepared by physical and chemical activation using the optimum conditions presented surface areas of 1090 m^2/g and 1740 m^2/g, respectively [146] . Further studies [147] indicate high adsorbed amounts of Methyleen Blue dye (340-390 mg/g and 190 mg/g by the chemically and physically activated carbons, respectively). Chemical activation demonstrated to produce activated carbons with better performances than three commercial samples. Higher pore size and more functional groups were appointed as the probable explanation for their better behaviour.

Polyethyleneterephtalate (PET) wastes (post-consumer bottles and plastics from industrial residues) have been also applied as raw materials for activated carbons preparation [13, 15, 148-152]. Dye adsorption results are still few but, at the moment, they seem to reveal good sorption properties. Methylene Blue basic dye adsorbed amounts were verified to be up to 1 mmol/g (about 320 mg/g) using a carbon obtained by combining chemical and steam activation [149]. Malachite green basic dye was also well removed (adsorption capacity of 169.49 mg/g, 26 °C) by a chemically (NaHO) activated carbon [151]. Activated carbons prepared by only steam activation and also using a chemical pre-treatment of waste PET bottles were found to have surface areas of 1200 m^2/g, highly mesoporous structure and an hydrophobic nature surface, being appropriate for reactive dye adsorption [152].

A carbonaceous slurry waste generated in fertilizer plants was also tested as precursor for carbon-based adsorbents for dye removal. This waste results from the partial oxidation of fuel oil/low sulphur heavy stock as feed stock in fertilizer plants [136]. Thermal activation followed by acid treatment (to remove the ash content) was found to produce carbon adsorbents with surface areas in the range 280-380 m^2/g and performances of 80-90 %, equal to the presented by a commercial activated carbon used for comparison [136, 142]. At 25 °C, dyes adsorption capacities were in the range 198.4-221.2 mg/g for anionic classes [136] and 85.0 and 91.1 mg/g for the cationic basic class [142].

Red mud is a residue generated in the alumina manufacture by the alkaline leaching of bauxite in Bayer process [153]. Literature reports some studies on reusing waste red mud for dyes adsorption, but results clearly indicate limited capacities. At 30 °C and natural pH 7.3, maximum adsorption capacity for Congo Red was measured as 4.05 mg/g. Desorption studies indicate ion-exchange as the predominant adsorption mechanism [144]. Results reported in reference [153] showed a higher but still low value for the same dye (7.08 mg/g) using a waste red mud after acid activation. According to Wang [135], thermal and acid treatments reduce Methylene Blue adsorption capacity of waste red mud.

Table 10 shows some capacities reported in the literature for the above mentioned adsorbents, including the respective treatments and operating conditions.

12. Conclusion

A wide variety of industrial wastes as adsorbents for textile dyes was presented. The sorption capacity was taken as the main criterion for measuring the potential of the studied materials. The most important advantage of using industrial wastes as adsorbents is, of course, their availability at low or no cost.

From the literature cited, preparation of activated carbons or simply carbon adsorbents from non-conventional raw materials appears to be a promising alternative. Some agricultural wastes, newspapers and sewage sludges were successfully applied as raw materials, under certain activation conditions. In general, chemical activation showed better results than the physical one. For dye adsorption purposes, the huge importance of the surface chemistry and physical properties for the effectiveness of the waste-based activated carbons was focused here. Several studies showed that activation does not always improve the adsorption properties of a raw material.

Chitin and chitosan-based materials derived from seafood wastes presented markedly high adsorption capacities for the most problematic dye class, the reactive dyes. These types of adsorbents, although they require purifying chemical treatments, emerge as an attractive possibility. The pH values frequently reported in literature (<4) were, however, impracticable in real situations, since textile effluents are generally alkaline and if the adsorption is to be used as a tertiary treatment (after a biological process, for example) the effluent pH would be neutral. So, it would be preferable to use more modest estimates obtained at a pH of around 7.

Other wastes, such as hydroxide metal sludges, fly ashes and leather wastes from chromium tanning, presented reasonable adsorption capacities. Considering that their cost is almost zero and that there is no need for any kind of treatment, these wastes also show a high potential of application.

Specific surface area, pore structure developed (size affecting the penetration of the dyes' large molecules into pores), presence of functional groups and pH_{ZPC} (affecting electrostatic interactions) are decisive factors in explaining the performance of an adsorbent.

In the authors' opinion, research on this topic should continue by applying the waste adsorbents in real textile effluents and evaluating the feasibility of using them in fixed bed columns or as an alternative to powdered activated carbon.

Acknowledgements

The authors are grateful to the Portuguese Foundation for Science and Technology (FCT) for S. Santos' scholarship (SFRH/BD/18477/2004).

References

[1] Gonçalves, I. M., Ferra, M. I. & Amorim, M. T. P. (1996). Processos de remoção biológica de corantes nos efluentes de indústria têxtil. *Tecnologias do Ambiente*, **11** 35-8.

[2] Crini, G. & Badot, P. M. (2008). Application of chitosan, a natural aminopoly-saccharide, for dye removal from aqueous solutions by adsorption processes using batch studies: A review of recent literature. *Progress in Polymer Science*, **33 (4)**, 399-447.

[3] Forgacs, E., Cserhati, T. & Oros, G. (2004). Removal of synthetic dyes from wastewaters: a review. *Environment International*, **30 (7)**, 953-71.

[4] Waters, B. D. (1995). The regulator's view. In P. Cooper, (Ed.), *Colour in Dyehouse effluent* (22-9). Oxford, The Society of Dyers and Coulorists, The Alden Press.

[5] Kouba, J. F. & Zhuang, P. (1994). Color removal for textile dyeing wastewater *Fluid / Particle Separation Journal*, **7** (**3**), 87-90.

[6] Easton, J. R. (1995). The dye maker's view. In P. Cooper, (Ed.), *Colour in Dyehouse Effluent* (9-21). Oxford, The Society of dyers and Colourists, The Alden Press.

[7] Anliker, R., Clarke, E. A. & Moser, P. (1981). Use of the Partition-Coefficient as an Indicator of Bioaccumulation Tendency of Dyestuffs in Fish. *Chemosphere*, **10 (3)**, 263-74.

[8] Crini, G. (2006). Non-conventional low-cost adsorbents for dye removal: A review. *Bioresource Technology*, **97 (9)**, 1061-85.

[9] Hazel, B. G. (1995). Industry evaluation of colour reduction and removal - the DEMOS project. In P. Cooper, (Ed.), *Colour in Dyehouse Effluent* (59-72). Oxford, The Society of dyers and Colourists, The Alden Press.

[10] Naim, M. M. & El Abd, Y. M. (2002). Removal and recovery of dyestuffs from dyeing wastewaters. *Separation and Purification Methods*, **31 (1)**, 171-228.

[11] Churchley, J. H. (1995). The water company's view. In P. Cooper, (Ed.), *Colour in Dyehouse Effluent* (31-43). Oxford, The Society of dyers and Colourists, The Alden Press.

[12] Rodrígues-Reinoso, F. (1997). *Introduction to Carbon Technologies*. Spain: Publicaciones de la Universidad de Alicante.

[13] Dias, J. M., Alvim-Ferraz, M. C. M., Almeida, M. F., Rivera-Utrilla, J. & Sanchez-Polo, M. (2007). Waste materials for activated carbon preparation and its use in aqueous-phase treatment: A review. *Journal of Environmental Management*, **85 (4)**, 833-46.

[14] Juang, R. S., Wu, F. C. & Tseng, R. L. (2000). Mechanism of adsorption of dyes and phenols from water using activated carbons prepared from plum kernels. *Journal of Colloid and Interface Science*, **227 (2)**, 437-44.

[15] Laszlo, K. (2005). Adsorption from aqueous phenol and aniline solutions on activated carbons with different surface chemistry. *Colloids and Surfaces a-Physicochemical and Engineering Aspects*, **265 (1-3),** 32-9.

[16] Banat, F. A., Al-Bashir, B., Al-Asheh, S. & Hayajneh, O. (2000). Adsorption of phenol by bentonite. *Environmental Pollution*, **107 (3)**, 391-8.

[17] Rengaraj, S., Moon, S.-H., Sivabalan, R., Arabindoo, B. & Murugesan, V. (2002). Agricultural solid waste for the removal of organics: adsorption of phenol from water and wastewater by palm seed coat activated carbon. *Waste Management*, **22 (5)**, 543-8.
[18] Vilar, V. J. P., Botelho, C. M. S. & Boaventura, R. A. R. (2007). Modeling equilibrium and kinetics of metal uptake by algal biomass in continuous stirred and packed bed adsorbers. *Adsorption-Journal of the International Adsorption Society*, **13 (5-6)**, 587-601.
[19] Bulut, Y. & Tez, Z. (2007). Removal of heavy metals from aqueous solution by sawdust adsorption. *Journal of Environmental Sciences*, **19 (2)**, 160-6.
[20] Qi, B. C. & Aldrich, C. (2008). Biosorption of heavy metals from aqueous solutions with tobacco dust. *Bioresource Technology*, **99 (13)**, 5595-601.
[21] Rozada, F., Otero, M., Morán, A. & García, A. I. (2008). Adsorption of heavy metals onto sewage sludge-derived materials. *Bioresource Technology*, **99 (14)**, 6332-8.
[22] Vilar, V. J. P., Botelho, C. M. S. & Boaventura, R. A. R. (2006). Equilibrium and kinetic modelling of Cd(II) biosorption by algae Gelidium and agar extraction algal waste. *Water Research*, **40 (2)**, 291-302.
[23] Martin, M. J., Artola, A., Balaguer, M. D. & Rigola, M. (2003). Activated carbons developed from surplus sewage sludge for the removal of dyes from dilute aqueous solutions. *Chemical Engineering Journal*, **94 (3)**, 231-9.
[24] Marsh, H., Heintz, E. A. & Rodríguez-Reinoso, F., eds. *Introduction to Carbon Technologies*. 1997, Universidade de Alicante, Secretariado de Publicaciones.
[25] Abrahart, E. N. (1977). *Dyes and their intermediates* (*Second Edition*). London: Edward Arnold
[26] O'Neill, C., Hawkes, F. R., Hawkes, D. L., Lourenco, N. D., Pinheiro, H. M. & Delee, W. (1999). Colour in textile effluents - sources, measurement, discharge consents and simulation: a review. *Journal of Chemical Technology and Biotechnology*, **74 (11)**, 1009-18.
[27] Araújo, M. & Castro, E. M. (1984). *Manual de Engenharia Têxtil, volume II*. Lisboa: Fundação Calouste Gulbenkian.
[28] Al-Degs, Y., Khraisheh, M. A. M., Allen, S. J., Ahmad, M. N. & Walker, G. M. (2007). Competitive adsorption of reactive dyes from solution: Equilibrium isotherm studies in single and multisolute systems. *Chemical Engineering Journal*, **128 (2-3)**, 163-7.
[29] Noroozi, B., Sorial, G. A., Bahrami, H. & Arami, M. (2008). Adsorption of binary mixtures of cationic dyes. *Dyes and Pigments*, **76 (3)**, 784-91.
[30] Allen, S. J., Mckay, G. & Porter, J. F. (2004). Adsorption isotherm models for basic dye adsorption by peat in single and binary component systems. *Journal of Colloid and Interface Science*, **280 (2)**, 322-33.
[31] Choy, K. K. H., Porter, J. F. & Mckay, G. (2004). Intraparticle diffusion in single and multicomponent acid dye adsorption from wastewater onto carbon. *Chemical Engineering Journal*, **103 (1-3)**, 133-45.
[32] Netpradit, S., Thiravetyan, P. & Towprayoon, S. (2004). Adsorption of three azo reactive dyes by metal hydroxide sludge: effect of temperature, pH, and electrolytes. *Journal of Colloid and Interface Science*, **270 (2)**, 255-61.
[33] Wang, S. B., Boyjoo, Y. & Choueib, A. (2005). A comparative study of dye removal using fly ash treated by different methods. *Chemosphere*, **60 (10)**, 1401-7.

[34] Al-Degs, Y. S., El-Barghouthi, M. I., El-Sheikh, A. H. & Walker, G. M. (2008). Effect of solution pH, ionic strength, and temperature on adsorption behavior of reactive dyes on activated carbon. *Dyes and Pigments*, **77 (1)**, 16-23.

[35] Santos, S. C. R., Vilar, V. J. P. & Boaventura, R. A. R. (2008). Waste metal hydroxide sludge as adsorbent for a reactive dye. *Journal of Hazardous Materials*, **153 (3)**, 999-1008.

[36] Pereira, M. F. R., Soares, S. F., Orfao, J. J. M. & Figueiredo, J. L. (2003). Adsorption of dyes on activated carbons: influence of surface chemical groups. *Carbon*, **41 (4)**, 811-21.

[37] Dhaouadi, H. & M'Henni, F. (2008). Textile mill effluent decolorization using crude dehydrated sewage sludge. *Chemical Engineering Journal*, **138 (1-3)**, 111-9.

[38] Walker, G. M. & Weatherley, L. R. (1997). Adsorption of acid dyes onto granular activated carbon in fixed beds. *Water Research*, **31 (8)**, 2093-101.

[39] Malik, P. K. (2004). Dye removal from wastewater using activated carbon developed from sawdust: adsorption equilibrium and kinetics. *Journal of Hazardous Materials*, **113 (1-3)**, 83-90.

[40] Boaventura, R. A. (1984). Characterization of industrial effluents in the Ave River basin, unpublished work.

[41] Langmuir, I. (1918). The adsorption of gases on plane surfaces of glass, mica and platinum. *Journal of the American Chemical Society*, **40** 1361-403.

[42] Freundlich, H. M. F. (1906). Over the adsorption in solution. *Journal of Physical Chemistry*, **57** 385-471.

[43] Moreno-Castilla, C. & Rivera-Utrilla, J. (2001). Carbon materials as adsorbents for the removal of pollutants from the aqueous phase *Materials Research Society Bulletin*, **26** 890-4.

[44] Moreno-Castilla, C. (2004). Adsorption of organic molecules from aqueous solutions on carbon materials. *Carbon*, **42 (1)**, 83-94.

[45] Al-Degs, Y., Khraisheh, M. A. M., Allen, S. J. & Ahmad, M. N. (2000). Effect of carbon surface chemistry on the removal of reactive dyes from textile effluent. *Water Research*, **34 (3)**, 927-35.

[46] Venkata Mohan, S., Chandrasekhar Rao, N. & Karthikeyan, J. (2002). Adsorptive removal of direct azo dye from aqueous phase onto coal based sorbents: a kinetic and mechanistic study. *Journal of Hazardous Materials*, **90 (2)**, 189-204.

[47] Faria, P. C. C., Orfao, J. J. M. & Pereira, M. F. R. (2004). Adsorption of anionic and cationic dyes on activated carbons with different surface chemistries. *Water Research*, **38 (8)**, 2043-52.

[48] Orfao, J. J. M., Silva, A. I. M., Pereira, J. C. V., Barata, S. A., Fonseca, I. M., Faria, P. C. C. & Pereira, M. F. R. (2006). Adsorption of a reactive dye on chemically modified activated carbons - Influence of pH. *Journal of Colloid and Interface Science*, **296 (2)**, 480-9.

[49] El Qada, E. N., Allen, S. J. & Walker, G. A. (2008). Adsorption of basic dyes from aqueous solution onto activated carbons. *Chemical Engineering Journal*, **135 (3)**, 174-84.

[50] Al-Duri, B. & Mckay, G. (1991). Prediction of Binary-System for Kinetics of Batch Adsorption Using Basic-Dyes onto Activated Carbon. *Chemical Engineering Science*, **46 (1)**, 193-204.

[51] Walker, G. M. & Weatherley, L. R. (1999). Kinetics of acid dye adsorption on GAC. *Water Research*, **33 (8)**, 1895-9.

[52] Choy, K. K. H., Porter, J. F. & Mckay, G. (2001). A film-pore-surface diffusion model for the adsorption of acid dyes on activated carbon. *Adsorption-Journal of the International Adsorption Society*, **7 (3)**, 231-47.

[53] Caballero, J. A., Front, R., Marcilla, A. & Conesa, J. A. (1997). Characterization of sewage sludges by primary and secondary pyrolysis. *Journal of Analytical and Applied Pyrolysis*, **40-1** 433-50.

[54] Bagreev, A., Bashkova, S., Locke, D. C. & Bandosz, T. J. (2001). Sewage sludge-derived materials as efficient adsorbents for removal of hydrogen sulfide. *Environmental Science & Technology*, **35 (7)**, 1537-43.

[55] Chiang, P. C. & You, J. H. (1987). Use of Sewage-Sludge for Manufacturing Adsorbents. *Canadian Journal of Chemical Engineering*, **65 (6)**, 922-7.

[56] Lu, G. Q. (1996). Preparation and evaluation of adsorbents from waste carbonaceous materials for SOx and NOx removal. *Environmental Progress*, **15 (1)**, 12-8.

[57] Martin, M. J., Artola, A., Balaguer, M. D. & Rigola, M. (2002). Towards waste minimisation in WWTP: activated carbon from biological sludge and its application in liquid phase adsorption. *Journal of Chemical Technology and Biotechnology*, **77 (7)**, 825-33.

[58] Tay, J. H., Chen, X. G., Jeyaseelan, S. & Graham, N. (2001). Optimising the preparation of activated carbon from digested sewage sludge and coconut husk. *Chemosphere*, **44 (1)**, 45-51.

[59] Otero, M., Rozada, F., Calvo, L. F., Garcia, A. I. & Moran, A. (2003). Kinetic and equilibrium modelling of the methylene blue removal from solution by adsorbent materials produced from sewage sludges. *Biochemical Engineering Journal*, **15 (1)**, 59-68.

[60] Calvo, L. F., Otero, M., Moran, A. & Garcia, A. I. (2001). Upgrading sewage sludges for adsorbent preparation by different treatments. *Bioresource Technology*, **80 (2)**, 143-8.

[61] Rozada, F., Calvo, L. F., Garcia, A. I., Martin-Villacorta, J. & Otero, M. (2003). Dye adsorption by sewage sludge-based activated carbons in batch and fixed-bed systems. *Bioresource Technology*, **87 (3)**, 221-30.

[62] Netpradit, S., Thiravetyan, P. & Towprayoon, S. (2003). Application of 'waste' metal hydroxide sludge for adsorption of azo reactive dyes. *Water Research*, **37 (4)**, 763-72.

[63] Namasivayam, C., Jeyakumar, R. & Yamuna, R. T. (1994). Dye Removal from Waste-Water by Adsorption on Waste Fe(III)/Cr(III) Hydroxide. *Waste Management*, **14 (7)**, 643-8.

[64] Netpradit, S., Thiravetyan, P. & Towprayoon, S. (2004). Evaluation of metal hydroxide sludge from reactive dye adsorption in a fixed-bed column system. *Water Research*, **38 (1)**, 71-8.

[65] Donald, M. (2000). A new process for electrocoagulation. *Journal of American Water Works Association*, **92** 34-43.

[66] Golder, A. K., Samanta, A. N. & Ray, S. (2006). Anionic reactive dye removal from aqueous solution using a new adsorbent - Sludge generated in removal of heavy metal by electrocoagulation. *Chemical Engineering Journal*, **122 (1-2)**, 107-15.

[67] Gan, Q. (2000). A case study of microwave processing of metal hydroxide sediment sludge from printed circuit board manufacturing wash water. *Waste Management*, **20 (8)**, 695-701.

[68] Lee, C. K., Low, K. S. & Chow, S. W. (1996). Chrome sludge as an adsorbent for colour removal. *Environmental Technology*, **17 (9)**, 1023-8.

[69] Namasivayam, C., Muniasamy, N., Gayatri, K., Rani, M. & Ranganathan, K. (1996). Removal of dyes from aqueous solutions by cellulosic waste orange peel. *Bioresource Technology*, **57 (1)**, 37-43.

[70] Annadurai, G., Juang, R. S. & Lee, D. J. (2002). Use of cellulose-based wastes for adsorption of dyes from aqueous solutions. *Journal of Hazardous Materials*, **92 (3)**, 263-74.

[71] Arami, M., Limaee, N. Y., Mahmoodi, N. M. & Tabrizi, N. S. (2005). Removal of dyes from colored textile wastewater by orange peel adsorbent: Equilibrium and kinetic studies. *Journal of Colloid and Interface Science*, **288 (2)**, 371-6.

[72] Hameed, B. H. & El-Khaiary, M. I. (2008). Batch removal of malachite green from aqueous solutions by adsorption on oil palm trunk fibre: Equilibrium isotherms and kinetic studies. *Journal of Hazardous Materials*, **154 (1-3)**, 237-44.

[73] Gupta, V. K., Jain, R. & Varshney, S. (2007). Removal of Reactofix golden yellow 3 RFN from aqueous solution using wheat husk - An agricultural waste. *Journal of Hazardous Materials*, **142 (1-2)**, 443-8.

[74] Gupta, V. K., Jain, R., Varshney, S. & Saini, V. K. (2007). Removal of Reactofix Navy Blue 2 GFN from aqueous solutions using adsorption techniques. *Journal of Colloid and Interface Science*, **307 (2)**, 326-32.

[75] Stavropoulos, G. G. & Zabaniotou, A. A. (2005). Production and characterization of activated carbons from olive-seed waste residue. *Microporous and Mesoporous Materials*, **82 (1-2)**, 79-85.

[76] Wu, F. C., Tseng, R. L. & Juang, R. S. (1999). Pore structure and adsorption performance of the activated carbons prepared from plum kernels. *Journal of Hazardous Materials*, **69 (3)**, 287-302.

[77] Tseng, R. L. (2007). Physical and chemical properties and adsorption type of activated carbon prepared from plum kernels by NaOH activation. *Journal of Hazardous Materials*, **147 (3)**, 1020-7.

[78] Guo, Y. P., Zhang, H., Tao, N. N., Liu, Y. H., Qi, J. R., Wang, Z. C. & Xu, H. D. (2003). Adsorption of malachite green and iodine on rice husk-based porous carbon. *Materials Chemistry and Physics*, **82 (1)**, 107-15.

[79] Guo, Y. P., Zhao, J. Z., Zhang, H., Yang, S. F., Qi, J. R., Wang, Z. C. & Xu, H. D. (2005). Use of rice husk-based porous carbon for adsorption of Rhodamine B from aqueous solutions. *Dyes and Pigments*, **66 (2)**, 123-8.

[80] Rahman, I. A., Saad, B., Shaidan, S. & Rizal, E. S. S. (2005). Adsorption characteristics of malachite green on activated carbon derived from rice husks produced by chemical-thermal process. *Bioresource Technology*, **96 (14)**, 1578-83.

[81] Attia, A. A., Girgis, B. S. & Fathy, N. A. (2008). Removal of methylene blue by carbons derived from peach stones by H3PO4 activation: Batch and column studies. *Dyes and Pigments*, **76 (1)**, 282-9.

[82] Guo, Y. P., Yang, S. F., Fu, W. Y., Qi, J. R., Li, R. Z., Wang, Z. C. & Xu, H. D. (2003). Adsorption of malachite green on micro- and mesoporous rice husk-based active carbon. *Dyes and Pigments*, **56 (3)**, 219-29.

[83] Zhang, H., Guo, Y. P., Liu, Y. H., Zhao, X., Deng, Y. H., Wang, Z. C. & Jiang, L. (2007). Adsorption of creatinine on porous carbon prepared from rice husk. *Acta Physico-Chimica Sinica*, **23 (6)**, 825-9.

[84] Kannan, N. & Sundaram, M. M. (2001). Kinetics and mechanism of removal of methylene blue by adsorption on various carbons - a comparative study. *Dyes and Pigments*, **51 (1)**, 25-40.

[85] Malik, P. K. (2003). Use of activated carbons prepared from sawdust and rice-husk for adsorption of acid dyes: a case study of Acid Yellow 36. *Dyes and Pigments*, **56 (3)**, 239-49.

[86] Kadirvelu, K., Karthika, C., Vennilamani, N. & Pattabhi, S. (2005). Activated carbon from industrial solid waste as an adsorbent for the removal of Rhodamine-B from aqueous solution: Kinetic and equilibrium studies. *Chemosphere*, **60 (8)**, 1009-17.

[87] Tan, I. A. W., Ahmad, A. L. & Hameed, B. H. (2008). Optimization of preparation conditions for activated carbons from coconut husk using response surface methodology. *Chemical Engineering Journal*, **137 (3)**, 462-70.

[88] Thinakaran, N., Baskaralingam, P., Pulikesi, M., Panneerselvam, P. & Sivanesan, S. (2008). Removal of Acid Violet 17 from aqueous solutions by adsorption onto activated carbon prepared from sunflower seed hull. *Journal of Hazardous Materials*, **151 (2-3)**, 316-22.

[89] Tsai, W. T., Chang, C. Y., Lin, M. C., Chien, S. F., Sun, H. F. & Hsieh, M. F. (2001). Adsorption of acid dye onto activated carbons prepared from agricultural waste bagasse by ZnCl2 activation. *Chemosphere*, **45 (1)**, 51-8.

[90] Valix, M., Cheung, W. H. & McKay, G. (2004). Preparation of activated carbon using low temperature carbonisation and physical activation of high ash raw bagasse for acid dye adsorption. *Chemosphere*, **56 (5)**, 493-501.

[91] Juang, R. S., Wu, F. C. & Tseng, R. L. (2002). Characterization and use of activated carbons prepared from bagasses for liquid-phase adsorption. *Colloids and Surfaces a-Physicochemical and Engineering Aspects*, **201 (1-3)**, 191-9.

[92] Garg, V. K., Amita, M., Kumar, R. & Gupta, R. (2004). Basic dye (methylene blue) removal from simulated wastewater by adsorption sawdust: a timber using Indian Rosewood industry waste. *Dyes and Pigments*, **63 (3)**, 243-50.

[93] Shukla, A., Zhang, Y. H., Dubey, P., Margrave, J. L. & Shukla, S. S. (2002). The role of sawdust in the removal of unwanted materials from water. *Journal of Hazardous Materials*, **95 (1-2)**, 137-52.

[94] Garg, V. K., Gupta, R., Yadav, A. B. & Kumar, R. (2003). Dye removal from aqueous solution by adsorption on treated sawdust. *Bioresource Technology*, **89 (2)**, 121-4.

[95] Ofomaja, A. E. & Ho, Y.-S. (2008). Effect of temperatures and pH on methyl violet biosorption by Mansonia wood sawdust. *Bioresource Technology*, **99 (13)**, 5411-7.

[96] Oliveira, L. S., Franca, A. S., Alves, T. M. & Rocha, S. D. F. (2008). Evaluation of untreated coffee husks as potential biosorbents for treatment of dye contaminated waters. *Journal of Hazardous Materials*, **155 (3)**, 507-12.

[97] Tseng, R. L., Wu, F. C. & Juang, R. S. (2003). Liquid-phase adsorption of dyes and phenols using pinewood-based activated carbons. *Carbon*, **41 (3)**, 487-95.

[98] Hamdaoui, O. (2006). Batch study of liquid-phase adsorption of methylene blue using cedar sawdust and crushed brick. *Journal of Hazardous Materials*, **135 (1-3)**, 264-73.
[99] Batzias, F. A. & Sidiras, D. K. (2004). Dye adsorption by calcium chloride treated beech sawdust in batch and fixed-bed systems. *Journal of Hazardous Materials*, **114 (1-3)**, 167-74.
[100] Batzias, F. A. & Sidiras, D. K. (2007). Simulation of methylene blue adsorption by salts-treated beech sawdust in batch and fixed-bed systems. *Journal of Hazardous Materials*, **149 (1)**, 8-17.
[101] Batzias, F. A. & Sidiras, D. K. (2007). Dye adsorption by prehydrolysed beech sawdust in batch and fixed-bed systems. *Bioresource Technology*, **98 (6)**, 1208-17.
[102] Batzias, F. A. & Sidiras, D. K. (2007). Simulation of dye adsorption by beech sawdust as affected by pH. *Journal of Hazardous Materials*, **141 (3)**, 668-79.
[103] Bailey, S. E., Olin, T. J., Bricka, R. M. & Adrian, D. D. (1999). A review of potentially low-cost sorbents for heavy metals. *Water Research*, **33 (11)**, 2469-79.
[104] Morais, L. C., Freitas, O. M., Goncalves, E. P., Vasconcelos, L. T. & Beca, C. G. G. (1999). Reactive dyes removal from wastewaters by adsorption on eucalyptus bark: Variables that define the process. *Water Research*, **33 (4)**, 979-88.
[105] Gan, Q., Allen, S. J. & Matthews, R. (2004). Activation of waste MDF sawdust charcoal and its reactive dye adsorption characteristics. *Waste Management*, **24 (8)**, 841-8.
[106] Sekaran, G., Shanmugasundaram, K. A. & Mariappan, M. (1998). Characterization and utilisation of buffing dust generated by the leather industry. *Journal of Hazardous Materials*, **63 (1)**, 53-68.
[107] Oliveira, L. C. A., Goncalves, M., Oliveira, D. Q. L., Guerreiro, M. C., Guilherme, L. R. G. & Dallago, R. M. (2007). Solid waste from leather industry as adsorbent of organic dyes in aqueous-medium. *Journal of Hazardous Materials*, **141 (1)**, 344-7.
[108] Yilmaz, O., Kantarli, I. C., Yuksel, M., Saglam, M. & Yanik, J. (2007). Conversion of leather wastes to useful products. *Resources Conservation and Recycling*, **49 (4)**, 436-48.
[109] Figueiredo, S. A., Loureiro, J. M. & Boaventura, R. A. (2005). Natural waste materials containing chitin as adsorbents for textile dyestuffs: Batch and continuous studies. *Water Research*, **39 (17)**, 4142-52.
[110] Chiou, M. S., Ho, P. Y. & Li, H. Y. (2004). Adsorption of anionic dyes in acid solutions using chemically cross-linked chitosan beads. *Dyes and Pigments*, **60 (1)**, 69-84.
[111] Crini, G. (2005). Recent developments in polysaccharide-based materials used as adsorbents in wastewater treatment. *Progress in Polymer Science*, **30 (1)**, 38-70.
[112] Wu, F. C., Tseng, R. L. & Juang, R. S. (2001). Enhanced abilities of highly swollen chitosan beads for color removal and tyrosinase immobilization. *Journal of Hazardous Materials*, **81 (1-2)**, 167-77.
[113] Prado, A. G. S., Torres, J. D., Faria, E. A. & Dias, S. C. L. (2004). Comparative adsorption studies of indigo carmine dye on chitin and chitosan. *Journal of Colloid and Interface Science*, **277 (1)**, 43-7.
[114] Annadurai, G., Ling, L. Y. & Lee, J. F. (2008). Adsorption of reactive dye from an aqueous solution by chitosan: isotherm, kinetic and thermodynamic analysis. *Journal of Hazardous Materials*, **152 (1)**, 337-46.

[115] Knorr, D. (1983). Dye Binding-Properties of Chitin and Chitosan. *Journal of Food Science*, **48 (1)**, 36-&.

[116] Juang, R. S., Tseng, R. L., Wu, F. C. & Lee, S. H. (1997). Adsorption behavior of reactive dyes from aqueous solutions on chitosan. *Journal of Chemical Technology and Biotechnology*, **70 (4)**, 391-9.

[117] Wong, Y. C., Szeto, Y. S., Cheung, W. H. & McKay, G. (2004). Adsorption of acid dyes on chitosan - equilibrium isotherm analyses. *Process Biochemistry*, **39 (6)**, 693-702.

[118] Bekçi, Z., Özveri, C., Seki, Y. & Yurdakoç, K. (2008). Sorption of malachite green on chitosan bead. *Journal of Hazardous Materials*, **154 (1-3)**, 254-61.

[119] Wong, Y. C., Szeto, Y. S., Cheung, W. H. & McKay, G. (2003). Equilibrium studies for acid dye adsorption onto chitosan. *Langmuir*, **19 (19)**, 7888-94.

[120] Cheung, W. H., Szeto, Y. S. & McKay, G. (2007). Intraparticle diffusion processes during acid dye adsorption onto chitosan. *Bioresource Technology*, **98 (15)**, 2897-904.

[121] Du, W. L., Xu, Z. R., Han, X. Y., Xu, Y. L. & Miao, Z. G. (2008). Preparation, characterization and adsorption properties of chitosan nanoparticles for eosin Y as a model anionic dye. *Journal of Hazardous Materials*, **153 (1-2)**, 152-6.

[122] Rosa, S., Laranjeira, M. C. M., Riela, H. G. & Fávere, V. T. (2008). Cross-linked quaternary chitosan as an adsorbent for the removal of the reactive dye from aqueous solutions. *Journal of Hazardous Materials*, **155 (1-2)**, 253-60.

[123] Wu, F.-C., Tseng, R.-L. & Juang, R.-S. (2000). Comparative adsorption of metal and dye on flake- and bead-types of chitosans prepared from fishery wastes. *Journal of Hazardous Materials*, **73 (1)**, 63-75.

[124] Figueiredo, S. A., Boaventura, R. A. & Loureiro, J. M. (2000). Color removal with natural adsorbents: modeling, simulation and experimental. *Separation and Purification Technology*, **20 (1)**, 129-41.

[125] Rinaudo, M. (2006). Chitin and chitosan: Properties and applications. *Progress in Polymer Science*, **31 (7)**, 603-32.

[126] Struszczyk, M. H. (2002). Chitin and chitosan - Part I. Properties and production. *Polimery*, **47 (5)**, 316-25.

[127] Agboh, O. C. & Qin, Y. (1997). Chitin and chitosan fibers. *Polymers for Advanced Technologies*, **8 (6)**, 355-65.

[128] Mckay, G., Blair, H. S. & Gardner, J. (1983). The Adsorption of Dyes in Chitin .3. Intraparticle Diffusion-Processes. *Journal of Applied Polymer Science*, **28 (5)**, 1767-78.

[129] Acemioglu, B. (2004). Adsorption of Congo red from aqueous solution onto calcium-rich fly ash. *Journal of Colloid and Interface Science*, **274 (2)**, 371-9.

[130] Dizge, N., Aydiner, C., Demirbas, E., Kobya, M. & Kara, S. (2008). Adsorption of reactive dyes from aqueous solutions by fly ash: Kinetic and equilibrium studies. *Journal of Hazardous Materials*, **150 (3)**, 737-46.

[131] Kumar, K. V., Ramamurthi, V. & Sivanesan, S. (2005). Modeling the mechanism involved during the sorption of methylene blue onto fly ash. *Journal of Colloid and Interface Science*, **284 (1)**, 14-21.

[132] Karagozoglu, B., Tasdemir, M., Demirbas, E. & Kobya, M. (2007). The adsorption of basic dye (Astrazon Blue FGRL) from aqueous solutions onto sepiolite, fly ash and apricot shell activated carbon: Kinetic and equilibrium studies. *Journal of Hazardous Materials*, **147 (1-2)**, 297-306.

[133] Hsu, T.-C. Adsorption of an acid dye onto coal fly ash. *Fuel, In Press, Corrected Proof*

[134] Lin, J. X., Zhan, S. L., Fang, M. H., Qian, X. Q. & Yang, H. (2008). Adsorption of basic dye from aqueous solution onto fly ash. *Journal of Environmental Management*, **87 (1)**, 193-200.

[135] Wang, S. B., Boyjoo, Y., Choueib, A. & Zhu, Z. H. (2005). Removal of dyes from aqueous solution using fly ash and red mud. *Water Research*, **39 (1)**, 129-38.

[136] Jain, A. K., Gupta, V. K., Bhatnagar, A. & Suhas (2003). Utilization of industrial waste products as adsorbents for the removal of dyes. *Journal of Hazardous Materials*, **101 (1)**, 31-42.

[137] Jha, V. K., Kameshima, Y., Nakajima, A. & Okada, K. (2008). Utilization of steel-making slag for the uptake of ammonium and phosphate ions from aqueous solution. *Journal of Hazardous Materials*, **156 (1-3)**, 156-62.

[138] Ramakrishna, K. R. & Viraraghavan, T. (1998). Use of slag for dye removal. *Waste Management*, **17 (8)**, 483-8.

[139] Zelic, J. (2005). Properties of concrete pavements prepared with ferrochromium slag as concrete aggregate. *Cement and Concrete Research*, **35 (12)**, 2340-9.

[140] Motz, H. & Geiseler, J. (2001). Products of steel slags an opportunity to save natural resources. *Waste Management*, **21 (3)**, 285-93.

[141] Maslehuddin, M., Sharif, A. M., Shameem, M., Ibrahim, M. & Barry, M. S. (2003). Comparison of properties of steel slag and crushed limestone aggregate concretes. *Construction and Building Materials*, **17 (2)**, 105-12.

[142] Jain, A. K., Gupta, V. K., Bhatnagar, A. & Suhas (2003). A comparative study of adsorbents prepared from industrial wastes for removal of dyes. *Separation Science and Technology*, **38 (2)**, 463-81.

[143] Gupta, V. K., Ali, I., Suhas & Mohan, D. (2003). Equilibrium uptake and sorption dynamics for the removal of a basic dye (basic red) using low-cost adsorbents. *Journal of Colloid and Interface Science*, **265 (2)**, 257-64.

[144] Namasivayam, C. & Arasi, D. J. S. E. (1997). Removal of congo red from wastewater by adsorption onto waste red mud. *Chemosphere*, **34 (2)**, 401-17.

[145] Bhatnagar, A. & Jain, A. K. (2005). A comparative adsorption study with different industrial wastes as adsorbents for the removal of cationic dyes from water. *Journal of Colloid and Interface Science*, **281 (1)**, 49-55.

[146] Okada, K., Yamamoto, N., Kameshima, Y. & Yasumori, A. (2003). Porous properties of activated carbons from waste newspaper prepared by chemical and physical activation. *Journal of Colloid and Interface Science*, **262 (1)**, 179-93.

[147] Okada, K., Yamamoto, N., Kameshima, Y. & Yasumori, A. (2003). Adsorption properties of activated carbon from waste newspaper prepared by chemical and physical activation. *Journal of Colloid and Interface Science*, **262 (1)**, 194-9.

[148] Kartel, M. T., Sych, N. V., Tsyba, M. M. & Strelko, V. V. (2006). Preparation of porous carbons by chemical activation of polyethyleneterephthalate. *Carbon*, **44 (5)**, 1019-22.

[149] Sych, N. V., Kartel, N. T., Tsyba, N. N. & Strelko, V. V. (2006). Effect of combined activation on the preparation of high porous active carbons from granulated post-consumer polyethyleneterephthalate. *Applied Surface Science*, **252 (23)**, 8062-6.

[150] Arenillas, A., Rubiera, F., Parra, J. B., Ania, C. O. & Pis, J. J. (2005). Surface modification of low cost carbons for their application in the environmental protection. *Applied Surface Science*, **252 (3)**, 619-24.

[151] Akmil-Basar, C., Onal, Y., Kilicer, T. & Eren, D. (2005). Adsorptions of high concentration malachite green by two activated carbons having different porous structures. *Journal of Hazardous Materials*, **127 (1-3)**, 73-80.

[152] Nakagawa, K., Namba, A., Mukai, S. R., Tamon, H., Ariyadejwanich, P. & Tanthapanichakoon, W. (2004). Adsorption of phenol and reactive dye from aqueous solution on activated carbons derived from solid wastes. *Water Research*, **38 (7)**, 1791-8.

[153] Tor, A. & Cengeloglu, Y. (2006). Removal of congo red from aqueous solution by adsorption onto acid activated red mud. *Journal of Hazardous Materials*, **138 (2)**, 409-15.

In: Industrial Waste...
Editor: John P. Samuelson, pp. 73-102
ISBN: 978-1-60692-720-5

Chapter 3

COST EFFECTIVE ADSORBED MATERIALS FOR INDUSTRIAL WASTEWATER TREATMENT

M. Nageeb Rashed[1]
Faculty of Science Aswan, South Valley University, 81528 Aswan, Egypt

Abstract

Developing countries suffer from water pollution, the high cost of contents and treatments make the major problems in these countries. Search for a cheaper alternative, reveals that utilizing natural processes and materials significantly reduce high cost. Adsorption as one of the effective technique widely used for treatment of industrial wastewater from organic and inorganic pollutants meet the great attention from researchers. In solving the problem of industrial solid waste disposal, adsorption technique using the treated solid waste also, meet the attention of researchers. Natural adsorbents such as ores, rocks, plant straw, soil, carbonaceous material developed from fertilizer waste slurry, activated carbons from date pits and the dried aquatic plants are of the effective adsorbent materials and environmentally sound low cost materials for treatment of industrial wastewater from organic pollutants and heavy metals. Also, ores (kaolin, talc, barite and chalcopyrite) and solid wastes from food industries and mining can be used successfully as low-cost materials for treatment of polluted water. This chapter presents recent researches on:-1-fundamental of adsorption. 2-recycle and reuse of industrial solid wastes for industrial wastewater treatment. 3- Naturally materials as adsorbents for the removal of pollutants from industrial wastewater.4- Case studies using adsorption technique

Introduction

The pollution of water with inorganic and organic pollutants has increased dramatically in the last 50 years as a consequence of the expansion of industrial activities. The well-established toxicity of inorganic pollutants (toxic and heavy metals) and organic pollutants (dyes, pesticides and pharmacological) in solution at sufficiently high concentrations affects humans, animals and vegetation.

---

[1] E-mail address: mnrashed@hotmail.com.

Industrial wastewater is one of the major sources of aquatic pollution. The type of pollutants present in industrial wastewater depends on nature of industry. However, some of the common pollutants generally present are metal ions, dyes, phenols, pesticides and a wide spectrum of aromatics. A number of methods such as coagulation, filtration with coagulation, precipitation, ozonation, adsorption, ion exchange, reverse osmosis and advanced oxidation processes have been used for the removal of pollutants from industrial wastewater. These methods have been found to be limited, since they often involve high capital and operational costs and may be associated with the generation of secondary waste which presents treatment problems, such as the large quantity of sludge generated by precipitation processes. On the other hand ion exchange and reverse osmosis are more attractive processes because the pollutant values can be recovered along with their removal from the effluents. Reverse osmosis, ion exchange and advanced oxidation processes do not seem to be economically feasible because of their relatively high investment and operational cost. Adsorption has advantages over the other methods because of simple design with a sludge free environment and can involve low investment in term of both initial cost and land required. The adsorption process is widely used for treatment of industrial wastewater from organic and inorganic pollutants and meet the great attention from the researchers. In recent years, the search for low-cost adsorbents that have pollutant –binding capacities has intensified. Materials locally available such as natural materials, agricultural wastes and industrial wastes can be utilized as low-cost adsorbents. Activated carbon produced from these materials can be used as adsorbent for industrial wastewater treatment, which improve economic value, helping industries reduce the cost of waste disposal and providing a potential alternative to activated carbon (Crini, 2005). Table 1 shows a non-exhaustive list of some selected examples of adsorbents used in wastewater treatment.

This chapter presents recent researches on:-

1) Fundamental of adsorption processes.
2) Recycle and reuse of agricultural and industrial solid wastes for industrial wastewater treatment.
3) Using adsorbent for removal of heavy metals and dyes from industrial wastewater.
4) Case studies using adsorption techniques.

Adsorption Theory

Adsorption is a common mechanism for organic and inorganic pollutants removal. It takes place at the solid-solute interface. It is a mass transfer process by which a substance is transferred from the liquid phase to the surface of a solid, and becomes bound by physical and/or chemical interactions. Other definition of adsorption is a mass transfer process by which a substance is transferred from the liquid phase to the surface of a solid, and becomes bound by physical and/or chemical interactions. Large surface area leads to high adsorption capacity and surface reactivity (Kurniawan et al., 2006).

The parameter which have been investigated for optimizing the use of adsorbent in industrial wastewater treatment include nature of adsorbent and adsorbate, pollutant concentration, solution pH, temperature, adsorbent dosage and the contact time.

An adsorption isotherm is the presentation of the amount of solute adsorbed per unit weight of adsorbent as a function of the equilibrium concentration in the bulk solution at constant temperature. Langmuir and Freundlich adsorption isotherms are commonly used for the description of adsorption data.

The Langmuir equation is expressed as:

$$Ce/qe = 1/bXm + Ce/Xm, \quad (1)$$

Where Ce is the equilibrium concentration of solute (mmol L^{-1}), qe is the amount of solute adsorbed per unit weight of adsorbent (mmol g^{-1} of clay), Xm is the adsorption capacity (mmol g^{-1}), or monolayer capacity, and b is a constant (L $mmol^{-1}$) (Faust and Aly, 1987).

The Freundlich equation is expressed as:

$$\log qe = \log k + 1/n \log Ce\,, \quad (2)$$

Where k (mmol g^{-1}) and $1/n$ are the constant characteristics of the system (Faust and Aly, 1987).

Table 1. Some low-cost adsorbents used in industrial wastewater treatment

Agricultural solid wastes	Industrial solid waste	Naturally materials
Hazelnut	Blast furnace sludge	Zeolites
Peanut hulls	Blast furnace slag	Clinoptilolite
Orange peel (citrus reticulate)	Iron slag	Montmorillonite
Grape stalk waste	Steel slag	Kaolinite
Sugar beet pulp	Red mud	Hematite
Wheat shell	Tea industry waste	Iron oxide
Saw dust	Red mud	Bentonite
Sugar beer pulp	Baggage fly ash	Dolomite
Coffee		China clay
Waste tea		Lignin
Peanut hulls		
Coffee residue –clay		
Nut shell		

Low Cost Adsorbents Divided into Several Adsorbent Types as Follow

1. Natural Materials

In recent years, increasing emphasis has been placed on environmental impact minimization and resulting from this the range and capability of natural and prepared materials capable of pollutant removal has seen steady development. In particular considerable work has been carried out on the use of both natural materials and their modifications. These natural materials, in many instances are relatively cheap, abundant in supply and have significant potential for modification and ultimately enhancement of their adsorption capabilities.

Natural materials locally available in certain areas are used as low-cost adsorbent for the removal of pollutants from industrial wastewater. The most common natural materials are:-

1.1. Natural Diatomite

It is attractive and adequate material that offers a great application in industry owing to both its high porosity and thermal resistance. Diatomite has its origin from a siliceous, sedimentary rock consisting principally of the fossilized skeletal remains of diatom, a unicellular aquatic plant related to the algae, during the tertiary and quaternary periods. The high porosity of diatomite and the charcoal affinity towards the organic and inorganic pollutants and also its important sorption properties (Hadjar, 2008).

1.2. Zeolites

Large amounts of fine zeolite by-product were produced when natural zeolite was processed into a powder with a specific particle size. Li and Bowman (2001) found that the surfactant-modified zeolite and kaolinite have the ability to remove chromate from solutions. (Volzone and Garrido, 2008). Also, the adsorption behavior of natural (clinoptilolite) zeolites with respect to Co(2+), Cu(2+), Zn(2+), and Mn(2+) has been studied in order to consider its application to purity metal finishing wastewaters (Erdem et al., 2004).

A granular material formulated by mixing zeolite by-product with Portland cement (ZeoAds) was applied for the removal of and heavy metal from aqueous solutions. The ZeoAds removed Pb and Cu in an aqueous solution up to 27.03 and 23.25 mg g (-1) (Ok et al., 2007).

Two natural zeolites found in the regions of Gordes and Bigadic, in western Turkey. Adsorption behavior of both zeolites is highly dependent on the pH. Adsorption dependence on lower pH values ($pH<4$) is explained by the dissolution of crystal structure and the competition of the zinc ions with the H+. Between pH 4 and 6, the basic mechanism is the ion exchange process. The results also showed that decrease in grain size does not increase the adsorption capacity of zeolite from Gordes, yet it increases that of zeolite from Bigadic about 23%.(Oren and Kaya, 2005).The adsorption of residual organic pollutants from flocculated printing ink wastewater onto several synthetic zeolites was investigated as a finishing method for additional reduction of TOC.

Adsorption of mercury by solid materials seems to be a suitable choice for final purification of wastewaters. The typical sorbents for mercury ions removal are activated carbons and zeolites. Mercury ion was removed by several sorbents, namely a non-impregnated activated carbon (Aquacarb 207EATM by Sutcliffe Carbon, is a commercially available non-impregnated granular activated carbon produced by Sutcliffe Carbon starting from a bituminous coal) and three natural materials: a char derived from a South African coal, a pozzolana and a tuff from a volcanic zone near Naples, in the South of Italy. These four materials have been chosen for their potential capture capacity as well as for their cost effectiveness and environmental safety which make them suitable sorbents for both natural and industrial water treatments. (Natale et al., 2006).

1.3. Calcite and Lignite

Removal of phosphate species from solution by adsorption onto calcite as natural adsorbent was applied. Significant amount of orthophosphates can be adsorbed onto relatively low mass of calcite. Besides, the calcite adsorbed-phosphate product is friendly to the environment, as it does not require further treatment for the phosphate species desorption because of its potential application to acid soils fertilization. (Karageorgiou et al., 2007).

The use of lignite in wastewater treatment has received increasing attention over the last several years (Mohan and Chander, 2006). Besides being plentiful, inexpensive lignite possesses several characteristics that make it effective media for the removal of aqueous metals from wastewater. Lignites possess a high oxygen content, which is fixed in carboxyl and hydroxyl groups. These groups are infecting the active centers of the ion exchange. So the lignite materials can be used as alternative cation exchangers. Carboxyl or hydroxyl groups are able to take part in the ion exchange reaction in the following manner (Eqs. 3–4).

$$n\text{L–COOH} + \text{M}n\text{+} = n\text{L–COOM} + n\text{H+} \quad (3)$$

$$n\text{L–COOH} + \text{M}n\text{+} = (n.1)\ \text{L–COOM (OH)}. \text{(Mohan and Chander, 2006)} \quad (4)$$

Remove and recover metal ions from acid mine drainage (AMD) by using lignite, a low cost sorbent, was investigated. Acid mine drainage (AMD), has long been a significant environmental problem resulting from the microbial oxidation of iron pyrite in presence of water and air, affording an acidic solution that contains toxic metal ions. Sorption of ferrous, ferric, manganese, zinc and calcium in multi-component aqueous systems was investigated (Mohan and Chander, 2006).

1.4. Bentonite

Bentonite was used successfully for the removal of dyes from industrial wastewater; Congo red (Bulut et al, 2008); xylene (Zitan et al, 2008); Direct Red 2 (Zohra et al.,2008); crystal violet (Eren and Afsin, 2008). The utilization of modified bentonite with a cationic surfactant (dodecyltrimethylammonium (DTMA) bromide) as an adsorbent was successfully carried out to remove a synthetic textile dye (Reactive Blue 19 (RB19)) by adsorption, from aqueous solutions. In this study, the immobilization of 8-hydroxy quinoline onto bentonite was carried out and it was then used to investigate the adsorption behavior of lead (II) ions from aqueous solutions. The changes of the parameters of pH, contact time, initial lead (II) ions concentration and temperature were tested in the adsorption experiments (Ozcan et al., 2008).

Adsorption of Cu(II) by raw bentonite (RB) and acid-activated bentonite (AAB) samples was investigated as a function of the initial Cu(II) concentration, solution pH, ionic strenght, temperature, the competitive and complexation effects of ligands (Cl^-, SO_4^{2-}, PO_4^{3-})(Eren and Afsin, 2008).

Modified OH-Al bentonites with HMP showed twice the levels of chromium retention as untreated bentonites. Use of modified hydroxy-aluminum bentonites for chromium (III) removal from solutions (Volzone and Garrido, 2008).

1.5. Clay Minerals (Kaolinite, Illite and Montmorillionite)

Typical clay minerals are kaolinite, illite and montmorillionite. These minerals were used for adsorption of heavy metals and organic pollutants from polluted water and industrial wastewater. The adsorption characteristics of heavy metals: cadmium (II), chromium (III), copper (II), nickel (II), lead (II), and zinc (II) ions by kaolin (kaolinite) and ballclay (illite) from Thailand were studied. This research was focused on the pH, adsorption isotherms of single-metal solutions at 30– 60 °C by batch experiments, and on ion selectivity in mixed and binary combination solutions. Adsorption of metals in the mixture solutions by kaolin was in the order: Cr > Zn > Cu ≈ Cd ≈ Ni > Pb, and for ballclay was: Cr > Zn > Cu > Cd ≈ Pb > Ni. The adsorption of metals was endothermic, with the exception of Cd, Pb and Zn for kaolin, Cu and Zn for ballclay. Kaolin and ballclay exhibited relatively hard Lewis base adsorption site (Chantawong et al., 2003).

The feasibility of using two important and common clay minerals, montmorillonite and kaolinite, as adsorbents for removal of toxic heavy metals has been reviewed. A good number of works have been reported where the modifications of these natural clays were done to carry the adsorption of metals from aqueous solutions. The modification was predominantly done by pillaring with various polyoxy cations of Zr^{4+}, Al^{3+}, Si^{4+}, Ti^{4+}, Fe^{3+}, Cr^{3+}or Ga^{3+}, etc. Preparation of pillared clays with quaternary ammonium cations, namely, tetramethyl-ammonium-, tetramethylphosphonium- and trimethyl-phenylammonium-, N′-didodecyl-N, N′-tetramethylethanediammonium, etc, are also common. Moreover, the acid treatment of clays often boosted their adsorption capacities. The adsorption of toxic metals, viz., As, Cd, Cr, Co, Cu, Fe, Pb, Mn, Ni, Zn, etc., have been studied predominantly. Montmorillonite and its modified forms have much higher metal adsorption capacity compared to that of kaolinite as well as modified-kaolinite (Krishna et al., 2008).

1.6. Talc

Because of its importance in the fields of environmental and mineral engineering, study of metal ion adsorption on silicate minerals has received extensive continuing interest. However, limited work has been done to compare the behavior of silicates that are hydrophilic with those that are naturally hydrophobic. It well understands the effect of hydrophobicity and surface heterogeneity on the adsorption of hydrolyzable metal ions (Pb and Cd) at the solid/liquid interface. Concurrent measurements of zeta potentials, the extent of the adsorption, and suspension stability were conducted on colloidal suspensions of two silicate minerals, namely naturally hydrophilic quartz and hydrophobic talc (Huang and Fuerstenau, 2000). Talic was used succuffuly as adsorbent for the removal of heavy metals. Rashed (2001) used talc for Pb removal from polluted water at pH7.

2. Agricultural Solid Wastes

Agricultural products and by-products is an abundant waste material, and needs proper disposal. When disposed by burning in situ, it generates CO_2 and other forms of pollutions. This creates a need for the conversion of agricultural products and by-products to useful, and hopeful, value-added products.

The application of low-cost adsorbents obtained from plant wastes as a replacement for costly conventional methods of removing heavy metal ions from wastewater has been reviewed. It is well known that cellulosic waste materials can be obtained and employed as cheap adsorbents and their performance to remove heavy metal ions can be affected upon chemical treatment. In general, chemically modified plant wastes exhibit higher adsorption capacities than unmodified forms. Numerous chemicals have been used for modifications which include mineral and organic acids, bases, oxidizing agent, organic compounds, etc.

Kou et al. (2001) and Houngshan Shang et al. (2003) have proposed the utilization of biomaterials or abandoned biomaterials (BIOM), new name for agricultural product and by-product of which the major component is cellulose $(C_6H_{10}\ O_5)n$. Cellulosic surface becomes partially negatively charged when immersed in water and, therefore, possess columbic interaction with cationic species in water. The high binding capacities of cationic species on the adsorbents are mainly the results of columbic interactions. Rashed (2006) used activated carbon produced from apricot and peach stones for the removal of Pb ion from wastewater and polluted water. The results reveal 93% and 97.6% Pb removal by apricot and peach stones.

Removal of lead, cadmium, zinc, and copper from industrial wastewater by carbon developed from walnut, hazelnut, almond, pistachio shell, and apricot stone was reported by Kazemipour et al. (2008). In their research adsorption of copper (Cu), zinc (Zn), lead (Pb), and cadmium (Cd) that exist in industrial wastewater onto the carbon produced from nutshells of walnut, hazelnut, pistachio, almond, and apricot stone has been investigated. All the agricultural shell or stone used were ground, sieved to a defined size range, and carbonized in an oven. Time and temperature of heating were optimized at 15 min and 800 °C, respectively, to reach maximum removal efficiency. Removal efficiency was optimized regarding to the initial pH, flow rate, and dose of adsorbent. The maximum removal occurred at pH 6–10, flow rate of 3 mL/min, and 0.1 g of the adsorbent.

Various carbonaceous materials, such as coal, lignite, coconut shells, pecan shell, wood and peat, are used in the production of commercial activated carbons. However, abundance and availability of agricultural byproducts make them good sources of raw materials for activated carbons (Bansode et al., 2003).

Low cost and non-conventional adsorbents include agricultural by-products such as nut shells, wood, bone, peat, coconut shells, processed into activated carbons (Bansode, et al., 2004), these have been reported to be important adsorbents for the removal of heavy metals and organics from municipal and industrial wastewater. Among these low cost adsorbents, chitosan has the highest sorption capacity for several metal ions (Nomanbhay and Palanisamy, 2005; Babel and Kurniawan, 2004). Coconut shell carbon was modified with chitosan and/or oxidizing agent (phosphoric acid) to produce composite adsorbent. The adsorption efficiency of the adsorbent was evaluated by measuring the extent of adsorption of zinc (II) in synthetic beverage industrial wastewater. (Amuda et al., 2007).

Rice husk (rice husk (RH) was from the local rice mill) as adsorbents was used as cheap agricultural adsorbent for color removal from wastewater using Rice husk activated carbon (Gupta et al., 2006).

Removal of mercury from aqueous solutions using activated carbon prepared from Ceiba pentandra hulls, Phaseolus aureus hulls and Cicer arietinum waste was investigated. The influence of various parameters such as effect of pH, contact time, initial metal ion concentration and adsorbent dose for the removal of mercury was studied using a batch process. The prepared adsorbents ACCPH, ACPAH and ACCAW had removal capacities of 25.88 mg/g, 23.66 mg/g and 22.88 mg/g, respectively, at an initial Hg (II) concentration of 40 mg/L. The order of Hg (II) removal capacities of these three adsorbents was ACCPH > ACPAH > ACCAW. The adsorption behavior of the activated carbon is explained on the basis of its chemical nature. The feasibility of regeneration of spent activated carbon adsorbents for recovery of Hg (II) and reuse of the adsorbent was determined using HCl solution (Rao et al., 2008).

Coffee husks (CH), a coffee processing residue, was used as untreated sorbents for the removal of heavy metal ions from aqueous solutions. Biosorption studies were conducted in a batch system as a function of contact time, initial metal ion concentration, biosorbent concentration and pH of the solution. A contact time of 72 h assured attainment of equilibrium for Cu (II), Cd (II) and Zn (II). The sorption efficiency after equilibrium was higher for Cu (II) (89–98% adsorption), followed by Cd (II) (65–85%) and Zn (II) (48–79%). Even though equilibrium was not attained in the case of Cr (VI) ions, sorption efficiency ranged from 79 to 86%. Sorption performance improved as metal ions concentrations were lowered. The effect of the initial pH in the biosorption efficiency was verified in the pH range of 4–7, and the results showed that the highest adsorption capacity occurred at distinct pH values for each metal ion (Waleska et al., 2008).

New sorbent potato peels, which is normally discarded as solid waste for removing toxic metal ion Cu (II) from water/industrial waste water have been studied. Potato peels charcoal (PPC) was investigated as an adsorbent of Cu (II) from aqueous solutions. Kinetic and isotherm studies were carried out by studying the effects of various parameters such as temperature, pH and solid liquid ratios. The optimum pH value for Cu (II) adsorption onto potato peels charcoal (PPC) was found to be 6.0. This work clearly indicates the potential of using PPC as an excellent adsorbent for the removal of Cu (II) from aqueous solutions. One gram of PPC is found enough to remove 99.8% of the copper (II) from an aqueous solution of 150 mg l^{-1} (100 ml) at pH 6.0 with a shaking time of 20 min (Amana et al., 2008).

An extensive list of plant wastes as adsorbents including rice husks, spent grain, sawdust, sugarcane bagasse, fruit wastes, weeds and others has been compiled. Some of the treated adsorbents show good adsorption capacities for Cd, Cu, Pb, Zn and Ni (Ngah and Hanafiah, 2008).

The feasibility of using various agricultural residues viz. sugarcane bagasse (SCB), maize corncob (MCC) and Jatropha oil cake (JOC) for the removal of Cd (II) from aqueous solution under different experimental conditions was reported. Batch experiments were carried out at various pH (2–7), adsorbent dose (250–2000 mg), Cd (II) concentration (5–500 mgl^{-1}) for a contact time of 60 min. The maximum cadmium removal capacity was shown by JOC (99.5%). Maximum metal removal was observed at pH 6.0 with a contact time of 60 min at stirring speed of 250 rpm with an adsorbent dose of 20 g l^{-1} of the test solution. The maximum adsorption of cadmium (II) metal ions was observed at pH 6 for all the adsorbents via; 99.5%, 99% and 85% for JOC, MCC, and SCB, respectively. Order of Cd (II) removal by various biosorbents was JOC >MCC> SCB. JOC may be an alternative biosorbent for the removal of Cd (II) ions from the aqueous solution (Garg et al., 2008).

Some granular activated carbons (GACs) were prepared from cherry stones (CS), an agricultural waste, by thermal methods following pyrolysis in nitrogen and subsequent activation (with air, carbon dioxide and steam) or single-step activation in steam. A GAC prepared by activation with carbon dioxide was further treated with several oxidizing agents (air, air–ozone mixture, nitric acid and hydrogen peroxide). Both, the non-oxidized and the oxidized GACs were used as adsorbent to study the adsorption of Cu (II) from aqueous solution. Adsorption isotherms were obtained at 20°C and data were well fitted to the Langmuir equation. The results show that the Cu (II) uptake of the non-oxidized GACs is not great, but similar to that of the commercial grade Filtrasorb 400 (Jaramilloa et al., 2008).

3. Industrial Solid Wastes

Rapid industrialization has led to increased disposal of solid wastes into the environment. Recycled or reuse these solid waste lead to solution of their environmental problems. One of these recycled methods is converting solid waste to activated carbon which used as adsorbent for the removal of toxic organic and inorganic pollutant from industrial wastewater. Activated carbons are carbonaceous materials with well developed porous structure consisting of hydrophobic graphene layer surface and hydrophilic surface functional groups. These characteristics make them useful as adsorbents, catalyst and catalyst support Activated carbon has undoubtedly been the most popular and widely used adsorbent in wastewater treatment applications throughout the world and has been successfully utilized for the removal of metal ions, dyes, phenols, detergents, pesticides, humic substances, chlorinated hydrocarbons and many other chemicals and organisms. (Marsh and Rodriguez-Reinoso, 2006). Activated carbon can be produced from a wide variety of raw materials, basically by two methods as: physical and chemical activation or a combination of both (Hayashi et al., 2002). Examples of commercial feedstock's are coal, wood and agricultural residues such as coconut shell, fruit stones (apricot and cherry stones, grape seeds), hard shells (almond, pistachio and pecan shell), sugar cane bagasse, etc (Philip and Girgis, 1996; Heschel and Klose, 1995; Aygun et al., 2003; Diaz-Diez et al., 2004; Soleimani and Kaghazchi, 2008).

Metal sludge (a waste product of electroplating industry) was used in removing vanadium from wastewater. The adsorption capacity of metal sludge for vanadium was found 24.8 mg/g at 25 °C (Bhatnagar et al., 2008).

The sugar industry is one of the most important agro based industries. The bagasse fly ash, an industrial solid waste generated from the sugar industry, causes a great disposal problem and it is currently being used as a filler in building material. The bagasse fly ash, an industrial solid waste of sugar industry, was used for the removal of cadmium and nickel from wastewater. As much as 90% removal of cadmium and nickel is possible in about 60 and 80 min (Guptaa et al., 2003). Also, a bagasse fly ash, a sugar industrial waste, was used as adsorbent for removal of Cr (III) and the maximum up take was 2–3 mg/g at pH 6. (Volzone and Garrido, 2008).

Adsorption of fluoride on waste carbon slurry was investigated. Waste carbon slurry was obtained from fuel oil based generators of a fertilizer industry. However, adsorption is a common technique used for fluoride removal from aqueous solution and applied frequently on large scale, as it can handle fairly large flow rates, producing a high quality of water without producing notorious sludge, residual contaminants, etc. Moreover, adsorption is universal, inexpensive and fast in nature and applicable for the removal of pollutants even at low concentration. The fluoride adsorption behavior of carbon slurry was investigated as it was of low cost, effective and capable to remove fluoride from aqueous solutions giving water suitable for drinking purpose (Gupta et al., 2007).

A large number of solid materials have been tested for fluoride removal, such as activated alumina (Rubel, 1983), amorphous alumina (Li et al., 2001), activated carbon (Srimurali et al., 1998), calcite (Yang et al., 1999), clay (Mahramanlioglu et al., 2002), zeolite (Onyango et al., 2004), charcoal (Bhargava and Killedar, 1992), bleaching earth (Mahramanlioglu et al., 2002) and red mud (Cengeloglu et al., 2002). It is interesting to note that some adsorbents are not capable to remove fluoride from water at low concentration ($2mgL^{-1}$) (Majima and Takatsuki, 1987; Wang and Reardon, 2001).

A comparative adsorption study was carried out using carbon slurry waste obtained from a fertilizer plant and blast furnace sludge, dust, and slag from steel plant wastes as adsorbents for the removal of two anionic dyes, methyl orange and brilliant blue G. It was found that carbonaceous adsorbent prepared from a fertilizer plant exhibits good potential for the removal of dyes as compared to the other three adsorbents prepared.

The use of a new activated carbon developed from date palm seed wastes, generated in the jam industry, for removing toxic chromium from aqueous solution has been investigated. The activated carbon has been achieved from date palm seed by dehydrating methods using concentrated sulfuric acid. The batch experiments were conducted to determine the adsorption capacity of the biomass. The effect of initial metal concentration (25–125 mg l^{-1}), pH, contact time, and concentration of date palm seed carbon have been studied at room temperature. A strong dependence of the adsorption capacity on pH was observed, the capacity increase as pH value decrease and the optimum pH value is pH 1.0. Kinetics and adsorption equilibrium were studied at different sorbent doses. The adsorption process was fast and the equilibrium was reached within 180 min. The maximum removal was 100% for 75 mg l^{-1} of Cr6+ concentration on 4 g l^{-1} carbon concentration and the maximum adsorption capacity was 120.48 mg g^{-1} (El Nemr Ahmed *et al.*, 2008).

An industrial waste sludge mainly composed by metal hydroxides (A metal hydroxide sludge generated in a Portuguese zinc electroplating industry by alkaline precipitation of the metal ions present in wastewater was used as adsorbent.) was used as a low-cost adsorbent for removing a reactive textile dye (*Remazol Brilliant Blue*) in solution. Characterization of this waste material included chemical composition, pHZPC determination, particle size distribution, physical textural properties and metals mobility under different pH conditions. Dye adsorption equilibrium isotherms were determined at 25 and 35 ^{o}C and pH of 4, 7 and 10 revealing reasonably fits to Langmuir and Freundlich models. At 25 ^{o}C and pH 7, Langmuir fit indicates a maximum adsorption capacity of 91.0 mg/g (Santos et al., 2008).

The adsorption of Cd^{2+} and Pb^{2+} on sugar beet pulp (SBP), a low-cost material, has been studied. The (SBP) an industrial by product and solid waste of sugar industry were used for the removal of Cd^{2+} and Pb^{2+} ions from aqueous water. The sorption process was relatively fast and equilibrium was reached after about 70 min of contact. As much as 70–75% removal of Cd2+ and Pb2+ ions for (SBP) are possible in about 70 min, respectively, under the batch test conditions. Uptake of Cd^{2+} and Pb^{2+} ions on (SBP) showed a pH-dependent profile. The overall uptake for the (SBP) is at a maximum at pH 5.3 and gives up to 46.1 mg g ˉ1 for Cd^{2+} and at pH 5.0 and gives 43.5 mg g^{-1} for Pb^{2+} for (SBP), which seems to be removed exclusively by ion exchange, physical sorption and chelation. A dose of 8 g L^{-1} was sufficient for the optimum removal of both the metal ions (Pehlivan et al., 2008).

The adsorption of Cr(VI) from aqueous solutions on sawdust (SD), base extracted sawdust (BESD) and tartaric acid modified sawdust (TASD) of Turkish red pine tree *(Pinus nigra),* a timber industry waste, was studied at varying Cr(VI) concentrations, adsorbent dose, modifier concentration and pH. Batch adsorption studies have been carried out. Sawdust was collected from waste timber industry and modified with various amount of tartaric acid (TA) (0.1–1.5 M). An initial pH of 3.0 was most favorable for Cr (VI) removal by all adsorbents. Maximum Cr (VI) was sequestered from the solution within 120 min after the beginning for every experiment. The experimental result inferred that chelation and ion exchange is one of the major adsorption mechanisms for binding metal ions to the SD. Percentage removal of Cr(VI) was maximum at the initial pH of 3.0 (87.7, 70.6 and 55.2% by TASD, BESD, and SD, respectively). Adsorption capacities range from 8.3 to 22.6 mg/g for SD samples (Godea et al., 2008). Cedar sawdust and crushed brick, low cost materials, have suitable adsorption capacity with regard to the removal of methylene blue from its aqueous solution (Hamdaoui, 2006).

The adsorption potential of metal sludge (a waste product of electroplating industry) in removing vanadium from water was investigated. The adsorption capacity of metal sludge for vanadium was found 24.8 mg/g at 25 ◦C. The adsorption was studied as a function of contact time (0.5–10 h), concentration (1.5×10^{-4} – 9.5×10^{-4} M) and temperature (25 and 45 °C) by batch method. The results clearly reveal that metal sludge can be fruitfully employed in treating industrial effluents containing vanadium and further safely dispose of by immobilizing it into cement. The proposed technology provides a two-fold advantage of wastewater treatment and solid waste management (Bhatnagar et al., 2008).

4. Biosorbents (or Bioadsorption)

Increasing industrial development and urbanization have led to continuously increasing production of toxic substances, discharged into the environment with risk for living organisms, and potentially constituting serious hazards to public health. Recently, bioadsorbents have emerged as an eco-friendly, effective and low cost material option. These bioadsorbents include some agricultural wastes, fungi, algae and bacteria. Biosorption (or bioadsorption) is capable of removing traces of heavy metals and other elements from dilute aqueous solutions.

Studies using bioadsorbents have shown that both living and dead microbial cells are able to adsorb metal ions and offer potential inexpensive alternative to conventional adsorbents. However, living cells are subject to toxic effect of the heavy metals, resulting in cell death Moreover, living cells often require the addition of nutrients and hence increase the BOD and COD the effluent . For these reasons, the use of nonliving biomaterials or dead cells as metal binding compounds has been gaining advantage because toxic ions do not affect them. In addition, dead cells require less care, maintenance and they are cheaper. Furthermore, dead biomass can be easily regenerated and reused. As a consequence, activated sludge produced the biological treatment facilities of the treatment plant is considered as a good bioadsorbent for the removal of heavy metal ions from industrial wastewater. Al-Qodah (2006)study the bioadsorption characteristics of the ions Cu^{2+}, Cd^{2+} and Ni^{2+} from synthetic water using activated sludge obtained from a treatment unit of a dairy factory.

Gadd (1988) and Brierley (1990) reviewed how bacteria, fungi and algae take up toxic metal ions. It is important to differentiate *biosorption* or sorption from *bioaccumulation...* Mechanisms of cell surface sorption are independent of cell metabolism; they are based upon physicochemical interactions between metal and functional groups of the cell wall. The microorganism's cell wall mainly consists of polysaccharides, lipids and proteins, which have many binding sites for metals. This process is independent of the metabolism and metal binding is fast. Bioaccumulation, in contrast, is an intracellular metal accumulation process which involves metal binding on intracellular compounds, intracellular precipitation, methylation and other mechanisms. Bioaccumulation can also be regarded as a second part of the metal sequestering process by living biomass. Sometimes, it is called *active biosorption* as the opposite to *passive biosorption*. Since it depends on the cell metabolism, it can be inhibited by metabolic inhibitors such as low temperature and lack of energy sources.

Fungi as one of biosorbents have been recently recognized as promising tools (Kapoor and Viraraghavan, 1997; Say et al., 2001). In fact, the purification of the water containing metals by fungal biomass is cheaper and has the following advantages: (i) production of small residual volume; (ii) possibility of valorisation of fungal waste biomasses from industrial fermentations; (iii) fast removal; (iv) easy installation of the process.

The uptake of metals by fungal biomass appears to involve a combination of two processes: bioaccumulation (i.e. active metabolism-dependent processes, which may include both transport into the cell and partitioning into intracellular components) and biosorption (i.e. the binding of metals to the biomass by processes that do not require metabolic energy). Several chemical processes may be involved in biosorption, including adsorption, ion exchange processes and covalent binding (Płaza et al., 1996).

Metal uptake by dead cells takes place by the passive mode. Living cells employ both active and passive modes for heavy metal uptake. Living and dead fungi cell removal of may offer an alternative method for wastewater remediation. The use of fungal biosorbents for heavy metals remediation has been reviewed (SAG, 221).Arange of equilibrium sorption models, and diffusion and sorption models in different reactor systems were reviewed to correlate fungal biosorption experimental data. Fungi are used in many industrial fermentation processes, which could serve as an economical source of biosorbent for arsenic removal. Fungi can also be easily grown in substantial amounts using unsophisticated fermentation techniques and inexpensive growth media. Lignin has also been used for the metal ions remediation (Mohan, 2006).

Sorption of Cr (VI) was carried out from dilute solutions using live and pretreated fungi biomass in a batch mode. Effects of agitation time, adsorbent dosage and pH were examined. The autoclaved biomass that showed maximum adsorption capacity was used as an adsorbent in column studies. The optimized flow rate of 2.5 ml min (-1) and bed height 10 cm were used to determine the effect of metal ion concentration on removal of Cr (VI). Batch sorption proved to be more efficient than the column sorption and hence batch sorption was used to remove Cr (VI) from a textile dyeing industry wastewater. The phytotoxic effect of treated and untreated wastewater was studied against Zea mays. Toxicity was reduced by 50% in the treated effluent. (Deepa et al., 2006).

The biosorption of cadmium (II) by the entrapped live and dead fungal biomass has been studied in a batch system. White rot fungus species *Lentinus sajor-caju* biomass was entrapped into alginate gel via a liquid curing method in the presence of Ca (II) ions. The heat-treatment process enhanced the biosorption capacity of the immobilized fungal biomass.The effect of initial cadmium concentration, pH and temperature on cadmium removal has been investigated. The maximum experimental biosorption capacities for entrapped live and dead fungal mycelia of *L. sajur-caju* were found to be 104.8"2. 7 mg Cd (II) gy1 and 123.5"4. 3 mg Cd (II) gy1, respectively.The kinetics of cadmium biosorption was fast, approximately 85% of biosorption taking place within 30 min. (Bayramoglu et al., 2002).

Fungi such as Agaricus macrosporus show potential for the removal of heavy metals from aqueous solutions contaminated by zinc, copper, mercury, cadmium or lead. This study investigated biosorption of these metals by living or non-living biomass of A. macrosporus from an acid solution, an acid solution supplemented with potassium and phosphorus, and an alkaline solution. Uptake showed a pH-dependent profile. Maximum percentage uptake of all metals was found to occur at alkaline pH (Cu 96%, Pb 89%). With living biomass, metal biosorption was greater and faster in K/P-supplemented acid medium than in non-supplemented acid medium, with equilibrium reached within 15 min for all metals, and the highest percentage uptake being of cadmium (96%).(Melgar et al., 2007).

Algae as one of bioadsorbent was used for industrial wastewater treatment. Continuous metal ions biosorption from Pb/Cu and Pb/Cd solutions onto seaweed *Gelidium sesquipedale* and a composite material prepared from an industrial algal waste was performed in a packed bed column. In the sorption process, Cd and Cu break through the column faster than Pb due to its lower affinity for the biosorbent. An overshoot in the outlet Cd concentration was observed and explained by competitive adsorption between Pb and Cd, whereby the higher Pb affinity for the biosorbent displaces bound Cd ions (Vilar et al., 2008).

Algal industrial waste from agar extraction process, algae *Gelidium* and a composite material obtained by immobilization of the algal waste with polyacrylonitrile (PAN) were physical characterized and used as biosorbents for dyes removal using methylene blue as model. The apparent and real densities and the porosity of biosorbents particles were determined by mercury porosimetry and helium picnometry. The methylene blue adsorption in the liquid phase was the method chosen to calculate the specific surface area of biosorbent particles as it seems to reproduce better the surface area accessible to metal ions in the biosorption process than the N2 adsorption-desorption dry method. The porous texture of the biosorbents particles was also studied. Equilibrium isotherms are well described by the Langmuir equation, giving maximum uptake capacities of 171, 104 and 74 mg g (-1), respectively for algae, algal waste and composite material (Vilar et al., 2007).

Laboratory investigations, of the potential use of dried Spirodela polyrrhiza biomass as an adsorbent for the removal of the basic dye methylene blue from aqueous solution were conducted. A series of experiments were undertaken in an agitated batch adsorber to assess the effect of the system variables, i.e. sorbent dosage, pH, and contact time. The results showed that as the amount of the dried S. polyrrhiza increased, the percentage of dye sorption increased accordingly. At pH 2.0 the sorption of dye was not favorable, while the sorption at other pHs (3.0-11.0) was remarkable. There was no significant difference in the dye concentration remaining when the pH was increased from 3.0 to 11.0. The dye removal time was influenced by the initial dye concentration, and the process followed the first-order rate kinetics. The rate constants for intraparticle diffusion were 1.00 and 3.27 $mg/g/min^{1/2}$ for 300 and 500 mg/l of dye, respectively. (Waranusantigu et al., 2003).

The adsorption behavior of mercury on Aspergillus versicolor biomass (AVB) has been investigated in aqueous solution to understand the physicochemical process involved and to explore the potentiality of AVB in pollution control management. This biomass has been successfully used for reducing the mercury concentration level in the effluent of chloralkali and battery industries to a permissible limit. The results establish that 75.6 mg of mercury is adsorbed per gram of biomass. The adsorption process is found to be a function of pH of the solution, with the optimum range being pH 5.0-6.0. The process obeys the Langmuir-Freundlich isotherm model (Das e al., 2007).

Activated carbon (AC) prepared from waste Parthenium was used to eliminate Ni (II) from aqueous solution by adsorption. Batch mode adsorption experiments are carried out, by varying contact time, metal ion concentration, carbon concentration, pH and desorption to assess kinetic and equilibrium parameters. The adsorption capacity (Q0) calculated from the Langmuir isotherm was 54.35 mg Ni (II)/g of AC at initial pH of 5.0 and 20 degrees C, for the particle size 250-500 microm. Increase in pH from 2 to 10 increased percent removal of metal ion. (Kadirvelu et al., 2002).

Biomass of Aspergillus niger, Penicillium chrysogenum, Rhizopus nigricans, Ascophyllum nodosum, Sargassum natans, Chlorella fusca, Oscillatoria anguistissima, Bacillus firmus and Streptomyces sp. have highest metal adsorption capacities ranging from 5 to 641 mg g^{-1} mainly for Pb, Zn, Cd, Cr, Cu and Ni. Biomass generated as a by-product of fermentative processes offers great potential for adopting an economical metal-recovery system (Ahluwalia and Goyal).

Unicellular microalgae Chlorella vulgaris and Scenedesmus dimorphus were reported to removing up to 55% of the phosphates from dairy industry and pig farming wastewaters (Gonzalez et al., 1997). Another strain of Scenedesmus, grown in artificial wastewater, also removed more than 50% of the phosphates (Voltolina et al., 1999).

An insoluble material of polysaccharidic nature has been obtained by thermal alkali treatment of the filamentous fungus Phomopsis sp. Information on Lewis base sites has also been obtained and used as a guideline in the evaluation of the complexing ability against a number of metal ions in aqueous media at pH in the range 4–6. Results indicate that after 24 h contact time, up to 870 mmol/g of lead, 390 mmol/g of copper, 230 mmol/g of cadmium, 150 mmol/g of zinc and 110 mmol/g of nickel ions are adsorbed into the material. After approximately 10 min, about 70% of the overall adsorption process has already been completed. Adsorbed metal ions can be recovered by washing with dilute acid. (Saianoa, et al., 2005).

Treatment of heavy metal wastewater by considering the competitive biosorption of two metal ions together, Cd (II) and Pb (II), by *Phanerochaete chrysosporium*, a filamentous fungus, under optimum biosorption conditions determined for each metal ion was studied. The effects of the presence of one metal ion on the biosorption of the other metal ion were investigated in terms of equilibrium isotherm and adsorption yield. Experimental results indicated that the uptake capacity and adsorption yield of one metal ion were reduced by the presence of the other metal ion. In addition, comparisons between biosorption of Pb (II) ions and Cd (II) ions by the biomass of *P. chrysosporium* in the binary solution could lead to the conclusion that biosorption of Pb (II) ions was preferential to that of Cd (II) ions. In the single-ion situation, biosorption of Cd (II) ions and Pb (II) ions had the optimum adsorption conditions in common, which were the solution pH 4.5, temperature 27 ^{o}C. The maximum uptake obtained at initial concentration of Cd (II) ions 50 mg l^{-1}, could reach 15.2 mg g^{-1}, for Pb (II) ions it could reach 12.34 mg g^{-1}. (Li et al., 2004).

Dead biomass can often be procured from industrial sources as a waste product of established fermentation processes (Kapoor and Viraraghavan, 1995). Systems using living cells are likely to be more sensitive to metal ion concentrations (toxicity effects) and adverse operating conditions (pH and temperature). Furthermore, a constant nutrient supply is required for systems using living cells, and recovery of metals and regeneration of biosorbent is typically more complicated. The uptake of metals by living cells depends on species, contact time, pH of the metal solution, culture conditions, initial metal ion concentration, and the concentration of cells in the solution. (Amudaa et al., 2007).

Adsorbents According to Pollutant Types

1. Adsorbents for Removal of Heavy Metals

Industrial uses of metals and other domestic processes (e.g. burning of fossil fuels, incineration of wastes, automobile exhausts, smelting processes and the use of sewage sludge as landfill material and fertiliser) have introduced substantial amounts of potentially toxic heavy metals into the atmosphere and into the aquatic and terrestrial environments.

Rapid industrialization has led to increased disposal of heavy metals into the environment. The tremendous increase in the use of the heavy metals over the past few decades has inevitably resulted in an increased flux of metallic substances in the aquatic environment. The metals are special because of their persistency in the environment. At least 20 metals are classified as toxic, and half of these are emitted into the environment in quantities that pose risks to human health. The ability of a water body to support aquatic life as well as its suitability for other uses, however, depends on many trace elements. (Amuda et al., 2007).

Heavy metal ions such as cobalt, copper, nickel, chromium and zinc are detected in the waste streams from mining operations, tanneries, electronics, electroplating and petrochemical industries, as well as in textile mill products. Industrial waste constitutes the major source of various kinds of metal pollution in natural waters. The important toxic metals i.e. lead (Pb), cadmium (Cd), zinc (Zn), and copper (Cu) find its way to the water bodies through wastewaters. The release of large quantities of heavy metals into the natural environment e.g. irrigation of agricultural fields by using sewage has resulted in a number of environmental problems and due to their non-biodegradability and persistence, can accumulate in the environment elements such as food chain, and thus may pose a significant danger to human health (Kazemipour et al., 2008).

Discharge of heavy metals from metal processing industries is known to have adverse effects on the environment. Table 2 show anthropogenic sources of metals in the environment. Conventional treatment technologies for removal of heavy metals from aqueous solution are not economical and generate huge quantity of toxic chemical sludge.

Table 2. Significant anthropogenic sources of metals in the environment

Industry	Metals	Pollution arising	Reference
Metallurgical industries Specialist alloys and steels	Pb, Mo, Ni, Cu, Cd, As, Te, U, Zn	Manufacture, disposal and recycling of metals. Tailings and slag heaps	Alloway and Ayres (1993) Rule et al. (2006) Cheng (2003)

Table 2. Continued

Industry	Metals	Pollution arising	Reference
Metal finishing industry Electroplating	Cr, Ni, Zn, Cu	Liquid effluents from plating processes.	Castelblanque and Salimbeni (2004) Zhao et al. (1999) Alvarez-Ayuso et al. (2003)
Miscellaneous sources Batteries	Pb, Sb, Zn, Cd,Ni, Hg	Waste battery fluid, contamination of soil and groundwater.	EU Directorate general of the Environment (2004)
Paints and pigments	Pb, Cr, As, Ti, Ba, Zn	Aqueous waste from manufacture, old paint deterioration and soil pollution.	Davis and Burns (1999) Barnes and Davis (1996) Monken (2000)

Several adsorbents such as saw dust, silica and iron oxide, sewage sludge ash, anatase-type titanium dioxide, olive mill residues, inorganic colloids, blast furnace sludge and activated carbon have been used for the treatment of heavy metal-rich effluents at the solid-solution interface.Cu (II) was adsorbed on potato peels, an abundantly available waste product is carried out along with studies of kinetics, isotherm and optimum conditions for adsorption. (Aman et al., 2008).

In industrial wastewater treatment technology, various methods have been used to remove heavy metals from aqueous solutions. The main techniques, which have been utilized to reduce the heavy metal ion content of effluents, include lime precipitation, ion exchange, adsorption into activated carbon, membrane processing, and electrolytic methods (Braukmann, 1990).

On the other hand ion exchange, reverse osmosis and adsorption are more attractive processes because the metals values can be recovered along with their removal from the effluents. Reverse osmosis and ion exchange do not seem to be economically feasible because of their relatively high investment and operational cost. These techniques can reduce heavy metal ions and do not appear to be highly effective due to the limitations in the pH range as well as high cost in the material and operation and may be associated with the generation of secondary waste which present treatment problems, such as the large quantity of sludge generated by precipitation processes. In recent years, considerable attention has been focused on the removal of heavy metals from aqueous solution using adsorbent derived from low-cost materials. Adsorption has advantages over the other methods because of simple design with a sludge free environment and can involve low investtment in term of both initial cost and land required (Viraraghaban et al., 1993). But the process has not been used by small and medium scale industries for the treatment of their metal bearing effluents, because of its high manufacturing cost. For this reason, the use of low cost materials as adsorbent for metal ion removal from the wastewater has been highlighted. However, efforts have been contributed to develop new adsorbent and improving the existing adsorbents to have an alternative to activated carbon. These materials range from industrial products such rubber tyres , industrial wastes and some natural material including agricultural product and by-product (Table 3).

Activated carbon has been the standard adsorbent for the reclamation of municipal and industrial wastewaters, however, it requires regeneration and its adsorption capacity for heavy metals is usually less than 50 mg/g. Furthermore, the adsorption density of inorganic species onto activated carbon surfaces varies significantly with pH, in accordance with its acid–base behavior (Chen et al., 2008).

Table 3. Heavy metal adsorption capacities for selected low cost adsorbents

Adsorbent	Metal	Uptake (mg g^{-1})	References
2-Industrial by-products			
Blast furnace sludge	Ni(II)	55.9	Feng et al. (2004)
	Cu(II)	133.4	
Metal sludge	V(II)	24.8	Bhatnagar et al.(2008)
Iron slag	Cu(II)	88.5	
	Pb(II)	95.2	
Steel slag	Cu(II)	16.2	Lopez et al. (1998)
	Pb(II)	32.2	
	Ni(II)	10.9	
Red mud	Cu(II)	19.7	
3-Agricultural waste			
Apricot stone	Pb(II)	85	Rashed (2006)
Hazelnut shell	Ni(II)	10.1	Demirbas et al. (2002)
Peanut hulls	Ni(II)	80.0	Periasamy and Namasivayam
Orange peel (citrus	Ni(II)	10.6	(1996) Ajmal et al. (2000)
reticulate)	Cu(II)	10.1	Martinez et al. (2006)
Grape stalk waste	Pb(II)	49.9	
	Pb(II)	73.8	Reddad et al. (2003)
Sugar beet pulp	Cd(II)	1.12	Zheng et al.(2008)
Areca waste	Cu(II)	2.84	

One of the cost effective biosorbent used for the adsorption of heavy metal-contaminated industrial wastewater is Coconut shell carbon modified with chitosan and/or oxidizing agent (phosphoric acid) to produce composite adsorbent. The adsorption efficiency of the adsorbent was evaluated by measuring the extent of adsorption of zinc (II) in synthetic beverage industrial wastewater (Amuda et al., 2007).

As a consequence, activated sludge produced the biological treatment facilities of the treatment plant is considered as a good bioadsorbent for the removal of Cu^{2+}, Cd^{2+} and Ni^{2+} ions from industrial wastewater (Al-Qodah, 2006).

Multi-component sorption studies were carried out for attenuation of divalent heavy metal cations (Pb^{2+}, Cu^{2+} and Zn^{2+}) by a low-cost mineral adsorbent from the aqueous solution. Kinetic and equilibrium batch-type sorption experiments were conducted under variable conditions for multicomponent using low-grade ($<12\%P_2O_5$) phosphate rock.

Percentage of multiple heavy metal species removal increases with decreasing initial metals concentration and particle size. The fixation capacity obtained at room temperature for Pb^{2+}, Cu^{2+} and Zn^{2+} was 227.2, 769.2 and 666.6 mol g^{-1}, respectively (Prasad et al., 2008).

Environmental friendly removal process of heavy metals by using biomass waste and natural product was reported. The process consists of the following three steps: (1) leaching of all metals including heavy metals by dilute sulfuric acid, (2) removal of turbid organic materials from the leach liquor by means of coagulation using astringent persimmon extract enriched with persimmon tannin leaving all heavy metal ions in the leach liquor and (3) adsorptive removal of heavy metals by means of adsorption onto the gel prepared from apple waste. Among which the pH adjustment is the key factor; i.e. the lower the pHs the higher the efficiency of the leaching, while reverse was the case for the adsorption (Ghimire et al., 2008).

Removal of lead (II) and copper (II) from aqueous solutions were studied using pomegranate peel (raw), activated carbon prepared from pomegranate peel (AC1) and activated carbon prepared from chemically treated pomegranate peel (AC2 and AC3). Batch adsorption experiments were performed as a function of pH, contact time, solute concentration and adsorbent dose. The optimum pH required for maximum adsorption were found to be 5.6 and 5.8, for lead and copper respectively.Results indicate the following order to fit the isotherms (El-Ashtoukhya et al., 2008).

Areca waste (AW) has been investigated as metal biosorbent for cadmium and copper from aqueous solution for its availability as food waste and also for its cellulosic matrix rich of potential metal binding active sites. The effect of various parameters on adsorption process such as contact time, solution pH, amount of AW and initial concentration of metal ions was studied at room temperature to optimize the conditions for maximum adsorption. Maximum metal sorption was found to occur at pH 5.6. Adsorption process revealed that the initial uptake was rapid and equilibrium was established about in 1 h for cadmium and copper. The highest value of Langmuir maximum uptake, (*b*), was found for cadmium (1.12 mg/g) and copper (2.84 mg/g) (Zheng et al., 2008).

Alum sludge was recycled using a chemical precipitation process to promote the removal of lead metal in wastewater. To make the process more cost-effective, two pH ranges were suggested for daily operation, depending on the involvement or otherwise of an aluminum regeneration process at a later stage. If aluminum regeneration was expected, an elevated pH (11.6) was suggested to optimize both the aluminum regeneration and lead removal. Lead hydroxide coprecipitated with alum sludge in alkaline conditions, where the high pH would also facilitate the alkaline regeneration of alum at a later stage. With the use of recycled alum sludge (RAS) in the coagulation process, lead removal rates increased from 79% to 96±98% with 100±180 mg/l of RAS, while the required fresh alum dosage was reduced from 175 mg/l to 50±12.5 mg/l only. If no fresh alum is added to the system, the RAS can still remove up to 94% of lead at low recycling doses (75±100 mg/l). However, if the regeneration of aluminum is not required, a lower initial pH (10.5) should be used to decrease chemical costs. (Chu, 1999).

Removal of copper ion using NaOH treated Rhizopus oryzae biomass was investigated in a batch reactor. The copper uptake exhibited substantial enhancement both in terms of kinetics of uptake as well as the loading capacity. The copper biosorption by viable and pretreated fungal biomass fit well to a Lagergren's pseudo second order reaction in comparison to pseudo first order kinetics. Investigation on effect of pH indicated improved performance in the range of pH 4–6 in alkali treated biomass. Copper uptake exhibited by viable biomass was highest at 21°C, unlike pretreated biomass that showed maximum uptake across the range of temperature 21–55 °C. The maximum copper loading capacity of the viable and pretreated biomass according to Langmuir isotherm was 19.4 and 43.7 mg/g, respectively. Microbial removal of metal ions from such wastewater has been indicated as being highly effective (Brierley, 1990).

Heavy metal resistant fungi and bacteria were isolated from the soil samples of an electroplating industry, and the bioaccumulations of Cr (VI) and Ni (II) by these isolates were characterized to evaluate their applicability for heavy metal removal from industrial wastewaters. The optimum pH and temperature conditions for both the growth and heavy metal removal were determined for each isolate. The optimal pH for fungal isolates was lower (5–5.2) than that for bacterial isolates. The observed effect(s) of pH was attributable mainly to organism-specific physiology because in all the tested cases the cellular growth positively correlated with heavy metal removal. Batch and tolerance experiments provided information for solid retention time (SRT) design and the lethal tolerance limits for the isolated microorganisms. Experimental results indicated that expanded SRTs (stationary phase) can be recommended while using the fungal and bacterial Cr-resistant isolates for removing chromium. In the case of Ni-resistant bacterial isolate, a non-expanded SRT was recommended for designing continuous-flow completely stirred (CFCS) bioreactor so that a mid-log phase of cellular growth can be kept during the bioaccumulation process. The tolerance data with a high range of heavy metal concentrations revealed the Cr-resistant isolates, especially the fungal one, could tolerate chromium toxicity at up to 10,000 mg L^{-1} chromium. (Congeevaram et al., 2007).

The removal of poisonous Cr (VI.) from industrial wastewater by different low-cost abundant adsorbents was investigated. Wool, olive cake, sawdust, pine needles, almond shells, cactus leaves and charcoal were used at different adsorbent/metal ion ratios (Dakiky et al., 2002).

2. Adsorbents for Removal of Dyes

Industrial wastewater containing dyes is a growing concern to the environment due to their color and accumulated toxic chemicals. These wastewaters cause many damages to the ecological system of the receiving surface water and create a lot of disturbance to the groundwater resources. Textile industries produce huge amounts of polluted effluents that are normally discharged to surface water bodies and groundwater aquifers. Approximately 10000 different dyes and pigments are used for industries. Dyes represent one of the problematic groups; they are emitted into wastewater from various industrial branches, mainly from the dye manufacturing and textile finishing and also from food coulring, cosmetics, paper and carpet industries. It is well known that the dye effluents from dyestuff manufacturing and textile industries, may exhibit toxic effects on microbial populations and can be toxic and/or carcinogenic to mammalian animal. Most dyes used in textile industries are stable to light and are not biologically degradable. Furthermore, they are resistant to aerobic digestion. (Ardejani, 2007).

Adsorption techniques are rapidly gaining prominence as treatment processes which produce high quality effluent that are law in concentration of dissolved organic compounds, such as dyes (Walker and Weatherly, 1998). On searching for economical and consistently available starting materials; sewage sludge has being identified as a potentially attractive material for the production of activated carbon as adsorbent for wastewater treatment plants. Sewage sludge are unavoidable by products of wastewater treatment plant in which it make a problem in its disposal, so activated or treated sewage sludge solve these problem and consider as low cost adsorbent for wastewater treatment. Research studies has been conducted to demonstrate the uses of treated sewage sludge for the removal of dyes from wastewater and polluted water (Rio et al,2005; Martin et al, 2003; Gulna et al, 2004, 2005; Martin et al, 2002; Chu and Chen, 2002; Rozada et al, 2003). In these studies activated carbons were produced that exhibited considerable adsorption capacity as dictated by the porous structure of the carbon. Other low cost adsorbent for the removal of dyes was activated rice husk (rice husk was from the local rice mill). It was used as cheap adsorbent for color removal from wastewater (Gupta et al., 2006).

Beer brewery waste has been shown to be a low-cost adsorbent for the removal of basic dye from the aqueous solution as compared to its precursor (i.e., diatomite) based on its physical and chemical characterizations. Removal of basic dye (methylene blue) from wastewaters utilizing beer brewery waste The results of preliminary adsorption kinetics showed that the diatomite waste could be directly used as a potential adsorbent for removal of methylene blue on the basis of its adsorption–biosorption mechanisms (Tasi et al., 2007). Table 4 show selected low cost adsorbent used for dyes removal from industrial wastewater.

Adsorption of basic dye, methylene blue, from aqueous solution (40 mg L^{-1}) onto cedar sawdust and crushed brick was studied as low-cost adsorbents for wastewater dye removal. The two adsorbents used in this study: brick and cedar sawdust were obtained from local industries. They were used directly for adsorption experiments without any treatment (Hamdaoui 2006).

Table 4. Recent reports on various adsorbents used for dyes removal from industrial wastewater

Dye	Adsorbent	Adsorption capacity	pH	Contact time	Reference
Methylene blue	Bagg1ase	77.9-69.8%	-	3 min	Reghuvanshi et al., 2004
Basic Red 16	Activated sludge	285 mg/g	7	45 min	Gulnaz et al., 2004
Basic Blue9	Activated sludge	256 mg/g	7	45 min	Gulnaz et al., 2004
Malachite green	Bagasse flash	170 mg/g	7	4 hr	Mall et al., 2005
	Activated carbon	8.27 mg/g	7	4 hr	Mall et al., 2005
Direct blue 2B	Sawdust	518 mg/g	3	2 hr	Malik, 2004
Direct green B	Activated carbon	327 mg/g	3	2 hr	Malik, 2004
Reactive blue2	Dried activated sluge	250 mg/g	7	36-60 min.	Aksu, 2001
Reactive yellow 2	Dried activated sluge	333 mg/g	5	15-30 min	Aksu, 2001
Congo red	Bagasse fly ash	2.07 mg/g	-	4 hr	Mall et al., 2005
Acid blue 40	Alunite	212 mg/g	1-2	90 min	Ozacar and Sengil, 2002
Acid yellow 17	Alunite	151 mg/g	1-2	90 min	
Congo red	Activated carbon	16.81 mg/g	6	1hr	Fu et al., 2002
	Fungal biomass	14.16 mg/g			

The utilization of wood-shaving bottom ash (WBA) for the removal of Red Reactive 141 (RR141), an azo reactive dye, was investigated. WBA/H_2O and WBA/H_2SO_4 were made by treating WBA with water and 0.1 M H_2SO_4, respectively, to increase adsorption capacity. Adsorption of RR141 from reactive dye solution (RDS) and reactive dye wastewater (RDW) by WBA/H_2O and WBA/H_2SO_4 involved the BET surface area and pore size diameter. Properties of adsorbents, effect of contact time, initial pH of solution, dissolved metals and elution studies indicated that the decolorisation mechanism involved both chemical adsorption and precipitation with calcium ions. The maximum dye adsorption capacities of WBA/H_2O, WBA/H_2SO_4 and activated carbon obtained from a Langmuir model at 30 ^oC were 24.3, 29.9, and 41.5 mg l^{-1}, respectively. In addition, WBA/H_2O and WBA/H_2SO_4 could reduce colour and high chemical oxygen demand (COD) of real textile wastewater. (Leechart et al., 2008).

3. Adsorbents for Removal of Other Organic Pollutants

Toxic organic pollutants cause several environmental problems if discharged to water body, so it is important to remove the toxic organic pollutants from industrial wastewater. Adsorption of organic pollutants using low-cost adsorbent is gaining interest as one of the effective processes of advanced wastewater treatment for treatment.

Phenol pollution is a serious problem in many countries. The major sources of phenolic waste are petroleum refineries, petrochemical, steel mills, coke oven plants, coal gas, synthetic resins, pharmaceuticals, paints, plywood industries and mine discharge. Phenolic waste imparts a carbolic odour to river water and is also toxic to fish and human beings.

In the study of Mukherjee (2007) three carbonaceous materials, activated carbon (AC), bagasse ash (BA) and wood charcoal (WC), was used as adsorbents for removal of phenol from water. The study was performed with two initial phenol concentrations, viz. 30 and 50 mg/L, with an equal amount of adsorbent dose (50 g/L). The effects of solution pH, concentrations of EDTA, anions, and dosages of adsorbent on removal of phenol were examined. Experimental results showed that for phenol–AC, phenol–WC and phenol–BA adsorption systems, approximately 98%, 90% and 90% removal efficiencies were achieved at given adsorption conditions. Removal efficiency of phenol slightly increased when the pH of adsorption system decreased (Mukherjee et al., 2007).

The wastewater with the highest concentration of phenol (>1000 mg/L) is typically generated from coke processing. Phonolic compounds are also emanated from resin plants with a concentration range of 12–300 mg/L. Environmental Protection Agency (EPA) has set a limit of 0.1 mg/L of phenol in wastewater. TheWorld Health Organisation (WHO) is stricter on phenol regulation. It sets a 0.001 mg/L as the limit of phenol concentration in potable water. Yapar and Yilmar (2004) explored the adsorptive capacity of some clays and natural zeolite materials found in Turkey for removal of phenol. They found calcined hydrotalcite was the best among the studied adsorbents. It can adsorb 52% of phenol from a solution with an initial phenol concentration of 1000 mg/L at the adsorbent/phenol ratio of 1:100 while the others could adsorb only 8% of phenol for the same operation conditions. Roostari and Tezel (2004) examined the adsorption of phenol from aqueous solution by silica gel, activated alumina, AC, fitrasorb 400, Hisir 1000. They found Hisir 1000 was the best among the tested materials in light of adsorption kinetics though the AC had the highest affinity and adsorption capacity of phenol... Das and Patnaik (2005) utilized blast furnace flue dust (BFD) and slag to investigate phenol adsorption through batch experiment (Mukherjee et al., 2007).

Removal of bromophenols from water using industrial wastes as low cost adsorbents was studied. A comparative study of the adsorbents prepared from several industrial wastes for the removal of 2-bromophenol, 4-bromophenol and 2, 4- dibromophenol has been carried out. The results show that maximum adsorption on carbonaceous adsorbent prepared from fertilizer industry waste has been found to be 40.7, 170.4 and 190.2 mg g−1 for 4-bromophenol 2-bromophenol and 2, 4-dibromophenol, respectively. As compared to carbonaceous adsorbent, the other three adsorbents (viz., blast furnace sludge, dust, and slag) adsorb bromophenols to a much smaller extent. This has been attributed to the carbonaceous adsorbent having a larger porosity and consequently higher surface area. The adsorption of bromophenols on this adsorbent has been studied as a function of contact time, concentration and temperature) Bhatnagar, 2007).

Parthenium hysterophorous (L) is a perennial weed. Carbonized parthenium activated with conc. H_2SO_4 and ammonium persulphate was effective in the removal of dyes, heavy metals and phenols. Variation in the percentage removal of adsorbates was observed with increase in the contact time (Rajeshwarisivaraj and Subburam, 2002).

Cost-Effectiveness of Low-Cost Adsorbents

Developing countries suffers from water pollution, the high cost of contents and treatments make the major problems in these countries. Search for a cheaper alternative, reveals that utilizing natural processes and materials significantly reduce high cost. Since the cost-effectiveness of an adsorbent is one of the important issues that must be considered when selecting an adsorbent, the price of low-cost adsorbents has to be compared. Expenditure on individual adsorbents varies, depending on the processing employed and its local availability (Hasan et al., 2000; Shukla and Pai, 2005). Low-cost adsorbents derived from agricultural waste, which are converted chemically or by heating into activated carbon, require extra processing cost. Improved adsorption capacities, however, may compensate the cost for such a process (Bailey et al., 1999; Babel and Kurniawan, 2003). Although expenditure on low-cost adsorbents may be negligible, further cost-benefit analysis needs to take into account any spending associated with regeneration or operation including chemicals, electricity, labor, transportation and maintenance (Krishnan and Anirudhan, 2003). If the overall cost remains economically attractive, low-cost adsorbents for metal uptake present promising alternatives to costly activated carbon, as their use can minimize treatment cost.

Refrences

Abollino O, Aceto M, Malandrino M, Sarzanini C, Mentasti E. (2003) Adsorption of heavy metals on Na-montmorillonite: effect of pH and organic substances. *Water Res*; **37**:1619–27.

Agrawal Archana , K.K. Sahu, B.D. Pandey (2004) Removal of zinc from aqueous solutions using sea nodule residue *Colloids and Surfaces A: Physicochem. Eng. Aspects* **237**, 133–140.

Ahluwalia SS; Goyal D. (2007) Microbial and plant derived biomass for removal of heavy metals from wastewater. *Bioresour. Technol.*; **98** (12):2243-57.

Ajmal M, Rao RAK, Ahmad R, Ahmad J. (2000) Adsorption studies on Citrus reticulata (fruit peel of orange) removal and recovery of Ni (II) from electroplating wastewater. *J Hazard Mater*;**79**:117– 31.

Aksu Zümriye (2001) Biosorption of reactive dyes by dried activated sludge: equilibrium and kinetic modeling. *Biochemical Engineering Journal* **7**, 79–84.

Allen S.J., L.J. Whitten, M. Murray, O. Duggan, P.A. Brown (1997) The adsorption of pollutants by peat, lignite and activated chars, *J.Chem. Technol. Biotechnol.* **68**, 442–452.

Al-Qodah Z. (2006) Biosorption of heavy metal ions from aqueous solutions by activated sludge. *Desalination* **196**, 164–176.

Alvarez-Ayuso E, Garcı´a-Sa´nchez A. (2003) Removal of heavy metals from wastewaters by vermiculites. *Environ. Technol.*; **24**:615– 25.

Amana Tehseen , Asrar Ahmad Kazi, Muhammad Usman Sabri, Qudsia Banoa. (2008) Potato peels as solid waste for the removal of heavy metal copper (II) from waste water/industrial effluent. *Colloids and Surfaces B: Biointerfaces* **63**, 116–121.

Amuda O.S., A.A. Giwa, I.A. Bello (2007) Removal of heavy metal from industrial wastewater using modified activated coconut shell carbon *Biochemical Engineering Journal* **36**, 174–181.

Ardejani F. Doulati , Kh. Badii, N. Yousefi Limaee, N.M. Mahmoodi, M. Arami, S.Z. Shafaei, A.R. Mirhabibi (2007) Numerical modelling and laboratory studies on the removal of Direct Red 23 and Direct Red 80 dyes from textile effluents using orange peel, a low-cost adsorbent. *Dyes and Pigments* **73**, 178-185.

Babel S. , T.A. Kurniawan (2004) Cr(VI) removal from synthetic wastewater using coconut shell charcoal and commercial activated carbon modified with oxidizing agent and/or chitosan, *Chemosphere* **54** (7) 951– 967.

Ballclay V. Chantawong, N. W. Harvey1 and V. N. Bashkin. (2003).Comparison Of Heavy Metal Adsorptions By Thai Kaolin .*Water, Air, and Soil Pollution* **148**: 111–125.

Bansal, R.C., Donnet, J.B., Stoeckli, F.(1988). Active Carbon. Marcel Dekker Inc, New York, USA.

Barnes, G.L., Davis, A.P.(1996). Dissolution of lead paint in aqueous solution. *Journal of Environmental Engineering ASCE*, **122**, 663–666.

Bayramoglu G., Adil Denizli, Sema Bektas, M.Yakup (2002) Aricac, Entrapment of *Lentinus sajor-caju* into Ca-alginate gel beads for removal of Cd (II) ions from aqueous solution: preparation and biosorption kinetics analysis. *Microchemical Journal* **72**, 63–76.

Berber-Mendoza, M.S., Leyva-Ramos, R., Alonso-Davila, P., Mendoza- Barron, J., Diaz-Flores, P.E.(2006). Effect of pH and temperature on the ion-exchange isotherm of Cd (II) and Pb (II) on clinoptilolite. *Journal of Chemical Technology and Biotechnology* **301**, 40–45.

Bhatnagar Amit , Ashwani Kumar Minoch, Deepak Pudasaine,Hyung-Keun Chung, Seong-Heon Kim, Hyoung-Soo Kim,Giehyeon Lee, Booki Min, Byong-Hun Jeo (2008) Vanadium removal from water by waste metal sludge and cement immobilization. *Chemical Engineering Journal.***144** (2)197-204.

Bhatnagar Amit (2007) Removal of bromophenols from water using industrial wastes as low cost adsorbents. *Journal of Hazardous Materials* **B139**, 93–102.

Bhattacharyy Krishna Gopal and Susmita Sen Gupta (2008) Adsorption of a few heavy metals on natural and modified kaolinite and montmorillonite: A review. *Advances in Colloid and Interface Science,* **140** (2),114-131.

Bowman,Li, Z., R.(2001). Regeneration of surfactant-modified zeolite after saturation with chromate and perchloroethylene. *Water Resources* **35** (1), 322–326.

Braukmann BM (1990). Industrial solution amenable to biosorption. In Biosorption (Edited by Volusky B), CRC Press, Boca Raton.

Brierley C.L. (1990) Bioremediation of metal-contaminated surface and groundwater. *J.Geomicrobiol*, **8**, 201–223.

Bulut Emrah, Mahmut Özacar and İ. Ayhan Şengil (2008) Equilibrium and kinetic data and process design for adsorption of Congo Red onto bentonite. *Journal of Hazardous Materials* **154** (1-3)613-622.

Castelblanque, J., Salimbeni, F.NF and RO (2004) Membranes for the recovery and reuse of water and concentrated metallic salts from waste water produced in the electroplating process. *Desalination* **167**, 65–73.

Cengeloglu, Y., Kir, E., Ersoz, M.(2002) Removal of fluoride from aqueous solution by using red mud. *Sep. Purif. Technol.* **28** (1), 81–86.

Chen Quanyuan , Colin D. Hills, Menghong Yuan, Huanhuan Liu, Mark Tyrer. (2008) Characterization of carbonated tricalcium silicate and its sorption capacity for heavy metals: A micron-scale composite adsorbent of active silicate gel and calcite. *Journal of Hazardous Materials* **153**, 775–783.

Cheng, S.P.(2003). Heavy metal pollution in China: origin, pattern and control. *Environmental Science and Pollution Research* **10** (3), 192–198.

Chu Wei (1999) Lead Metal Removal By Recycled Alum. Sludge. *Wat. Res.***33** (13) 3019-3025.

Crini Gre´gorio (2005) Recent developments in polysaccharide-based materials used as adsorbents in wastewater treatment. *Prog. Polym. Sci.* **30**, 38–70.

Dakiky M. , M. Khamis, A. Manassr, M. Merie (2002) Selective adsorption of chromium(VI) in industrial wastewater using low-cost abundantly available adsorbents *Advances in Environmental Research* **6**, 533-540.

Das N.C., M. Bandopadhyay (1993) Removal of nickel by low coast natural medium, *Indian J. Technol.* **31** 312–317.

Das SK; Das AR; Guha AK. A (2007) Study on the adsorption mechanism of mercury on Aspergillus versicolor biomass. *Environ. Sci. Technol.* **15**; 41(24):8281-7.

Das C.P., L.N. Patnaik (2005) Removal of phenol by industrial solid waste, Practice Period. Hazard. *Toxic Radioactive Waste Manage. ASCE* **9** (2) 135–140.

David William O'Connell, Colin Birkinshaw, Thomas Francis O'Dwyer (2008) Heavy metal adsorbent prepared from the modification of cellulose: A review. *Bioresource Technology* **99** (15) 6709-6724.

Davis, A.P., Burns, M.(1999). Evaluation of lead concentration in runoff from painted structures. *Water Research* **33** (13), 2949–2958.

Deepa KK; Sathishkumar M; Binupriya AR; Murugesan GS; Swaminathan K; Yun SE. (2006).Sorption of Cr(VI) from dilute solutions and wastewater by live and pretreated biomass of Aspergillus flavus. *Chemosphere*; **62** (5):833-40.

Demirbas¸ E, Kobya M, Oncel S, Sencan S. (2002) Removal of Ni (II) from aqueous solution by adsorption onto hazelnut shell activated carbon: equilibrium studies. *Biores. Technol.;* **84**:291–3.

Diaz-Diez, M.A., Go´mez-Serrano, V., Ferna´ndez Gonza´lez, C., Cuerda-Correa, E.M., Macı´as-Garcı´a, A.(2004). Porous texture of activated carbons prepared by phosphoric acid activation of woods. *Appl. Surf.Sci.* **238**, 309–313.

El-Ashtoukh E.-S.Z. , N.K. Amin, O. Abdelwahab (2008) Removal of lead (II) and copper (II) from aqueous solution using pomegranate peel as a new adsorbent. *Desalination* **223**, 162–173.

El Nemr Ahmed , Azza Khaled, Ola Abdelwahab, Amany El-Sikaily (2008) Treatment of wastewater containing toxic chromium using new activated carbon developed from date palm seed. *Journal of Hazardous Materials* **152**, 263–275.

Erdem E, Karapinar N, Donat R. (2004) The removal of heavy metal cations by natural zeolites. *J. Colloid Interface Sci.*; **280** (2):309-14.

Eren E. and B. Afsin (2008) Investigation of a basic dye adsorption from aqueous solution onto raw and pre-treated bentonite surfaces. *Dyes and Pigments* **76** (1)220-225.

Eren E. and B. Afsin (2008b) An investigation of Cu (II) adsorption by raw and acid-activated bentonite: A combined potentiometric, thermodynamic, XRD, IR, DTA study. *Journal of Hazardous Materials* **151**(2-3)682-691.

Evans, M.J.B., Halliop, E., MacDonald, J.A.F.(1999). The production of chemically-activated carbon. *Carbon* **37**, 269–274.

Feng D, Van Deventer JSJ, Aldrich C. (2004) Removal of pollutants from acid mine wastewater using metallurgical by-product slags. *Sep.Purif. Technol.*; **40** (1):61 – 7.

Fu Y, Viraraghavan T. (2002) Removal of Congo red from an aqueous solution by fungus Aspergillus niger. *Adv. Environ. Res.*; 7:239–47.

Gadd G.M.(1988), Accumulation of metal by micro-organisms and algae, in: H. Rehm (Ed.), Biotechnology: A Complete Treatise, vol. 6B, Special Microbial Processes, vol. 4, VCH, Verlagsgesellschaft,Weinheim, pp. 401–430.

Garg Umesh , M.P. Kaur, G.K. Jawa, Dhiraj Sud, V.K. Garg (2008) Removal of cadmium (II) from aqueous solutions by adsorption on agricultural waste biomass. *Journal of Hazardous Materials* **154** ,1149–1157.

Ghimire Kedar Nath , Huang Kai, Katsutoshi Inoue, Keisuke Ohto, Hidetaka Kawakita, Hiroyuki Harada, Minoru Morita (2008) Heavy metal removal from contaminated scallop waste for feed and fertilizer application. *Bioresource Technology* **99**, 2436–2441.

Gode Fethiye, Elif Dincturk Atalay, Erol Pehlivan (2008) Removal of Cr (VI) from aqueous solutions using modified red pine sawdust *Journal of Hazardous Materials* **152**, 1201–1207.

Gonzalez, L.E., Can˜ izares, R.O., Baena, S.(1997). Efficiency of ammonia and phosphorus removal from a Colombian agroindustrial wastewater by the microalgae Chlorella vulgaris and Scenedesmus dimorphus. *Bioresource Technol.* **60**, 259–262.

Gulnaz, O, A. Kayaa, F. Matyar, B.Arikan (2004) Sorption of basic dyes from aqueous solution by activated sludge', *Journal of Hazardous Materials B*, **108**, 183–188.

Gupta Vinod K., Alok Mittal, Rajeev Jain, Megha Mathur, Shalini Sikarwar (2006) Adsorption of Safranin-T from wastewater using waste materials- activated carbon and activated rice husks *Journal of Colloid and Interface Science* **303**, 80–86.

Hadjar H., B. Hamdi, M. Jaber, J. Brendle, Z. Kessaı¨ssia, H. Balard, J.B. Donnet (2008) Elaboration and characterisation of new mesoporous materials from diatomite and charcoal. *Microporous and Mesoporous Materials.* **107**, 219–226.

Hamdaoui Oualid (2006) Batch study of liquid-phase adsorption of methylene blue using cedar sawdust and crushed brick. *Journal of Hazardous Materials* **B135**, 264–273.

Harvey1 N. W. and V. N. Bashkin (2003) Comparison of heavy metal adsorptions by Thai kaolin and Ballclay ,*Water, Air, and Soil Pollution* **148**: 111–125.

Huang and Fuerstenau (2000) The effect of the adsorption of lead and cadmium ions on the interfacial behavior of quartz and talc. *Colloids and Surfaces A: Physicochemical and Engineering Aspects.***177**, Issues 2-3, 147-156.

Jaramilloa J., V. Go′ mez-Serranob, P.M. Alvareza (2008) Enhanced adsorption of metal ions onto functionalized granular activated carbons prepared from cherry stones. *Journal of Hazardous Materials* (**161**)2-3, 670-676.

Kadirvelu K; Senthilkumar P; Thamaraiselvi K; Subburam V.(2002). Activated carbon prepared from biomass as adsorbent: elimination of Ni (II) from aqueous solution. *Bioresour .Technol.***81** (1):87-90.

Karageorgiou Kostantinos , Maximos Paschalis, Georgios N. Anastassakis.(2007) Removal of phosphate species from solution by adsorption onto calcite used as natural adsorbent *Journal of Hazardous Materials* **A139**, 447–452.

Kazemipour Maryam , Mehdi Ansari, Shabnam Tajrobehkar, Majdeh Majdzadeh, Hamed Reihani Kermani) (2008) Removal of lead, cadmium, zinc, and copper from industrial wastewater by carbon developed from walnut, hazelnut, almond, pistachio shell, and apricot stone. *Journal of Hazardous Materials* **150**, 322–327.

Kou Y, Shang HS, Yang F, Yan Z (2001). *China Pat. Appl. Number*: 01115505.

Knocke WR, Hemphill LH (1981). Mercury sorption by waste rubber. *Water Res.* **15**: 275-282.

Krishnan, K.A., Anirudhan, T.S.(2003). Removal of cadmium(II) from aqueous solutions by steamactivated sulphurised carbon prepared from sugar-cane bagasse pith: kinetics and equilibrium studies. *Water SA* **29**, 147.

Kurniawan T. A. , Gilbert Y.S. Chan, Wai-hung Lo, Sandhya Babel (2006) .Review, Comparisons of low-cost adsorbents for treating wastewaters laden with heavy metals *Science of the Total Environment* **366**, 409– 426.

Leechart Piyawan, Woranan Nakbanpote, Paitip Thiravetyan (2008) Application of ‘waste’ wood-shaving bottom ash for adsorption of azo reactive dye. *Journal of Environmental Management* **1–9.**

Li Q. , Songtao Wu, Gang Liu, Xinkai Liao, Xu Deng, Daohua Sun, Yuelin Hu, Yili Huang (2004) Simultaneous biosorption of cadmium (II) and lead (II) ions by pretreated biomass of *Phanerochaete chrysosporium*. *Separation and Purification Technology* **34**, 135–142.

Li, Y.H., Wang, S., Cao, A., Zhao, D., Zhang, X., Xu, C., Luan, Z., Ruan, D., Liang, J., Wu, D., Wei, B.(2001). Adsorption of fluoride from water by amorphous alumina supported on carbon nanotubes. *Chem. Phys. Lett.* **350** (5 and 6), 412–416.

Lopez, E., Soto, B., Arias, M., Nunez, A., Rubinos, D., Barral, T. (1998) Adsorbent properties of red mud and its use for wastewater treatment. *Water Research* **32** (4), 1314–1322.

Mahramanlioglu, M., Kizilcikli, I., Bicer, I.O.(2002). Adsorption of fluoride from aqueous solution by acid treated spent bleaching earth. *J. Fluorine Chem.* **115** (1), 41–47.

Malik, P.K. (2004) Dye removal from wastewater using activated carbon developed from sawdust: adsorption equilibrium and kinetics, *Journal of Hazardous Materials B.*, **113**, 81–88.

Mall, I.D., Srivastava, V.C., Agarwal, N.K., Mishra, I.M.(2005). Removal of congo red from aqueous solution by bagasse fly ash and activated carbon: kinetic study and equilibrium isotherm analyses. *Chemosphere* **61**, 492–501.

Marsh, H., Rodriguez-Reinoso, F.(2006) Activated Carbon, first ed., Elsevier Ltd.

Martinez, M., Miralles, N., Hidalgo, S., Fiol, N., Villaescusa, I., Poch, J.(2006). Removal of lead (II) and cadmium (II) from aqueous solutions using grape stalk waste. *Journal of Hazardous Materials* **133** (1–3), 203–211.

Melgar M.J. , J. Alonso, M.A. García. Melgar (2007) Removal of toxic metals from aqueous solutions by fungal biomass of Agaricus macrosporus. *Bioresource Technology,* **98** (12),2745-59.

Mohan D. , C.U. Pittman Jr., P.H. Steele (2006) Single, binary and multicomponent adsorption of copper and cadmium from aqueous solutions on Kraft lignin—a biosorbent, *J. Colloid Interf. Sci.* **297** (2) 489–504.

Mohan Dinesh , Subhash Chander (2006) Removal and recovery of metal ions from acid mine drainage using lignite—A low cost sorbent. *Journal of Hazardous Materials* **B137**, 1545–1553.

Monken Monken, A.(2000). Water pollution control for paint booths. *Metal Finishing* **98** (6), 464–471.

Mukherjee Somnath , Sunil Kumarb, Amal K. Misra , Maohong Fan (2007) Removal of phenols from water environment by activated carbon, bagasse ash and wood charcoal *Chemical Engineering Journal* **129**, 133–142.

Murari Prasad, Huan-yan Xub, Sona Saxena (2008) Multi-component sorption of Pb(II), Cu(II) and Zn(II) onto low-cost mineral adsorbent. *Journal of Hazardous Materials* **154**, 221–229.

Natale F. Di , A. Lancia , A. Molino, M. Di Natale, D. Karatza, D. Musmarr (2006) Capture of mercury ions by natural and industrial materials. *Journal of Hazardous Materials* **B132**, 220–225.

Nemerrow (1971).Liquid wastes of industry. Addison Wesley Publishing, California, USA.

Ngah W.S. Wan , M.A.K.M. Hanafiah. (2008) Removal of heavy metal ions from wastewater by chemically modified plant wastes as adsorbents: A review, *Bioresource Technology* **99**, 3935–3948.

Ng C., J.N. Losso,W.E. Marshall, R.M. Rao (2000) Physical and chemical properties of selected agricultural by products based activated carbons and their ability to adsorb geosim, *Bioresour. Technol.* **84**, 177–185.

Nomanbhay S.M. , K. Palanisamy (2005) Removal of heavy metal from industrial wastewater using chitosan coated oil palm shell charcoal, Electron. *J. Biotechnol.* **8** (1) 43–53.

Onyango, M.S., Kojima, Y., Aoyi, O., Bernardo, E.C., Matsuda, H.(2004). Adsorption equilibrium modeling and solution chemistry dependence of fluoride removal from water by trivalentcation- exchanged zeolite F-9. *J. Colloid Interface Sci.* **279** (2), 341–350.

Oren AH, Kaya A. (2006) Factors affecting adsorption characteristics of Zn^{2+} on two natural zeolites. *J Hazard Mater.***17**;131(1-3):59-65.

Ozacar, M., Sengil, A. (2002) Adsorption of Acid dyes from aqueous solution by calcined alunite and granular activated carbon, *Adsorption*, **8,** 301-308.

Ozcan A, Omeroğlu C, Erdoğan Y, Ozcan AS (2006) Modification of bentonite with a cationic surfactant: An adsorption study of textile dye Reactive Blue 19. *J. Hazard. Mater.* **140** (1-2):173-9.

Ozcan AS, Gök O, Ozcan A.(2009) Adsorption of lead(II) ions onto 8-hydroxy quinoline-immobilized bentonite. *J. Hazard. Mater.***161**(1),499-509

Pehlivan E. , B.H. Yanık, G. Ahmetli, M. Pehlivan. (2008) Equilibrium isotherm studies for the uptake of cadmium and lead ions onto sugar beet pulp.*Bioresource Technology* **99**, 3520–3527.

Periasamy and Namasivayam (1996) Periasamy K, Namasivayam C. Removal of copper (II) by adsorption onto peanut hull carbon from water and copper plating industry wastewater. *Chemosphere*; **32**(4):769– 89.

Peric′ J, Trgo M, Medvidovic NV. (2004) Removal of zinc, copper, and lead by natural zeolite—a comparison of adsorption isotherms. *Water Res* 2004; **38** (7):1893–9.

Philip, C.A., Girgis, B.S. (1996). Adsorption characteristics of micro porous carbon from apricot stones activated by phosphoric acid. *J. Chem. Tech. Biotechnol.* **67**, 248–254.

Płaza G, ŁukasikW, Ulfig K. (1996) Sorption of cadmium by filamentous soil fungi. *Acta Microbiol. Pol.*; **45**:193–201.

Rajeshwarisivaraj; Subburam V. (2002) Activated parthenium carbon as an adsorbent for the removal of dyes and heavy metal ions from aqueous solution. *Bioresour. Technol. Nov.* ; **85** (2):205-6.

Rao M. Madhava , D.H.K. Kumar Reddy, Padala Venkateswarlu, K. Seshaiah (2008) Removal of mercury from aqueous solutions using activated carbon prepared from agricultural by-product/waste. *Journal of Environmental Management*,**1-10.**

Raghuvanshi S. P., singh R., C. P. Kaushik (2004) Kinetics study of methylene blue dye bioadsorption on baggase. *Applied ecology and environmental research* **2**(2): 35–43.

Rashed M.N. (2001) Lead removal from contaminated water using mineral adsorbents. *The Environmentalist,* **21**, 187-195.

Rashed M.N. (2006). Fruit stones from industrial waste for the removal of lead ions from polluted water. *Environmental monitoring and Assessment*, **119**, 31-41.

Reddad Z, Gerenete C, Andres Y, Le Cloeiric P. (2003) Mechanisms of Cr (III) and Cr (VI) removal from aqueous solutions by sugar beet pulp. *Environ. Technol* .2003; 24:257–64.

Roostari N., F.H. Tezel (2004) Removal of phenol from aqueous solution by adsorption, *J. Environ. Manage.* **70**, 157–164.

Rubel, J.F. (1983). In: Shupe, J.L., Peterson, H.P., Leone, N.C. (Eds.), The Removal of Excess Fluoride from Drinking Water by the Activated Alumina Method. Paragon Press, Salt Lake City, p. 345.

Rule, K.L., Comber, S.D.W., Ross, D., Thornton, A., Makropoulos, C.K., Rautiu, R. (2006). Diffuse sources of heavy metals entering an urban wastewater catchment. *Chemosphere* **63** (1), 64–72..

Sag Y. (2001) Biosorption of heavy metals by fungal biomass and modeling of fungal biosorption: a review, *Sep. Purif. Methods* **30** (1) 1–48.

Saianoa F, Maurizio Ciofaloa, Santa Olga Cacciolab, Stefania Ramirez. (2005) Metal ion adsorption by Phomopsis sp. biomaterial in laboratory experiments and real wastewater treatments. *Water Research* **39**, 2273–2280.

Shankar Congeevaram, Sridevi Dhanarani a, Joonhong Park b, Michael Dexilin a, Kaliannan Thamaraiselvi (2007) .Biosorption of chromium and nickel by heavy metal resistant fungal and bacterial isolates. *Journal of Hazardous Materials* **146**, 270–277.

S′ılvia C.R. Santos, V′ıtor J.P. Vilar, Rui A.R. Boaventura (2008) Waste metal hydroxide sludge as adsorbent for a reactive dye. *Journal of Hazardous Materials* **153**, 999–1008.

Soleimani Mansooreh , Tahereh Kaghazchi. (2008) Adsorption of gold ions from industrial wastewater using activated carbon derived from hard shell of apricot stones – An agricultural waste. *Bioresource Technology*, **99**(13) 5374-5383.

Tsai Wen-Tien , Hsin-Chieh Hsub, Ting-Yi Su b, Keng-Yu Lin b, Chien-Ming Lin (2007) Removal of basic dye (methylene blue) from wastewaters utilizing beer brewery waste. *Journal of Hazardous Materials* **154** (1-3)73-78.

Vilar VJ; Botelho CM; Boaventura RA. (2007) Methylene blue adsorption by algal biomass based materials: biosorbents characterization and process behaviour. *J. Hazard. Mater;* **147**(1-2):120-132.

Vilar V. J.P., Jos′e M. Loureiro, Cid′alia M.S. Botelho, Rui A.R. Boaventur.(2008) Continuous biosorption of Pb/Cu and Pb/Cd in fixed-bed column using algae *Gelidium* and granulated agar extraction algal waste. *Journal of Hazardous Materials* **154**, 1173–1182.

Vinod Kumar Guptaa, Imran Ali, Vipin Kumar Saini (2007) Defluoridation of wastewaters using waste carbon slurry. *Water Research* **41**, 3307 – 3316.

Vinod K. Guptaa, C.K. Jainb, Imran Alib, M. Sharmaa, V.K. Saini (2003) Removal of cadmium and nickel from wastewater using bagasse fly ash—a sugar industry waste *Water Research* **37**, 4038–4044 .

Viraraghavan T, Dronamrajum (1993). Removal of copper, nickel and zinc from wastewater by adsorption using peat. *J. Environ. Sci. Health Part A* **28**: 1261.

Voltolina, D., Cordero, B., Nieves, M., Soto, L.P. (1999). Growth of Scenedesmus sp. in artificial wastewater. *Bioresource Technol.* **68**, 265–268.

Volzone Cristina, Liliana Beatriz Garrido (2008) Use of modified hydroxy-aluminum bentonites for chromium (III) removal from solutions. *Journal of Environmental Management* **88** (4)1640-1648.

Waleska E. Oliveira , Adriana S. Franca, Leandro S. Oliveira, Sonia D. Rocha.(2008) Untreated coffee husks as biosorbents for the removal of heavy metals from aqueous solutions. *Journal of Hazardous Materials* **152** (2008) 1073–1081.

Wang, Y., Reardon, E.J.(2001) Activation and regeneration of a soil sorbent for defluoridation of drinking water. *Appl. Geochem.* **16**, 531–539.

Waranusantigul P; Pokethitiyook P; Kruatrachue M; Upatham ES. (2003) Kinetics of basic dye (methylene blue) biosorption by giant duckweed (Spirodela polyrrhiza). *Environ. Pollut.*; **125** (3):385-92.

Yang, M., Hashimoto, T., Hoshi, N., Myoga, H. (1999). Fluoride removal in a fixed bed packed with granular calcite.*Water Res.* **33** (16), 3395–3402.

Xiang-Yang Gu , Jonathan W.C. Wong (2007) Degradation of inhibitory substances by heterotrophic microorganisms during bioleaching of heavy metals from anaerobically digested sewage sludge. *Chemosphere* **69**, 311–318.

Yavuz O, Altunkaynak Y, Guzel F. (2003) Removal of copper, nickel, cobalt, and manganese from aqueous solution by kaolinite. *Water Res.***37**:948– 52.

Yapar S., M. Yilmar (2004) Removal of phenol by using montmorillonite, clinoptilolite and hydrotalcite, *Adsorption* **10**, 287–298.

YS Ok, Yang JE, Zhang YS, Kim SJ, Chung DY.(2007) Heavy metal adsorption by a formulated zeolite-Portland cement mixture. *J. Hazard. Mater*; **147** (1-2):91-6.

Zaitan Hicham, Daniel Bianchi, Ouafae Achak and Tarik Chafik (2008) A comparative study of the adsorption and desorption of *o*-xylene onto bentonite clay and alumina . *Journal of Hazardous Materials*, **153** (1-2)852-859.

Zheng Wei , Xiao-ming Li, Fei Wang, Qi Yang, Pin Deng, Guang-ming Zenga (2008) Adsorption removal of cadmium and copper from aqueous solution by areca—A food waste . *Journal of Hazardous Material,* **157**(2-3) 490-495.

Zohra Bouberka , Khenifi Aicha, Sekrane Fatima, Bettahar Nourredine and Derriche Zoubir (2008) Adsorption of Direct Red 2 on bentonite modified by cetyltrimethylammonium bromide .*Chemical Engineering Journal*, **136**, 2-3, 295-305.

In: Industrial Waste...
Editor: John P. Samuelson, pp. 103-119
ISBN: 978-1-60692-720-5

Chapter 4

VALORIZATION OF AGRO-INDUSTRIAL WASTES BY BIOLOGICAL TREATMENT

Nikolay Vassilev[1,3,*]***, Iana Nikolaeva***[2]***, Monica Odlare***[2]***, Miguel Garcia Roman***[3]***, Mercedes Serrano***[3]***, Encarnación Jurado***[3] ***and Maria Vassileva***[3]

[1]Institute of Biotechnology, Faculty of Sciences, University of Granada, c/Fuentenueva s/n, Granada-18071, Spain
[2]Department of Public Technology, Malardalen University, Vasteras, Sweden
[3]Department of Chemical Engineering, University of Granada, Spain

Abstract

Agro-industry generates a wide number of waste products. As their disposal and landfill present serious environmental concerns, research studies are performed proposing value-adding technologies. This paper describes a number of biotechnological processes aimed at treatment and valorisation of wastes with high organic load derived from the industrial production of olive oil, sugar, fruit juices, etc. Employing solid-state fermentation technology, microorganisms are able to degrade solid wastes simultaneously, producing metabolites that can be further used. A special emphasis is given to the recent achievements leading to formulation of bio-control products, production of some important microbial metabolites, and biological organic-matter–based fertilizers. Concomitant solubilization of low-grade rock phosphates (introduced into solid waste microorganism systems), which are considered waste materials from the mining industry, is also discussed. The review deals with the possibility of creating a sustainable practice in industries processing agricultural products.

1. Introduction

Agriculture is an important area that can benefit from microbial biotechnology. We are able to feed far more people using far less land thanks to the use of fertilizers and pesticides.

[*] E-mail address: nbvass@yahoo.com. Tlf: + 34647462581; Department of Chemical Engineering, Faculty of Sciences, University of Granada, c/Fuentenueva s/n, Granada-18071, Spain; Corresponding author.

However, a lot of farming systems manifest signs of deterioration and lack of sustainability, mainly due to intensive applications of chemical products such as fertilizers and pesticides, and numerous environmental problems exist for both chemical products. Agricultural runoff has led to eutrophication of bodies of water, while the use of broad-spectrum pesticides may harm beneficial organisms along with target pests, thus decreasing soil biodiversity. On the other hand, disposing of untreated agro-industrial wastes directly in landfills for improving soil physical/chemical properties and plant growth, or their incineration, present another unsustainable practice. The negative effects of such lack of environmental sustenance has resulted in the search for alternative management schemes that are capable of maintaining agro-systems with high productivity and, simultaneously, ensuring sustainability. Accordingly, the main tasks of sustainable agriculture are to minimize the existing adverse impacts of the chemical farming practice by implementing soil fertility restoring alternatives. It is now well known that agronomists are looking for biological approaches that can solve at least partly the problems of conventional agriculture.

Green chemistry and biotechnology are accepted as important tools in achieving sustainability. Their implementation, and the design of chemical products and microbial processes that reduce or eliminate the use and generation of hazardous substances and abundant waste materials, is essential if the expanding global population is to enjoy an increased standard of living without having a negative impact on the health of the planet (Kirchhoff, 2005). The complete recycling of agro-industrial wastes is now accepted as an important part and essential element of sustainable agriculture and agro-industry (Tengerby and Szakacs, 1998; Saddler, 1993). Microbially-treated agricultural residues may restore soil fertility and microflora. Alternatively, excess agro-industrial wastes may be converted to valuable feed supplements and bio-industrial products. In recent years, there has been an increasing trend towards more efficient utilization of agro-industrial residues based on processes that employed solid-state fermentation. This review will focus on research work employing solid state fermentation in solubilization of insoluble phosphates and mass-production of biocontrol agents

2. Solid State Fermentation Processes: Definition

Solid-state fermentation (SSF) is defined as a fermentation process in which microorganisms grow on solid materials without the presence of free water (Cannel and Moo-Young, 1980). We should distinguish between solid-state and solid-substrate fermentation, although many researchers consider these two processes essentially one and the same. In fact, Pandey et al. (2001a) defined solid-substrate fermentation as a process in which the substrate itself acts as a carbon (and energy) source, occurring in the absence or near-absence of free water. Solid-state fermentation was defined by the same authors as any fermentation process occurring in the absence or near-absence of free water, employing a natural substrate or an inert substrate used as solid support. We would prefer the definition according to which SSF can be understood as any process in which substrates in a solid particulate state are utilized (Mitchell et al., 2000).

As Holker et al. (2004) have recently mentioned that free water does not appear to be the natural environment for the great part of microorganisms. Even in the marine environment, 98% of the microorganisms have been found in/on solid surfaces (Kelecom, 2002). Therefore, the cultivation of microorganisms freely suspended in nutritive liquids may rather impair their development and metabolic efficiency. This statement is not surprising, bearing in mind that immobilized (attached or entrapped in polymers) cell systems are often reported to demonstrate higher stability and catalytic longevity compared to the same cultures in free state (Vassilev and Vassileva, 1992; Vassilev et al., 2001, 2005a). It should be mentioned, however, that many microorganisms are well adapted to submerged liquid fermentations (in some cases by genetic engineering) and very efficiently produce a wide range of bioproducts on an industrial scale. Nevertheless, SSF is the cultivation method of choice in cases in which the biotechnological process is required to include morphological and metabolic differentiation into substrate-penetrating and aerial hyphae (such as the production of conidiospores). When compared with the SSF processes that have long been established in Asian countries, SSF technology in western countries appears to still be in its infancy. However, today, SSF is given more emphasis on the development of wide range of bioprocesses which include bioremediation, biodegradation of hazardous wastes, biodetoxification of agro-industrial wastes, biotransformation of crops and crop residues for nutritional enrichment, large-scale production of secondary metabolites, enzymes, antibiotics, organic acids, biopesticides, biosurfactants, biofuels and aromatic compounds (Holker and Lenz, 2005; Holker et al., 2004; Mitchell et al., 2002; Pandey, 2003; Pandey et al., 1999; Tengerby and Szakacs, 2003). During the period 1991–2000, more than 1,000 publications have appeared in various journals, proceedings and books.

Using SSF technology, we can bring the cultivated microorganisms into tight contact with agro-industrial wastes and thus achieve the highest substrate concentrations for fermentation. It is important to note that this type of biotreatment of waste materials requires smaller fermenters with subsequent less effort for downstream operations, less energy and cost requirements. Another environment-protective advantage of SSF is that the growth of contaminating microorganisms is minimized and, therefore, sterilization processes can be eliminated. Among the important features of SSF is the possibility of coexistence of more than one microorganism in commensal or symbiotic associations (Rayner and Boddy, 1988). Therefore, the overall environmental impact of biotreatment of agro-industrial wastes can be considered in several inter-related directions.

In nature, there are many microorganisms, mainly bacteria and fungi, that live on natural, solid supports that in the majority of cases are wastes. Similarly, the major microbial groups used in solid-state fermentations are fungi and bacteria. However, it is well established that fungi are the most important and the best adapted for SSF because of their mode of growth and high tolerance to low water activity (Himmel et al., 1999).

All solid substrates that include starch, cellulose, lignocellulose, pectin and other polysaccharides in their structure can be used in the solid-state fermentations as substrates. Products from agriculture and agro-industry provide excellent solid matrices for SSF. The most commonly-used type of solid-state cultivation includes natural solid wastes or by-products directly or after pre-treatment.

In this chapter, we will review studies on biotreatment of agro-industrial wastes that use solid-state fermentation processes. These wastes are derived from agro-industrial activities such as the production of sugar, fruit juices, various oils, wine, grains, coffee, etc.

Bioreactors commonly used in SSF are flat trays, packed-bed bioreactors, drum bioreactors, and fluidized-bed bioreactors. Their advantages and disadvantages were recently discussed in detail (Holker and Lenz, 2005; Couto and Sanroman, 2006). It is now believed that the adaptation of the koji bioreactor—widely used in Southwest Asia on a large scale—to various industrial applications could facilitate and stimulate the use of SSF in the western world (Nagel, 2002).

Here, we will present two main fields of research and application: solubilization of insoluble phosphates, and processes aimed at the production of biocontrol agents using agro-industrial wastes.

3. Solubilization of Insoluble Inorganic Phosphate Sources in SSF and Further Use of the Resulting Products

In general, current management of phosphate-bearing resources, which in fact are finite non-renewable sources, is not fully in accord with the principles of sustainability. All this may lead to a progressive increase in prices mainly because of increases in processing costs. In order to reduce the depletion of global phosphorus reserves, research should be oriented towards more effective utilization of phosphates. For these reasons, solubilization of insoluble inorganic phosphates has been actively studied but also because of its environmental and significance (Khan et al., 2007 and references therein). A biological approach for extracting phosphate from rock phosphate (RP) was proposed as a less expensive and lower-energy technique compared with the conventional processes (Goldstein and Rogers 1999). Alternatively, direct application of RP in soils would minimize pollution and decrease the costs of chemical treatment (Iyamuremye and Dick 1996). The latter approach is, however, not always effective without previous treatment. Various techniques for RP solubilization have been proposed, with increasing emphasis on application of P-solubilizing microorganisms (Rodriguez and Fraga, 1999; Whitelaw 2000; Vassilev et al., 2001; Vassilev and Vassileva, 2003) employing low-grade RP which is often considered as a waste material from the mining industry.

A number of in vitro studies have shown that bacteria, fungi and actinomycetes are able to liberate phosphate ions from sparingly soluble inorganic P-bearing compounds. The P-solubilizing activity is determined by the microbial biochemical ability to produce and release metabolites such as organic acids that, through their hydroxyl and carboxyl groups, chelate the cations (mainly calcium) bound to phosphate, the latter being converted into soluble forms. However, metabolizable C compounds must be applied as energy source to the microbial solubilizers to ensure their growth, organic acid production, and, simultaneously, RP solubilization.

Approaches to increasing the agronomic value of low-grade RP using the advantages of SSF can be split into two categories: the application of agro-wastes in SSF where both organic acid production and RP solubilization occur simultaneously, and, rarely, use of waste materials as a support for P-solubilizing microorganisms for their further agronomic use. A typical example of the latter approach has been described in the early work of Kucey (1987), in which wheat straw amended with glucose was preincubated with *Penicillium bilaji*. Similarly, a biofertilizer containing *P. bilaji* was proposed where peat was the carrier also for the nitrogen-fixing *Rhizobium meliloti* (Rice et al. 1994). After incubation at 30°C for 96 h, the mature peat was found to meet the requirements of a common delivery system for both microorganisms, containing 2.3×10^8 *Penicillium* cfu/g and 6.3×10^9 *Rhizobium* cfu/g.

SSF is accepted as a promising technology for production of organic acids on agro-industrial wastes (Pandey et al., 2000). Organic acids are excellent P-solubilizing agents and, therefore, biotreatment of low-grade RP by SSF can be accepted as an attractive approach within microbially-mediated solubilization. The use of SSF technology for RP fungal solubilization using agro-industrial wastes has been for the first time proposed by Vassilev et al. (1994) and further investigated with *A. niger* grown on wastes derived from the sugar-producing industry from sugar beet (Vassilev et al. 1995). Sugar beet waste material (SB) was selected as it provided acceptable mycelial growth, organic acid production and a level of mineralization between 56% and 69% depending on the substrate concentration. This system was able to solubilize 76% of the RP supplemented to the medium at concentration of 15 g/kg SB. The presence of RP affected the behaviour of *A. niger*, particularly its growth and citric acid production.

A similar biotechnological scheme has been applied using olive cake (OC)-based and dry olive cake (DOC)-based media and *A. niger* (OC) and *Phanerochaete chrysosporium* (DOC) (Vassileva et al. 1998; Vassilev et al. 2002; Vassilev et al., 2006b). Depending on the olive oil industrial technology used, several types of wastes can be produced in different ratios and physico-chemical characteristics, such as wastewaters, OC, and DOC. The degree of OC and DOC mineralization, mycelial growth, and organic acid production was lower in comparison to the SB-based process, which resulted in lower solubilization of the total P present in the RP. Obviously, the insoluble and crystalline nature of cellulose, associated with a higher content of lignin and the low content of oxidizable carbon in the OC/DOC experiments decreased the overall process efficiency. In addition, the presence of polyphenolic compounds in olive mill wastes could affect the physiological and biochemical characteristics of the fungal P-solubilizers.

Agro-industrial wastes have been used as amendment to increase soil organic matter content and improve plant growth (Huang and Huang, 1993). Organic soil amendment may induce chemical and physical changes that also affect soil microflora. Introduction of the resulting bio-treated products, containing mineralized organic matter, plant-available soluble phosphate and fungal mycelium, into plant-soil systems in greenhouse and field conditions resulted in a significant plant growth enhancement (Vassilev et al., 1996; Vassileva et al., 1998; Vassilev et al., 2006b). In addition, sugar beet wastes, OC and DOC, processed under the conditions of the above described scheme, increased soil β-glucosidase and phosphatase enzyme activities (Medina et al., 2004). Enhancement of soil aggregate stability, water soluble C, and water-soluble carbohydrates was also registered in typical, stressed Mediterranean soils. It is important to mention that bio-treated agro-industrial wastes with simultaneously solubilized RP have successfully been used in aforestation of desertified areas and restoration of metal-contaminated soils (Caravaca et al., 2004; Medina et al., 2005).

4. Application of Agro-industrial Wastes in Phytase Production

Apart of the inorganic fraction of phosphate, which is formed by various reactions with soil components, a great part (50-80 %) of total phosphate in soil exists as organic phosphate compounds unavailable to plants. This organic fraction, 50 % of which is phytate, should be mineralized by organic phosphate mineralisers like phytases. On the other hand, phytases are considered of significant importance in supplementation the nutritional value of feed that is rich in phytate and due to their multifunctional properties, it is now a common practice to include microbial phytases in animal (mainly monogastric animals) feed and human nutrition in order to enhance nutrient utilization (Feil, 2001; Selle and Ravindran, 2007). Although phytases can be found in various plants and animals, microbial sources are more promising for the production of phytases on a commercial level (Pandey et al., 2001b; Vats and Banerjee, 2004).

Techniques using solid-state fermentation and submerged fermentation have been employed for microbial phytases production. However, there is considerable greater interest for production of this kind of enzymes by SSF as the results, obtained in laboratory conditions have shown increased enzyme activity (Gowthman et al., 2001). Filamentous fungi, mainly *Aspergillus* spp., are widely employed in phytase production by SSF technique although other microorganisms such as *Rhizopus, Mucor* and *Peniophora* spp. have been also used.

The production of phytase from *A. ficuum* (AF) NRRL 3135 has been achieved by three different cultivation methods, namely, solid state (Elbune et al., 1995), semi-solid (Han et al., 1997) and submerged fermentations (Vats and Banerjee, 2004). However, Nair and Duvnjak (1990) reported a complete reduction of phytic acid content in canola meal using SSF by *A. niger* NRRL 3135. Ebune et al. (1995) studied the phytase production by *A. ficuum* using canola wastes and found that the age of inoculum had profound effect on enzyme synthesis. Krishna and Nokes (2001) studied the effect of culture conditions, particularly inoculum age, media composition (wheat bran and full-fat soybean flour) and duration of SSF on the phytase production by *A. niger.* Bogar et al. (2003) reported phytase production by *A. ficuum* NRRL3135, *Mucor racemosus* NRRL1994 and *Rhizopus oligosporus* NRRL 5905 on canola meal, cracked corn, soybean meal and wheat bran in SSF. Optimisation studies were carried out using various types of experiment-optimization designs, showing the improved production when wheat bran was supplemented with starch and ammonium sulphate.

During the last ten years, biotechnological applications of residues derived from the production of different kind of oils have been increased (Ramachandran et al., 2007). Particularly oil cakes, obtained after extraction of oil from the seeds, were experimented in conditions of SSF for the production of phytase. For example, the fermentation conditions of extracellular phytase production in SSF using coconut oil cake as substrate with no additional nutrients by *R. oligosporus* were recently reported in literature (Sabu et al., 2002). Maximal enzyme production (14.29 U per g of dry substrate) was registered at pH 5.3, 30°C, and 54.5% moisture content after 96 h of incubation. The addition of other nutrients to the substrate resulted in inhibition of product formation. In another study, Ramachandran et al. (2005) confirmed the excellent properties of coconut oil cake as a substrate employing three strains of *Rhizopus* spp. while application of groundnut oil cake and palm kernel cake in the medium resulted in comparable phytase titres with one filamentous fungus, *R. oryzae* NRRL 1891. However, mixed substrate fermentation using coconut cake and sesame oil cake (1:1, w/w) further enhanced enzyme production which reached 64 U per g dry substrate when 1% glucose was supplemented to the fermentation medium that was a two-fold increase in comparison to the yield by individual substrate. Similarly, Roopesh et al. (2006) used mixed sesame oil cake with wheat bran to produce phytase with *Mucor racemosus* NRRL 1994. Optimizing the most important parameters of the SSF (incubation time, initial moisture content, and inoculum concentration) the production of phytase reached 44.5 U per g dry substrate which was 1.5-fold higher than the highest yield obtained with any individual substrate. These fermented agro-industrial wastes could be directly used as animal feed enriched with phytase.

Dry olive waste (DOW), an agro-industrial material derived from the two-phase-manufacture of olive oil, was used as a substrate for phytase production by *Aspergillus niger* (Vassilev et al., 2007a). A series of experiments were performed in conditions of solid-state fermentation in the presence of rock phosphate. Both enzyme production and phosphate solubilization depended on water medium content, type of nitrogen source, inoculum size, and the presence and initial concentration of phosphate in the medium. It was found that at optimized process conditions (moisture 70%; corn steep liquor as a nitrogen source; inoculum size of 3-4 ml; presence of slow release phosphate), the filamentous fungal culture was able to produce 58 U phytase/g dry substrate and 31 mg soluble phosphate per flask. Further research should be carried out to evaluate the final bio-product in soil-plant systems.

5. Application of Solid-State-Processed Agro-industrial Wastes in Production of Biological Control Agents

Until recently, the major disinfestant used has been the chemical fumigant methyl bromide, which is now being phased out of use due to its negative environmental effects (Ristiano and Thomas, 1997). Steaming soil by portable steam devices is already widely used in glasshouses and similar techniques are developed for use in field situations as well (White et al., 2000). Although soil sterilisation targets deleterious microorganisms, other beneficial microorganisms are also destroyed in the process, resulting in a biological disturbance being created in the treated soil. This process negatively affects the ecology and, in turn, allows other microorganisms (including pathogens) to recolonise the substrate easily, as there is a high nutrient availability and little competition. It is believed that the solution of this problem coiuld be to directly introduce into the sterile soil or substrate (in case of ornamental and vegetable crops) biocontrol agents or plant beneficial microorganisms.

Microbially-mediated treatment of agro-industrial wastes by SSF is the process of choice for mass production of biocontrol agents. Using this biotechnological approach, the solid wastes materials serve as carriers for the biocontrol agents and, therefore, it is not necessary to develop a sophisticated formulation process. Such type of formulation is adapted for horticultural usage where the final SSF-produced biomaterial is mixed with potting mixture. For example, following this scheme, *Fusarium oxyporum* Fo47 has been produced in SSF using either sterilized peat or calcinated clay as the solid support. The resulting product was stored at room temperature without loss of activity (Olivian et al., 1999).

More recently, Vassilev et al. (2005b; 2008a) reported an efficient SSF-based process for preparing a material with biocontrol and plant growth promoting functions using sugar beet wastes and wastes from the olive oil producing industry. These waste materials were selected as they constitute a major environmental problem especially for Mediterranean countries.

Sugar beet wastes were mineralized by an acid-producing strain of *A. niger* with a simultaneous solubilization of rock phosphate in conditions of solid-state fermentation process (Vassilev et al., 2005b). Soil amendment with the resulting material resulted in 347% and 467% (vs. un-amended control) higher plant biomass in plant-soil experiments contaminated or not with *F. oxisporum* f.sp. *licopersici*, respectively. Disease severity and number of *F. oxysporum* CFU reached the lowest levels, particularly when plants were mycorrhized with *Glomus deserticola*. In vitro studies proved the biocontrol activity of *A. niger* which was partly attributed to siderophore production on modified chrome azurol S (CAS) medium.

When describing valorization of agro-industrial wastes for their further application in plant-soil systems contaminated with phytopathogens, special attention deserve olive mill wastes. Olive oil mill residues vary depending on the technology (2-phase or 3-phase) used. In general, disposal on agricultural land by field spreading is the best of all solutions so far proposed to solve the problem of whichever olive mill waste disposal (Federici, 2006). Field spreading particularly of olive mill wastewaters is now allowed by the Italian legislation while simply accepted in Spain. It is widely accepted that these wastes have a high fertilizer potential. However, there is serious concern about their introduction into soil because of their proved antimicrobial and phytotoxic properties. In a recent study, Bonanomi et al. (2006) investigated the effects of dry olive mill wastes (DOW) derived from 2-stage olive oil production technology on growth of the four crop species and, simultaneously on saprophytic growth of the four phytopathogenic fungi. Depending on their concentrations, residues resulted phytotoxic, both in laboratory and greenhouse bioassays, for all plant species. In contrast with the results observed with plant species, the radial growth and hyphae density of the tested phytopathogenic fungi were positively affected by DOW which caused high *Lactuca sativa* mortality in greenhouse bioassays and foliar disease dramatically increased on *L. sativa* and *Lycopersicon esculentum* plants grown on soil amended with DOR at all used concentrations. Therefore, untreated DOW demonstrated well pronounced phytotoxic effect on the growth of plants concomitantly stimulating growth of phytopathogenic fungi in DOW-amended soil thus increasing fungal disease incidence. On the other hand, one of the main disadvantages of applying olive mill wastes in general is their low content of plant available phosphate.

Bering the above in mind, in our studies we proved the positive effect of microbial treatment of olive mill wastes in on soil-plant system infected by a phytopathogenic fungus (Vassilev et al., 2008a). A microcosm studies was carried out to determine the effect of *Aspergillus niger*-treated mixture of two agro-industrial waste materials, dry olive wastes and sugar beet press mud, on tomato (*Lycopersicon esculentum*) plants grown in a soil inoculated with *Fusarium oxysporum* f. sp. *lycopersici* (*Fol*). Agro-wastes were treated in conditions of solid-state fermentation in the presence of Morocco apatite (RP) and further applied at a rate of 50 g/kg soil. Soil-plant systems were additionally inoculated or not with the arbuscular mycorrhizal (AM) fungus *Glomus intraradices*. Plant growth and nutrition, symbiotic developments and soil enzymatic activities were stimulated in *Fol*-free soil supplemented with treated agro-wastes and significantly greater in treatments where AM fungus was introduced compared with the non-amended control. The introduction of *Fol* into the soil-plant system adversely influenced all studied parameters. AM fungus alone reduced the effect of the plant pathogen on tomato plants and significantly decreased (by 2.3×10^3) the number of *Fol*-CFU compared with the non-mycorrhizal control grown in non-amended soil. Higher levels of pathogen control were achieved in *A. niger*/Agro-Industrial-Wastes/RP-amended mycorrhized plant-soil system. In separate, *in vitro* studies, *A. niger* demonstrated a strong suppressive effect on *Fol* with a 5-fold reduction of the pathogen colony diameter compared to the control.

Similar biotechnological scheme was employed in DOW treatment by the white-rot fungus *Phanerochaete chrysosporium* in conditions of SSF where the basal medium was enriched with rock phosphate (Vassilev et al., 2007b). The results of this study showed that DOW-based medium provided necessary nutrients for manganese peroxidase production by fungal culture of *P. chrysosporium* in conditions of SSF. Optimizing several process parameters such as solid:liquid ratio, temperature and pH resulted in enhanced enzyme production which reached 1100 U/l in the presence of insoluble inorganic phosphate source (rock phosphate). Solubilization of RP added to the medium was achieved which was probably due to organic acid release by the fungus. It is well established that various carboxylic acids are produced by white-rot and brown-rot fungi which serve as organic chelators thus stabilizing the formation of Mn^{3+} (Hakala, 2005). The secretion of organic (mainly oxalic) acids is probably in equilibrium related to the production of H_2O_2. Milagres et al. (2002) demonstrated that acidic compounds present in fermentation broth of several of these fungi showed metal-chelating ability. In a separate (CAS) experiment, we confirmed this activity of *P. chrysosporium* which most likely stimulated RP dissolution. SSF produced viable cultures of *P. chrysosporium* with efficient biocontrol activity which was capable of suppressing growth of *Fusarium oxisporum* in experiments on solid agar media where at the end of the studied period the white-rot fungus invaded the colony of the pathogenic fungus.

In another study, a chitinase-producing filamentous fungus, *Penicillium janthinellum*, was tested for its ability to grow and produce chitinolytic metabolites in conditions of SSF on DOW-based medium (Vassilev et al., 2008b). Chitinolytic activity was significantly improved in media supplemented with chitin from crab shell although chitinases are known to be produced constitutively. The maximum chitinase yield reached 1612.9 U per kg initial dry matter in the presence of insoluble inorganic phosphate source. Solubilization of RP added to the medium was achieved which was probably due to proton release by *P. janthinellum*. Fresh preparations of SSF-grown *P. janthinellum* were efficient antifungal agents as proved in a microcosm soil-plant experiment inoculated with the phytopathogen fungus *Fusarium oxisporum*. Microbial chitinases attracted the attention as one of the potential agents for control of phytopathogenic fungi and insect pests (Stleger et al. 1986; Roco and Perez 2001). Production of lytic enzymes can be stimulated by the presence of organic substances rich in ligno-cellulose used in solid-state fermentations. It was recently reported that soil filamentous fungi (*Aspergillus and Penicillium*) were capable of producing chitinases in solid-state fermentation conditions using mouldy bran (Patidar et al. 2005) or corncob/shrimp shellfish wastes (Rattanakit et al. 2002) as substrates.

In the last two studies based on dry olive wastes, *P. chrysosporium* and *P. janthinellum* were entrapped in alginate carriers in order to compare their biocontrol effectiveness with that of final products obtained in conditions of SSF. The results showed that both approaches are equally effective but further studies are necessary to test the usefulness of these products after storage to determine their final economic parameters.

For several decades, formulations containing *Bacillus thuringiensis* have been accepted as the most efficient biocontrol material used in agriculture as this microorganism is environmentally safe and could be mass-produced at low cost employing agro-industrial wastes. Production of *Bacillus thuringiensis* has been reported in conditions of solid state fermentation using agro-industrial wastes such as wheat bran, rice bran, rice husks, soybean powder, fish wastes, molasses and protein hydrolysates (Morris et al., 1996; Suyanandana et al., 1996; Capalbo and Morales, 1997; Morales et al., 1998; Fernendez-Larrea Vega, 1999).

Broiler litter has also been reported as a particularly attractive substrate for solid state production of *B. thuringiensis*. This biologically produced material, which contains crude protein, stimulates microbial growth (Summers and Leeson, 1985) although some previous treatment is needed. Introduction into soil of methanol extracted litter fermented with *B. thuringiensis* resulted in over 90% mortality in three weeks for first instars of Japanese beetles when compared to a control treatment using compost without fermented litter (Adams et al. (2002).

Agro-industrial wastes used as substrates in SSF are accepted as the best biotechnological method for production of high quality spores of biocontrol microorganisms (Holker et al., 2004). In some cases, SSF is carried out using substrates previously processed in conditions of submerged fermentation (Tengerdy and Szakacs, 2003). The resulting spore-containing material is used for seed coating or, alternatively, co-encapsulated with micropropagated plantlets in natural polysaccharide gels, preferably calcium alginate. Another scheme, which includes further introduction of SSF-produced biocontrol monocultures at a rate of about 10^9 CFU/g into a compost pile of steam-treated agro-industrial residues, resulted in the formation of directed compost products.

Spores of *Penicillium oxalicum* grown in SSF were more effective than those from submerged cultivation in reducing rhizosphere population and disease symptoms caused by *Fol* (Pascual et al., 2000). Spores obtained by SSF have higher surface hydrophobicity which affect the level of adhesion to plant roots and their survival was similar at 4° or 25° C thus avoiding the need of refrigeration. On the other hand, conidia of *Trichoderma harzianum* produced on solid medium have been found to persist longer under harsher environmental conditions that those produced in conditions of submerged process (Muñoz et al., 1995). Wall thickness of conidia obtained by this technique was twice that of submerged formed conidia. Using SSF-based technology, Cliquet and Scheffer (1997) obtained spores of *Trichoderma* spp. which were used in industrial film coating of seeds.

Microorganisms with biocontrol activity can be employed in SSF for production of antimicrobial metabolites. The effectiveness of SSF for iturin production using agro-industrial wastes (Ohno, 1996; Makkar and Comeotra, 1999). Iturin production in SSF was found to be up to 20 times higher in a 3-kg SSF bioreactor as compared with a 5-l submerged agitated bioreactor using synthetic media. Bacterially produced anti-fungal peptide antibiotic iturin is a potent biocontrol agent which has been proved in field conditions against plant pathogens (Balakrishnan and Pandey, 1996).

Solid-state fermentation has been proved as an excellent tool for cultivation of *Trichoderma viride* using wastes from the wine industry (Bai et al., 2008). *Trichoderma* species are widely used in biocontrol of plant pathogens and, in the same time, recognized as plant growth promoters. While the majority of studies are stressed on *Trichoderma* spore production few reports focused on the production of extracellular enzymes which, in fact, play important role in biocontrol mechanisms. *T. viride* was reported to produce chitinase and β-glucanase which can hydrolyze plant pathogen cell wall, and pectinase (Markovich and Kononova, 2003) which can elicit plant disease resistance. Mixture of grape marc, spent grape marc and wine lees can be suitable substrates for solid state fermentation employing *T. viride* which produced 47,8 U/g initial dry substrate (IDS), 8.32 U/g IDS, and 9.83 U/g IDS of chitinase, β-glucanase, and pectinase, respectively.

Biocontrol microorganisms cultivated on agro-industrial waste materials and further introduced into soil, present a number of advantages including soil enrichment with organic matter, and can be incorporated into soil maintenance routines. Greenhouse conditions offer good opportunities for the application of microorganisms with biocontrol activity, mostly due to their more uniform and controlled environment. Antagonists to be used in greenhouses should be selected in order to fit the set of conditions, such as pH, temperature and matrix potential, of the cultivation environment. Moreover, cultural practices and amendments inducing disease suppressiveness in soil are more effective in greenhouses than in the open field (Spadaro and Gullino, 2005). An increasing number of research projects are focusing on using plant residues and organic matter to increase the population of microorganisms antagonistic to plant pathogens. Organic amendments (including microbially-treated agro-industrial wastes) have long been recognized to facilitate biological control if applied before planting. Moreover, biologically transformed organic matter can serve as an ideal source of nutrient for antagonists into soils and offer an opportunity to introduce and establish specific biocontrol agents into soils, which in turn leads to a sustainable disease control based on the activities of microbial communities (Hoitink and Boehm, 1999).

6. Conclusion

Biological treatment of agro-industrial wastes by solid state fermentation is one of the oldest bioprocesses known and its advantages are now proved in many biotechnological applications. Solid state fermentations involve many natural microbiological processes including composting and ensiling which were consciously avoided in this mini-review. Critical analysis of the literature shows that SSF offers environmental friendly and sustainable alternatives in the transformation of wide variety of agro-industrial wastes into value-added products. The resulting final biomaterials can be applied in the sustainable agriculture in order to minimize use of chemicals. In support of the latter we should mentioned that the European Union scheduled a gradual 56-percent reduction of synthetic pesticides in the period 1991-2008 (Montesinos, 2003). Similar restrictions are now acting in the USA and Canada (Hynes and Boyetchko, 2006). While preparation of plant protection (biocontrol) agents, employ SSF as a rule, in the near future scaling-up of processes of RP solubilization, protein enrichment, and production of various microbial metabolites will be expected. In some cases, future research should be carried out on SSF processes for the preparation of microbial inoculants with complex action modes. As was recently reported, many microorganisms which are employed in SSF processes show simultaneous P-solubilizing, biocontrol, and plant growth promoting activity (Vassilev et al., 2006a; 2007). Bioconversion of agro-industrial wastes by SSF will have an important role in future biotechnologies focussed on agriculture, mainly because of its favourable economy, and ease of *on site* operation. However, the research focus in this field will be on searching for efficient engineering developments.

Acknowledgements

This work is supported by Project MEC CTM2005-06955 (CICYT), R&C Programme-NV (co-financied by the European Social Fund), and EC Project "Waste to Products".

References

Adams, T.T., Eiteman, M.A. & Hanel, B.M. (2002) Solid state fermentation of broiler litter for production of biocontrol agents. *Bioresource Technology*, **82**, 33-41.

Bai, Z, Jin, B., Li, Y., Chen, J. & Li, Z. (2008) Utilizatiion of winery wastes for Trichoderma viride biocontrol agent production by solid state fermentation. *Journal of Environmental Sciences,* **20**, 353-358.

Balakrishnan, K. & Pandey, A. (1996) Production of biologically active secondary metabolites in solid state fermentation. *Journal of Science and Industry Research*, **55**, 365-372.

Bogar, B., Szakacs, G., Linden, J.C., Pandey, A. & Tengerdy, R.P. (2003) Optimization of phytase production by solid substrate fermentation. *Journal of Industrial Microbiology and Biotechnology,* **30**, 183–189.

Bonanomi, G., Giorgi, V., Del Sorbo, G., Neri, D. & Scala, F. (2006) Olive mill residues affect saprophytic growth and disease incidence of foliar and soilborne plant fungal pathogens. *Agriculture, Ecosystems and Environment,* **115**, 194-200.

Cannel, E. & Moo-Young, M. (1980) Solid-state fermentation systems. *Process Biochemistry,* **4**, 2-7.

Capalbo, D.M.F. & Morales, I.O. (1997) Use of agro-industrial residues for bioinsecticidal endotoxin production by *Bacillus thuringiensis* var. israelensis or kurstaki in solid state fermentation. In: Roussos S (ed.) *Advances in Solid State Fermentation 2nd International Symposium on Solid State Fermentation FMS-95*. Montpellier, France, Norwell, MA.

Caravaca, F., Alguacil, M.M., Vassileva, M., Diaz, G. & Roldan, A. (2004) AM fungi inoculation and addition of microbially treated dry olive cake-enhanced afforestation of a desertified Mediterranean site. *Land Degradation and Development,* **15**, 153-161.

Cliquet, S. & Scheffer, R.J. (1997) Influence of culture conditions on growth and survival of conidia of *Trichoderma* spp. coated on seeds. *Biocontrol Science and Technology,* **7**, 171-181.

Couto, S. & Sanroman, A. (2006) Application of solid state fermentation to food industry. *Journal of Food Engineering,* **76**, 291-302.

Ebune, A., Al-Asheh, S. & Duvnjak Z. (1995) Production of phytase during solid-state fermentation using *Aspergillus ficuum* NRRL 3135 in canola meal. *Bioresource Technology,* **53**, 7–12.

Federici, F. (2006) Wastewaters from the olive-oil extraction process: disposal or valorization? *Pomologia Croatica,* **12**, 15-27.

Feil, B. (2001) Phytic acid. *Journal of New Seeds*, **3**, 1–35.

Fernendez-Larrea Vega, O. (1999) A review of *Bacillus thuringiensis* (Bt) production and use in Cuba. *Biocontrol News Information,* **20**, 47N-48N.

Goldstein, A.H. & Rogers, R.D. (1999) Biomediated continuous release phosphate fertilizer. US Patent 5,912,398.

Gowthaman, M.K., Krishna, C. & Moo-Young M. (2001) Fungal solid state fermentation-An overview. In: Applied Mycology and Biotechnology, vol 1 *Agriculture and Food Productions*. pp. 305-352. Khachatourians G.G., Arora D.K., Eds. Elsevier Science, The Netherlands.

Hakala, T.K. (2005). Manganese peroxidases, laccases and oxalic acid from the selective white-rot fungus *Physisporinus rivulosus* grown on spruce wood chips. *Enzyme and Microbial Technology,* **36**, 461-468.

Han, Y.W., Gallagher, D.J. & Wilfred, D.J. (1997) Phytase production by *Aspergillus ficuum* on semisolid substrate. *Journal of Industrial Microbiology,* **2**, 195–200.

Harman, G.E., Hayes, C.K., Lorito, M., Broadway, R.M., Di Pietro, A., Peterbaues, C. & Tronsmo, A. (1993) Chitinolytic enzymes of *Trichoderma harzianum*: purification of chitobiosidase and endochitinase. *Phytopathology,* **83**, 313-318.

Himmel, M.E., Ruth, M.F. & Wyman, C.E. (1999) Cellulase for commodity products from cellulosic biomass. *Current Opinion in Biotechnology,* **10**, 358-364.

Hoitink, H.A.J. & Boehm, M.J. (1999) Biocontrol within the context of soil microbial comunities: a substrate dependent phenomenon. *Annual Reviews in Phytopathology,* **37**, 427–446.

Holker U. & Lenz J. (2005) Solid-state fermentation – are there any biotechnological advantages. *Current Opinion in Microbiology,* **8**, 301-306.

Holker, U., Hofer, M. & Lenz J. (2004) Biotechnological advantages of laboratory-scale solid-state fermentation with fungi. *Applied Microbiology and Biotechnology,* **64**, 175-186.

Huang, H.C. & Huang, J.W. (1993) Prospects for control of soilborne plant pathogens by soil amendment. *Current Topics in Botany Research,* **1**, 223– 235.

Hynes, R.K. & Boyetchko S.M. (2006) Research initiatives in the art and science of biopesticides formulations. *Soil Biology and Biochemistry,* **38**, 845-849.

Iyamuremye, F. & Dick, R.P. (1996) Organic amendments and phosphorus sorption by soils. *Advances in Agronomy,* **56**:139–185

Kelecom, A. (2002) Secondary metabolites from marine microorganisms. *Annales Academia Brasileña de Ciencias,* **74**, 151-170.

Khan, M.S., Zaidi, A. & Wani P.A. (2007) Role of phosphate-solubilizing microorganisms in sustainable agricultura-A review. *Agronomy for Sustainable Development,* **27**, 29-43.

Kirchhoff, M. (2005) Promoting sustainability through green chemistry. *Research in Conservation and Recycling,* **44**, 237-243.

Krishna, C. & Nokes, S.E. (2001) Predicting vegetative inoculum performance to maximize phytase production in solid-state fermentation using response surface methodology. *Journal of Industrial Microbiology and Biotechnology,* **26**, 161–170.

Kucey, R.M.N. (1987) Increased phosphorus uptake by wheat and field beans inoculated with a phosphorus-solubilizing *Penicillium bilaji* strain and with vesicular-arbuscular mycorrhizal fungi. *Applied and Environmental Microbiology,* **55**, 2699–2703

Makkar, R.S. & Comeotra S.S. (1999) Biosurfactant production by microorganisms on unconventional carbon sources. *Journal of Surfactants and Detergents,* **2**, 237-241.

Markovich, N. A. & Kononova, G. L. (2003) Lytic enzymes of *Trichoderma* and their role in plant defense from fungal diseases: A review. *Applied Biochemistry and Microbiology,* **39**, 341–351.

Medina, A., Vassilev, N. & Alguacil, M.M. (2004) Increased plant growth, nutrient uptake, and soil enzymatic activities in a desertified Mediterranean soil amended with treated residues and inoculated with native mycorrhizal fungi and a plant growth promoting yeast. *Soil Science,* **169**, 260-270.

Medina, A., Vassilev, N., Barea, J.M. & Azcon, R. (2005) Application of *Aspergillus niger*-treated agrowaste residue and *Glomus mossae* for improving growth and nutrition of *Trifolium repens* in a Cd-contaminated soil. *Journal of Biotechnology,* **116**, 369-378.

Milagres, A.M. (2002) Production of metal chelating compounds by white and brown-rot fungi and their comparative abilities for pulp bleaching. *Enzyme and Microbial Technology,* **30**, 562-565.

Mitchell D.A., Berovic M., Krieger N. (2000) Biochemical engineering of solid state bioprocessing. *Adv. Biochem. Eng. Biotechnol.* **68**, 61-138.

Mitchell, D.A., Berovic, M. & Krieger N. (2002) Overview of solid state bioprocessing. *Biotechnology Annual Reviews,* **8**, 183-225.

Montesinos, E. (2003) Development, registration and commercialization of microbial pesticides for plant production. *International Microbiology,* **6**, 245-252.

Morales, I.O., Capalbo, D.M.F., Arruda, R.O.M., Bianchi, V.L. & Ascher, K.R.S. (1998) *Bacillus thuringiensis* development from 1971 to 1996: Cases of a research group in Brazil. *Israel Journal of Entomology,* **32**, 45-48.

Morris, O.N., Knagaratnam, P. & Converse, V. (1996) Suitability of 30 agricultural products and by-products as nutrient sources for laboratory production of *Bacillus thuringiensis* subsp. *aizawai* (HD133). *Journal of Invertebrate Pathology,* **70**, 113-120.

Muñoz, G., Agosin, E., Cotoras, M., San Martin, R. & Volpe, D. (1995) Comparison of aerial and submerged spore properties of *Trichoderma harzianum*. *FEMS Microbiology Letters,* 125, 63-69.

Nagel, F.J.I. (2002) Process control of solid state fermentation. PhD thesis, Wageningen University, ISBN 90-5808-513-9.

Nair, V.C. & Duvnjak Z. (1990) Reduction of phytic acid content in canola meal by *Aspergillus ficuum* in solid-state fermentation process. *Applied Microbiology and Biotechnology,* **34**, 183–188.

Ohno, A., Ano, T. & Shoda M. (1996) Use of soybean curd residue, okara, for the solid state substrate in the production of a lipopeptide antibiotic iturin A by *Bacillus subtilis* NB22. *Process Biochemistry,* **31**, 801-806.

Olivian, C., Steinberg, C., Dreumont, C., Stawiecki, K., Pouhair, D., Wadoux, P. & Alabouvette C. (1999) Production of a mixed inoculum of *Fusarium oxisporum* Fo47 and *Pseudomonas fluorescens* C7 in peat. In: Alabouvette C (Ed.) Second International Fusarium Biocontrol Workshop. INRA, Dijon, France.

Ordentlich, A., Elad, Y. & Chet, I. (1988) The role of chitinase of *Serratia marcescens* in biocontrol of *Sclerotium rolfsii*. *Phytopathology,* **78**, 48-88.

Pandey, A. (2003) Solid state fermentation. *Biochemistry Engineering Journal,* **13**, 81-84.

Pandey, A., Soccol, C.R. & Mitchell D. (2000) New developments in solid state fermentation: I-bioprocesses and products. *Process Biochemistry,* **35**, 1153-1169.

Pandey, A., Selvakumar, P., Soccol, C.R. & Nigam, P. (1999) Solid state fermentation for the production of industrial enzymes. *Current Science,* **77**, 149-162.

Pandey, A., Soccol, C.R., Rodriguez-Leon, J.A. & Nigam P. (2001a) Solid-State Fermentation in Biotechnology-Fundamentals and Applications. Asiatech Publishers, Inc. New Delhi, pp. 100-221.

Pandey, A., Szakacs, G., Soccol, C.R., Rodriguez-Leon, J.A. & Soccol V. (2001b) Production, purification and properties of microbial phytase. *Bioresource Technology,* **77**, 203-214.

Pascual, S., de Cal A., Magan N. & Melgarejo, P. (2000) Surface hydrophobicity, viability and efficacy in biological control of *Penicillium oxalicum* spores produced in aerial and submerged culture. *Journal of Applied Microbiology,* **89**, 847-853.

Patidar, P., Agrawal, D., Banerjee, T. & Patil, S. (2005) Optimization of process parameters fro chitinase production by soil isolates of *Penicillium chrysogenum* under solid substrate fermentation. *Process Biochemistry,* **40**, 2962-2967.

Ramachandran, S., Roopesh, K., Nampoothiri, K.M., Szakacs, G. & Randey A. (2005) Mixed substrate fermentation for the production of phytase by *Rhizopus* spp. Using oilcakes as substrates. *Process Biochemistry,* **40**, 1749-1754.

Ramachandran, S., Singh, S.K., Larroche, Ch., Soccol, C.R. & Randey A. (2007) Oil cakes and their biotechnological applications. *Bioresource Technology,* **98**, 2000-2009.

Rattanakit, N., Plikomol, A., Yano, S., Wakayama, M. & Tachiki T. (2002) Utilization of shrimp shellfish waste as a substrate for solid-state cultivation of *Aspergillus* sp. S1-13: Evaluation of a culture based on chitinase formation which is necessary for chitin-assimilation. *Journal of Bioscience and Bioengineering,* **93**, 550-556.

Rayner, A.D.M. & Boddy, L. (1988) Fungal communities in the decay of wood. *Advances in Microbial Ecology,* **10**, 115–166.

Rice W.A., Olsen P.E., Leggett M.E. (1994) Co-culture of *Rhizobium meliloti* and a phosphorus-solubilizing fungus (*Penicillium bilaji*) in sterile peat. *Soil. Biol. Biochem.* **27**, 703–705.

Ristaino, J.B. & Thomas, W. (1997) Agriculture, methyl bromide, and the ozone hole: can we fill the gaps? *Plant Disease,* **81**, 964–977.

Roco, A. & Perez, L.M. (2001) In vitro biocontrol activity of *Trichoderma harzianum* on *Alternaria alternata* in the presence of growth regulators. *Electronic Journal of Biotechnology,* **4** (http://www.ejbiotechnology.info/content/vol4/issue2/full/1/).

Rodriguez, H. & Fraga R. (1999) Phosphate solubilizing bacteria and their role in plant growth promotion. *Biotechnology Advances,* **17**, 319–339.

Roopesh, K., Ramachandran, S., Nampoothiri, K.M., Szakacs, G. & Pandey, A. (2006) Comparison of phytase production on wheat bran and oilcakes in solid-state fermentation by *Mucor racemosus*. *Bioresource Technology,* **97**, 506–511.

Sabu, A., Sarita, S., Pandey, A., Bogar, B., Szakacs G. & Soccol R.C. (2002) Solid-state fermentation for production of phytase by *Rhizopus oligosporus*. *Applied Biochemistry and Biotechnology,* **102–103**, 251–260.

Saddler, J.N. (1993) *Bioconversion of Forest and Agricultural Plant Residues*. CAB International, UK.

Selle, P.H. & Ravindran, V. (2007) Microbial Phytase in poultry nutrition. *Animal Feed Science and Technology,* **135**, 1-41.

Spadaro, D. & Gullino, M.L. (2005) Improving the efficacy of biocontrol agents against soilborne pathogens. *Crop Protection* **24**, 601–613.

Stleger, R.J., Cooper, R.M. & Charnley A.K. (1986) Cuticle degrading enzyme of entomopathogenic fungi: cuticle degradation in vitro by enzymes from entomopathogens. *Journal of Invertebrate Pathology,* **47**, 167-177.

Summers, J.D. & Leeson, S. (1985) In: *Poultry Nutrition Handbook*. Department of Animal and Poultry Science, Ontario Agricultural College, University of Guelph, Guelph, Ontario, Canada, pp. 139-155.

Suyanandana, P., Potacharoen, W., Arunpairojana, V., Boonsong, P. & Vattanakul, J. (1996) Solid state cultivation of *Bacillus thuringiensis* for insect larvicide in agriculture and public health. *Khon Kaen Agricultural Journal,* **24**, 190-196.

Tengerby, R.P. & Szakacs, G. (1998) Perspectives in agrobiotechnology. *Journal of Biotechnology,* **66**, 91-99.

Tengerby, R.P. & Szakacs, G. (2003) Bioconversion of lignocellulose in solid substrate fermentation. *Biochemical Engineering Journal,* **13**, 169-179.

Vassilev, N. & Vassileva, M. (1992) Production of organic acids by immobilized filamentous fungi. *Mycological Research,* **96**, 563-570.

Vassilev, N. & Vassileva, M. (2003) Biotechnological solubilization of rock phosphate on media containing agro-industrial wastes. *Applied Microbiology and Biotechnology,* **61**, 435-440.

Vassilev, N., Baca, M.T. & Vassileva, M. (1994) Lignocellulose and Fungi: From nature to industrial use. *Mycologist,* **8**, 113-114.

Vassilev, N., Nikolaeva, I. & Vassileva M (2005a) Polymer-based preparation of soil inoculants: Applications to arbuscular mycorrhizal fungi. *Reviews in Environmental Science and Bio/Technology,* **4**, 235-243.

Vassilev, N., Nikolaeva, I. & Vassileva, M. (2005b) Biocontrol properties of microbially-treated sugar beet wastes in presence of rock phosphate. *Journal of Biotechnology,* **118**, S1: S177

Vassilev, N., Nikolaeva, I. & Vassileva, M. (2006a) Simultaneous P-solubilizing and biocontrol activity of microorganisms: potentials and future trends. *Applied Microbiology and Biotechnology,* **71**, 137-144.

Vassilev, N., Vassileva, M. & Medina, A. (2006b) Microbial solubilization of rock phosphate on media containing agro-industrial wastes and effect of the resulting products on plant growth and P-uptake. *Plant and Soil,* **287**, 77-84.

Vassilev, N., Baca, M.T., Vassileva, M. & Franco, I. (1995) Rock phosphate solubilization by *Aspergillus niger* grown on sugar-beet waste medium. *Applied Microbiology and Biotechnology,* **44**, 546–549.

Vassilev, N., Franco, I., Vassileva, M. & Azcon, R. (1996) Improved plant growth with rock phosphate solubilized by *Aspergillus niger* grown on sugar-beet waste. *Bioresource Technology,* **55**, 237–241.

Vassilev, N., Vassileva, M., Azcon, R. & Barea, J.M. (2002) The use of ^{32}P dilution techniques to evaluate the effect of mycorrhizal inoculation on plant uptake of P from the resulting products of fermentation mixtures including agrowastes, *Aspergillus niger* and rock phosphate. In: The use of Nuclear and Related Techniques for Evaluating the Agronomic Effectiveness of Phosphate fertilizers. International Atomic Energy Agency Technical Document. FAO/IAEA, Vienna, Austria.

Vassilev, N., Vassileva, M., Fenice, M. & Federici, F. (2001) Immobilized cell technology applied in solubilization of insoluble inorganic (rock) phosphates and P plant acquisition. *Bioresource Technology,* **79**, 263-271.

Vassilev, N., Vassileva, M., Bravo, V., Fernandez-Serrano, M. & Nikolaeva I. (2007a) Simultaneous phytase production and rock phosphate solubilization by *Aspergillus niger* grown on dry olive cake. *Industrial Crops and Products,* **26**, 332-336.

Vassilev, N., Fenice, M., Jurado, E., Reyes, A., Nikolaeva, I. & Vassileva, M. (2008a) Antagonistic effect of microbially-treated mixture of agro-industrial wastes and inorganic insoluble phosphate to *Fusarium* wilt disease. In: *Progress in Environmental Microbiology*, Ed. Myung-Bo Kim, NOVA Science Publishers, pp. 223-234.

Vassilev, N, Nikolaeva, I, Someus, E, Martinez-Nieto, L. & Vassileva, M. (2007b) Manganese peroxidase production by *Phanerochaete chrysosporium* in different modes of fermentation. *Paper presented at the XV International Workshop on Bioencapsulation*, Vienna, Austria, 6-8 September. Proceedings, P4-02, 1-4.

Vassilev, N, Reyes, A, García, M & Vassileva, M (2008b) Production of chitinase by free and immobilized cells of *Penicillium janthinellum.* A comparison between two biotechnological schemes for inoculant formulation. Paper accepted for presentation at the XVI International Conference on Bioencapsulation, Dublin, Ireland.

Vassileva, M., Vassilev, N. & Azcon, R. (1998) Rock phosphate solubilization by *Aspergillus niger* on olive cake-based medium and its further application in soil-plant system. *World Journal of Microbiology and Biotechnology,* **14**, 281–284.

Vats, P. & Banerjee U. (2004) Production studies and catalytic properties of phytases (myo-inositolhexkisphosphate phosphohydrolases. *Enzyme and Microbial Technology,* **35**, 3-14.

White, G., Bond, B. & Pinel, M. (2000) Steaming ahead. *Grower,* **134**, 19–20.

Whitelaw, M.A. (2000) Growth promotion of plants inoculated with phosphate-solubilizing fungi. *Advances in Agronomy,* **69**, 99–151.

In: Industrial Waste...
Editor: John P. Samuelson, pp. 121-204
ISBN 978-1-60692-720-5

Chapter 5

INDUSTRIAL POLLUTION MANAGEMENT: A SUSTAINABILITY PERSPECTIVE

Y. Shastri*
Vishwamitra Research Institute and University of Illinois at Chicago,
Department of Bioengineering,
851 S. Morgan Street, Chicago, IL 60607
U. Diwekar
Vishwamitra Research Institute, Center for Uncertain Systems:
Tools for Optimization and Management,
368 56th Street, Clarendon Hills, IL-60514.

Abstract

With ever increasing focus on industrial pollution and environment, traditional industrial design framework needs to be extended from green process design, to industrial ecology, to sustainability. The concept of overall sustainability goes beyond industrial ecology and brings in time dependent nature of the ecosystem. Decisions regarding regulations and human interactions with ecosystem come in picture. It involves dealing with various time scales and time dependent uncertainties. The available management options and decision making change from the single industrial production plant to the larger scales of an integrated industrial park, community, and natural systems. Further, green engineering concepts change the goal of engineering design from profitability to include a number of different and conflicting objectives that can define sustainability in the end. This integration poses challenging problem of discrete and continuous decisions, nonlinearities and uncertainties. Systems theory approach deals with similar ideas of disparate temporal and spatial systems, their interactions and interdependencies. This prompts one to extend systems theory approach for decision making in sustainability. This chapter presents such an idea through a case-study problem of mercury waste management in a watershed. The harmful effects of mercury pollution are well documented. Mercury in water particularly presents a grave danger because of its bioaccumulative nature. It is further complicated by the complex mercury cycling in the environment. As a result, successful control of mercury

*Current affiliation: Energy Biosciences Institute, Department of Agricultural and Biological Engineering, University of Illinois at Urbana-Champaign.

pollution in water depends on actions at various points in the mercury cycle, creating an appropriate case study for the sustainability initiative. This work looks at three levels of pollution control: individual industry level, industrial sector level and ecosystem level. At the individual industry level, the goal is to optimize mercury treatment technology selection and design. At the industrial sector level, the option of pollutant trading to balance ecological and economic objectives is analyzed, and an optimization based decision making framework is proposed. Once mercury enters the water bodies, controlling the bioaccumulation of mercury can further reduce the harmful effects and affect the regulatory decisions. This work investigate the idea of lake pH control through time dependent liming, constituting work at the ecosystem level. Uncertainties increase as we move from the individual industry level, where model and tools are well developed, to ecosystem level. Although the application is specific to mercury pollution, the goal is to understand the generalized integrated approach for sustainable waste management from industrial perspective. To achieve these goals, appropriate modeling of the objectives, risks and uncertainties is necessary, followed by the use of appropriate tools to optimize various decision. This work takes a systems theory approach, employing techniques such as optimization, optimal control and real options theory, to come up with efficient management strategies.

1. Introduction

The focus on industrial pollution and environment has been mounting in recent times, not only from social perspective but also from industrial perspective. There are concerns over the harmful effects of a lot of pollutants and waste streams leading to strict environmental regulations. Since industries are the primary source of majority of pollutants, industrial sector is simultaneously tackling the issue of rising waste management and treatment costs. It has been proposed that the ultimate solution to this problem lies in simultaneous consideration of all relevant aspects in decision making, also known as the sustainability approach.

Sustainable development is defined as "the development that meets the needs of the present without compromising the ability of the future generations to meet their own needs" (Tomlinson, 1987). The concept of overall sustainability goes beyond industrial ecology and brings in the complex interactions between a single industrial production plant with the larger scales of an integrated industrial park, community, and natural systems. It also brings in the time dependent nature of the ecosystem. Consequently, the scope of available management options and decision making widens, and decisions regarding regulations, human interactions with ecosystem come in picture. Further, green engineering concepts change the goal of engineering design from profitability to include a number of different and often conflicting objectives that can define sustainability in the end. Sustainability therefore calls for a synergistic approach to technological, social and regulatory decision making.

However, this integration poses challenging problem of discrete and continuous decisions, nonlinearity, disparate spatial and temporal scale interactions and uncertainties. Static and dynamic uncertainties, which become more pronounces as the envelop is extended, need to be accounted for. Systems theory approach offers a suitable methodology to tackle such complex issues. It employs efficient techniques to model objectives, risks and uncertainties to pervade inter-system boundaries and allow communication between various aspects. This is followed by the use of mathematical theories such as optimization, optimal control and real options theory, to come up with efficient management strategies. The approach is presented here through a case-study problem of mercury waste management in a watershed.

The harmful effects of mercury pollution are well documented. Mercury in water presents a grave danger because of its bioaccumulative nature. It is further complicated by the complex mercury cycling in the environment. As a result, successful control of mercury pollution in water depends on actions at various points in the mercury cycle, creating an appropriate case study for a sustainability initiative. This work looks at three levels of pollution control: individual industry level, industrial sector level and ecosystem level. At the individual industry level, this work analyzes various mercury control technologies for industries, where the goal is to optimize technology selection and design. Strict regulations for environmental cause need to be balanced with the profitability and economic sustainability for the industries. Therefore, at the industrial sector level, this work analyzes the option of pollutant trading, and proposes a decision making framework to balance the economic and ecological objectives. Once mercury enters the water bodies, controlling the bioaccumulation of mercury can further reduce the harmful effects and affect regulatory decisions. This in turn can affect the industries responsible for the pollution. This work investigates the idea of lake pH control through time dependent liming, which constitutes the work at the ecosystem level (Shastri & Diwekar, 2008b). Uncertainties increase as we move from the individual industry level, where model and tools are well developed, to ecosystem level.

To summarize, the chapter investigates the application of systems theory approach in decision making for waste management with sustainability as the overall objective. Although the case study example deals with mercury pollution, the goal is to understand the generalized integrated approach for sustainable waste management from industrial perspective. The chapter is arranged as follows: The next section elaborates the idea of sustainability, followed by section 3 discussing systems theory approach and its correlation with sustainability. Sections 4, 5, and 6 present tools for uncertainty modeling, nonlinear stochastic programming and stochastic optimal control, respectively. These systems theory based innovative tools are used at various stages of the work. Section 7 emphasizes the importance of mercury as a pollutant and sets the stage for the case study analysis. This is followed by sections 8, 9 and 10 elaborating the work done at industry, inter-industry and ecosystem level to manage mercury pollution. The last section presents a summary and important conclusions from the work.

2. Sustainability

The topic of sustainability goes beyond industrial ecology and is, perhaps, operationally and conceptually one of the most complex that modern science has faced as it involves socio-economic interactions and its effect on the overall ecosystem. Sustainability must embody, in some form, elements of physics, engineering, ecology, law, economics, sociology, and politics and cannot be investigated successfully within a confines of a single discipline. This gives rise to severe operational difficulties in cross-disciplinary communication. It is not easy to capture economic demands, legal issues, and ecosystem functions in a single framework. Fig. 1 presents the extension of framework from process design, to industrial ecology leading to socio-economic sustainability. At the center of this new framework is the green chemical plant engineered with clean products, clean processes, and green

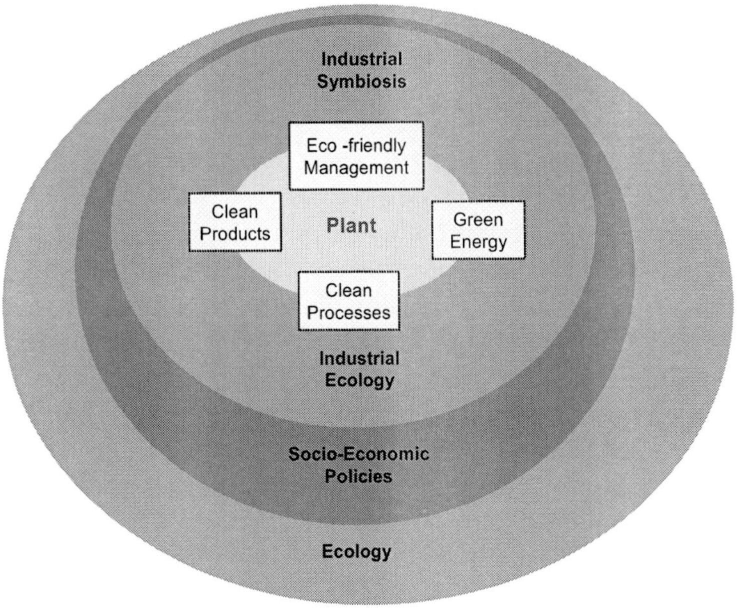

Figure 1. From Green Engineering to industrial Ecology to Sustainability.

energy, and eco-friendly management and planning. The decision making is then extended to industrial symbiosis and industrial ecology, time-dependent socio-economic policies and together their effect on ecological systems leading to long term sustainability of the planet. Fig. 2 indicates the various objectives that need to be considered while making decisions in this field.

However, it is necessary to provide a metric for sustainability that can address the cross disciplinary nature of sustainability where conceptual difficulties arise due to the lack of a general underlying theory. Sustainability has different perspectives depending on the nature of the system at hand. For industrial sector, sustainability means economic viability and profitability while following the regulations. For the society, maintenance of clean environment and its services is a sustainability metric. Sustainability of the natural systems on the other hand refers to biodiversity and species survival. Moreover, the criteria that identify a sustainable, as opposed to an unsustainable system, are also unknown.

It is evident from this discussion that sustainability has a wide scope. Many aspects related to it are still not clearly understood. Decision making with a sustainability perspective is therefore quite difficult. It can be easily realized that logic and heuristics based decisions might be sub-optimal because such approaches often do not account for the interactions between various sub-systems. A systematic approach is thus needed to resolve these issues. Systems theory is an option to achieve these objectives. The next section gives an overview of this approach and justifies its use for the concerned problem.

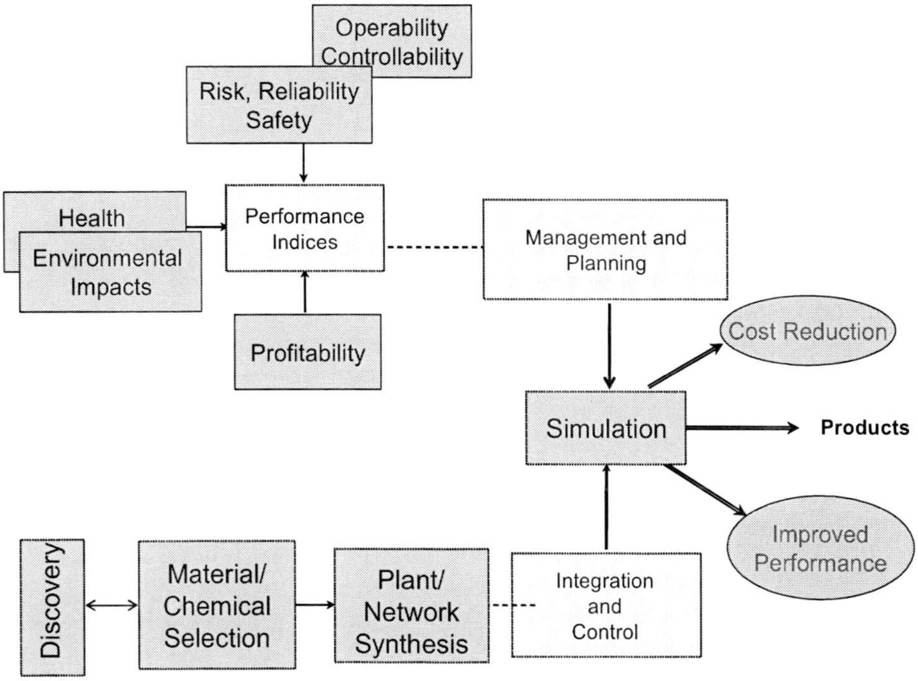

Figure 2. From Green Engineering to industrial Ecology to Sustainability.

3. Systems Theory Approach

Systems theory is defined as the trans-disciplinary study of the abstract organization of phenomena, independent of their substance, type, or spatial or temporal scale of existence. It investigates both, the principles common to all complex entities, and the (usually mathematical) models which can be used to describe them. Systems analysis applies systems principles to aid a decision-maker with problems of identifying, reconstructing, optimizing, and controlling a system (usually a socio-technical organization), while taking into account multiple objectives, constraints and resources. It aims to specify possible courses of action, together with their risks, costs and benefits.

The definitions of systems theory and systems analysis clearly indicate that there is a close correlation between systems approach and sustainability. Like sustainability, systems theory aims to connect various disciplines through a common base and account for the interdependencies between these disciplines. Hence it is prudent to apply this theory towards the proposed sustainability analysis.The theory is primarily based on development in mathematics. The advantage of dealing with all aspects in mathematical realm is that disparate systems can be made to communicate with each other. Once a mathematical representation is given, one can account for the interactive effects between the systems. Modeling forms the basis of this approach since an efficient model to represent a given system renders it amiable for further analysis. Models can be of various types, such as: first principle models, regression (data based models), conceptual models etc. Models can also be classified as: linear vs nonlinear, discrete time vs continuous time, static vs dynamics etc. Another very important model classification is deterministic vs stochastic. Each type of model has differing development requirements, complexities and applications, and the suitable model must be identified based on the particular application.

It has been emphasized before that the consideration of uncertainty is very important in sustainability analysis, primarily due to the wide scope covered by this field, leading to lack of complete understanding of the interdependencies. If the scope includes natural systems such as lakes, forests, predator-prey interactions, then the uncertainty increases substantially since many of these natural systems are poorly understood. Consequently, models incorporating uncertainty i.e. stochastic models, are valuable. Uncertainties can be classified as time independent (static) and time dependent (dynamic). Static uncertain parameters have a fixed distribution function which does not change with time, while for dynamic uncertain parameters, the parameter value depends on time along with its distribution function. Static uncertainties have been frequently incorporated in various engineering applications (e.g. Monte Carlo simulations). Since sustainability focuses on the long term effects along with the short term goals, forecasting is essential. This means that time dependent stochastic models are needed, highlighting the importance of dynamic uncertainty modeling.

The model development stage is followed by systems analysis to achieve desired objectives such as parameter optimization, benefit maximization, or goal attainment. The tools available to perform these tasks are again based on the theory of mathematics. Two widely established tools in the field of engineering are optimization and optimal control theory. The theory of optimization and control has been extensively developed with numerous successful applications. A review of the theoretical basics can be found in standard literature. Here, we propose to discuss those aspects of optimization and optimal control theory that are important for the considered application.

The work, as stated before, deals with industrial process models for process selection and design optimization. Many of these models are nonlinear in nature. When uncertainty is added to this, optimization of such stochastic nonlinear models is a difficult task, numerical complexity being one of the major hurdles. We review a recent nonlinear stochastic optimization algorithm that has been proposed by the authors which will be useful in solving problems encountered in this analysis. Similarly, optimal control theory, when extended to include uncertainty, becomes complicated to deal with. Here we review a recent development by Rico-Ramirez & Diwekar (2004) to solve stochastic optimal control problems in an efficient manner. The next few sections, therefore, deal with various aspects of systems theory that are deemed to be important for the proposed analysis. These are: time dependent uncertainty modeling, stochastic nonlinear programming algorithm and stochastic optimal control solution methodology.

4. Time Dependent Uncertainty Modeling

It is a common practice to use probability distribution functions like normal, lognormal or uniform, to model uncertainties. However, these distributions are used for scalar parameter uncertainties. Modeling dynamic or time-dependent uncertainties is a difficult task. Recently Diwekar (2003b) presented the basic concepts for modeling time dependent uncertainties. These concepts, derived from finance literature where time dependent uncertainties dominate, are presented here briefly.

Finance literature presents different ways to represent and forecast uncertainty using the stochastic processes. In any stochastic process, there is a variable evolving over time t, and both, time t as well as the variable itself, can assume discrete as well as continuous nature. Wiener process, also known as Brownian motion, is a simple continuous time and continuous state stochastic process. It can be used to model a variety of continuous stochastic processes (Dixit & Pindyck, 1994; Diwekar, 2003b). The Wiener process is represented as:

$$dz = \epsilon_t \sqrt{dt} \tag{1}$$

where, dz is the random variable, and ϵ_t is a normally distributed random variable with zero mean and unit standard deviation. Random variable dz has the property that the expectation is zero ($E[dz] = 0$) and the variance is dt ($var[dz] = dt$). Using this definition of the Wiener process, the general form of the Ito process is given as:

$$dx = a(x,t)\,dt + b(x,t)\,dz \tag{2}$$

Here, $x(t)$ is the continuous time stochastic variable that is to be modeled, and dz is the Wiener increment, as defined by Eq. (1). $a(x,t)$ (drift parameter) and $b(x,t)$ (variance parameter) are known (nonrandom) functions. Many stochastic processes, such as the simple Brownian motion with drift, the geometric Brownian motion and the mean reverting Ito process are derived from Eq. (2). In this work, the mean reverting Ito process is used to model the natural variation in lake pH for lake liming operation. The mean reverting process has been previously used to model many stochastic variables, such as crude oil and copper prices (Dixit & Pindyck, 1994), relative volatility of non-ideal mixtures (Ulas & Diwekar, 2004) and human mortality rate (Diwekar, 2005). The characteristic of the mean reverting Ito process is that, although it models the random variable fluctuations for a short time, in the long run, the variable is drawn back to the mean value. The exact form of the mean reverting Ito process is presented later while discussing the lake liming operation in section 11.

5. Optimization: L-shaped BONUS Algorithm

Various optimization problem types encountered at different stages of the proposed work include simple mixed integer linear programming problems (pollutant trading), mixed integer nonlinear programming problems (technology selection and design), and the stochastic variants of both of them. To solve the deterministic problems, well established methods from optimization literature can be used (e.g. branch and bound, Generalized Bender's Decomposition, Outer-Approximation etc). However, consideration of uncertainty complicates the solution of the resulting stochastic programming problems. If there is nonlinearity in the model in addition to uncertainty, then result problem falls into the class of stochastic nonlinear programming (SNLP) problem. Such problems are difficult to solve. A general stochastic nonlinear programming problem can be represented as follows:

$$\begin{aligned} \text{Optimize } J &= P_1(f(\theta, x, u)) && (3) \\ \text{such that} & && \\ P_2(g_1(\theta, x, u)) &= 0 && (4) \\ P_3(g_2(\theta, x, u) \leq 0) &\geq \alpha && (5) \end{aligned}$$

where, θ is the set of decision variables, x is the set of system parameters, u is the set of uncertain variables, and P_1, P_2 and P_3 are probabilistic measures such as expected value and variance. The uncertainty may affect the objective function and/or any of the constraints to make it a stochastic programming problem.

The field of process systems engineering is replete with the applications of stochastic programming, many of which are nonlinear. Numerous well known tasks in this field, such as project planning and scheduling, chemical synthesis, process design and optimization, and some new fields, such as computer aided molecular design, use stochastic programming. An extensive review of the stochastic programming methods and their applications in the process engineering field is given in Diwekar (2003c) and Sahinidis (2004). Some of the recent applications include enterprise-wide process network (Rya et al., 2004), planning and scheduling related tasks (Jung et al., 2004; Lin et al., 2004) and environment related applications (Diwekar, 2003a, 2005; Kheawhom & Hirao, 2004). Many of these problems are nonlinear, complicating the problem solution.

However, despite the relative importance of such problems, comparatively few solution methodologies have been found to be successful over the entire domain of problems in this class. Computational efficiency is a major issue with the existing algorithms to solve such problems. Hence efficient techniques are called for. From optimization perspective, there are two more issues that are often encountered in such problems:

- The models are large scale, resulting in high computational requirements
- The models are black-box models, which need to be integrated with the optimization algorithm

The examples of these include, WASP5 model for the pollutant fate and transport simulation and LakeWeb model for aquatic biota and bioaccumulation simulation. Such large scale stochastic nonlinear programming problems are computationally difficult to solve. Discussed below is a new algorithm called the L-shaped BONUS, which can solve such problems in a computationally efficient manner (Shastri & Diwekar, 2006a). The proposed algorithm is an integration of the traditional sampling based L-shaped method with a new algorithm BONUS (Better Optimization of Nonlinear Uncertain Systems), proposed to solve the SNLP problems. The algorithm is discussed in the following sections, starting with a brief overview of the existing solution methods, followed by the discussion of the standard L-shaped method, BONUS and the proposed L-shaped BONUS algorithm.

5.1. Methods for SNLP Problem Solution

5.1.1. General Overview

Over the years, a lot of research has gone in devising strategies to solve the SNLP problems. One kind of solution methods, such as chance constraint programming method (Charnes & Cooper, 1959), convert these problems into deterministic equivalents. The deterministic optimization methods can then be applied. But these methods are restricted to problems with known and stable density functions of the random variables. The second kind of solution methods are aimed at extending the deterministic nonlinear programming methods to included uncertainty (Elmaghraby, 1960; Ziemba, 1970, 1972). For the optimization problems that can be decomposed into two or multiple stages, decomposition based stochastic programming methods, such as the L-shaped method, are developed (Birge & Louveaux, 1997). The basic L-shaped method is modified for various problem types, which includes the regularized decomposition method (Ruszczyński, 1986) and the piecewise quadratic form of the L-shaped method (Louveaux, 1978). The methods based on the stochastic programming Lagrangian include the basic Lagrangian dual ascent method (Birge & Louveaux, 1997), the Lagrangian finite generation method for linear quadratic stochastic programs (Rockafellar & Wets, 1986) and the progressive hedging algorithm (Rockafellar & Wets, 1991). These decomposition based methods require convexity of the problem and/or a dual block angular structure. The stochastic quasi-gradient methods (SQG) are less specialized than the other algorithms, but are useful to solve problems having complex objective functions and constraints (Louveaux, 2001). The SQG methods represent one of the first computational developments in stochastic programming. An exhaustive review of all these methods is omitted for brevity, and the reader is referred to the respective references made for these solution algorithms. Even though important, application of these methods to solve the real life problems has always been restricted. This is due to the various limitations in the form of functional requirements (convexity, differentiability etc.) or distribution of uncertain variables (stable).

5.1.2. Sampling Based Methods

In stochastic programming problems, it is common to use sampling approximations when the probability distributions of the uncertain parameters are known. The aim is to model the complete uncertain parameter space as closely as possible through a sufficient number of samples. Fig. 3 represents the generalized solution procedure for the sampling based approach. The structure is similar to that for a deterministic problem, apart from the fact that the deterministic model is replaced by a stochastic model with the (shaded) sampling loop representing the discretized uncertainty space. The goal in stochastic programming is to improve the probabilistic objective function value with each iteration. In the set up of Fig. 3, the calculation of these terms requires the simulation of the stochastic modeler at each iteration. In the traditional sampling based methods, this is achieved by model simulations for a given number of samples, and the subsequent computation of the probabilistic function (e.g. expected value of the objective function). Two such methods are the sampling based L-shaped method (Dantzig & Glynn, 1990; Dantzig & Infanger, 1992) and the Stochastic Decomposition algorithm (Higle & Sen, 1991). The L-shaped method is a scenario based

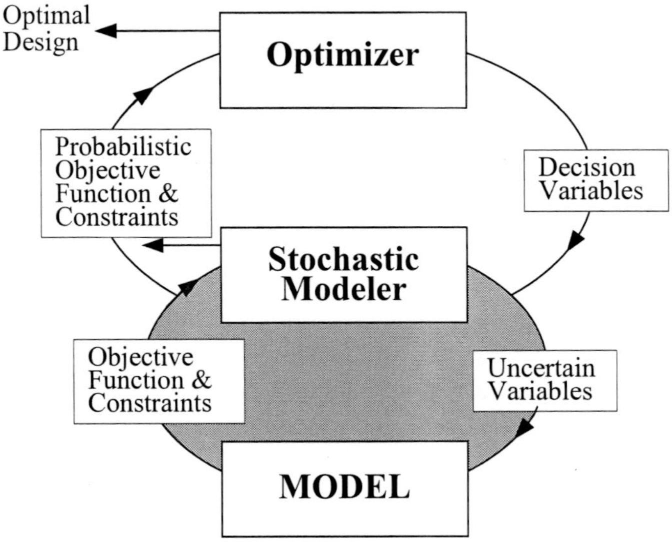

Figure 3. Optimization under uncertainty.

method. But the number of scenarios increases exponentially as the number of uncertain variables increases. Monte Carlo sampling avoids this problem, and hence, Monte Carlo sampling based approximations have been incorporated in the L-shaped method. The key feature is the use of statistical estimates to obtain confidence intervals on the results.

The simulation of the stochastic modeler at each iteration is a major drawback of the sampling based methods. For a sample size of n, the model needs to be simulated n times in each iteration as a part of the stochastic modeler. With a larger sample size, required for better approximation, the computational load increases tremendously. The next section explains BONUS, an algorithm that overcomes this problem through the use of a reweighting scheme.

5.2. BONUS

The sampling based approaches to solve the stochastic nonlinear programming problems suffer from the main drawback of computational complexity, as mentioned in the previous section. Recently, a new method has been proposed to solve the SNLP problems that holds its advantage in circumventing the problem of repeated model simulations. This new method is Better Optimization of Nonlinear Uncertain Systems (BONUS) (Sahin & Diwekar, 2004). Fig. 4 shows the BONUS algorithm structure.

Fig. 3 shows the standard stochastic programming algorithm structure with the sampling loop requiring repeated model simulations. Compared to this, the BONUS algorithm uses reweighting approach to skip these repeated model simulations. This reweighting scheme is central to the BONUS algorithm. An initial, uniform base distribution of the uncertain parameters is generated. For the first iteration, the algorithm emulates the standard sampling based algorithm in that the model is simulated for each sample to determine the output distribution. At the subsequent iterations, when the optimizer needs new estimates of the probabilistic values of the objective function, a new set of samples is taken. But this time the model is not re-run. Instead, the reweighting approach is applied to approximate the probabilistic behavior of the new output distribution. Fig. 4 illustrates this step of the algorithm. The reweighting scheme uses the initial sample set, the initial output distribution and the new sample set data to estimate the information about the new output distribution. Owing to the importance of reweighting in the proposed algorithm, it is explained in the following section.

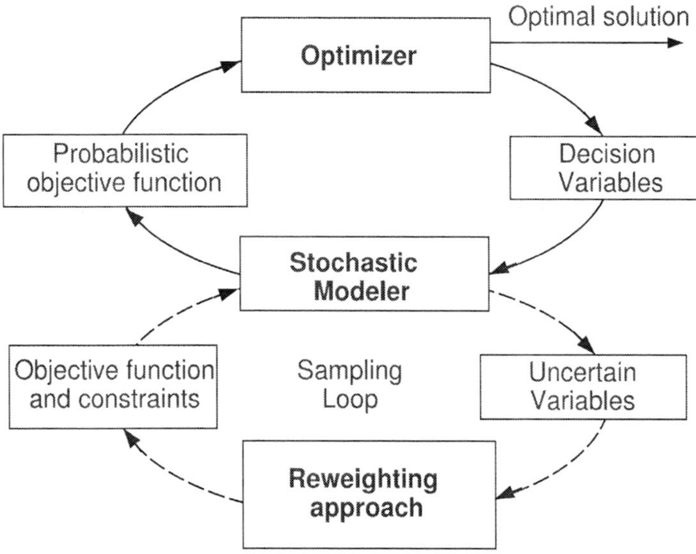

Figure 4. BONUS algorithm.

5.2.1. The Reweighting Approach

The reweighting approach is based on the various reweighting schemes proposed by Hesterberg (1995). It is an extension of the importance sampling concept of estimating something about a distribution (target distribution $f(x)$) using observations from a different distribution (design distribution $\hat{f}(x)$), where these distributions are represented by the respective probability density functions. Let X be a random variable with the probability density function $f(x)$, and $Q(X)$ be a function of X. Then, to estimate a certain property of $Q(X)$, such as the expected value $\mu = E_f[Q(X)]$, importance sampling solves a different problem of estimating $E_{\hat{f}}[Y(X)]$, where

$$Y(x) = Q(x)\frac{f(x)}{\hat{f}(x)} \tag{6}$$

and the samples X_i are now drawn from $\hat{f}(x)$. The distribution $\hat{f}(x)$ can be designed to achieve the desired results (e.g. reduced variance, better representation of the rare events). The weight function $W(x)$ is defined as

$$W(x) = \frac{f(x)}{\hat{f}(x)} \tag{7}$$

which gives the likelihood ratio between the target and the design distributions and weighs observations of $Q(x)$. To perform this estimation effectively, Hesterberg (1995) proposed various design distributions $\hat{f}(x)$ (e.g. defensive mixture distributions) and estimation schemes (integration estimate, ratio estimate). In the ratio estimate, the weights W_i are normalized to avoid problems when they do not sum to 1. The normalized weights V_i and the estimate μ is given as

$$V_i = \frac{W_i}{\sum_{j=1}^{n} W_j} \tag{8}$$

$$\mu = \sum_{i=1}^{n} V_i Q(X_i) \tag{9}$$

where, n is the sample size. Means, higher moments and percentiles can be computed using such relations. The reweighting scheme in the proposed algorithm is based on the ratio estimate just explained.

The reweighting approach, as used in the BONUS algorithm, is schematically shown in Fig. 5. Suppose X represents the uncertain variable in the stochastic programming problem, and $Q(X)$ is the output of the stochastic modeler. For the first iteration, the base case samples X_i^* with a uniform distribution ($\hat{f}(x)$) are drawn and the model is simulated for each sample to get the complete model output distribution $Q(X_i^*)$. During the subsequent iterations, new samples X_i of the required distribution ($f(x)$) are drawn. Having known the model response $Q(X_i^*)$ for the sample set X_i^* from the distribution $\hat{f}(x)$, it is possible to use Eq. 9 to estimate the expected value of the model response $Q(X_i)$ for the new sample set X_i from the distribution $f(x)$. The expected value of the stochastic model response $Q(X_i)$ for the new sample set X_i is therefore given as

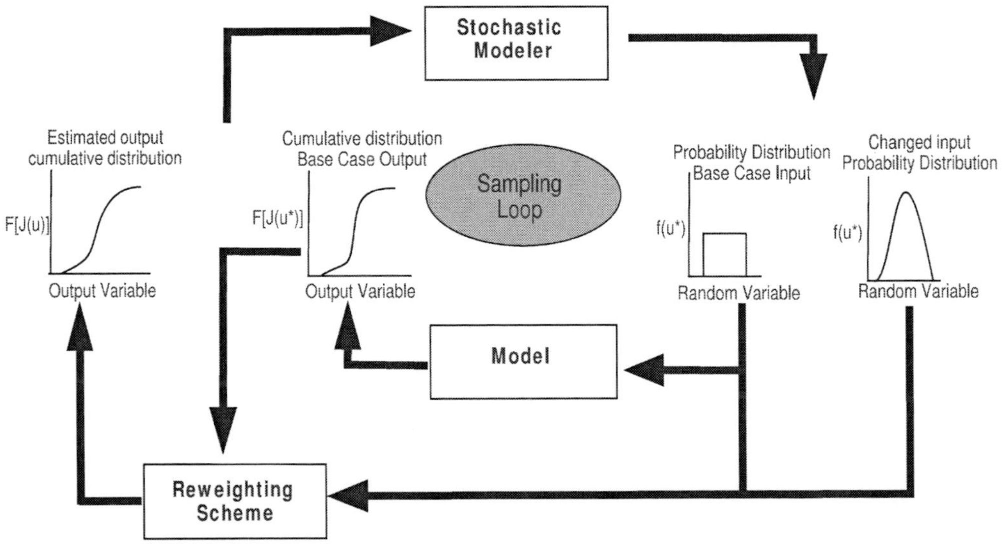

Figure 5. Reweighting approach in the BONUS algorithm.

$$E_f[Q(X_i)] = \sum_{j=1}^{n} \frac{\frac{f(X_j)}{\hat{f}(X_j^*)}}{\sum_{i=1}^{n} \frac{f(X_i)}{\hat{f}(X_i^*)}} Q(X_j^*) \tag{10}$$

In a sampling based algorithm, this procedure requires determining the probability density function from the available sample set. This is carried out using the Gaussian Kernel Density Estimation technique (Silvermann, 1986), which is a nonparametric density estimation technique. The basic idea behind this technique is to place a bin of a certain width $2h$ around every sample X and weigh that sample by the number of other samples X_i in the same bin. If this bin is replaced by a kernel function, such as the normal density function, the density function for the sample set X_i is calculated using Eq. 11.

$$f(X) = \frac{1}{n.h} \sum_{i=1}^{n} \frac{1}{\sqrt{2\pi}} e^{-\frac{1}{2}\left(\frac{X-X_i}{h}\right)^2} \tag{11}$$

where, h is the window width, also called the smoothing parameter or bandwidth. The value of h decides the fineness of the density estimation. For this work, it is taken as the standard deviation of the sample set. Thus, given the two sample sets, Eq. 11 is used to determine the density function at each sample point for both the distributions, which are then used in Eq. 10 to find out the output distribution for the second sample set.

For the proposed algorithm, this idea of reweighting is used in the sampling based L-shaped method to solve the decomposable SNLP problems in a computationally efficient manner. Before describing the proposed algorithm however, it is prudent to understand the L-shaped method, which is explained in the next section.

5.3. L-shaped Method with Sampling

The basic L-shaped method is a scenario based method applicable for discrete distributions to solve two or multi stage stochastic programming problems (Van Slyke & Wets, 1969). The basic idea of the L-shaped method is to approximate the nonlinear term in the objective function of the problem. The principle behind this approach is that, since the nonlinear objective function term (recourse function) involves the solution of all the second stage recourse problems, numerous function evaluations for it are avoided. This term therefore is used to build a master problem (with first stage decision variables) and the recourse function is exactly evaluated only as a subproblem, referred to as the second stage problem. Fig. 6 shows the algorithm structure.

The method is essentially a Dantzig-Wolfe decomposition (inner linearization) of the dual or a Benders decomposition (outer linearization) of the primal. The problem is decomposed into two or multiple stages. The first stage problem (master problem) uses a linear approximation (also the lower bound) of the second stage recourse function to fix the first stage decision variables. These first stage decisions are passed on to the second stage (sub problem), where the dual of the second stage problem is solved for different scenarios. The solution of all the dual problems is used to calculate the expected value of the recourse function, which is also its upper bound. Two kinds of cuts are sequentially generated, the

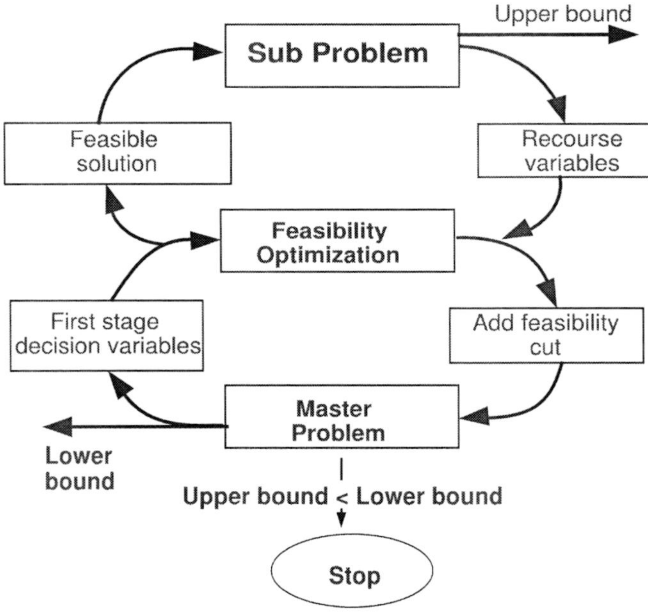

Figure 6. L-shaped algorithm structure.

feasibility cut and the optimality cut. These cuts are added to the master problem for a better approximation of the recourse function in the subsequent iterations. The algorithm terminates when the upper bound from the sub problem is less than or equal to the lower bound from the master problem (Diwekar, 2003b).

For continuous distributions, sampling is used to approximate the distribution. The use of importance sampling as a variance reduction technique was proposed by Dantzig & Glynn (1990), while the use of sampling in the basic L-shaped method was proposed by Dantzig & Infanger (1992). Samples, instead of scenarios, are used in the second stage recourse function calculations of the L-shaped method. Statistical approximations of the recourse function and the simplex multipliers are used to generate the cuts and the bounds. This method is known as the internal sampling method. In the context of the sampling based algorithms explained in section 5.1, the sub-problem solution for each realization of the uncertain space can be compared with the stochastic modeler in Fig. 3.

Thus to summarize, the decomposition strategy of the L-shaped method offers computational savings, but is still not very efficient for the SNLP problems. Therefore, we are proposing an integration of the decomposition structure and the reweighting scheme. In the next section, the proposed L-shaped BONUS algorithm is detailed which achieves the same.

5.4. Proposed Algorithm: L-shaped BONUS

The sampling based algorithm suffers from the computational bottleneck of repeated model simulations, while BONUS uses the reweighting approach to bypass this problem. The proposed algorithm is an integration of the sampling based L-shaped method with BONUS.

The central idea of reweighting in BONUS is utilized in this algorithm. The modification is in the second stage recourse function calculation procedure of the L-shaped method. Since the structure of the algorithm is based on the L-shaped method and the application of the reweighting concept is similar to that in BONUS, the mathematical details are not reproduced here, and those can be found in sections 5.2. and 5.3.. The proposed algorithm is shown schematically in Fig. 7 and explained below.

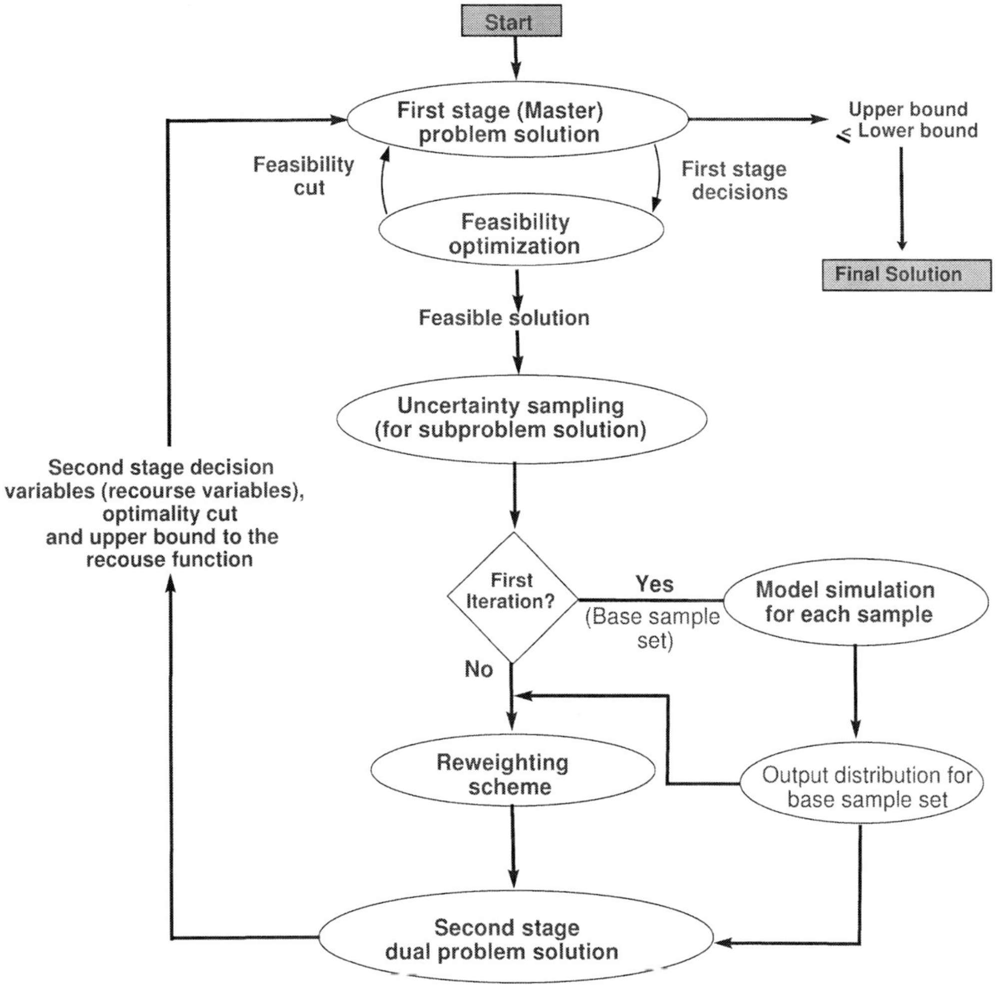

Figure 7. Proposed L-shaped BONUS algorithm structure.

5.4.1. Algorithm Details

The given stochastic programming problem is first converted into a two stage stochastic programming problem with recourse. The first stage decisions are taken using a linearized approximation of the second stage nonlinear recourse function and utilizing the feasibility and the optimality cuts, if generated. This also determines the lower bound for the objective function value. The second stage objective is the expected value of the recourse function, which depends on the first as well as the second stage decision (recourse) variables. Following the sampling based L-shaped method structure, the first stage decisions are passed on to the second stage, where the sub-problem is solved for each uncertainty realization. The idea of the proposed algorithm is to reduce the computations at the sub-problem solution stage by using the reweighting scheme to bypass the nonlinear model computations. The reweighting scheme, as mentioned in section 5.2, needs the model output distribution for a base case uniform input distribution. For this purpose, during the first optimization iteration, the nonlinear model is simulated, and the sub-problem is solved for each sample. The model simulation results for the base case constitute the base case output distribution. The sub-problem solution for each sample is used to derive the optimality cut for the master problem and generate an upper bound for the objective function value according to the L-shaped algorithm. The second optimization iteration solves the first stage master problem using these cuts. The new first stage decisions along with an updated lower bound are passed on to the second stage problem. During this iteration, when the new set of samples is taken by the stochastic modeler, the model simulation and the sub-problem solution steps are not performed for each sample. The reweighting scheme, with Gaussian Kernel Density Estimation, as explained in section 5.2, is used to predict the probabilistic values (expectation) of the model output. The base case output distribution along with the two sample sets are used for this prediction. The expected value of the model output is used to solve only one second stage dual sub-problem to generate the cuts and to update the upper bound on the objective function value. It should be noted that for the second iteration, only one sub-problem is solved. Thus, not only the nonlinear model simulation time but also the sub-problem solution time is saved. This procedure of reweighting based estimation is continued in every subsequent iteration till the termination criteria based on the L-shaped method is encountered.

The primary advantage of the proposed L-shaped BONUS algorithm, as has been repeatedly stressed, is its computational efficiency. Repeated model simulations, which are a bottleneck in the stochastic optimization procedure, being avoided, the problem solution becomes faster. The effect is expected to be more pronounced in the case of nonlinear and/or high dimensional models, as the ones often encountered in the real life applications. Another advantage of this algorithm is its ability to convert an SNLP problem into an SLP problem by using reweighting to approximate the nonlinear relationships. The disadvantage is that reweighting is an approximation and the quality of this approximation is a point of contention. It has been shown that the estimation accuracy of reweighting improves with the increasing sample size, which also increases the computational load to a certain extent. The exact quantitative nature of this relationship is difficult to establish. It is thought that it will depend on the particular nonlinear system. For the details, the readers are referred to Sahin & Diwekar (2004). Finally, as with any sampling based optimization technique, the sampling properties are very important for this algorithm. The accuracy of the reweighting scheme depends on the number and uniformity of the samples. For this algorithm, we propose to use the Hammersley Sequence Sampling (HSS), which is shown to be very efficient (Kalagnanam & Diwekar, 1997; Wang et al., 2004). The sampling technique is based on the generation and inversion of the Hammersley points, and is shown to have k dimensional uniformity property.

5.5. Algorithm Results and Discussion

The proposed L-shaped BONUS algorithm has been used to solve different stochastic nonlinear programming problems (Shastri & Diwekar, 2006a, 2008a). These problems include an illustrative farmer's problem, a process systems engineering relevant chemical blending problem (Shastri & Diwekar, 2006a) and a large scale problem of sensor placement in water distribution network (Shastri & Diwekar, 2006b). Another application to water pollutant trading is discussed later in the chapter. Table 1 gives the important information about these problems. The detailed problem description, found in Shastri & Diwekar (2006a), and Shastri & Diwekar (2006b), has been omitted here for the sake of brevity. A summary of the results for these problems is reported in Table 2. The important conclusions that can be drawn from the results are:

- The computational time increases exponentially for the sampling based L-shaped method while it increases linearly for the proposed algorithm.
- The results for the two methods differ by less than 5% in all problems, indicating that the accuracy is not compromised.
- Hammersley Sequence sampling technique is better than Monte Carlo sampling technique for this algorithm.
- The algorithm converts the stochastic nonlinear programming problem into a stochastic linear programming problem by using reweighting to bypass the model nonlinearity.
- Off-line model simulations can be used, thereby allowing the use of black-box model simulation results in the optimization algorithm.

Table 1. Problems solved using L-shaped BONUS algorithm

	Farmer's problem	Chemical blending problem	Sensor placement problem
Aim	Maximize profit	Minimize blending cost	Minimize risk due to contamination attack
Decisions	Planting strategy for the crops	Blending strategy of the basic components	Sensor locations in the network
Nonlinearity	Relationship between atmospheric parameters and seasonal yield	Relationship between fraction of impurity and total impurity parameter	Relationship between nodal demands and network flow patterns
Uncertainty	Atmospheric parameters	Individual impurity fraction	Nodal demands in the water network

Table 2. Comparison of the results of different problems solved using the L-shaped method and the L-shaped BONUS algorithm

Problem	Results					Other advantages
	Sample Size	Number of iterations/CPU time (sec)		Objective function value		
		L-shaped	L-shaped BONUS	L-shaped	L-shaped BONUS	
Farmer's problem	100	26/NA	7/NA	−122804	−123513	Converts a stochastic nonlinear programming (SNLP) problem into a stochastic linear programming (SLP) problem
	200	38/NA	7/NA	−122997	−125840	
	300	45/NA	7/NA	−123012	−125872	
	400	25/NA	7/NA	−123077	−125356	
	500	54/NA	7/NA	−123109	−125224	

Table 2. Continued

Problem	Results					Other advantages
	Sample Size	Number of iterations/CPU time (sec)		Objective function value		
		L-shaped	L-shaped BONUS	L-shaped	L-shaped BONUS	
Chemical blending problem	100	11/16.32	2/1.98	64980	66054	Converts a stochastic nonlinear programming (SNLP) problem into a stochastic linear programming (SLP) problem
	200	9/25.84	2/4.54	64979	66027	
	300	6/26.59	2/7.12	64977	66010	
	400	5/29.64	2/10.03	64977	66006	
	500	6/43.71	2/13.28	64976	66003	
Sensor placement problem	100	NA[1]	11	NA[1]	1.145×10^7	Allows the use of black-box simulation models in optimization algorithm

[1]Not solved using the L-shaped method: Black-box EPANET model simulation is needed for the sub-problem solution for each sample

In conclusion, the better computational properties, aided with the advantages of allowing black-box model simulations, make the L-shaped BONUS algorithm a valuable tool in the field of stochastic optimization.

6. Stochastic Optimal Control Theory

In general, control refers to a closed loop system, where the desired operating point is compared with an actual operating point and the knowledge of difference is fed back to the system. Conventional frequency domain techniques are then used to design a controller. Optimal control problems on the other hand are defined in time domain, and their solution requires establishing an index of performance for the system and designing the course of action (future) so as to optimize the performance index (Diwekar, 1996).

For the deterministic case, Pontryagin's maximum principle can be used to formulate the optimal control problem. For the stochastic case though, methods based on the Ito's lemma need to be used. Stochastic dynamic programming, an extension of the dynamic programming technique (Hamilton-Jacobi-Bellman equation), has been proposed but leads to a complex partial differential equation. Instead, this work applies recently proposed stochastic maximum principle (Rico-Ramirez et al., 2003; Rico-Ramirez & Diwekar, 2004). The main advantage of using this approach is that the solution to the partial differential equations in dynamic programming formulation is avoided. Instead, a set of ordinary differential equations needs to be solved as a boundary value problem. The problem formulation is briefly explained below.

Consider a system represented by the following set of differential equations.

$$dx = f(x, u, t)\, dt + g\, dz \tag{12}$$

where, x is the state variable vector of dimension n ($x(t) \in R^n$), and u is the control variable vector of dimension m ($u(t) \in R^m$). The starting condition for the state vector is given by $x(t_0) = x_0$, and the final condition at time T is $x(T)$. Let $1, ..., n_k$ be the set of deterministic states, and $n_{k+1}, ..., n$ be the set of uncertain states. The second part of Eq. (12) models the uncertainty. For deterministic states, the function $g = 0$. In optimal control, there is a time dependent performance index, which, in this case, is represented as:

$$J(t_0) = \int_{t_0}^{T} F(x(t), u(t), t)dt \tag{13}$$

where, F is the function to be optimized over the time interval of $[t_0, T]$. The Hamiltonian for this stochastic case is defined as:

$$H(x, u, t) = F(x, u, t) + \lambda'\, f(x, u, t) + \frac{1}{2}\, g^2\, w \tag{14}$$

where, $\lambda(t)$ is the set of costate or adjoint variables ($\lambda(t) \in R^n$) (the first derivatives of the objective function F with respect to state variables), and λ' represents the matrix transpose. $\omega(t)$ represent the second derivatives of the objective function F with respect to the state variables. This term is included due to the Ito process contribution. The optimal control law is then given by the solution of the following set of equations:

State Equation

$$\dot{x}_i = \frac{\partial H}{\partial \lambda_i} = f \qquad i = 1, ..., n \tag{15}$$

Costate Equation

$$-\dot{\lambda}_i = \frac{\partial H}{\partial x_i} = \frac{\partial f'}{\partial x_i}\lambda_i + \frac{\partial F}{\partial x_i} \qquad i = 1, ..., n \tag{16}$$

$$\frac{dw_j}{dt} = -2\, w_j \frac{\partial}{\partial x_j} f_j - \frac{1}{2} w_j \frac{\partial^2}{\partial x_j^2}(g_j^2) - \lambda_j \frac{\partial^2}{\partial x_j^2} f_j \qquad j = n_{k+1}, ..., n \tag{17}$$

Stationarity Condition

$$0 = \frac{\partial H}{\partial u_p} = \frac{\partial F}{\partial u_p} + \frac{\partial f'}{\partial u_p}\lambda \qquad p = 1, ..., m \tag{18}$$

This is a set of $2n+(n-n_k)$ ordinary differential equations (state and costate equations) and m algebraic equations (stationarity condition), and it is solved as a boundary value problem. The boundary values of the state and costate variables depend on the problem specification, while the boundary values for w are given as $w(T) = 0$ (Rico-Ramirez & Diwekar, 2004). The control trajectory obtained is optimal for the considered objective function and starting conditions. Previous applications of stochastic maximum principle to solve optimal control problems can be found in Shastri & Diwekar (2006d); Ulas & Diwekar (2004); Rico-Ramirez & Diwekar (2004).

The preceding sections reviewed important systems theory based techniques that are useful to deal with problems in sustainability. The remaining part of the chapter presents the application of these techniques to industrial waste management using mercury waste in water as the case study. The work is explained in the following sections.

7. Sustainable Waste Management: Case of Mercury Pollution

Mercury is recognized internationally as an important pollutant and is fast becoming a major concern to the environmentalists, primarily due to better understanding of its harmful environmental and health impacts. For humans, the primary targets for toxicity of mercury and mercury compounds are the nervous system, kidney, and developing fetus. Other systems in the human body that may be affected include respiratory, cardiovascular, gastrointestinal, hematologic, immune and reproductive (USEPA, 1997b).

Mercury can cycle in the environment in all media as part of both natural and anthropogenic activities. Fig. 8 gives the pictorial representation of the atmospheric mercury cycling. Majority of mercury is emitted in air, where coal fired power plants are the major sources. Other polluters include waste incinerators, chlor-alkali plants etc. However, most of the mercury present in air (in elemental or inorganic form) is deposited into various water bodies such as lakes, rivers and oceans. This transfer of mercury is attributed to the processes of dry and wet deposition. In addition, the water bodies are enriched in mercury due to direct industrial waste water discharge, storm water runoffs, agricultural runoffs etc. Once present in water, mercury is highly dangerous not only to the aquatic communities but also to humans through direct and indirect effects (USEPA, 2000). Inorganic mercury in water is converted into organic methyl mercury through the process of methylation. Methyl mercury accumulates up the aquatic food chains, so that organisms in higher trophic levels have higher mercury concentrations (Jensen & Jernelöv, 1969; DeSimone et al., 1973). The consumption of these aquatic animals by humans and wild animals further aids bioaccumulation along the food chain. As a result, contaminated fish consumption is the most predominant path of human exposure to mercury. This has resulted in fish consumption advisories at various water bodies throughout the US.

Owing to such a complex cycling, successful management of mercury pollution should consider management strategies at various stages of the cycle. While doing so, it is essential to juxtapose the environmental, economic and social objectives. However, simultaneous consideration of such disparate aspects while devising strategies is a challenging task. The concept of sustainability and sustainable development can contribute towards achieving this objective. The presented work is a step in this direction, where the aim is to address the mercury cycling issues through the multidisciplinary concept of sustainability, targeting overall sustenance. Since the idea of sustainability changes with the system being considered, different objectives, management options and tools are applied for different applications. The work proposes management strategies at following three different stages of the mercury cycle.

- Industry level management: Simultaneous selection and design of mercury treatment technologies to minimize treatment cost
- Inter-industry (industrial sector) level management: Technology selection and pollutant trading to achieve environmental objectives at reduced overall cost, and
- Ecosystem level management: Control of water pH using liming to reduce the bioaccumulative effects of mercury present in water bodies

Due to the stated importance of uncertainty consideration for a robust analysis, each stage of the work incorporates uncertainty through effective modeling and efficient solution techniques.

8. Industrial Level Management

The major contribution of mercury to the environment comes from industries, which includes air polluters like coal fired power plants, and liquid waste dischargers, like Public Owned Treatment Works (POWT). Regulations on the permitted mercury discharge are getting increasingly strict, thereby increasing the waste treatment costs. It is important therefore for the industries to optimize their waste treatment option by selecting the right technology and the correct design and operating specifications.

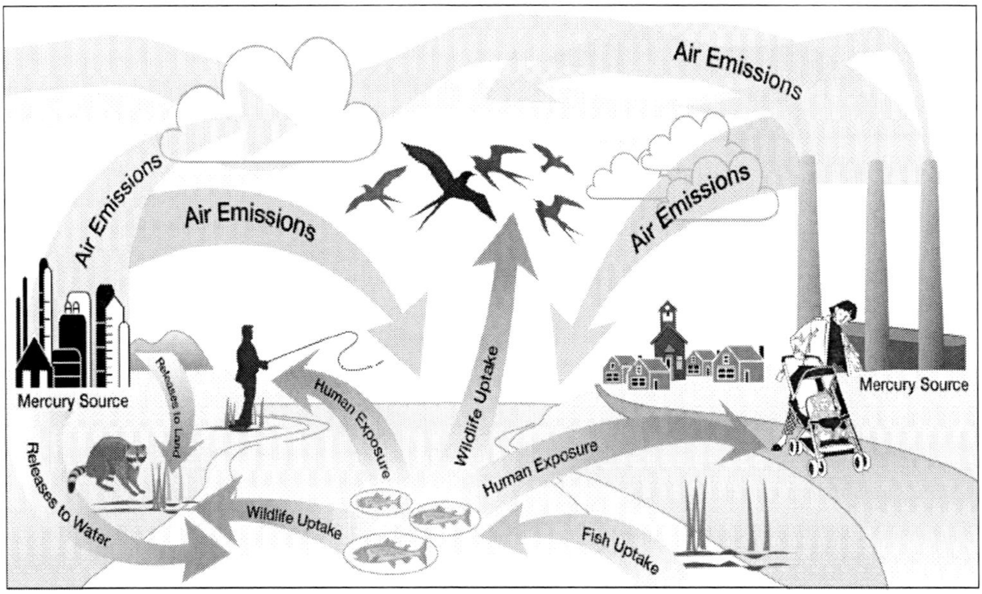

Figure 8. Environmental cycling of mercury.

The idea envisioned here for industry level management is the simultaneous optimization of technology selection and design decisions for mercury waste treatment. The primary task is to model the treatment technologies and decision making aspects of the problem. This includes the modeling of process dynamics and costs for various mercury treatment technologies, followed by the identification of appropriate decision variables and objectives. The resulting framework will involve discrete decisions, such as technology selection, as well as continuous decisions, such as design parameters. Given that there are new technologies emerging in this field, limited performance data results in uncertainties in parameters. These uncertainties need to be properly characterized, quantified and included in decision making. The continuous design decisions might be governed by nonlinear relationships, further complicating the task. The challenging task is to integrate these aspects to achieve overall optimization through simultaneous decision making. The second aspect of the work is solve the resulting optimization problem using appropriate algorithmic tools. In optimization terminologies, the proposed problem is a stochastic mixed integer nonlinear programming problem. Section 5 describes a tool that will aid in efficiently solving these computationally challenging problems.

9. Industrial Sector Level Management: Pollutant Trading

In the wake of increasingly stringent discharge regulations, selection of the right technology by the individual industry is not sufficient, and innovative methods are required that will analyze the problem from industrial as well as environmental perspective, achieving simultaneous sustainability. Pollutant trading is a possible option.

Pollutant trading is a market based strategy to economically achieve environmental resource management. The overall goal is to attain the same or better environmental performance with respect to pollution management at a lower overall cost. Since the U.S. Environmental Protection Agency (USEPA) issued its Emissions Trading Policy in 1986, the most frequent application of this approach has been in SOx trading (Solomon, 1998). The strategy has been found to achieve desirable cost reductions or environmental benefits in many applications (Kling, 1994; Bohi & Burtraw, 1998; Farrell, 2000). On March 15, 2005, USEPA issued the federal rule allowing cap and trade policy to reduce mercury emissions from coal-fired power plants. This relative success of trading to deal with air pollutants has encouraged the introduction of effluent trading concept for water pollution control, making it a possible option for managing mercury waste. Pollutant trading adds flexibility and introduces new options to the policy makers and industries alike. A policy maker, while aiming for a better environment and efficient waste management, has to understand the industry constraints, since the regulations need to be satisfied by the industries. With increasingly stringent environmental regulations, waste management has become an important financial consideration for the industries. Trading offers these industries an option to satisfy the regulations at reduced costs. Individual industry level decisions are thus affecting the overall goal. Such decision are beyond heuristics and a systematic approach is called for.

The presented work uses optimization techniques to develop a decision making framework that will guide industries in taking optimal decisions in wake of the added flexibility due to trading. The model will also guide regulators in developing optimal regulations in different situations. Work in Farrell et al. (1999) presents one such analysis for NOx pollution. It analyzes the effect of trading on overall cost and emission reductions, and suggests industry level decisions. This work performs a similar analysis for mercury trading in watershed management. Compared to the model presented in Farrell et al. (1999), this model also accounts for the health care costs, arising through pollutant exposure, to assess the benefits of various decisions. For a watershed, the framework can be used to develop efficient regulations such as TMDLs (Total Maximum Daily Load), given the watershed, industries and technology specific details, optimizing the industry decisions. The model is then extended to include health care cost as an objective in decision making, which further distinguishes it from the model in Farrell et al. (1999). The effect of uncertainty, prevalent in most situations, on model solutions is also analyzed using chance constrained programming.

The proposed model is applied on a Savannah River watershed mercury waste management case study. It is opined that the application of trading for mercury is not advisable since mercury cycle in the environment is still not fully understood, partly due to its bioaccumulative potential and slow, long term effects (USEPA, 1996). Consideration of human health care costs resulting from mercury exposure partially addresses this issue by quantifying the long term effects of mercury pollution. The problem formulation also ensures that the discharge conditions can not deteriorate over the existing ones, and hence trading will not compromise the environment.

Currently, trading is practiced mostly in a decentralized manner. Given the trading parameters, each participant strives to maximize its benefits. The role of the regulatory agency is then restricted to setting the trading parameters. Another way of practicing trading is centralized, where the regulatory agency has a major role to play. In the presented work, the conventional economic goals of trading are merged with the environmental and social goals. This being the case, the centralized trading mechanism has greater potential, since it can simultaneously consider the various aspects while taking decisions. This will mean that the role of the regulatory authorities will be paramount to ensure the success of pollutant trading. This work therefore might be viewed from two perspectives. The possibility of using trading for mercury waste management is explored through the Savannah river case study. Consideration of aquatic mercury treatment technologies and health care costs at the watershed level helps in properly assessing the perceived benefits of this technique for mercury. The basic optimization model on the other hand, is applicable to any pollutant and media. It is thus a step towards an integrated policy making framework for sustainable waste management.

9.1. Pollutant Trading

Environmental credit trading is an approach to environmental protection that uses market based mechanisms to efficiently allocate emission or pollutant reductions among sources with different marginal control costs. The concept is attributed to Crocker (1966), Dales (1968) and Montgomery (1972). Early applications of trading were designed to provide greater flexibility for emission sources to meet air quality standards in a cost-effective manner (Tietenberg, 1985). Recently, Bush administration has approved a cap and trade policy for mercury air pollution in coal-fired power plants, which aims to reduce mercury loading by about 70% by 2018.

Apart from the environmental gains from practising trading, considerable focus has been given to the economic aspects, understanding the various economic implications of trading (Tietenberg, 1995). This includes the study of overall economics of pollutant trading (Newell & Stavins, 2003; Tietenberg, 2000; Hahn & Hester, 1989; Boumol & Oates, 1988). Research has also focused on various specific aspects of pollutant trading and their impact on the economic performance. This includes the analysis of trading ratio (Woodward, 2001; Horan, 2001), transaction costs (Cason & Gangadharan, 2003; Gangadharan, 2000) and initial allocation of the permits (Kampas & White, 2003). McKitrick (1999) analyzed the dependence on the relationship between the technology of the point source and the marginal abatement costs. The focus of many of these articles has been to understand the economic consequences by studying the trading equilibrium.

Despite the extensive study of trading economics, work on the actual implementation of trading and its effect on industry level decisions has been relatively scarce. Selection of the right technology has always been complicated for an industry, and trading further complicates it due to added flexibility. Oates et al. (1989) and Atkinson & Tietenberg (1982) have also highlighted the importance of including emission level differences while comparing trading with the traditional command and control approaches. This work tries to include both these aspects in understanding the trading mechanism. The work integrates the major economic interactions of individual industries, trading and the environment, and thereby studies the trading implications. The next section describes watershed based trading, which is the focus of this work.

9.2. Watershed Based Trading

9.2.1. Overview

Following on the lines of air pollutant trading, trading principles have also been sanctioned by USEPA for water pollution problems on a limited basis since the early 1980s. USEPA has since then formalized the concept into a framework to guide effective implementation of water pollutant trading (USEPA, 1996, 2003). Trading is based on the fact that sources in a watershed can face very different costs to control the emission of the same pollutant. Trading programs allow facilities facing higher pollution control costs to meet their regulatory obligations by purchasing environmentally equivalent (or superior) pollution reductions from another source at lower cost. Firms having financial capability and infrastructure to perform pollutant reduction below the required limit get credits for it. These credits can be sold to another firm to gain monetary benefits or banked for future use, if that is a possibility.

Pollutant dischargers are mainly classified as either point or non-point sources. Point sources are defined as those having direct and measurable emissions (e.g. industries, municipal waste treatment plants etc.), while non-point sources are those with diffused emissions that are difficult to measure (e.g. agricultural or storm-water runoffs). Amongst the various possibilities of carrying out trading, the one between point sources is thought to be simpler and achievable. This is owing to their measurable discharges in terms of quality and quantity, and also due to the measurable assessment of pollution reduction techniques.

Point sources are regulated under the National Pollution Discharge Elimination System (NPDES) established under the Clean Water Act (CWA). Under NPDES, all facilities which discharge pollutants from any point source into waters of the United States are required to obtain a permit allowing them to discharge only a certain amount of pollutant. The permit provides two levels of control: technology-based limits (based on the ability of the dischargers in the same industrial category to treat wastewater), and water quality-based limits (if technology-based limits are not sufficient to provide protection of the water body). The permits can be individual (specific to a company) or general (applicable to a group of companies). The existence of general water quality-based permits in a watershed is equivalent to the concept of Total Maximum Daily Load (TMDL) in the watershed. TMDL is established by the state for waterbodies or watersheds where technology-based requirements alone are not sufficient to attain water quality goals. TMDL establishes the loading capacity of a defined watershed area, identifies reductions or other remedial activities needed to achieve water quality standards, identifies sources, and recommends waste load allocation for point (and non-point) sources. EPA has proposed two possible trading mechanisms for point sources (USEPA, 1996).

- Trades can occur in the context of individual point source permits. In this case, different point sources have individual (technology or water-quality based) pollutant permits and there is no common water quality based limit. Treatment cost differentials encourage trading between various point sources.

- Trades can also occur through the development of TMDL or other equivalent analytical framework. In this case, water quality-based limits, common to all the point sources in the watershed, are established. This provides a starting point to compare the costs of the baseline responsibilities necessary to achieve water quality goals with alternative allocations. Parties to the trade then negotiate within the loading capacity determined under the TMDL so as to satisfy the common discharge objective.

Various parameters that affect the economics of trading are trading ratio (how many units of pollutant reduction a source must purchase to receive credit for one unit of load reduction), transaction costs (expenses for trading participants that occur only as a result of trading), number of participants, availability of cost data and uncertainties related to continued industry participation and data availability.

9.2.2. Implementation Perspective

Once the TMDL has been established, the point source is assigned a particular load allocation and may need to reduce its discharge to satisfy the allocation. It has two options to accomplish this:

- Implementing end of pipe treatment methods, which entail certain capital and operating costs depending on the existing technology, the amount being treated and level of reduction being achieved. These costs will vary for different point sources
- Trading a particular amount of pollutant to another point source in the watershed that is able to reduces its discharge more than that specified by the regulation

The major reason for pollutant trading being attractive is the flexibility it offers, not only for the polluters, but also for the policy makers. From the polluter's perspective, multiple waste treatment options are typically available for implementation. Therefore, in the presence of multiple polluters, multiple technologies and option of trading, decisions such as if and how much to trade become difficult for the polluters. For policy makers, trading opens up new avenues to achieve better environmental goals at lower overall cost, for which the policy maker has various parameters like trading ratio, trading transaction cost, number of participants at his disposal. The policy maker, although driven by environmental objectives, must also consider industry capabilities and limitations. It is also important to understand the effect of trading on the final reductions. Whatever be the discharge level of a pollutant, even below the TMDL regulation, it adversely affects the environment and humans. Consideration of such effects is important in decision making. Finally, information about of the watershed and trading, if available while finalizing the regulation, might influence the regulation development itself. In this wake, a formal decision making framework that takes into account these aspects and guides the policy maker and polluters in decision making is desirable.

Use of optimization techniques offers a way to achieve this, where the objective of overall cost minimization can be achieved by incorporating the watershed and pollutant specific details through constraints. Once the modelling framework has been developed, it also opens up the option of exploring various scenarios to finalize the best settings. The next section explains a basic optimization model which is a step towards this objective.

9.3. Optimization Based Decision Making Model

The formulation considers that TMDL regulation has already been developed by the state in consultation with USEPA. This translates into a specific load allocation for each point source. The model includes only point sources, primarily due to two reasons. Firstly, trading between point sources is thought to be easier for implementation. Secondly, majority of the industries, which are the focus of this chapter, can be classified as point sources.

Consider a set of point sources (PS_i), $i = 1, ..., N$, disposing pollutant containing waste water to a common water body or watershed. The various point source specific parameters are:
D_i = Discharge quantity of polluted water from PS_i [volume/year]
c_i = Current pollutant discharge concentration for PS_i [mass/volume]
a_i = Current pollutant discharge quantity for PS_i [mass/year]
L_i = Load allocation to PS_i based on TMDL regulation [mass/year]
red_i = Desired pollutant quantity reduction in discharge of PS_i [mass/year]
P_i = Treatment cost incurred by PS_i to reduce pollution when trading is not possible.

Here, the value of red_i is given by

$$red_i = D_i.c_i - L_i = a_i - L_i \tag{19}$$

Every PS has the option of trading or implementing a particular waste reduction technology. Let $j = 1, ..., M$ be the set of reduction technologies available to the point sources for implementation. The technology specific parameters are:

TC_j= Total treatment plant cost [$/volume]
q_j = Pollution reduction possible from the process [mass/volume]

The total plant cost TC_j is the sum of the annualized capital cost and annual operating and maintenance cost. The annualized capital cost in turn depends on the total equipment and setup cost (USDOI, 2001). Trading is possible between all point sources. For simplicity, a single trading policy exists between all possible pairs of point sources, and a single trading ratio and transaction fee is applicable to all the trades. Let r be the trading ratio and F be the transaction cost (in $/mass) to be paid by the point source trading its pollutants. The objective of the model is to achieve the desired TMDL goal at *minimum overall cost.*

Let b_{ij} be the binary variables representing the point source-technology correlation. The variable is 1 when PS i installs technology j, and 0 otherwise. Let t_{ik} (mass/year) be the amount of pollutant traded by PS i with PS k, i.e. PS i pays PS k to take care of its own pollution. All the parameters are on annual basis. The problem is then formulated as follows:

Objective :

$$\text{Minimize} \qquad \sum_{i=1}^{N}\sum_{j=1}^{M} TC_j.D_i.\, b_{ij} \tag{20}$$

Constraints :

$$t_{ii} = 0 \qquad \forall i = 1, ..., N \tag{21}$$

$$red_i \le \sum_{j=1}^{M} q_j.D_i.\, b_{ij} + \sum_{k=1}^{N} t_{ik} - r\sum_{k=1}^{N} t_{ki} \qquad \forall i = 1, ..., N \tag{22}$$

$$P_i \ge \sum_{j=1}^{M} b_{ij}.TC_j.D_i + F\Big(\sum_{k=1}^{N} t_{ik} - \sum_{k=1}^{N} t_{ki}\Big) \qquad \forall i = 1, ..., N \tag{23}$$

The objective function gives the sum of the technology implementation cost for all point sources. Although each PS will also spend or gain from practicing trading, expense for one PS in a watershed is earning for one or more PS in the same watershed. As a result, for the complete watershed, trading does not contribute to the cost objective. The first set of constraints eliminates trading within the same PS. The second set of constraints ensures that all the regulations are satisfied. The reduction of the pollutant discharge at the end of technology implementation and/or trading must be at least equal to the targeted reduction, for all the PS. The trading ratio r is usually set higher than 1, which requires the buyer to purchase more units of reduction than they have to achieve without trading, to include a margin of safety in trading mechanism. This is done to account for uncertainties in the level of control needed to attain water quality standards, and to provide a buffer in case traded reductions are less effective than expected (USEPA, 1996). Consequently, the PS taking the pollutant has to reduce the mercury by an amount equal to the actual quantity traded times the trading ratio. Accordingly, in Eq. 22, the reduction achieved by industry i reflects the amount traded by industry k (t_{ki}) modified by the trading ratio r. The last constraint ensures that the expenses incurred by each PS with trading are not more that those without trading. This is because the trading framework by USEPA mentions that no polluter can be forced to trade. As a result, a polluter will participate in trading only if there is a financial incentive for it. In the absence of this constraint, the problem solution might force some industries to spend more when trading is practiced, if such as solution results in lower overall cost. This clearly is an unlikely scenario which is avoided by this constraint.

The problem given by Eqs. 20-23 is a mixed integer linear programming problem (MILP). The decision variables in the problem are binary variables b_{ij} and continuous variables t_{ik}. In the subsequent text, this formulation is referred to as 'formulation A'. It should be noted that the reduction capabilities of different treatment processes are considered to be fixed at q_j. Therefore, selection of a technology by a PS to achieve reduction targets red_i is a discrete process. For comparison, the optimization problem is also solved when the industries can not trade, and hence have to implement a technology. For this case, the optimization problem is constituted by Eqs. 20-22 and forcing t_{ik} and t_{ki} equal to zero. P_i in Eq. 23 for the trading problem solution is provided by the solution of the optimization problem without trading.

It is important to ascertain that the trading solution achieves cost-effectiveness. The literature on pollutant permit trading has extensively discussed the aspects of market equilibrium under trading systems (Hung & Shaw, 2005; Weber, 2001). For the proposed optimization model in the absence of trading (equivalent to the model in the presence of environmental constraints only), the constraint is written as:

$$red_i \leq \sum_{j=1}^{M} q_j.D_i.\, b_{ij} \qquad \forall i = 1, ..., N \tag{24}$$

Comparing Eqs. 22 and 24, it is obvious that the solution to the problem without trading (satisfying the constraint of Eq. 24) is a subset of the solution with trading. Furthermore, since each industry has to achieve the minimum reduction target red_i, with or without trading, it is not possible to violate the constraint while minimizing the cost. In the limiting case, a PS acts as a middle agent in pollutant trading (i.e., $\sum_{k=1}^{N} t_{ik} = r \sum_{k=1}^{N} t_{ki}$). Since the minimum trading ratio value in this setting is 1, such an action by a PS will result in no profit out of trading, and will result in a loss if the trading ratio is greater than 1. These things indicate that the solution to the problem with trading has to be at least as effective as the one without trading.

The model in Farrell et al. (1999) for NOx budget is also a mixed integer problem with a similar objective of cost minimization for all sources. It selects the combination of control technologies for each plant to yield the lowest cost. However, there are a few differences in the two models. The NOx model takes decisions at the individual boiler level, while the proposed model takes decisions at the point source (company) level. The NOx model is a dynamic model and the decisions are made over a period of time to maximize the net present value. This permits the analysis of banking of trades over time. The proposed model on the other hand is not dynamic, since the EPA framework does not mention any option of banking of trades for water effluent trading. The structure of the NOx model is such that, although emission goals over a period of time are achieved, there is a possibility of excessive emissions at localized space or time instants. This possibility does not exist in the proposed model, due to the constraint by Eq. 22. Further differences arise when the proposed model is extended to include health care cost, first for solution comparison and later in the objective function. Although Farrell et al. (1999) discuss various sources of uncertainty, the optimization model does not solve the problem considering uncertainty. The proposed model, on the other hand, is extended to systematically include uncertainty in decision making through chance constrained programming. These differences will become evident as the extended models are discussed later in the article.

The problem formulated above is quite general, applicable to any watershed and any pollutant. The next section discusses the application of the model on a case study of mercury waste management in Savannah River basin.

9.4. Savannah River Case Study

9.4.1. Savannah River Watershed Details

TMDL has been established for five contiguous segments of the Savannah River. These segments are included in the list of impaired water bodies since the tissue mercury concentration in certain fish species exceeds the Georgia Department of Natural Resources (GDNR) Fish Consumption Guidelines. Although the mid-line of the Savannah River serves as the east-west boundary between the states of Georgia and South Carolina, the TMDL does not provide wasteload allocations to South Carolina NPDES facilities. This TMDL reflects the assumption that the concentrations of mercury in the South Carolina portion of Savannah River will meet the applicable Georgia water quality standards at the South Carolina-Georgia border. In order to develop the TMDL, the applicable water quality standard, which gives the maximum safe concentration of mercury in water, must be determined. EPA determined the applicable water quality standard for total mercury in the ambient water of the Savannah River Basin to be 2.8 ng/l (parts per trillion). At this concentration, or below, fish tissue residue concentrations of mercury will not exceed 0.4 mg/kg, which is protective of the general population from the consumption of freshwater fish. This interpretation of Georgia's water quality standard was based on site-specific data gathered for the Savannah River in 2000 specifically for the purpose of this TMDL. The loading of mercury from the watershed into the Savannah River was simulated using a Watershed Characterization System (WCS) model developed by EPA Region 4 (USEPA, 2001b). The water quality model known as WASP5 (Ambrose et al., 1993) is used to simulate mercury fate and transport in Savannah River. WASP5 is a general dynamic mass balance framework for modeling contaminant fate and transport in surface waters. The calculated allowable load of mercury that can come into the Savannah River without exceeding the applicable water quality standard of 2.8 ng/l as interpreted by EPA is 32.8 kilograms/year, which represents the TMDL for the watershed. Since about 99% of the mercury loading is through atmospheric sources, only 1% of the allowable load is assigned to the wasteload allocation for NPDES sources. This amounts to the combined allowable discharge of 0.33 Kg/year. Based on the current volumetric discharge of each of the NPDES sources, wasteload allocation (permitted discharge of mercury by each source) is carried out.

The impact of seasonal and spatial variations in the discharge as well as that of mercury transportation and fate on wasteload allocations is expected. This will consequently affect the trading mechanism. For the case of Savannah River basin, these issues were considered while simulating the mercury fate and transport in the concerned river basin segments using the WASP5 simulator. However, with the argument that seasonal variations are expected to be dampened due to the long process of bioaccumulation, the seasonal variations were ignored while developing the TMDL (USEPA, 2001a). Furthermore, the spatial characteristics of the basin were effectively overlooked by considering the highest mercury concentration among the various segments of the basin for TMDL calculation to provide a conservative estimate.

In all, there are 29 significant point sources discharging mercury in the Savannah River watershed. The point sources represent a wide spectrum and include 13 major municipal polluters, 12 major industrial polluters, 2 minor municipal polluters and 2 minor industrial polluters. It should be noted that there are more point sources in this region. But the discharges from those are negligible owing to their relatively low discharge volumes. One of the options of implementing the TMDL is to apply a common WQS of 2.8 ng/lit to all the point source discharges across the watershed. Therefore, under this option, the wasteload allocation for each NPDES point source identified in this TMDL would be the product of 2.8 ng/l and its permitted or design flow rate. The sum of these individual wasteload

Table 3. Point source data for the Savannah river basin

Industry	Total Volumetric Discharge (MGD-Million Gallons per Day)	Current Discharge concentrations (ng/lit)	Targeted reduction (g/year)	Treatment cost without trading ($/year)
I_1	46.1	4.65	0.1178	1.68×10^7
I_2	1.5	3.7	0.0018	355 875
I_3	4.6	4.3	0.0095	1679 000
I_4	1.5	3.4	0.0012	355 875
I_5	2.0	3.88	0.0029	730 000
I_6	2.24	3.7	0.0027	531 440
I_7	1.2	3.9	0.0018	438 000
I_8	27.0	4.83	0.0757	1.53×10^7
I_9	4.5	4.0	0.0075	1642 500
I_{10}	1.0	3.1	0.00041	237 250
I_{11}	1.0	3.06	0.00036	237 250
I_{12}	1.0	3.22	0.00058	237 250
I_{13}	2.0	3.31	0.0014	474 500
I_{14}	3.765	4.8	0.0104	2130 049
I_{15}	18.0	4.33	0.0381	6570 000
I_{16}	7.2	5.1	0.0229	4073 400
I_{17}	58.6	4.87	0.1676	3.32×10^7
I_{18}	23.0	4.52	0.0546	8395000
I_{19}	1.152	5.05	0.0036	651 744
I_{20}	0.362	4.14	0.00067	132 130
I_{21}	108.0	4.58	0.2656	3.94×10^7
I_{22}	4.68	5.2	0.0155	2647 710
I_{23}	28.09	4.41	0.0625	1.03×10^7
I_{24}	1.921	3.9	0.0029	701 165
I_{25}	0.544	4.5	0.0013	198 560
I_{26}	0.5	3.95	0.0008	182 500
I_{27}	0.003	3.72	3.81×10^{-6}	711.75
I_{28}	1.246	4.1	0.0022	454 790
I_{29}	0.054	3.4	4.47×10^{-5}	12 811.5

allocations is 0.001 kg/year, which is significantly less than the 0.33 kg/year cumulative wasteload allocation provided to all NPDES facilities (USEPA, 2001a). Current discharge concentrations for the 29 point sources are not reported in literature. For this case study, they are not considered to be related to their discharge quantities. The overall reduction needed to achieve the TMDL criteria is about 44% (USEPA, 2001a). The targeted overall reduction for the PS is therefore taken to be 40%, and the individual discharge concentrations are adjusted accordingly. Further details about the TMDL development can be found in USEPA (2001a).

Table 3 gives the various parameter values related to the point sources. The table also gives the values of red_i (targeted reduction) and P_i (treatment cost without trading) for each PS at TMDL 32 Kg/year, for which the problem is solved below.

Table 4. Data for the various treatment technologies

Process	Mercury reduction capability (ng/lit)	Capital requirement ($/1000 gallons)
Coagulation and Filtration (A)	2.0	1.0
Activated carbon adsorption (B)	3.0	1.5
Ion exchange (C)	1.0	0.6

9.4.2. Technology Details

Three treatment technologies are considered for this problem and they are available to all point sources for implementation. These include: coagulation and filtration, activated carbon adsorption and ion exchange process. The capital requirement and reduction capability of any process is expected to be (nonlinearly) related to the capacity of the treatment plant and the form and concentration of the waste to be treated, amongst many other factors. For this analysis though, such complex relationships are ignored for simplicity, and the treatment cost is only linearly related to the volume of the waste. Total plant cost data for the treatment methods is reported in USDOI (2001) as a function of the waste volume. The total plant cost includes capital as well as annual operating cost per unit volume of waste treated, calculated using the following equations (USDOI, 2001):

$$\begin{aligned}\text{Annualized capital cost} &= [\text{Total capital equipment cost} + \\ &\quad \text{Project related spacial cost}] \\ &\quad \times (\text{capital recovery factor for 30 years} \\ &\quad \text{at 3.89\% real annual interest based} \\ &\quad \text{on lagged impact on interest} = 0.057) \qquad (25)\\ \text{Total annual cost}(TC) &= \text{Annualized capital cost+ Total annual} \\ &\quad \text{operations and maintenance cost} \qquad (26)\end{aligned}$$

Since waste volumes encountered in this case study are mostly greater that 1 MGD, asymptotic values reported in USDOI (2001) are used. The treatment efficiencies depend on the waste composition and concentration. In general though, a more efficient treatment is likely to be more expensive. This criteria, along with data given in (USEPA, 1997a), is used to decide the treatment efficiencies. Table 4 gives the technology data.

9.4.3. Trading Details

Trading parameters needed for the problem solution are the trading ratio and the trading transaction cost. For a point source-point source trading, as considered in this problem, a study conducted for the USEPA for trading of toxic pollutants in Delaware River recommends a trading ratio between 1.1-1.25 (Marshall, 1999). For this problem, the ratio r is 1.1. It should be noted here that since mercury trading in water polluters has not been practiced yet, transaction fee is not easy to decide. However, an EPA document gives a hypothetical example of water quality trading (USEPA, 1996) in which the transaction fee is taken to be in the range of the per kg treatment cost of the pollutant. This is also observed for SO_2 trading, which is already practiced. The values of the per unit marginal abatement costs and trading transaction cost for SO_2 trading are reported in Burtraw et al. (2005). For this work, using the average volumetric discharge and the average desired discharge reduction for all the point sources, the average treatment cost is calculated. To calculate this cost, data for the three treatment processes reported in Table 4 is used. This gives the treatment cost in '\$/Kg of mercury'. Based on this value, the transaction cost is fixed at 4.0×10^8 \$/Kg.

9.4.4. Health Care Cost Consideration

The discharge of mercury to the watershed, although below TMDL limit, is still harmful to the humans and is consequently associated with some health care cost. From a social perspective, it is preferable to reduce the discharge as much as possible to avoid any harmful health effects. This model considers the health care cost to compare the solutions obtained for various scenarios. The cost is not considered in the objective function for the optimization problem, rather it is added to the final cost.

The bioaccumulative nature of mercury and its slow dynamics make the long term effects of mercury exposure important. Hence, it is essential to account for such effects while quantifying the health care costs. Majority of mercury accumulates in the food chain as methyl mercury. Therefore, quantification of the health care costs based on methyl mercury concentration is most appropriate. IRIS (Integrated Risk Information System) database reports the methyl mercury reference dose for chronic oral exposure (RfD), which is the highest dose of methyl mercury without any harmful effects. However, since the TMDL for the Savannah River is developed on the discharge of total mercury to the watershed, not only methyl mercury, the problem needs a quantifying measure based on total mercury. IRIS database does not report the RfD value for mercury (elemental).

Given these considerations, this work quantifies the health care costs through the LC50 (Lethal Concentration 50%) value for mercury, which quantifies the harmful effects of long term exposure to mercury. The health care cost is a function of the final overall mercury discharge. This discharge value is used to calculate the average mercury consumption by humans. When compared with the LC50 value, this gives the approximate mortality rate of humans. The parameters required to calculate this cost are:

LC50 value for mercury = LC50 = 350 μg/liter
Average fish consumption = F_{avg}= 17.5 grams per person per day,
Average water consumption = W_{avg}= 2 liters per person per day,
Safe concentration of mercury in fish = Hg_{safe}= 0.4 mg/Kg,
Population affected by the consumption = P = 10000,
Compensation for the health cost = C_{health} = \$ 1 Million per person, and
Final average water quality standard = WQS_{final}

Values of F_{avg}, W_{avg} and Hg_{safe} are taken from USEPA (2001a). The LC50 value is reported in the material and safety data sheet for mercury by Fisher Scientific. The compensation amount is based on the typical insurance compensations received in road accident fatalities. Another method of calculating the approximate health care compensation is through a quick present value calculation. Assume the average salary of a person to be $ 60,000 per annum and the average period of employment to be 30 years. Then, considering the discount rate of 5% per annum, the present value for the person is around $ 1 Million. The watershed population affected by the polluted water is based on the data reported by US Census Bureau. The overall health care cost is then given as

$$\text{Health Cost} = \left(\frac{WQS_{final}.Hg_{safe}}{WQS} \frac{1000.F_{avg}.P}{2.LC50.W_{avg}} \right).C_{health} \tag{27}$$

where, WQS is the targeted water quality standard resulting in safe desired level of mercury in fish and the term in the brackets represents the mortality rate. The mortality rate is calculated by calculating the total amount of mercury consumed per day by a human, and comparing it with the average per day mercury intake by a human through drinking, assuming that health effects due to drinking represent the long term effects quantified by the LC50 value.

9.5. Savannah Case Study: Results and Discussion

The problem formulation is given by Eqs. 20-23. In the subsequent text, analysis when trading is not permitted is referred to as 'technology option', while the analysis when trading is allowed is referred to as 'trading option'. Following different studies are conducted for the problem:

- Comparing the solutions for technology option and trading option to quantify the impact of trading on the cost to satisfy regulations
- Analyzing the dependence of the solutions for technology option and trading option on the TMDL value to determine the effect of regulations on the problem that will guide the policy makers

The next two sections report the results.

9.5.1. Comparison between Trading and Non-trading Solutions

TMDL for this analysis is 32 Kg/year and the total mercury reduction target for all the point sources is 0.911 g/year. A comparison of the solutions obtained for the two cases is given in Table 5. It can be seen that there is about 18% reduction in the treatment cost as a result of implementing trading, which amounts to around 27 Million $ annually. This supports the expected result that trading will reduce the overall expenditure. However, while satisfying TMDL, the total mercury discharge for the trading option is higher by about 17%. This becomes significant when one considers the health care costs associated with the mercury discharge. A comparison of the health care costs shows that the cost for technology option is about 14% less than that for trading option, a difference of 7 Million $. Thus, considering only the reduction costs, trading option appears to be economically beneficial, but it is not necessarily so in the wake of the ensuing health care costs.

The technology only solution results in the implementation of technology A by 14 polluters, technology B by 6 polluters and technology C by 9 polluters. In comparison, when

Table 5. Comparison of costs for two options

	Technology option	Trading option
Reduction cost (Million $)	147.99	121.10
Health care cost (Million $)	41.59	48.57
Total cost (Million $)	189.59	169.68
Total mercury discharge reduction (grams)	1.102	0.911

trading is available, 14 industries implement technology A, while technology B and C are not implemented at all. 15 polluters trade all their reduction quantity to some other industry, while 6 polluters implement technology and also trade some portion of their discharge. Total quantity of mercury traded is 0.053 g/year, which is about 5% of the desired reduction. The results show a trend towards avoiding expensive technology options and satisfying part of the pollutant reduction through trading. Also observed is a significant preference towards one technology (technology A) after trading is allowed.

9.5.2. Solution Dependence on the TMDL Regulation

To analyze and compare the trading and technology options with respect to the TMDL regulation, the problem is solved for various TMDL values between 23 Kg/year and 40 Kg/year. Plots showing the treatment, health care and total costs for both the options as a function of TMDL are shown in Fig. 9. The plots reveal a familiar trend with respect to the reduction costs i.e. reduction costs with trading are always lower than those without trading. But the variation in these costs, particularly the technology option costs, is not linear, leading to varying differences in the reduction costs for trading and technology options. This means that the exact benefits of trading depend on the TMDL value. It can also be observed that the health care costs for trading option are always higher due to lesser discharge reductions. The relationship and variation between the technology and trading option for the total cost is similar to that for the reduction cost, but the difference is reduced due to the opposite trend in health care cost. The advantages offered by trading option are therefore subject to the correct relative evaluation of these two opposing trends. For this problem, with the given considerations, it can be concluded that trading becomes economically attractive, even after considering health care costs, at lower TMDL values. In future, regulations are expected to get more stringent making trading more beneficial. Another important aspect to consider in Fig. 9 is the nature of variation of different plots. The nonlinearity of these variations puts forth certain points where reductions in TMDL are associated with relatively smaller increase in the total cost. This observation is quite pronounced for technology option where the integrality of decisions related to technology implementation (decision variables b_{ij}) makes a strong impact. For policy makers, such information can be invaluable to extract maximum environmental benefits under given financial constraints. For trading option, although these effects are observed, they are considerably reduced due to the flexibility in decision making offered by trading. Fig. 10 plots the reductions achieved by both options as a function of TMDL. It can be seen that technology option always achieves higher reductions and trading option is able to exactly achieve the desired reductions on many occasions. But the difference between the reductions and subsequent environmental benefits for technology option over trading option varies, depending on the TMDL value.

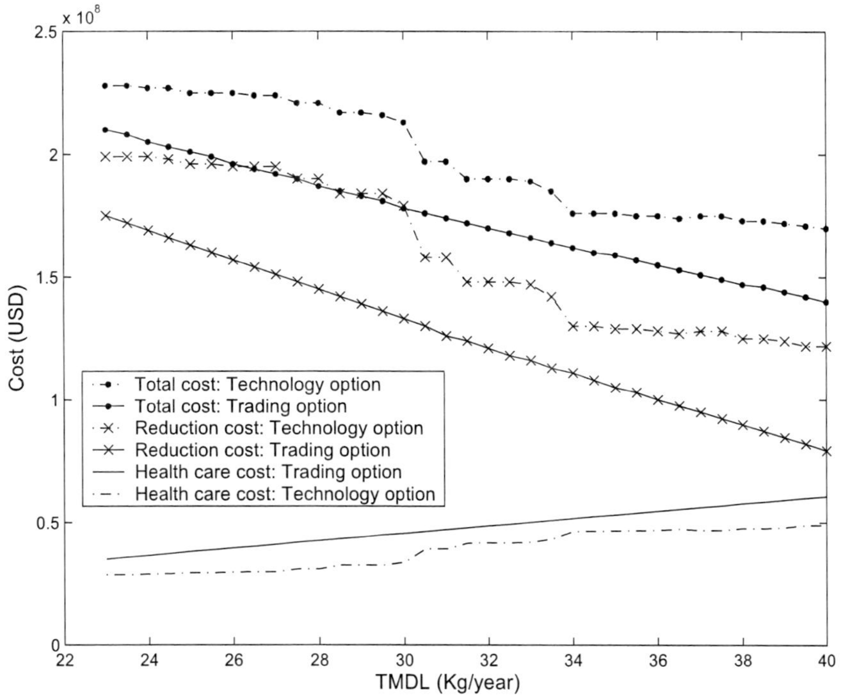

Figure 9. Cost variations with TMDL for the two solution options.

9.6. Health Care Cost Consideration

The previous formulation minimized the discharge reduction cost while considering health care cost only for comparison and indicated that this cost affects the overall cost trends. This section therefore studies the impact of health care cost on the problem solution more extensively. First, a sensitivity analysis for the health care cost is carried out for the model. The health care cost depends on the mortality rate and monetary compensation per affected person, both of which are approximated here. The mortality rate is a function of the LC50 value, which is assumed to model the long term effects of mercury contamination. The monetary compensation puts a value on human life and can be a difficult parameter to estimate, being dependent on the region, country, and policies. The sensitivity analysis accounts for some of these unknown factors to a certain extent. Then, formulation A is modified to include the health care cost as a part of the objective function, the new objective being the minimization of the total (reduction and health care) cost. The new formulation is solved for various cases and sensitivity analysis for it is also performed. The next two sections discuss these cases.

9.6.1. Sensitivity Analysis for Health Care Cost

The sensitivity analysis is conducted by varying the compensation per affected person (C_{health}) in formulation A. Previously reported results considered C_{health} to be 1 Million \$. The problem is now solved for two more values of C_{health}, 3 Million \$ and 5 Million \$. The high value of \$ 5 Million is chosen to compensate for the uncertainty in the effect of mercury, since the LC50 values are only a rough estimates of the possible harmful effects.

The results are shown in Fig. 11, which plots the total costs for various cases. The plots show that increase in health care cost increases the total cost at any TMDL. But more important is the dependence of these costs on TMDL at various compensation values. For C_{health}=1 Million \$, increase in TMDL reduces the total cost and hence higher TMDL appears to be more optimal. Since the reduction costs are also decreasing with increasing TMDL, while the health care costs are simultaneously increasing, this means that health care costs are not as significant as the reduction costs. This trend is seen to be changing when C_{health} is increased to 3 Million and 5 Million \$. For C_{health}= 3 Million \$, the plots are almost flat, meaning that increase in health care costs is equivalent to decrease in reduction costs as TMDL is increased. When C_{health}= 5 Million \$, the overall cost increases with increasing TMDL. The health care costs become more prominent than the reduction costs, thereby dominating the trend in the overall cost. Thus, at C_{health} = 5 Million \$, a lower TMDL appears to be more optimal. It should be noted that since the health care cost is not an objective, the decisions do not change. Changes in the overall cost trends are therefore a results of the varying significance of health care costs in the problem. Since the parameter C_{health} quantifies the value of human life in some sense, the results point

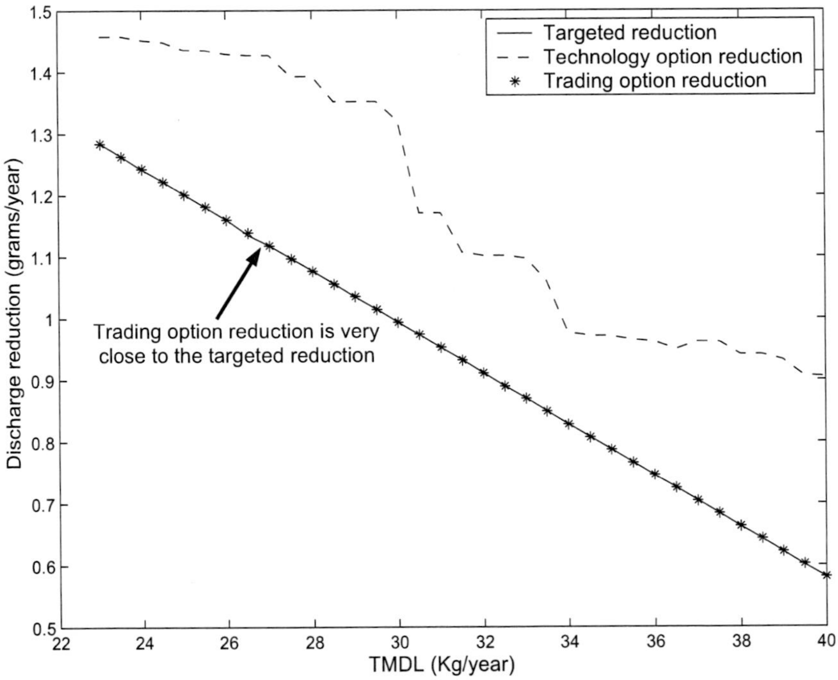

Figure 10. Final discharge reduction variation with TMDL for the two solution options.

that economies which value human life more (signified by higher value of C_{health}), will be putting stricter regulations in place (lower TMDL). The contrary holds true for economies putting lesser value on human life. One can also analyze these trends in conjuncture with the current TMDL for Savannah river basin set by USEPA. The current approved TMDL is 32.8 Kg/year. Assuming that USEPA has set this regulation optimally (through consideration of other parameters like safe fish tissue mercury concentration), one can get a crude estimate on the value that has been put on human life. For this case, this will come out to be around 3 Million \$ per affected person. Another interesting observation from Fig. 11 is that at lower C_{health} values, trading option is more economical than technology option. But as C_{health} values are increased, the advantages diminish, and ultimately technology option becomes more economical than trading option. Simulations reveal that the trends change at the compensation value of 3.75 Million \$. At this C_{health} value, the trading option is more economical at lower TMLDs but expensive at higher TMDLs. This change in the trend can be linked to the higher reductions achieved by the technology option than trading option and the increasing importance of these higher reductions with increasing C_{health} values. The sensitivity analysis results for health care cost have thus thrown light on some interesting aspects of the optimization model, most importantly emphasizing the role of valuation of human life in regulation development.

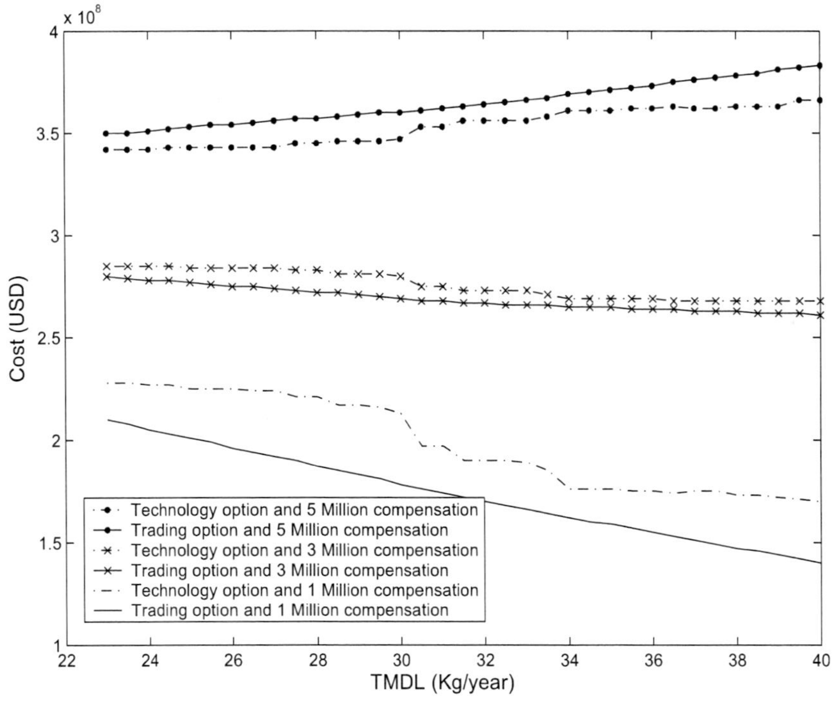

Figure 11. Total cost variations with TMDL for different health care compensations for original problem formulation.

9.6.2. Modified Problem Formulation

As illustrated in the previous section, health care cost is an important parameter while evaluating different options to achieve regulation compliance. The ignorance of health care costs while making decisions can result in a trading mechanism that is more expensive than the technology only solution. This motivates the inclusion of these health care costs in the objective for minimization.

The new problem formulation, referred hereafter as 'formulation B', with health care cost as a part of the objective, is given by Eqs. 28-35. The objective function of formulation A has been appended to include the health care cost in Eq. 28. This health care cost is calculated as a function of the mortality rate as mentioned before. Accordingly, additional equations are included in the problem to calculate the mortality rate. red_i^{final} is the final reduction achieved by point source i. All other symbols have their previously assigned meanings.

Objective :

$$\text{Minimize} \quad \sum_{i=1}^{N}\sum_{j=1}^{M} TC_j.D_i.\, b_{ij} + \text{Mortality}.C_{health} \tag{28}$$

Constraints :

$$t_{ii} = 0 \quad \forall i = 1, ..., N \tag{29}$$

$$red_i \leq \sum_{j=1}^{M} q_j.D_i.\, b_{ij} + \sum_{k=1}^{N} t_{ik} - r\sum_{k=1}^{N} t_{ki} \quad \forall i = 1, ..., N \tag{30}$$

$$P_i \geq \sum_{j=1}^{M} b_{ij}.TC_j.D_i + F\Big(\sum_{k=1}^{N} t_{ik} - \sum_{k=1}^{N} t_{ki}\Big) \quad \forall i = 1, ..., N \tag{31}$$

$$red_i^{final} = \sum_{j=1}^{M} q_j.D_i.\, b_{ij} + \sum_{k=1}^{N} t_{ik} - r\sum_{k=1}^{N} t_{ki} \quad \forall i = 1, ..., N \tag{32}$$

$$WQS_i = c_i - \frac{red_i^{final}}{D_i} \quad \forall i = 1, ..., N \tag{33}$$

$$WQS_{final} = \frac{\sum_{i=1}^{N} WQS_i.D_i}{\sum_{i=1}^{N} D_i} \tag{34}$$

$$\text{Mortality} = \left(\frac{WQS_{final}.Hg_{safe}}{WQS}\,\frac{1000.F_{avg}.P}{2.LC50.W_{avg}}\right) \tag{35}$$

The objective is to minimize the total cost rather than just the compliance cost as in formulation A. This formulation is appropriate under the centralized trading mechanism envisioned in this work. With this formulation, given the data such as industry willingness to spend on pollution abatement, available technologies, actual values of discharges etc., the regulators will be able to evaluate different possible scenarios, considering various health care compensations and TMDL values, and arrive at the optimal regulatory decisions. Next section presents the results for the modified problem.

9.6.3. Model Results: Savannah River Basin

The model is applied to the previously analyzed case study of mercury waste management in Savannah River basin. Data reported in Tables 3 and 4 is applicable here, and the problem studies variation in the reduction, health care and total cost with TMDL. The problem is solved for various TMDL values between 23 Kg/year and 40 Kg/year.

The variation of the total cost with TMDL is shown in Fig. 12 for trading option. For comparison, variation of formulation A results is also shown. It can be seen that at 1 Million \$ compensation (C_{health}), there is little to no difference in the overall cost for the two formulations. Analysis of the results indicates that health care costs are not significant enough to change the optimal decisions when these costs are included in the objective. To determine the dependence of these trends on the health care compensation, simulations were carried out for different C_{health} of 3 Million \$ and 5 Million \$, the results for which are also plotted in Fig. 12. While the results at C_{health}=3 Million \$ are the same as those at 1 Million \$, there is a significant difference in the results for C_{health}=5 Million \$. At this compensation value, health care costs in the objective become more significant than the reduction costs. As a result, the optimization model tries to minimize the mercury discharge under the given constraints by changing the optimal solutions of formulations A and B. It is also seen that the optimal solution in this case is independent of TMDL. Analysis of technology allocations and pollutant trading reveals that point sources are made to reduce more than their load allocations. This reflects in increased treatment costs which are compensated for by reduced health care costs leading to the reduced total costs. These results might change when one considers the financial constraints (availability of limited funds) on individual point source. Such a constraint will reduce the possible options with each point source, thereby affecting the overall problem solution.

Table 6 compares the results for formulations A and B at TMDL of 32 Kg/year for C_{health}=5 Million \$. It should be noted that at this high health compensation cost, the health care costs become very important. Consequently, ignoring these costs in decision making, as done in formulation A, results in extremely high health care costs. However, when these costs are considered in the objective by formulation B, a preference towards more efficient and expensive technologies and consequent reduction in the health care and overall cost is evident. This further emphasizes the importance of health care costs in decision making.

The results for the sensitivity analysis of formulation B (Eqs. 28-35) are shown in Fig. 13. As for formulation A, one can see that at every TMDL, increase in health care cost increases the total cost. However, the trends in the variation of the total cost with TMDL for this formulation do not change with C_{health} value. The slope of the curves reduces as C_{health} is increased, as was observed for formulation A, and the optimal values for the technology and trading option approach each other. At C_{health}=5 Million \$, the solutions are identical and no pollutant is traded. One can also see that the trading option is never expensive than the technology option, contrary to the results for formulation A, because in this case the total cost is optimized as compared to only the reduction cost in formulation A.

Fig. 14 plots the variations in the total cost for different problem formulations and both technology and trading options. These plots are for C_{health} value of 3.75 Million \$. For formulation A, trading option is economical at lower TMDL (below 33 Kg/year), but becomes uneconomical above this TMDL. When variations for formulation B are plotted though, trading option is economical at all TMDL values. Moreover, the total costs for formulation B with trading option are the least amongst all. It can therefore be argued that considering health care cost in the objective of trading option is important to accurately assess the relative economy of the two options.

Table 6. Solution comparison with and without health care cost as objective

	Formulation A	Formulation B
Reduction cost (Million $)	121.10	281.15
Health care cost (Million $)	242.88	53.46
Total cost (Million $)	363.98	334.61
Total mercury discharge reduction (Kg)	0.911	1.945
No. of Technology A implemented	14	3
No. of Technology B implemented	0	29
No. of Technology C implemented	0	13
No. of companies with multiple technologies	0	16

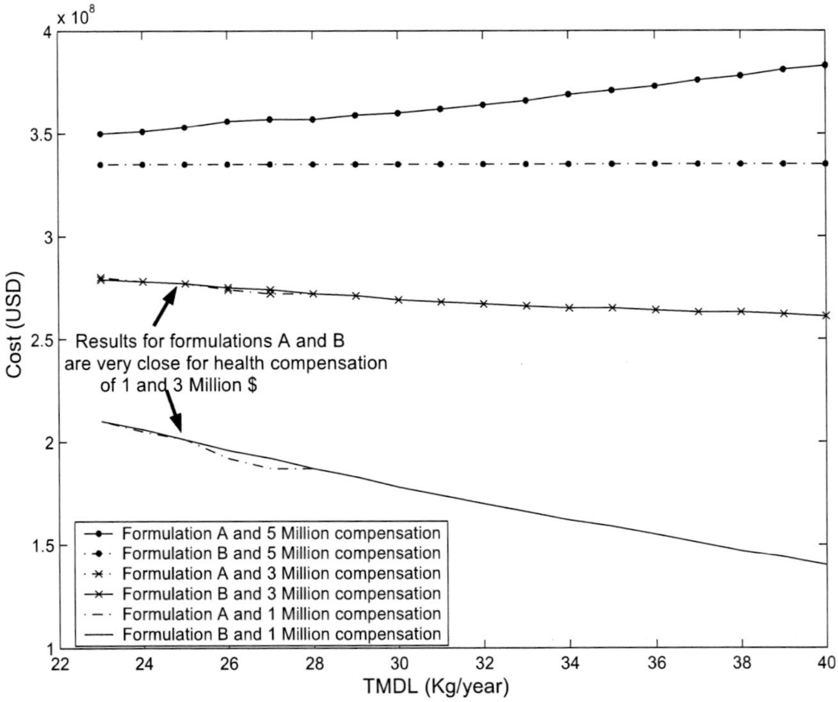

Figure 12. Cost variations with TMDL for objectives with and without health cost.

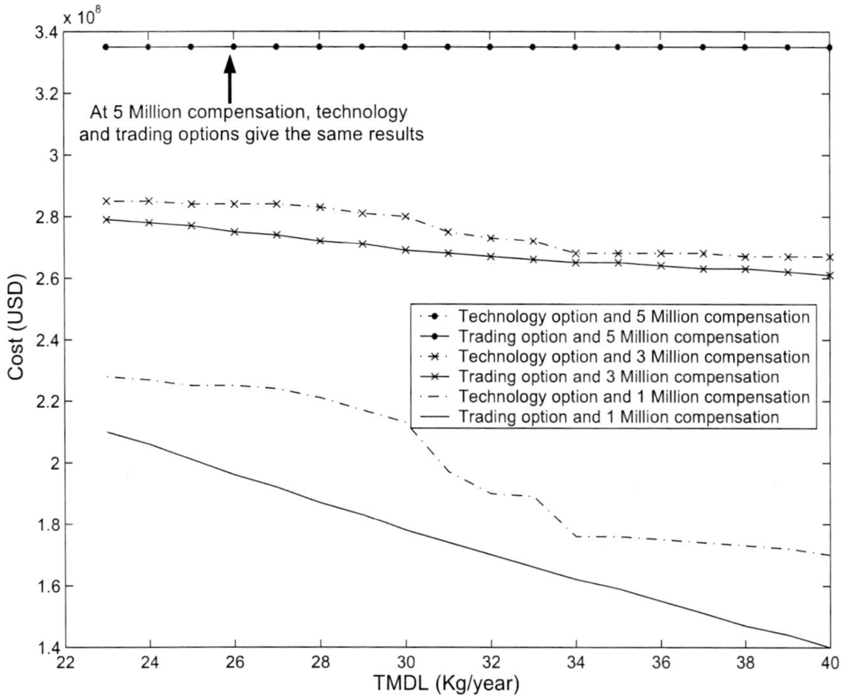

Figure 13. Total cost variations with TMDL for different health care compensations for modified problem formulation.

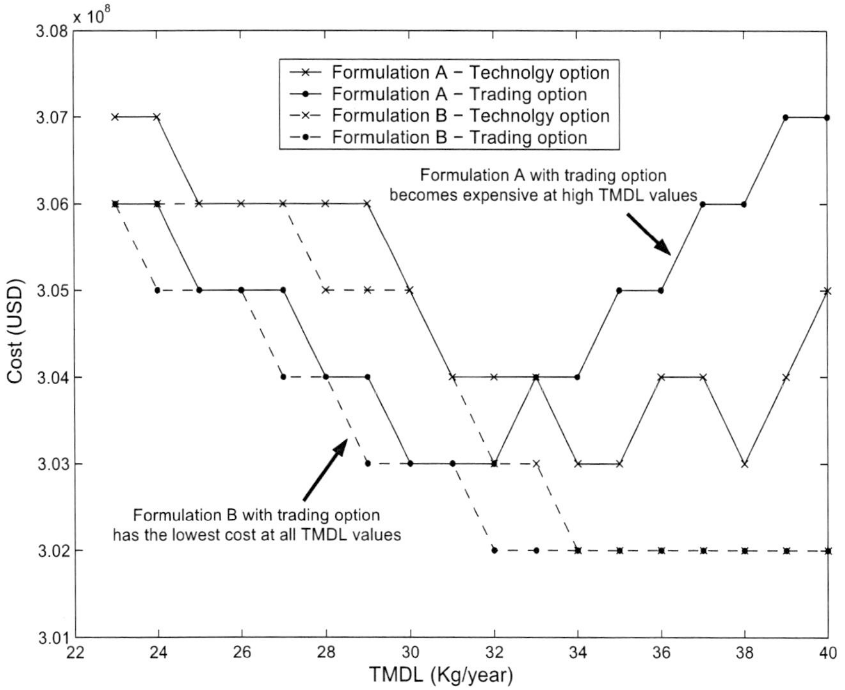

Figure 14. Total cost variations with TMDL for different formulations for C_{health}=3.75 Million \$.

9.7. Chance Constrained Formulation and Analysis

The previous two formulations, given by Eqs. 20-23 and Eqs. 28-35, assumed that data is known deterministically, without any uncertainty. However in many cases, data is known only in terms of the parameter ranges and sometimes their distributions. For example, the Mercury Study Report to Congress (USEPA, 1997b), reports that the uncertainty in the point estimates of the anthropogenic mercury emissions ranges from medium ($\pm 25\%$) to high ($\pm 50\%$). Under these circumstances, one has to work with the available data to arrive at the optimal decisions.

For the problem of pollutant trading, there are various possible sources of uncertainty. At the TMDL development step, the discharge from the individual point sources and the fate and transportation of mercury is variable. These will affect the final bioaccumulation results, and hence the regulations and discharge allocations. Also uncertain are the efficiencies of the mercury treatment technologies implemented by the industries, affecting the actual reduction achieved. As an initial step in incorporating uncertainty in trading mechanism, this work models the current discharge of mercury by each industry (a_i) to be uncertain, normally distributed around a mean value. As explained before, once the TMDL has been developed, each industry is assigned a specific load reduction target based on the current discharge levels. Since the current discharge values are uncertain, load allocations and subsequent decisions are affected by the uncertainty. This converts the optimization problem from deterministic to stochastic, a class of optimization problems dealing with uncertainty. The next section explains the model for this stochastic case and suggests a solution strategy.

9.7.1. Problem Formulation

The formulation is an extension of formulation A given by Eqs. 20-23, and does not consider health care costs in the objective function. Eqs. 20-23 again constitute the stochastic problem, the difference being that in the stochastic case parameter red_i is uncertain. Since TMDL value is fixed and red_i is a linear function of TMDL and a_i, red_i and a_i have the same distribution. a_i is considered to be normally distributed with standard deviation σ_i and value used in previous deterministic analysis as mean. This means that parameter red_i in Eq. 22 is also normally distributed.

Different solution techniques are available to solve stochastic programming problems, such as chance constrained programming, decomposition techniques, various sampling based techniques etc. (Birge & Louveaux, 1997; Diwekar, 2003b). Since the constraint represented by Eq. 22 is linear and the distribution of uncertain parameter red_i is stable (normal), chance constrained method is used in this work to solve the stochastic programming problem (Charnes & Cooper, 1959).

In chance constrained programming, some of the constraints are to be satisfied with a certain probability, which allows the conversion of the problem into a deterministic equivalent. Consider the constraint given by following equation

$$P(g(x) \geq u) \geq \alpha \tag{36}$$

Here, u is the uncertain variable with cumulative distribution function F, and $g(x) \geq u$ is the constraint that is to be satisfied with a probability of α. This constraint corresponds to Eq. 22 in trading problem formulation (Eqs. 20-23). $g(x)$ and u are the right hand side and left hand side of Eq. 22, respectively. Then the deterministic equivalent of constraint by Eq. 36, due to chance constrained programming, is

$$g(x) \geq F^{-1}(\alpha) \tag{37}$$

The chance constrained formulation of Eq. 22 is therefore given as

$$\sum_{j=1}^{M} q_j.D_i.\, b_{ij} + \sum_{k=1}^{N} t_{ik} - r \sum_{k=1}^{N} t_{ki} \geq F_i^{-1}(\alpha) \qquad \forall i = 1, ..., N \tag{38}$$

Here, F_i is the cumulative distribution function of uncertain variable red_i with mean red_i^* and standard deviation σ_i. The variations in the aggregate load will be a function of the variations in the individual loads. Also, the actual required reductions red_i for various point sources might not be correlated. However, incorporating the constraint given by Eq. 38 for all the point sources ensures that the worst case scenario under the given constraint satisfaction probability (α) is accounted for. This will guarantee that there are no localized "hotspots" due to discharge uncertainties. The constraint represented by Eq. 38 is used in deterministic optimization techniques to solve the chance constrained programming problem, the results of which are reported in the next section.

9.7.2. Model Results: Savannah River Basin

The desired reduction red_i has a constant standard deviation of 5% for all the point sources i.e. $\sigma_i = 0.05(red_i^*)$. This simulates $\pm 15\%$ uncertainty in discharge concentration. The problem is solved for various values of α, the probability of constraint satisfaction. The values of α used in this problem are 0.5, 0.9 and 1.0. $\alpha = 1$ (equivalent to 0.999) simulates the case when the constraint is always satisfied. This case is approximated by solving the problem for $red_i = red_i^* + 3\sigma_i$. The problem solution (i.e. technology selection and trading policy) is then used to estimate the actual health care costs by taking into consideration the uncertain mercury discharge values. Since the technology decisions are already made, expected values of the health care costs are directly calculated for each TMDL. The analysis is performed for various values of TMDL, ranging from 23 Kg/year to 40 Kg/year, to explore the dependence of the results on the regulation.

The variations in the total cost for different values of α are plotted in Fig. 15, while those for the health care costs are plotted in Fig. 16. The plots indicate that higher values of α, by increasing the constraint satisfaction probability, reduce the health care costs. This increases the total costs due to the increased reduction costs. The total cost fluctuates a little for $\alpha = 0.5$, but is almost linear for α=0.9 and 1.0. Since the flexibility offered by trading diminishes the integrality effects as discussed before, these results indicate that trading becomes more important at larger uncertainties. The nonlinearity in the trends also puts forth certain interesting points such as the TMDL of 35 Kg/year. At this TMDL, cost difference in 50% and 90% constraint satisfaction is relatively small as compared to other TMDLs. This can be explained from the variation of the reduction cost for this range in Fig. 9. The reduction cost does not increase appreciably for the TMDL between 35 Kg/year and

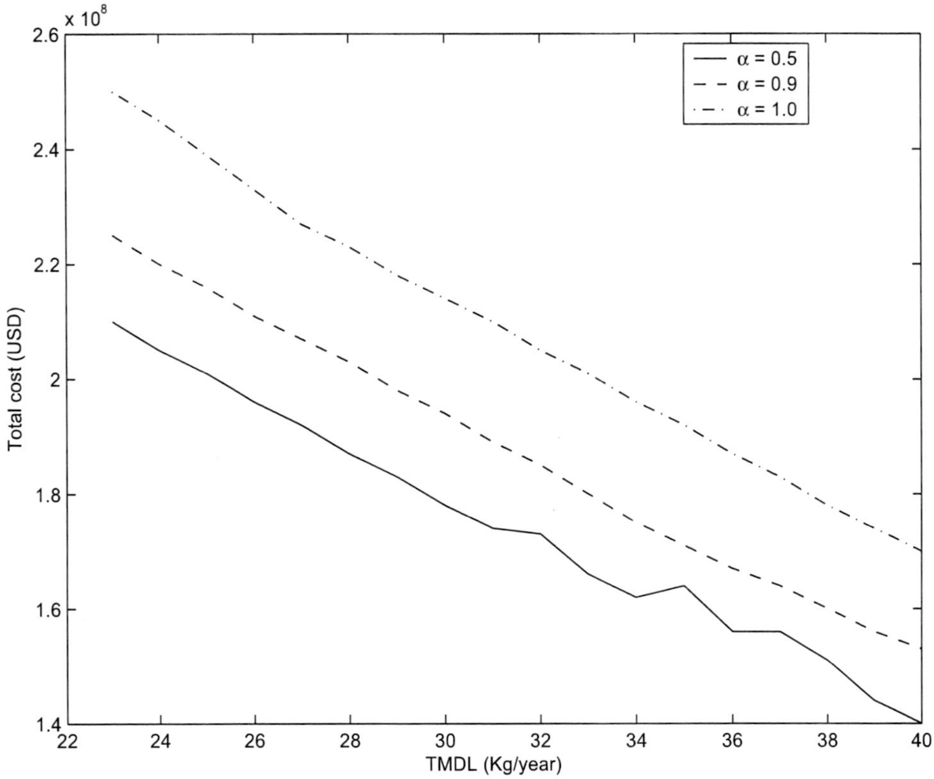

Figure 15. Total cost variation of the stochastic problem using chance constrained programming.

33 Kg/year. This means that the solutions obtained for TMDL 35 Kg/year can also satisfy lower TMDLs. The solutions are therefore overachieving the targets in this region, and hence can take care of some uncertainties, resulting in smaller differences in the cost for 50% and 90% constraint satisfaction. Such results help in finding the TMDL values for which the solution is more robust.

To analyze the effect of the degree of uncertainty on problem solution, the chance constraint analysis is carried out for 16.67% standard deviation, i.e. $\sigma_i = 0.167(red_i^*)$, for various values of α. This simulates $\pm 50\%$ uncertainty in the current discharge concentrations. The results indicate that increase in uncertainty increases the total cost. Table 7 compares the solutions for the two cases of uncertainty ($\pm 15\%$ and $\pm 50\%$) for 90% constraint satisfaction ($\alpha = 0.9$) at TMDL 32 Kg/year. It can be seen that cost increase is also accompanied by higher discharge reduction and additional implementation of expensive technology (technology B). Simulations were carried out for $\sigma_i = 16.67\%$ and when point sources can not implement more that one technology. This restriction can possibly be due to the internal financial constraints. The simulations showed that the problem is infeasible below TMDL 30 Kg/year, even if trading is an option. These results show that presence of uncertainty causes additional cost burden to ensure load reduction satisfaction.

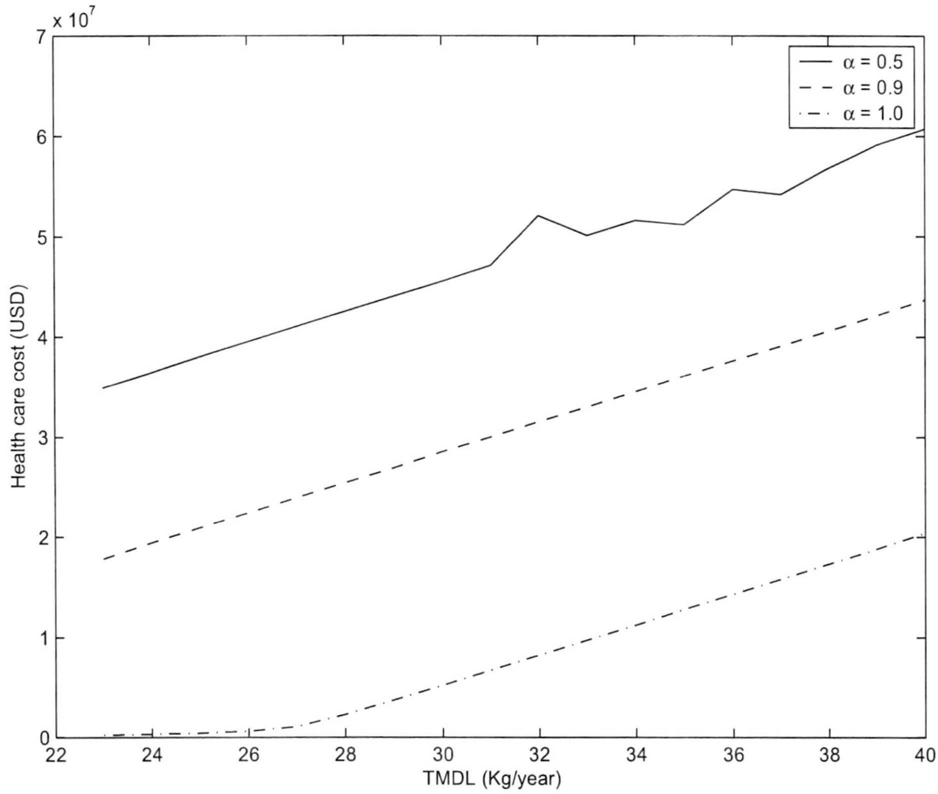

Figure 16. Health care cost variation of the stochastic problem using chance constrained programming.

9.8. Stochastic Programming Pollutant Trading Problem

In the mercury trading problem discussed here, the technology implementation cost is computed using simple linear models. This results in the formulation of a mixed integer linear programming problem (deterministic and stochastic). However, cost functions for various technologies are typically nonlinear. Inclusion of these nonlinear models leads to nonlinear programming problem formulations which are more complicated than their linear counterparts. Furthermore, data related to many mercury treatment technologies are scarce leading to considerable uncertainties about the technology performance. The addition of uncertainty further complicates the optimization problem formulation.

The idea proposed in this work is an extension of the mixed integer linear programming problem of simultaneous technology section and pollutant trading to include nonlinear technology models. The new formulation also considers the uncertainties associated with various technology models. This results in the formulation of a mixed integer stochastic nonlinear programming problem. The following sections briefly discuss the problem formulation and solution methodology.

Table 7. Solution comparison for different levels of uncertainty.

	$\sigma_i = 5\%$	$\sigma_i = 16.67\%$
Total cost (Million \$)	184.52	219.81
Total mercury discharge reduction (Kg)	1.051	1.378
No. of Technology A implemented	17	17
No. of Technology B implemented	4	9
No. of Technology C implemented	3	0

9.8.1. Basic Problem Formulation

The general stochastic programming problem for pollutant trading can be represented as:

$$\begin{aligned}
&\textbf{Objective}: \\
&\quad \text{Minimize} \qquad E[f_1(\zeta) + f_2(\phi, u)] \qquad (39) \\
&\textbf{Constraints}: \\
&\qquad \text{Environmental constraints (regulations)} \\
&\qquad \text{Technological constraints (technology models)} \\
&\qquad \text{Trading constraints} \qquad (40)
\end{aligned}$$

where, f_1 is the cost function corresponding to trading decisions ζ. f_2 represents the technology cost models, where f_2 depends on design variables ϕ and uncertain parameters u. E represents the expectation operator. The specific problem formulation is given as:

Objective :

$$\text{Minimize} \qquad E\Big[\sum_{i=1}^{N}\sum_{j=1}^{M} f_j(\phi, u)\, b_{ij}\Big] \qquad (41)$$

Constraints :

$$t_{ii} = 0 \qquad \forall i = 1, ..., N \qquad (42)$$

$$red_i \leq \sum_{j=1}^{M} q_j.D_i.\, b_{ij} + \sum_{k=1}^{N} t_{ik} - r\sum_{k=1}^{N} t_{ki} \qquad \forall i = 1, ..., N \qquad (43)$$

$$P_i \geq \sum_{j=1}^{M} f_j.b_{ij} + F\Big(\sum_{k=1}^{N} t_{ik} - \sum_{k=1}^{N} t_{ki}\Big) \qquad \forall i = 1, ..., N \qquad (44)$$

where, all notations have their previously assigned meanings. Since cost related to trading does not contribute to the objective function (for the complete watershed), function f_1 represented in the generalized formulation is eliminated. The decision variables in the problem are: b_ij representing the selection of a technology j by industry i, t_{ik} representing the amount of pollutant traded between industries i and k, and ϕ representing the design parameters for the technology. Since cost models f_j are nonlinear, this represents a stochastic nonlinear programming problem. Such problems are known to be computationally very difficult to solve.

9.8.2. Two Stage Problem Formulation

Literature discusses various approaches to solve stochastic programming problems. One such approach is decomposition strategy that reduces the computational requirements for problem solution. The decomposition method is discussed in section 5 and hence will not be reviewed here. In summary, the given stochastic programming problem is decomposed into two or multiple stages, where the nonlinear recourse of the objective function is computed accurately in the second stage. The principle behind this approach is that, since the nonlinear objective function term (recourse function) involves the solution of all the second stage recourse problems, numerous function evaluations for it are avoided. The first stage problem, known as the master problem, uses a linear approximation of the recourse function to fix the first stage decision variables, and the recourse function is exactly evaluated only as a subproblem, referred to as the second stage problem. The stochastic pollutant trading problem is to be solved using the decomposition method.

The first stage problem uses linear approximations of the nonlinear technology models. These approximations are similar to those used in the linear problem discussed earlier in this chapter. Technology selection b_{ij} and trading amount t_{ik} represent the first stage decision variables. The first stage problem is represented as:

$$\textbf{Objective}:$$

$$\text{Minimize} \quad \theta \tag{45}$$

$$\textbf{Constraints}:$$

$$t_{ii} = 0 \quad \forall i = 1, ..., N \tag{46}$$

$$red_i \leq \sum_{j=1}^{M} q_j.D_i.\, b_{ij} + \sum_{k=1}^{N} t_{ik} - r\sum_{k=1}^{N} t_{ki} \quad \forall i = 1, ..., N \tag{47}$$

$$P_i \geq \sum_{j=1}^{M} TC_j.D_i.\, b_{ij} + F\big(\sum_{k=1}^{N} t_{ik} - \sum_{k=1}^{N} t_{ki}\big) \quad \forall i = 1, ..., N \tag{48}$$

$$\theta \geq \sum_{i=1}^{N}\sum_{j=1}^{M} TC_j.D_i.\, b_{ij} \tag{49}$$

$$g \leq G.\zeta + \theta \tag{50}$$

where, θ represents the first stage objective function. Constraints represented by Eqs.46-48 are explained before. Eq.49 puts a lower bound on the linear approximation of the nonlinear cost models, while Eq.50 represents the optimality cut which is introduced after the solution of the second stage problem. This optimality cut includes the first stage decision variables b_{ij} and t_{ik} represented collectively here as ζ.

The first stage decisions are passed on to the second stage where the recourse function is computed using nonlinear models. In the second stage, the uncertain variables are sampled $Nsamp$ times. The second stage sub-problem is solved for each sample to calculate the expected value of the nonlinear recourse function. The second stage problem is thus given as:

$$\textbf{Objective}: \quad \text{Minimize} \quad \sum_{n=1}^{Nsamp} \sum_{i=1}^{N} \sum_{j=1}^{M} C_j(n).\, b_{ij} \tag{51}$$

$$\textbf{Constraints}: \quad C_j(n) = f_j(n) \tag{52}$$

where, $C_j(n)$ represents the exact cost computed using the nonlinear cost models f_j for the particular sample n of the uncertain parameter. The solution of the second stage problem results in possible generation of an optimality cut which is included in the subsequent master problem solution through the computation of G and g matrices.

The details of problem solution using decomposition strategy are omitted for the sake of brevity and can be found in many stochastic programming textbooks such as Birge & Louveaux (1997). Since the second stage problem solution requires nonlinear model simulation for each sample n, computational requirements can be substantial. The L-shaped BONUS algorithm discussed in section 5 of this chapter can be used to solve this problem efficiently by reducing the computational requirements. The algorithm details are discusses in section 5. This constitutes an area of further research in this field.

9.9. Discussion: Industrial Sector Level Management

The work presented in the last few sections analyzed water pollutant trading as an industrial sector level option to achieve sustainable waste management of mercury. Pollutant trading complicates decision making for policy makers and polluters alike, and hence a centralized trading mechanism can be an effective way of optimizing the mutual benefits. To efficiently carry out this task, an optimization model was formulated which minimized the overall cost taking into account the watershed and technology specific aspects. The model formulation suggests optimal decisions to individual polluters and allows one to take optimal policy decisions as well. The proposed optimization model was applied to the Savannah River basin mercury waste case study. Following important conclusions can be drawn from the results:

- Trading reduces the overall treatment cost for pollution reduction. However, this is accompanied by comparatively higher, although less than permitted, mercury emissions. This means that the advantages offered by reduced treatment cost with trading should be carefully weighed against increased risk of adverse health effects of mercury.
- The results after the inclusion of health care cost in the objective of minimization depend on the significance of health care cost. For high compensation values, higher

emission reductions are achieved, offsetting the increased treatment costs, thereby arriving at lower overall costs.

- Sensitivity analysis for the health care cost highlights the importance of correct valuation of human life in regulation development. It suggest that higher valuation will typically lead to stricter regulations.
- Problem analysis in the presence of uncertainty indicates that the impact of uncertainty is more pronounced as the regulations become stricter.
- When the dependence of the solutions on TMDL regulation was analyzed, a non-linear variation was observed in all cases. This indicates that the consideration of trading and watershed specific details during regulation development might help develop better regulations, achieving an effective tradeoff between cost and reductions. This points towards the centralized trading mechanism. For this particular problem of Savannah river basin, trading became more attractive with more stringent regulations.

When these results are compared with the results presented in Farrell et al. (1999) for NOx, one observes that the effect of trading on the overall cost is more for mercury waste management. This difference is particularly obvious for the total reductions achieved with and without trading, the difference being larger for the case of mercury.

9.10. Future Extensions

The work can be extended further by considering a larger watershed (scaling up) and including non-point sources in the analysis. Also important is the dry and wet deposition of mercury from the atmosphere to the watershed, which accounts for a substantial amount of mercury present in water bodies. Non-point sources are believed to have considerable potential for pollution reduction (Jarvie & Solomon, 1998), but also complicate the problem as they are associated with uncertainties, static as well as time dependent (e.g. seasonal variations). Stochastic programming techniques will be needed to solve such cases. As regards to mercury, better characterization of mercury cycling and its harmful effects is needed to generate confidence in trading. Bioconcentration factors (BCF) and bioaccumulation factors (BAF) are possible options. The lack of reliable models have restricted their use in such situations. Better characterization of their uncertainties can resolve some of the issues and would be looked into in future.

10. Water Pollutant Trading: Application of L-shaped BONUS

The previous section presented the application of pollutant trading concept for the management of mercury waste in a watershed. The work analyzed different deterministic formulations, a chance constrained programming formulation and a two stage stochastic programming formulation, all for point sources. Two possible extensions to the optimization framework are:

- More extensive uncertainty analysis for the problem considering point sources

- Inclusion of non-point source in trading mechanisms.

Both cases will lead to the formulation of a stochastic programming problem. It is also possible that the problem might be nonlinear in nature. In section 5, L-shaped BONUS algorithm was introduced which solves stochastic nonlinear programming problem in a computationally efficient manner. That algorithm can be a useful tool to solve stochastic nonlinear pollutant trading problems.

To illustrate the application and usefulness of the algorithm in pollutant trading field, this work discusses a water pollutant trading problem involving non-point sources. The example does not discuss mercury waste management for a specific watershed, as done in the previous section. The purpose to show the applicability of L-shaped BONUS algorithm in solving pollutant trading problems. The problem is discussed in the next section.

10.1. Non-point Pollutant Trading

As mentioned before, trading between various point sources is considered to be the easiest one to implement. However, trading between point and non-point sources is considered to have the greatest potential for water pollution control (Jarvie & Solomon, 1998). This is because non-point sources are the primary polluters from a volumetric perspective in most water bodies. Moreover, pollution abatement costs for the non-point sources (through Best Management Practices) are quite often considerably lower than those for the point sources. The non-point sources are though associated with uncertainties which complicate the problem solution. For starters, contributions from non-point sources are not accurately measurable (at a reasonable cost), and since the contributions are diffused, the effect of the pollutant reduction strategies is also difficult to estimate. Furthermore, contributions from many of the non-point sources depend on essentially stochastic factors, such as rainfall, weather conditions etc. (Jarvie & Solomon, 1998; USEPA, 1996).

The next section proposes an optimization based model to make trading decisions in the presence of non-point sources. The consideration of non-point sources leads to the formulation of a stochastic nonlinear programming problem. The solution of this stochastic programming problem is achieved using the recently proposed L-shaped BONUS algorithm.

10.2. Trading Problem Formulation

The literature on modeling of point/non-point water pollutant trading is diverse with the focus varying from regulations to economics (Malik et al., 1993; Shortle et al., 1998; Horan et al., 2001). The trading model used in this work is based on the models proposed in Malik et al. (1993); Zhang & Wang (2002). The problem considers trading between a set of point sources (PS), which includes one Public Owned Water Treatment plant (POWT), and one non-point source (NPS), representing a farm. A POWT is a state or municipality owned waste water treatment facility, designed to treat a variety of pollutants such as sewage waste, water run-offs etc. All these sources discharge the pollutants into a common water body such as a lake. Regulations have been placed on the discharge of a set of chemicals, specifying the maximum amount that is allowed to be discharges per day into the water body. Currently, the non-point sources are not regulated, as it is not possible to accurately measure the emissions from these sources. However, EPA is thinking of including these non-point sources in future. For simplicity, this model does not consider regulations on non-point source emission. Depending on the current discharge amounts of each chemical, each point source has different levels of reduction to achieve. It has two options to achieve the targets:

- Implementation of advanced end of pipe treatment methods.
- Trading a particular amount of reduction by implementing and operating better treatment methodologies on a piece of land from the NPS (farm). The reduction techniques for the NPS nonlinearly depend on the type and quantity of chemical being treated.

There are three sources of uncertainty in this setup. As mentioned previously, the non-point sources have uncertain emission quantities and uncertain reduction efficiencies for a particular technology. These two factors combine to give variability in the actual reduction achieved for a particular chemical. Another source of uncertainty is through the POWT. Although the source is a point source in terms of the output concentration of the chemicals, the total amount of waste being discharged is subject to variability. This is because a POWT often treats wastes from many disparate sources, some of which are stochastic in nature. As a result, the amount of input and consequently the output varies within certain limits.

Let $i = 1, \ldots, P$ represent the set of PS and $j = 1, \ldots, M$ be the indicator of the chemical. For the PS, the current pollutant discharge levels are considered as known, along with the discharge reduction (treatment) cost for both the chemicals. The model parameters associated with the PS are:

$D(i)$ = Total volumetric discharge from PS i ($[L^3][s]^{-1}$)
$e_{p0}(i,j)$ = Pre-treatment discharge quantity of chemical j from PS i ($[M][L^3]^{-1}$)
$c_p(i,j)$ = Cost for the point source discharge reduction of chemical j by PS i ($[\$][M]^{-1}$)

where, the terms in bracket represent the dimensions of various quantities, with $[M]$=Mass, $[L]$=length, $[s]$=time and $[\$]$=cost in dollars.

For the NPS, it is considered that a fixed amount of land is available, divided among all the PS, to implement the NPS treatment technologies (Best Management Practice). Similar to the PS, current discharge quantities, discharge reduction costs, and abatement efficiencies are known for the NPS. The model parameters associated with the non-point source are:

L_{max} = Maximum amount NPS land available from trading ($[L^2]$)
$e_{n0}(j)$ = Pre-treatment discharge quantity of chemical j from non-point source ($[M][L^2][s]^{-1}$)
$c_n(j)$ = Cost for the non-point source discharge reduction of chemical j ($[\$][L^2]^{-1}$)
$b_{NPS}(j)$ = Non-point source abatement efficiency of chemical j
$q_n(j)$ = Abatement in non-point source discharge of chemical j ($[M][L^2][s]^{-1}$)

Here, $q_n(j) = e_{n0}(j).b_{NPS}(j)$.

From the regulatory perspective, it is assumed that the targeted discharge rate of both chemicals for each PS has been decided through the development of some regulatory mechanism like TMDL (Total Maximum Daily Load) (USEPA, 1991). In addition to these limits, there are restrictions on the maximum amount of any chemical that can be discharged into the water body at a particular location. Enforcing this limit ensures that the implementation of pollutant trading does not result in the creation of 'hot-spots', defined as the localized points with high pollutant concentration. Accordingly, the regulatory parameters of the model are:

$z_{red}(i,j)$ = Targeted reduction in emission of pollutant j by PS i ($[M][s]^{-1}$)
$z_{allowed}(j)$ = Maximum permitted emission of pollutant j at any location ($[M][s]^{-1}$)

In the context of sources of uncertainty described before, parameters $e_{p0}(i,j)$, $b_{NPS}(j)$ and $D(i)$ are uncertain and are considered to be normally distributed around a known mean value with a known standard deviation. The decisions to make are: if and how much pollutant quantity should be traded with the land (non-point source) and how much reduction should be achieved by end of pipe treatment, with the aim to achieve the reduction targets at the lowest overall cost. Accordingly, the decision variables in the model are:

$L(i,j)$ = Land allocated for trading ($[L^2]$), and
$q_p(i,j)$ = Abatement in point source emission quantity ($[M][L^3]^{-1}$)

The uncertainties associated with the problem necessitate the formulation of a stochastic programming problem, given as follows:

$$\text{Minimize} \qquad f_1(c_n, L) + E[f_2(D, c_p, q_p)] \tag{53}$$

where, E is the expectation operator over the uncertain parameters, and f_1 and f_2 are linear/nonlinear functions of the respective variables. The first term in the objective represents the cost incurred due to trading through the allocation of land for each PS from the NPS. The second term represents the expected value of the total treatment cost incurred by the PS to satisfy the regulations in the presence of uncertainty. This objective is subjected to the following constraints.

$$\sum_{i,j} L(i,j) \leq L_{max} \tag{54}$$

$$E[D(i,j).q_p(i,j) + L(i,j).q_n(j)] \geq z_{red}(i,j) \qquad \forall(i,j) \tag{55}$$

$$E[D(i,j).(e_{p0}(i,j) - q_p(i,j))] \leq z_{allowed}(j) \qquad \forall(i,j) \tag{56}$$

$$0 \leq q_p(i,j) \leq e_{p0}(i,j) \qquad \forall(i,j) \tag{57}$$

$$q_n(j) = e_{n0}(j).b_{NPS}(j) \qquad \forall(j) \tag{58}$$

The first constraint ensures that the total land allocation does not exceed the available land with the NPS. The second set of constraints ensures that each industry achieves the individual reduction target of each chemical, with or without trading. The third set ensures that the emission of a pollutant j at any location does not exceed the permitted value. Constraint 57 bounds the PS reduction. The last constraint models the effect of uncertainty on the problem. $e_{n0}(j)$ and $b_{NPS}(j)$ are uncertain parameters, affecting pollutant reduction by the non-point source $q_n(j)$.

The problem can be converted into a two stage stochastic programming problem with recourse, where the first stage decisions are the land allocations between the various point sources and the non-point source $L(i, j)$, and the second stage decision variables are the point source abatement $q_p(i, j)$ to be achieved by end of pipe treatment to take care of the uncertainties. The two stage formulation, where specific definitions of functions f_1 and f_2 are also included is given below:

First stage problem:

$$\text{Minimize} \quad \sum_{i=1}^{P}\sum_{j=1}^{M} c_n(i,j)L(i,j)^{\alpha_j} \; + \; E[R(L,q_p,q_n,D)] \tag{59}$$

$$\sum_{i=1}^{P}\sum_{j=1}^{M} L(i,j) \; \leq \; L_{max} \tag{60}$$

where, R is the recourse function. α_j is the constant for chemical j, representing the nonlinear relationship between the land allocation and pollutant reduction.

Second stage problem:

$$\text{Minimize} \quad E[R(L,q_p,q_n,D)] \; = \; \sum_{n=1}^{N}\sum_{i=1}^{P}\sum_{j=1}^{M} D(i,j,n)c_p(i,j)q_p(i,j,n) \tag{61}$$

$$D(i,j,n).q_p(i,j) + L(i,j).q_n(j,n) \; \geq \; z_{red}(i,j,n) \qquad \forall(i,j,n) \tag{62}$$

$$D(i,j,n)[e_{p0}(i,j) - q_p(i,j,n)] \; \leq \; z_{allowed}(j) \qquad \forall(i,j,n) \tag{63}$$

$$0 \leq q_p(i,j) \; \leq \; e_{p0}(i,j) \qquad \forall(i,j) \tag{64}$$

$$q_n(j,n) \; = \; e_{n0}(j).b_{NPS}(j,n) \qquad \forall(j,n) \tag{65}$$

where, N is the sample size used to represent the uncertain space in the optimization algorithm, and n is a particular sample from the uncertain space. Eqs. 59 through 65 represent the general two stage stochastic programming problem with recourse. This work considers the problem with four point sources (PS) ($P = 4$), comprising of three industries, one Public Owned Water Treatment plant (POWT), and regulations in place for two chemicals, A and B ($M = 2$). The next section gives the numerical results for the problem.

10.3. Results of the Trading Problem

Tables 8 and 9 give the data for the point and non-point sources, respectively. The total quantity of land available for trading (L_{max}) is 200 units. The regulation has assigned a 40% reduction target on the discharge of each chemical by each of the point sources. The regulation for the POWT is based on the mean value of the discharge. To avoid occurrence of ‘hot-spots’ in the lake, no PS is allowed to discharge more than 200 and 170 units per unit time of chemical A and B, respectively ($z_{allowed}(j)$). Since pollutant trading for water pollutants, particularly between point and non-point sources is not practiced extensively yet, data availability is scarce. The data reported in this work, although not taken from an actual problem, is comparable to the typical values and reductions encountered in practical implementations (USEPA, 2001a).

Apart from discussing the solution of trading problem, the objective of this work is to perform a comparative study of the L-shaped BONUS algorithm with the standard L-shaped method. Since the sampling technique plays an important role in the sampling based algorithm, a comparative analysis between the Hammersley Sequence Sampling technique (HSS) and the Monte Carlo sampling technique (MCS) is also performed. Accordingly, the problem is solved using four different methodologies:

- Standard L-shaped method with HSS
- Standard L-shaped method with MCS
- L-shaped BONUS method with HSS
- L-shaped BONUS method with MCS

The four methodologies are compared by studying the problem solutions as a function of the sample size, for which the sample size is varied from 25 to 500. The variation of the objective function for the four methods is shown in Fig. 17, while the typical variation of the decision variables is shown in Fig. 18.

Table 8. Details of point source emission and treatment

Point Source	Total discharge	Standard deviation in discharge amount	Concentration in untreated waste		Treatment Cost	
			Chemical A	Chemical B	Chemical A	Chemical B
1	100	0.0	3.35	2.15	10.0	11.5
2	120	0.0	2.20	2.95	9.0	13.0
3	110	0.0	1.60	1.90	11.9	8.8
4	150	0.1	1.99	2.10	8.9	8.1

Table 9. Details of non-point source emission and treatment

	Chemical A	Chemical B
Mean value of emission quantity	5.0	5.5
Standard deviation in emission quantity	0.2	0.2
Technology implementation cost	0.91	1.0
Mean value of reduction efficiency	0.6	0.8
Standard deviation in reduction efficiency	0.06	0.06

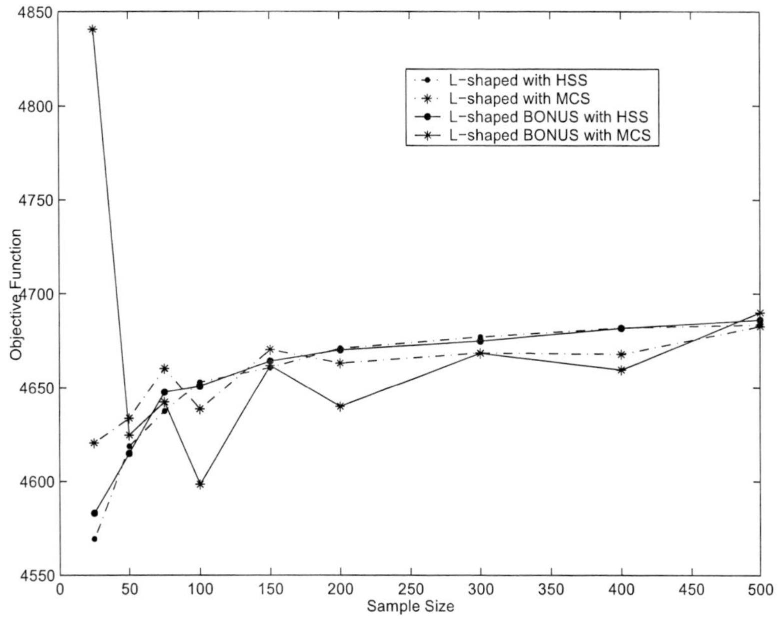

Figure 17. Variation of the objective function result with sample size.

Fig. 17 indicates that the difference in the results for the two methods is within reasonable limits, 0.05% for the maximum sample size, indicating that the reweighting approximation does not sacrifice accuracy. Also evident from the plots is the fact that the two solutions approach a steady state value as the sample size is increased. The plot for the decision variable shows a similar qualitative behavior. A comparison of the two sampling techniques indicates that the results with HSS technique reach the steady state value faster as the sample size is increased than for the MCS technique. Also the error between the results for the two methods is smaller for the HSS technique, particularly at smaller sampling sizes.

An important advantage of the L-shaped BONUS algorithm is the computational efficiency due to the reweighting scheme. Computational times for the four methods have been plotted in Fig. 19. The plots show that the computational requirements go down considerably for the L-shaped BONUS algorithm, with the reduction of computational time to almost 10% of that for the standard L-shaped method for the maximum sample size of 500. This advantage becomes more obvious at larger sample sizes. Also observed is the fact that there is not much difference in the computational times for the different sampling techniques when used in the same solution method.

As regards to the trading problem, the solution indicates that all the point sources need to achieve part of the required reductions through trading with the NPS to minimize the overall cost. Such a decision will be hard to take, if the industries decide on their own, without consideration of the overall cost. The results therefore strongly support the use of rigorous and systematic mathematical analysis to perform decision making in trading, since the decisions based on heuristics are likely to be sub-optimal.

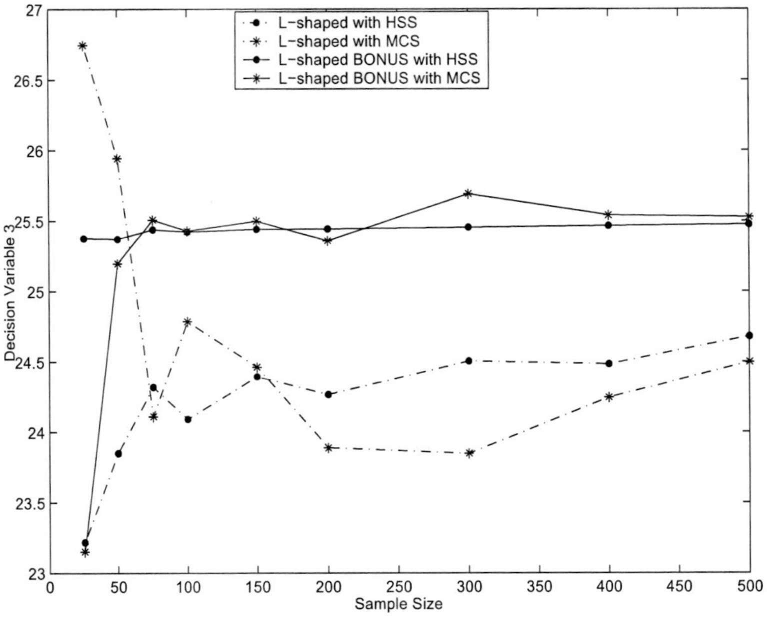

Figure 18. Variation of the decision variable 3 result with sample size.

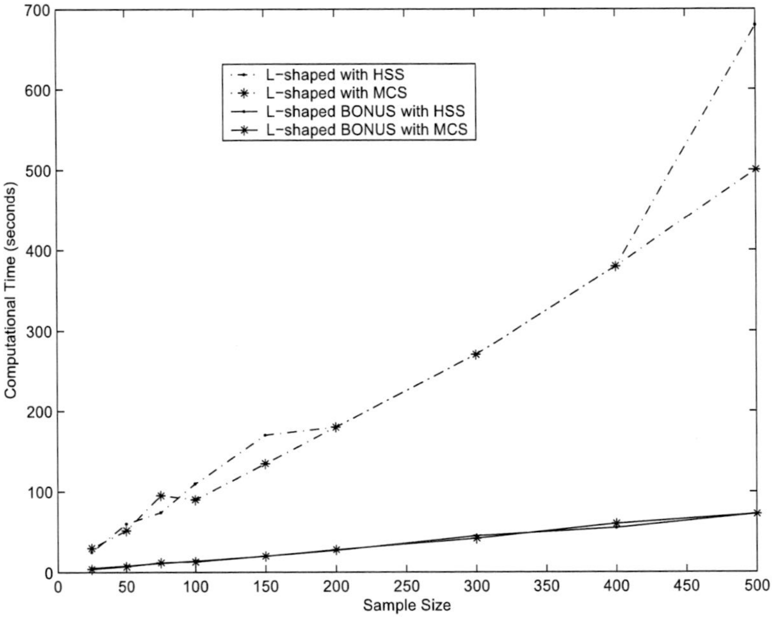

Figure 19. Variation of the computational time with sample size.

10.4. Discussion

The environmental trading problem is formulated as two-stage stochastic programming problem making it amenable to specialized stochastic programming solution methods like the L-shaped method. The properties of the environmental trading problem solution lend support to the argument that the L-shaped BONUS algorithm is computationally much more efficient as compared to the sampling based L-shaped method. From trading point of view, the potential of water pollutant trading between point and non-point sources is well known. But as the problem formulation and the results suggest, decision making in point/non-point trading is complex and requires a systematic analysis. The existence of various tradeoffs and restrictions make it an ideal problem for the optimization theory. The presented work is the first instance of the formulation of a point/non-point trading problem using stochastic programming ideas. Advances in the field of stochastic programming offer an exciting avenue to make better decisions in the goal of environment preservation. In this regard, further investigations into a more detailed analysis of stochastic programming applications in environmental trading problems offer considerable prospects.

11. Ecosystem Level Management: Lake Liming

Section 7 explains the complicated mercury cycling depicted in Fig. 8. It is mentioned that mercury cycles in all media as a part of natural as well as anthropogenic effects. An important aspect of this cycling is the bioaccumulation of mercury along the aquatic food chain. Mercury, in the form of methylmercury, bioaccumulates up the aquatic food chains so that organisms in higher trophic levels have higher mercury concentrations (DeSimone et al., 1973). As a result, contaminated fish consumption is the most predominant path of human exposure to mercury. The primary targets for toxicity of mercury and mercury compounds are the nervous system, kidney, and developing fetus (USEPA, 1997b).

Since methyl mercury is the primary bioaccumulative form of mercury, controlling the conversion of mercury in water bodies to methyl mercury is a possible option to control bioaccumulation. This work looks into the option of using time dependent liming strategy of lakes and rivers to control water pH. To make the analysis more robust, it proposes to derive liming strategies in the presence of time dependent uncertainty. Optimal control theory, in combination with the real options theory to effectively deal with various sources of uncertainty, is used for the derivation of the liming stratcgics. Such an analysis is expected to make the liming operation more reliable, thereby presenting one more tool to manage the harmful effects of mercury pollution. The work is explained in the subsequent sections in detail.

11.1. Mercury Bioaccumulation Overview

Fig. 20 represents the overall mercury cycling in water bodies such as lakes and rivers. As shown in the schematic, mercury can exists in various forms, mainly, elemental mercury (Hg), inorganic mercury (Hg(II)), organic methyl mercury (CH_3Hg) and complexes of these with dissolved organic carbon. There are number of pathways by which mercury can enter freshwater environment: Hg(II) and methylmercury from atmospheric deposition (wet

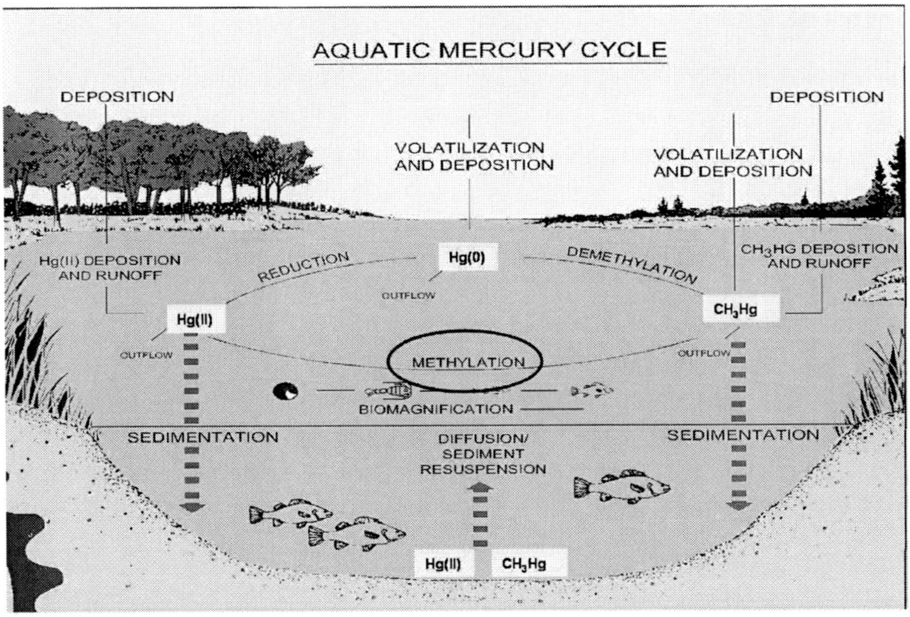

Figure 20. Mercury cycling in water.

and dry) can enter bodies directly; Hg(II) and methylmercury can be transported to water bodies in runoff (bound to suspended soil/humus or attached to dissolved organic carbon); or Hg(II) and methylmercury can leach into the water body from groundwater flow in the upper soil layers (USEPA, 1997b). Mercury and its compounds exist in different segments of the water body such as the water column, sediment (active and passive), and the biota (fish).

Once in the water bodies, mercury can exist in dissolved as well as particulate form and it can undergo following different transformations: Elemental Hg can be oxidized to Hg(II) or volatilized to the atmosphere; Hg(II) can be methylated in sediments and water column to form methylmercury; methylmercury can be alkylated to form dimethylmercury; and Hg(II) and methylmercury can form organic and inorganic complexes with sediment and suspended particular matter. All these transformations are simultaneously observed in a given water body. The concentration of each chemical form depends on the extent of various reactions, which can differ for different water bodies.

Of the various chemical forms of mercury, methylmercury (MeHg) is considered to be the most dangerous due to the following reasons:

- All forms of mercury can be converted to methylmercury by natural processes in the environment.
- Methylmercury bioaccumulates and biomagnifies in aquatic food webs.
- Methylmercury is the most toxic form of mercury.

Owing to its high bioaccumulative potential, the concentration of methylmercury in large aquatic animals (such as predatory fishes) is many times more than the water column or sediment concentration. Generally, methylmercury fraction in a water column is not more than 25% of the total mercury content. In lakes without point source discharges, this fractions is typically less than 10% (USEPA, 1997b). However, it has been observed that almost all of the mercury present in fish tissues is as methylmercury, confirming its preferential bioaccumulation.

"Bioaccumulation" refers to the net uptake of a contaminant from the environment into biological tissue via all pathways. It includes the accumulation that may occur by direct contact of skin or gills with mercury-contaminated water as well as ingestion of mercury-contaminated food. "Biomagnification" refers to the increase in chemical concentration in organisms at successively higher trophic levels in a food chain as a result of the ingestion of contaminated organisms at lower trophic levels. The reason for the strong toxicity of methylmercury is its strong affinity for sulfur containing compounds (e.g. proteins). Biological membranes, including the blood-brain barrier and the placenta, that tend to discriminate against other forms of mercury allow relatively easy passage of methylmercury and dissolved mercury vapor (Eisler, 1987). This leads to the harmful effects that have been previously mentioned.

Thus, methylation of mercury to MeHg is a key step in the bioaccumulation of mercury in aquatic food chains (Sorensen et al., 1990). The concentration of MeHg in water depends on the equilibrium between the methylation and demethylation reactions, which occur in the water column as well as the sediments. The exact mechanism of the methylation reaction is not well understood. Various mechanisms have been proposed in the literature for the formation of methylmercury. The proposed mechanisms include abiotic and biotic. Abiotic methylation includes photochemical and due to the presence of humic and fulvic acids in the solution (Nagase et al., 1982). However abiotic methylation is considered to be insignificant in most cases. Biotic methylation is due to the presence of bacteria that excrete methylcobalamine (Vitamin B_{12}) (Jensen & Jernelöv, 1969). It can be enzymatic (Wood et al., 1968) as well as non-enzymatic (Bertilsson & Neujahr, 1971). It has also been shown that biotic methylation can take place in sediment (Winfrey & Rudd, 1990) and water column (Xun et al., 1987). It has also been proposed that sulfate reducing bacteria mediate mercury methylation (Gilmour et al., 1992). This illustrates that there is considerable uncertainty over the exact methylation mechanism. Studies have also been carried out to understand the effect of physical and chemical conditions such as pH, dissolve oxygen, dissolved organic carbon (DOC), temperature, salinity etc., on methylation (Winfrey & Rudd, 1990; Driscoll et al., 1995). These studies have shown a strong correlation between acidic conditions, i.e. low pH values and high mercury bioaccumulation in fish. This correlation could be due to higher concentration of bioavailable methylmercury in the ecosystems caused by altered chemical partitioning of methylmercury across the sediment water interface, increased inputs of methylmercury to lakes from the terrestrial ecosystems or precipitation, or increased in-lake production of methylmercury (Winfrey & Rudd, 1990). Even though the exact contribution of each of these factors is not accurately known, the strong correlation between pH and mercury concentration suggests a possible option to reduce mercury bioaccumulation. This idea forms the basis of lake liming operation which is described in the next section.

11.2. Lake Liming

The process of lowering of the pH of a lake is known as lake acidification. The primary reason for lake acidification is burning of different fossil fuels (coal, oil, petrol). The resulting emissions of different types of compounds (sulphur dioxide and nitrogen oxides) oxidize in the atmosphere giving rise to the deposition of hydrogen, that is, to acidification. Acidifying substances, which are mainly S and N, are deposited on land and water as wet and dry deposition. Some other natural processes such as degradation of organic matter or weathering of base rock can also affect lake acidity. However, these are shown to be secondary in most cases. Since gases emitted due to anthropogenic activities can travel a long distance before getting deposited, acidification problem as assumed global scale.

The previous section discussed the correlation between pH and mercury bioaccumulation. There are also direct effects of the acidic conditions on lake biota. Very acidic lakes (i.e. pH below 4) are often devoid of fish, frogs, salamanders, crayfish, insects, and plankton. Certain fish types are sensitive to changes in pH and prefer lake waters with pH values ranging from 7 to 9. Because very acidic surface waters can have toxic concentrations of aluminum in solution, aquatic animals may be subjected to a potentially lethal double dose of poisonous acid and metals. Acidic conditions have also been shown to affect the reproductive processes in aquatic animals. Therefore, control of lake pH and hence its acidification can be an effective method to avoid such undesirable effects.

Although, reduction in the emission of these gases is the ultimate solution of acidification, this will be a slow process and will take time to implement. Therefore, more immediate methods need to be resorted to. One such method is lake liming. Liming is the addition of a base, such as limestone, to the water body to neutralize acid waters and soils and buffer them from rapid fluctuations in pH. Liming is an established practice with its origins in the Roman era. Soil liming has been extensively practiced in the U.S. and U.K. The treatment of surface waters to control acidity (surface water liming) began in Norway in 1920s (Porcella et al., 1989). Since then, the use of liming as a mitigative techniques has been widespread in the Scandinavian countries. Approximately 8000 lakes and 6000 km of running water have been limed in Sweden during the last two decades at an annual cost of about USD 20 million. This makes liming one of the most extensive activities applied to preserve and restore threatened environments (Håkanson, 2003b). Considerable research on lake liming has been conducted in United State, Canada, United Kingdom, Finland and Sweden in the past few years, increasing the overall understanding about the liming operation.

Base treatment neutralizes acidity in surface water and increases the supply of basic cations, thereby improving water quality and fostering the presence of a broader array of organisms than found in acidic waters. The possible liming agents include: limestone minerals such as calcite ($CaCO_3$) and dolomite ($CaMg(CO_3)_2$), hydrated lime ($Ca(OH)_2$) and soda ash (Na_2CO_3). Limestone has been the preferred material because: (1) it neutralizes acidity without causing excessively high or rapidly changing pH; (2) it increases acid neutralizing capacity (ANC) with high buffer capacity at typical ambient pH; (3) it increases the amount of divalent cations, which are physiologically important to fish and may competitively inhibit the uptake of toxic metal ions; (4) it provides flexibility in treatment design owing to its pH solubility relationship; and (5) it is inexpensive, readily available, and usually contains few toxic contaminants (Porcella et al., 1989). In Sweden, more than 95%

of liming operations are carried out using limestone and dolomite (Henrikson & Brodin, 1995). Finely ground calcite, usually less than 0.2 mm in diameter, is an effective liming agent. Porcella et al. (1989) discuss the water chemistry related to lake liming operation in detail. In summary, the de-acidification is due to the rapid reaction of calcite with hydrogen ions and aqueous carbon-di-oxide, increasing the pH value.

Two important questions related to liming operation are:

- What are the cost effective ways of liming?
- What are the ecological effects of liming?

The cost depends primarily on the method of liming. There are various ways to carry out liming operation: (1) direct addition of lime to the water body; (2) wetland liming, generally by helicopter, to create a smoother influence on lake pH and prolong the duration of the liming; (3) full scale drainage area liming to influence the entire catchment area and obtain long-term effects; and (4) by dosers in the tributaries to the lake (Håkanson, 2003b). Although the last three methods often result in more effective liming, they are considerably more expensive than direct addition of lime in water. There are various options to carry out direct lime addition, such as: boat or barge, surface ice (snowmobile), shore land (tractor), feeder stream, and air (helicopter, plane). Liming by boat or barge is the most popular liming method. Lime is usually distributed in the form of slurry which aids its dissolution leading to more immediate impact on the lake pH. The selection of a particular method often depends on criteria such as budget, accessibility to the site, weather conditions, geographical details etc.

To answer the second question, many experimental studies have been conducted to determine the effect of lime addition on aquatic biota (Håkanson, 2003a; Morabito, 2001; Prepas et al., 2001; Erikkson, 1998; Hincks & Mackie, 1997; Findlay, 1996; Hindar et al., 1996). Since these case studies and observations are site specific, it is difficult to make generalized conclusions about the effect of liming on lake biota. The important conclusions can be summarized as:

- Response to acidic conditions and liming is species specific. There are species that thrive on acidic conditions.
- The biomass and diversity of a particular trophic level may not always be correlated while liming (Findlay, 1996).
- Recovery to pre-acidification stage might take a long time after the initial improvements.
- pH value above 6 often ensures the survival and reproduction of most species.

This information is used in the subsequent sections to formulate the lake liming problem.

11.3. Systematic Approach to Lake Liming

Since liming entails considerable costs, it is essential that the liming operation is optimized so as to reduce expenses. Even though the liming technique is the major factor deciding the expenses, efficient implementation of the selected technique can reduce expenses. Previous work in this area includes Håkanson (2003a) and Riely & Rockland (1988). Currently, most of the liming decisions are based on rule of thumb. The amount of lime to be added is decided using parameters such a lake volume, current lake pH, targeted pH, water salinity etc. (Håkanson & Boulion, 2002). These are mostly static decisions and do not take into account the dynamic nature of the natural system (lake). It is obvious that such heuristics based decisions do not maximize the benefit. An effective approach is to use time dependent liming where liming decisions (amount of lime to be added) change with time based on the current lake conditions. In engineering field, such problems are commonly solved and come under the aegis of control theory applications. Control theory has been developed with engineering applications in mind. However, natural systems also present an interesting avenue to implement these ideas. Recent examples of such applications include Shastri & Diwekar (2006c,d), Ludwig et al. (2003), Chukwu (2001) and Kolosov (1997) among others. This work proposes to use control theory to achieve time dependent liming of a lake. There are different types of control techniques that might be applicable to a control problem (e.g. optimal control, model predictive control, linearized control etc.). In this work, optimal control theory has been used. The fundamentals of optimal control theory have been well established. The main advantages of using optimal control theory are: it does not make any assumption about the structure of the controller (control law), and it theoretically works for all types of systems, including nonlinear. Due to these advantages, most of the control theory applications in natural systems use optimal control theory. The theoretical details are skipped here for brevity and interested readers are referred to texts such as Kirk (1970) and Lewis (1986).

There are other issues that complicate the liming operation. The primary issue is the presence of various kinds of uncertainties, such as lack of information on the exact pH of the lake, seasonal variations in lake pH, and topological effects of liming. Moreover, the spatial and temporal effects of liming on lake biota are subjective. As a result, lake liming has not been a widespread practice in north America even though it has been relatively successful in Scandinavian countries. In order to make liming implementable, one needs to incorporate these uncertainties in the analysis. Uncertainty can be of two types: time independent (static) and time dependent (dynamic). Many of the uncertainties associated with natural systems are dynamic in nature. Moreover, static uncertainties in dynamic systems can often lead to time dependent uncertainties. Hence this work will focus on incorporating dynamic uncertainty in control problem formulation.

The primary requirement to formulate and solve a control problem is a model for the liming operation. Ottosson & Håkanson (1997) present a simplified deterministic model to simulate lake pH in the presence of natural inputs and liming actions. The inclusion of uncertainty in the analysis however necessitates uncertainty modeling. Real options theory has dealt with time dependent uncertain variables and proposes efficient methods to model those. The modeling basics are discussed in section 4 of this chapter. Those ideas are used here to model uncertain parameters associated with the liming operation.

The next section presents the liming model in detail. The basic deterministic model is first discussed followed by the stochastic model which is later used to formulate and solve the optimal control problem. The control problem solution has been achieved using stochastic maximum principle which is presented in section 6.

11.4. Lake Liming Model

11.4.1. Basic Deterministic Model

The basic lake liming model is presented in Ottosson & Håkanson (1997) and further discussed in Håkanson & Boulion (2002) and Håkanson (2003b). It is a mixed model consisting of both statistical regression and dynamic interactions. An empirical model is used to predict the initial pH (mean annual pH). The model also includes a regression that predicts natural pH. In addition to these empirical submodels, the lake liming model consists of dynamic (time dependent) interactions. It is a compartmental model with three different compartments, namely, water, active sediment and passive sediment. Accordingly, the three model variables are: lime in water, lime in active sediment and lime in passive sediment. Four continuous flows of lime connect the three compartments: sedimentation to active sediments, internal loading from active sediments to water, outflow from the lake water and transport from active to passive sediments. In addition, two flows give the inflow of lime from the liming, one to the lake water and one directly to the active sediments. The model is easy to handle since all input data can be obtained from maps and no field measurements are necessary. The necessary input parameters in the equation are: lime distribution coefficient (Dc), internal loading rate (ILR), dynamic ratio (Dr), lake water retention time (Rt), sedimentation rate (Sr) and active sediment age (ASA). These parameters are dependent on other basic lake chemical and physical properties such as: lake area, lake mean depth, lake maximum depth, lake color, lake total phosphorous concentration, drainage area, mean annual precipitation. The model variables are:

y_1: Lime in water
y_2: Lime in active sediment
y_3: Lime in passive sediment
The governing ordinary differential equations for the model are:

$$f_1 = \frac{dy_1}{dt} = \text{Lime input}.Dc.(0.422 * 0.712) + y_2.ILR.Dr - \frac{y_1}{Rt.52} - y_1.Sr \quad (66)$$

$$f_2 = \frac{dy_2}{dt} = y_1.Sr + u.(1 - Dc).(0.422 * 0.712) - y_2.ILR.Dr - \frac{y_2}{ASA} \quad (67)$$

$$f_3 = \frac{dy_3}{dt} = \frac{y_2}{ASA} \quad (68)$$

The change in lake pH value is computed from the lime in water variable of the model using a logical condition given below.

$$\Delta pH = \begin{cases} log_{10}(LC + 0.01) - AC & \text{if } log_{10}(LC + 0.01) - AC > 0 \\ 0 & \text{if } log_{10}(LC + 0.01) - AC \leq 0 \end{cases}$$

where,
AC = Additive constant,

LC = Lime concentration = Lime in water/Lake volume, and
Lake pH = Lake initial pH + ΔpH

Extensive discussion of the model and its validation can be found in Ottosson & Håkanson (1997) and Håkanson & Boulion (2002) and hence skipped in this text. The next section explains a modification to the basic model in order to make it suitable for optimal control application.

11.4.2. Modified Nonlinear Lake Liming Model

In the deterministic lake liming model, the computation of the lake pH from lime in water is through a logical condition. However, the presence of a logical condition is problematic in the formulation and solution of the control problem. The logical condition therefore must be eliminated, which is done here through a nonlinear approximation.

The modified model assumes, for the sake of mathematical representation, that the base pH value of the lake is a result of the presence of lime in water. Any additional lime, which represents the actual lime added in the liming operation, increases the lake pH above the base pH value. A continuous function relates the lake pH value and the total lime quantity in water (which includes the lime quantity corresponding to the base pH value). The variation in pH with base concentration is often represented in chemistry literature by a well known sigmoidal function. The general equation for the sigmoidal curve is given as:

$$f(x) = Vmin + \frac{Vmax - Vmin}{1 + 10^{[log_{10}(E_{50}) - x]}} \tag{69}$$

Here, $Vmin$ and $Vmax$ are the minimum and maximum values of the dependent variable, respectively, E_{50} is the 50% value in the given range (where the steepest part of the curve is situated), and x is the value of the independent variable. It is assumed that such a sigmoidal curve governs the relationship between lake pH and lime in water (the steepest part of the sigmoidal curve being around pH 7). Accordingly, the dependent variable is lake pH while the independent variable is lime quantity in water. The pH range of interest for liming operation is take as 0-10. Thus, $Vmin = 0$, $Vmax = 10$. Let u be the quantity of lime in water, u_{max} be the maximum quantity of lime expected in water, u_{min} be the minimum quantity of lime in water (taken as zero), and u_{mean} be the lime quantity in water corresponding to the initial lake pH. The range of independent variable in the actual sigmoidal equation is $v_{min} = -10$ and $v_{max} = +10$, which corresponds with u_{min} and u_{max}, respectively and therefore $u_{mean} = 0$. The resulting pH equation is given as:

$$pH = \frac{10}{1 + 10^{[k.(\frac{u_{mean}}{u})^4.(log_{10}(E_{50}) + 10 - 20.\frac{u}{u_{max}})]}} \tag{70}$$

where, k is a tuning parameter for better approximation. The value of E_{50} is adjusted to ensure that simulation results for the approximate model match those for the original model reasonably well. Values for u_{mean} and u_{max} will depend on the properties of the lake being limed. The lake pH is thus calculated as a function of u, and the actual lime addition is given by $u - u_{mean}$.

11.4.3. Modified Linear Lake Liming Model

The modification to the lake liming model presented in the previous section included a nonlinear sigmoidal function to approximate lake pH as a function of the lime quantity in water. It must be noted that the results of the control problem solution (lime addition profile) must be optimal for the basic model. This means that the results obtained using the approximation must also hold for the actual model. The nonlinear model discussed in the last section gives a reasonable approximation over the considered pH horizon of 0-10. However, in most cases, the actual lake pH values are not less than 4. Only in extreme cases of lake acidification, the values are found to be below 4. Hence, the pH range of interest can be reduced to 4-10. This allows one to use a linear approximation for lake pH instead of a nonlinear one. A linear relationship gives a better approximation for the considered range and also reduces numerical complexities. It must however be ensured that the lake pH values are within the range of interest. Outside this range, the linear variation will be much worse than the nonlinear one considered before.

The basis of the linear model is same as that for the nonlinear model. It assumes that the lake base pH is due to the presence of lime in water. For linear approximation, only the linear portion of the nonlinear sigmoidal curve is considered. The linear model is thus given as:

$$pH = \frac{k}{Vmax - Vmin}\left[v_{min} + \frac{v_{max} - v_{min}}{u_{max}}\right].u \tag{71}$$

where, all symbols have their previously assigned meanings. For linear approximation, based on the smaller pH range of interest, $Vmin = 4$ and $Vmax = 10$. The sigmoidal relationship for the independent variable value between -3 and +3 shows a fairly linear relationship. This portion of the curve is considered here and approximated using a linear relationship. Accordingly, with reference to the notations used in the previous section, $v_{min} = -3$ and $v_{max} = +3$. k is again a constant used as a tuning parameter for better approximation. The values of u_{mean} and u_{max} again depend on the lake characteristics. The linear model can therefore be written as:

$$pH = \frac{k}{6}\left[-3 + \frac{6}{u_{max}}\right].u \tag{72}$$

The quality of approximation using the linear liming model has been verified through test simulations and found to be quite good. The results of the test simulations are omitted here for the sake of brevity. The subsequent discussion in this work refers to the linear model presented in this section.

11.4.4. Stochastic Lake Liming Model

It has been mentioned before that the lake liming operation is complicated due to the presence of various uncertainties. Natural pH of a lake varies seasonally and hence constitutes an uncertain parameter. Typical seasonal fluctuations in natural lake pH are reported in Ottosson & Håkanson (1997). The data gives fractional change in lake pH for a particular month over the annual average lake pH value. To include these variations, the modified lake pH equation is written as:

$$pH = \frac{k}{6}\Big[-3 + \frac{6}{u_{max}}\Big].u.\Big(1 + \delta pH\Big) \tag{73}$$

where, δpH is the fractional variation in lake pH value for that time period.

However, fluctuations represented by δpH can vary for different lakes (due to geographic location, land use practices and other geochemical properties). Determination of the exact values of δpH for each lake is not practical. Since this is a parameter that changes randomly with time, time dependent uncertainty modeling techniques can be used. A generalized representation of δpH in Eq. 73 using a stochastic process will also make the liming model flexible for application to different lakes. In this work, mean revering Ito process is used to model δpH owing to its success in modeling various time dependent stochastic parameters (Diwekar, 2005; Shastri & Diwekar, 2006d; Ulas & Diwekar, 2004). The stochastic liming model is then represented as:

$$\begin{aligned}
f_1 = \frac{dy_1}{dt} &= (u - u_{mean}).Dc.(0.422 * 0.712) \\
&+ y_2.ILR.Dr - \frac{(y_1 - u_{mean})}{Rt.52} - (y_1 - u_{mean}).Sr && (74) \\
f_2 = \frac{dy_2}{dt} &= (y_1 - u_{mean}).Sr + (u - u_{mean}).(1 - Dc).(0.422 * 0.712) \\
&- y_2.ILR.Dr - \frac{y_2}{ASA} && (75) \\
f_3 = \frac{dy_3}{dt} &= \frac{y_2}{ASA} && (76) \\
pH = y_4 &= \frac{k}{6}\Big[-3 + \frac{6}{u_{max}}\Big].y(1).(1 + y_5) && (77) \\
f_{ito} = \frac{dy_5}{dt} &= \eta(d\bar{pH} - y_5) + \frac{\sigma\,\epsilon}{\sqrt{\Delta t}} && (78)
\end{aligned}$$

Here, the parameter representing natural pH fluctuations (δpH) becomes the additional state variable (y_5), which is modeled as a stochastic process represented by equation f_{ito}. $d\bar{pH}$ is the mean value of fractional natural pH variation. The stochastic lake liming model is used further in this work to formulate and solve optimal control problem.

11.5. Optimal Control Problem

Optimal control problems require establishing an index of performance for the system and designing the course of action so as to optimize the performance index. The goal in lake liming operation is to maintain the pH value at some desired level or within a desired range. Let $\bar{pH}$ represent the targeted pH value, which can be a constant or a time dependent parameter. The objective is to achieve target pH value as closely as possible, or alternately, to minimize the variance of the actual lake pH value around the targeted pH value. Accordingly, the time dependent objective is mathematically represented as:

$$J = \int_0^T (y_4 - p\bar{H})^2 \, dt \tag{79}$$

where, T is the simulation time horizon. The lake liming problem is thus a stochastic optimal control problem. Stochastic maximum principle, as describe in section 6, is used to formulate and solve the control problem. Since the resulting boundary value problem is complex, analytical solution is very difficult. Hence, computational technique of steepest ascent of Hamiltonian is used. The problem details and equations are omitted here for the sake of brevity. The next section discusses the important results of the problem.

11.6. Lake Liming Problem Results

The methodology explained in the previous sections is applied on a case study lake liming problem. The parameter values for the case study lake, based on the values reported in Håkanson & Boulion (2002), are:

Initial lake pH = 6.15
Lake area = 1.26 km^2
Lake mean depth = 8.5 m
Lake maximum depth = 26.2 m
Drainage area = 51.5 km^2
Mean annual precipitation = 602 mm/year
Active sediment age = 519.6 weeks
Internal loading rate = 0.001 (1/month)
Distribution coefficient = 0.5
Settling velocity = 0.074 meter/week
Additive constant = 2.375

The other parameters are computed using these basic parameters as per the following relationships Håkanson & Boulion (2002):

Lake volume = Lake area * Lake mean depth
Water discharge = 0.01*DrainageArea*Precipitation/600
Lake water retention time = Lake volume/(Water discharge*60*60*24*365/7) (weeks)
Sedimentation rate = Settling velocity/Lake mean depth (1/week)
Dynamic ratio = Lake area$^{0.5}$/Lake mean depth

Ottosson & Håkanson (1997) report the typical natural lake pH variation for a lake. As described in section 11.4, a mean revering Ito process is used to model the natural lake pH variation for the stochastic optimal control problem. Fig. 21 shows a comparison between the predefined natural variation and the variation modeled using mean reverting Ito process. The targeted pH value to be used in the time dependent objective function is 7.0. The simulation time horizon is 1000 weeks. The upper limit on the control variable (lime addition rate) is 100 T/week. Such an upper limit is expected to be present in actual implementations due to logistic or equipment limitations.

Fig. 22 presents the result of the control problem solution, indicating that the targeted lake pH is effectively achieved. The lake pH rises very quickly within about one year and then fluctuates around the targeted pH for the remaining time horizon. The control variable profile for this result is shown in Fig. 24. Lime input for the initial simulation horizon is high to raise the lake pH value to the targeted value as quickly as possible. After the targeted range has been achieved, lime addition settles at a non-zero value for the remaining time horizon. The value fluctuates continuously to account for the variation in natural lake pH. The application of the optimal control variable profile on the basic lake liming model leads to pH variations illustrated in Fig. 23. The plot shows that fluctuations for the basic model are much less than those for the approximate model. There is a slight offset for the complete time horizon due to the approximation. Such an error can however be effectively taken care of by reducing the actual lime additions by a constant amount throughout.

To ascertain the importance of using stochastic maximum principle, the stochastic model is solved using deterministic optimal control technique. Thus, the natural lake pH variation is represented using the mean reverting Ito process. However, the control problem is solved using the deterministic techniques, i.e. the natural pH variations do not affect the optimal control problem solution. This is equivalent to ignoring natural pH variations while taking liming decisions. The result for this problem, along with the result for the stochastic problem is shown in Fig. 25. The use of stochastic optimal control (stochastic maximum principle) leads to better lake pH control. Similar comparison for the basic liming model is illustrated in Fig. 26. The results for the basic model clearly illustrate the advantage of stochastic control over deterministic control. This emphasizes that taking liming decisions while ignoring natural pH variations will lead to suboptimal results, thereby highlighting the importance of uncertainty consideration in decision making.

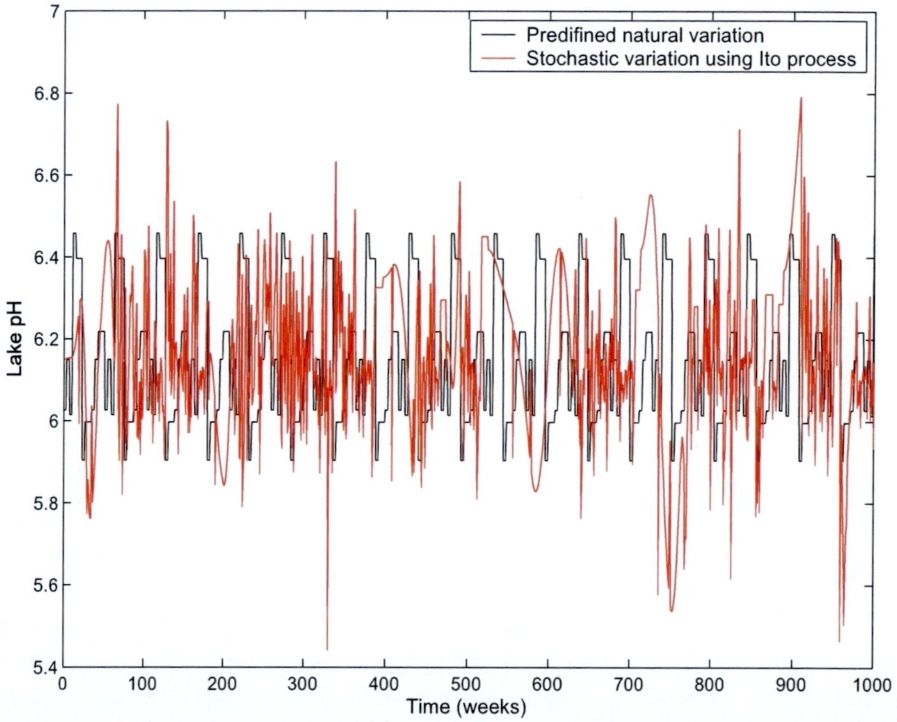

Figure 21. Natural lake pH variation: Comparison between predefined and Ito process variation.

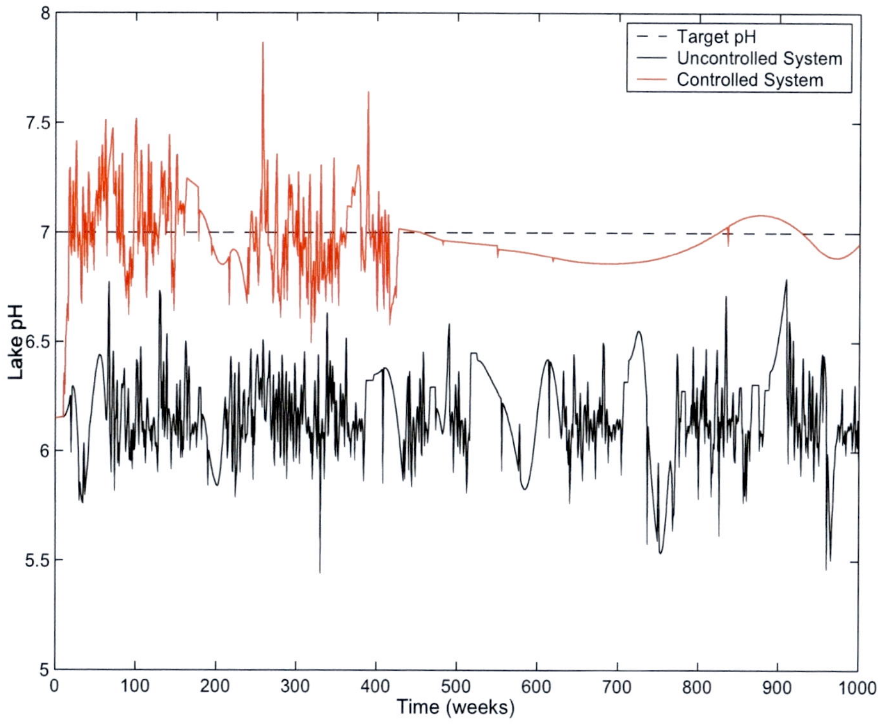

Figure 22. Lake pH variation for approximate model: Controlled and uncontrolled system.

11.7. Multi-objective Liming Problem

The previous section presented the optimal control analysis of a stochastic lake liming problem. The objective for the control problem is to achieve the target pH as closely as possible for the complete time duration under consideration. The only restriction on the control variable is an upper bound to account for logistical and equipment limitations. However lake liming is expensive. Hence, financial aspect will have a definite impact on the decisions related to liming. This represents an additional objective for the control problem. Such problems appear in almost all real life decisions and are known as multi-objective problems in optimization and control theory terminology. Quite often these multiple objectives are conflicting and a trade off exists between them. The decision maker first has to decide the importance of each of the objectives and then solve the control problem that optimizes the objectives according to the given weights. The more difficult task for the decision maker is then to decide the appropriate weights for the objectives. Quite often the approach to solve the problem for different weights on the objective and generate a Pareto surface (Diwekar, 2003b). For the lake liming problem, the objectives are to achieve the target lake pH and minimize the liming expenses, which are conflicting. To formulate the multi-objective optimization problem though, the liming cost function has to be finalized.

The cost of liming depends on the method used for liming. The costs incurred for liming using different methods have been reported in Riely & Rockland (1988) based on experimental studies. Since liming by boat or barge is the most popular method, it is assumed to be the method employed in this analysis. Based on the values reported in Riely & Rockland

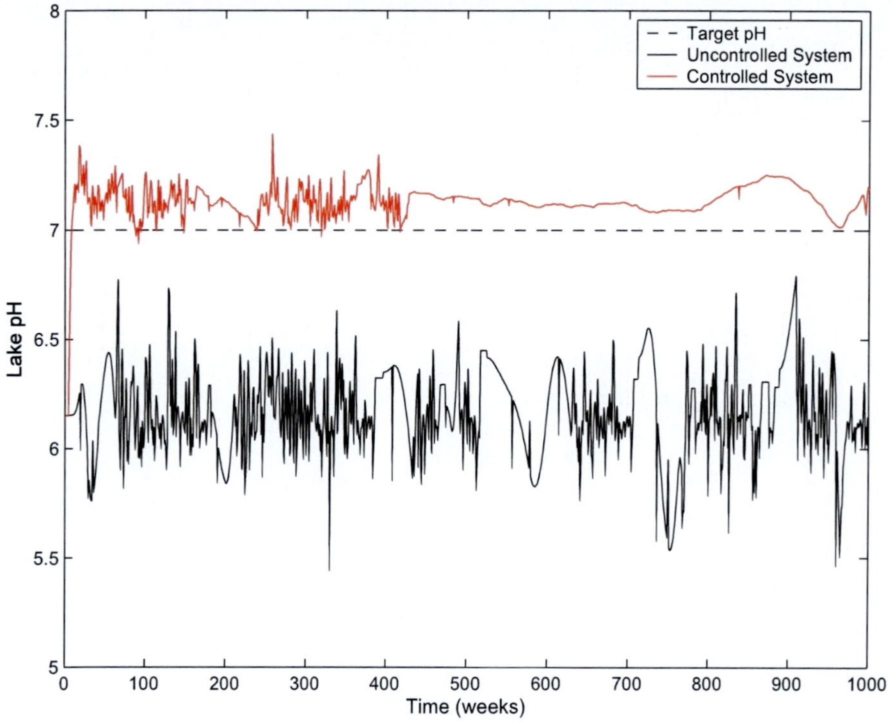

Figure 23. Lake pH variation for basic model: Controlled and uncontrolled system.

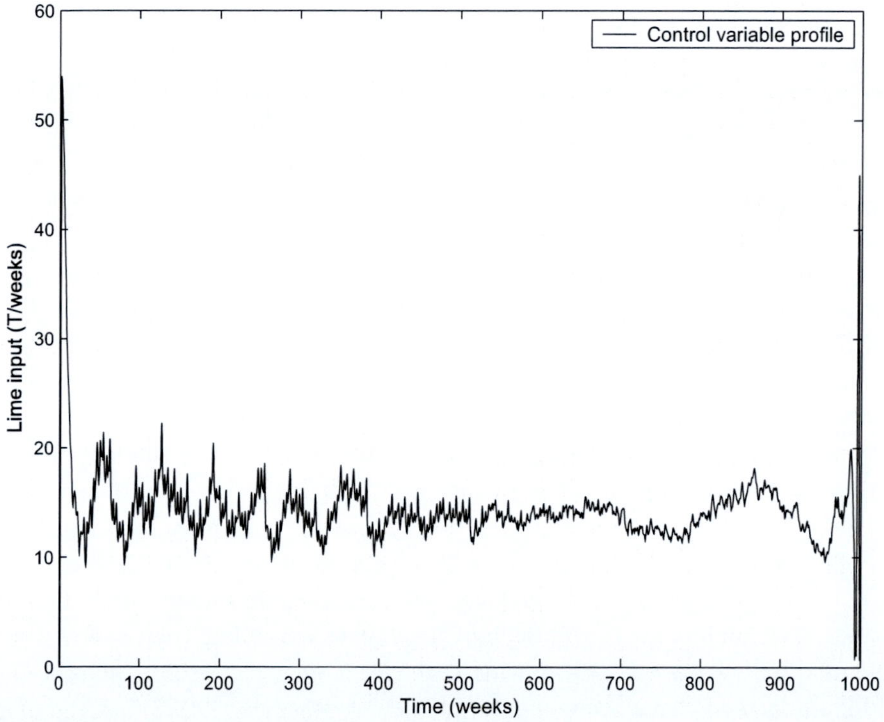

Figure 24. Control variable (lime addition) profile for the stochastic lake pH control.

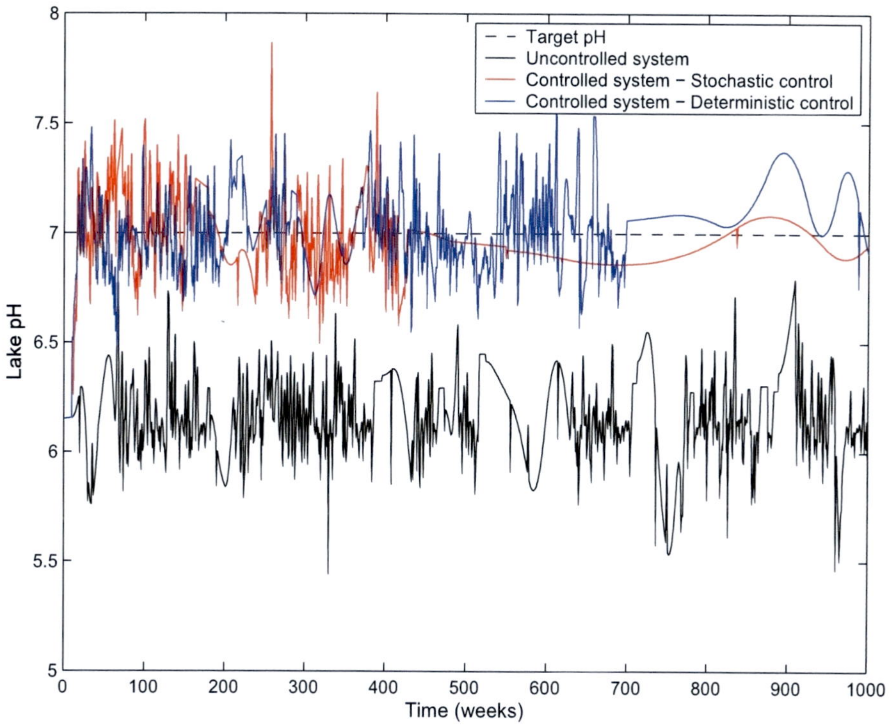

Figure 25. Comparison between stochastic and deterministic control problem result: Modified model.

(1988), the average liming cost is \$236 per unit Ton of lime. However, it must be noted that there is a minimum cost of liming irrespective of the quantity of lime added. This represents the basic logistic and equipment expense. The fraction of this cost in the total expenditure typically reduces as the quantity of lime added is increased. Thus, cost and lime quantity are nonlinearly related. Here, \$236 is assumed to be the minimum required cost for any non-zero lime addition. This discontinuous function is approximated using a power law as given below:

$$\text{Liming cost} = 236.u(t)^{0.25} \tag{80}$$

where, u is the quantity of lime added at any instant. This equation is the mathematical representation of the second objective function. For the multi-objective problem, the combined objective function is therefore defined as:

$$J = \int_0^T \left[w_1.(y_4 - \bar{pH})^2 + w_2.236.u(t)^{0.25}\right] dt \tag{81}$$

where, w_1 and w_2 represent the weights given to each objective function.

The stochastic optimal control problem solved in the previous section represents the solution with no importance to the economic objective, i.e. with $w_2 = 0$. To find the trade off between the two objectives, the problem is solved with a non-zero weight on the economic objective function. The comparative results for the modified model are shown

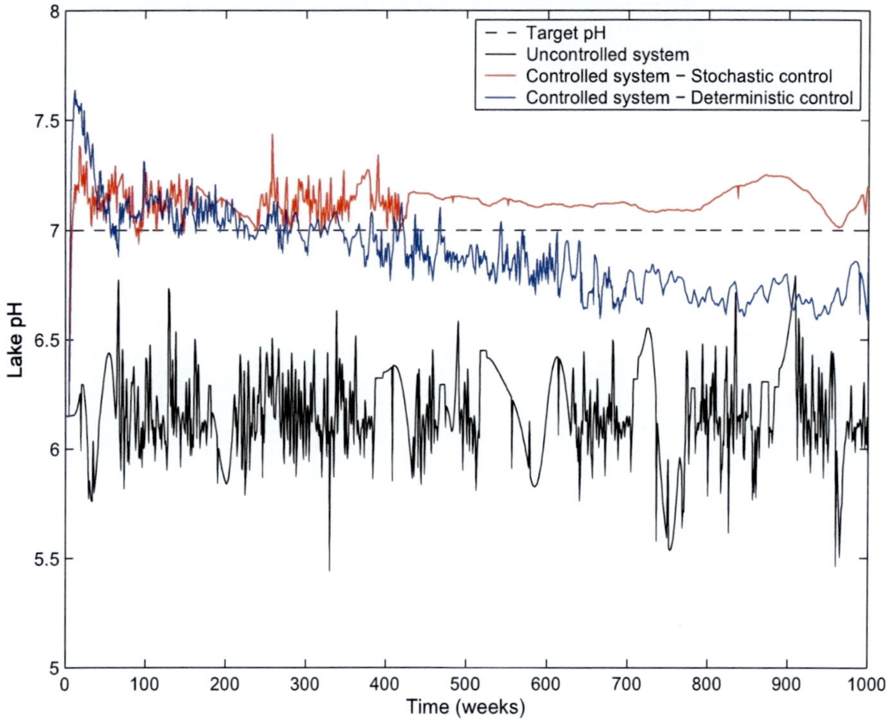

Figure 26. Comparison between stochastic and deterministic control problem result: Basic model.

in Fig. 27. Similar comparison for the basic model, using the optimal control profiles for the model simulation, is shown in Fig. 28 The comparison shows that the inclusion of economic objective degrades the quality of pH control. In the multi-objective case, the lake pH is constantly fluctuating and never settles around the targeted pH value. A comparison of the numerical values suggests that there is about 50% reduction in the liming cost from the single objective solution. However, this is accompanied by about 50% deterioration in the liming objective (variance around the target value).

This comparison thus illustrates the trade off between accurate pH control and liming cost. Problem solution for different weights on the objectives will generate the complete trade off surface (Pareto surface). The decision maker can then chose the right combination of the weights and use the corresponding lime input profile for implementation.

11.8. Discussion: Ecosystem Level Management

Mercury methylation is an important step in the bioaccumulation process. Since acidic conditions aid methylation, control of lake pH through external lime addition, known as lake liming, is a possible solution to minimize the adverse effects. To maximize the benefits from lake liming, time dependent liming, and use of systems theory based technique of optimal control for decision making is proposed. The basic lake liming model from literature is modified to formulate the optimal control problem. However, there are considerable uncertainties associated with the liming operation, mainly due to the unpredictable natural

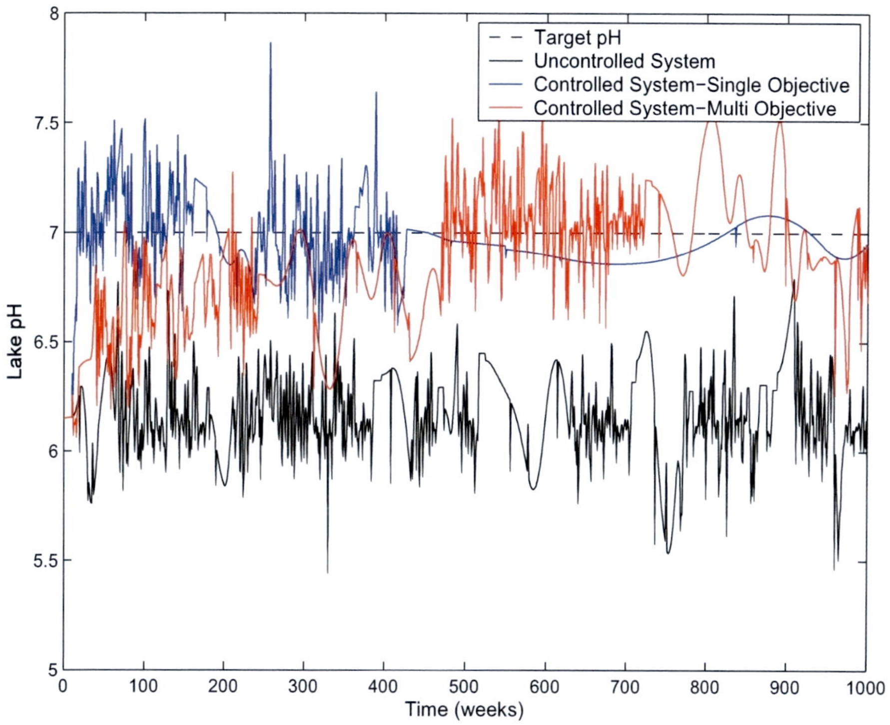

Figure 27. Comparison between single objective and multi-objective liming problem solution: Modified model.

fluctuations in lake parameters. This work proposes to use efficient uncertainty modeling techniques from finance literature, resulting in a stochastic optimal control problem. A case study problem, considering natural lake pH variation as an uncertain parameter, is discussed. The results show that use of time dependent liming is a worthwhile option and highlight the importance of considering uncertainty in taking liming decisions. The multi-objective problem results illustrate the trade-off between the liming and financial objectives. The results should generate more interest in lake liming operation, thereby presenting an additional tool to minimize the harmful effects of mercury pollution.

12. Conclusion

Pollution in all media such as air, water and soil, has been a concern for the last few decades due to valid reasons. Continuous degradation of the natural environment starting mainly with the industrial revolution has manifested itself quite severely in the last few years. Changes in the environment, some of them apparently non-reversible, have been alarming, forcing humans to take immediate corrective measures. Industries, being the major polluters, attract much attention. Earlier efforts to achieve environmentally friendly industrial sector went from waste treatment and management, to green engineering and green process design, coming under the auspice of industrial ecology. However, it has been realized that the aspects of society and overall ecosystem need to be considered more strongly. Further-

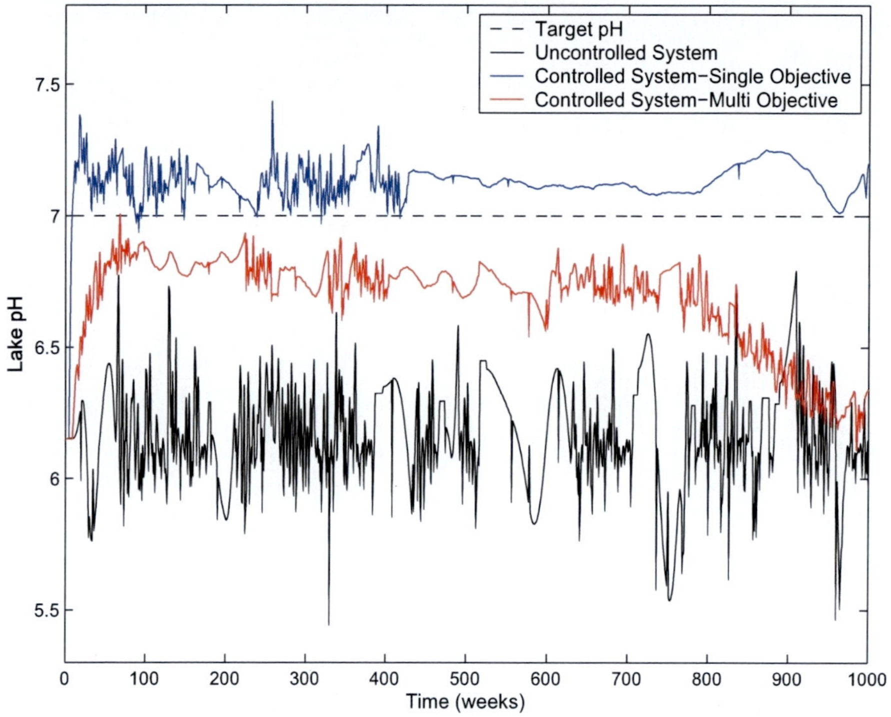

Figure 28. Comparison between single objective and multi-objective liming problem solution: Basic model.

more, since future is quite uncertain, forecasting and decision making under uncertainty is necessary for a realistic analysis. This brings in the concept of sustainability for industrial waste management. Sustainability is a relatively new concept in the field of environmental management. Sustainable development refers to management decisions that target long term goals and consider disparate systems interacting in a complex manner. Such an integrated approach is thought to be the key to future developments. The inter-disciplinary nature of the work however poses serious problems for efficient decision making. The concepts and tools of systems theory provide a means to resolve this issue. This forms the central idea of this chapter.

The main idea presented in this chapter is sustainable industrial pollution management using systems theory based tools. The scope of sustainability and systems theory are shown to be overlapping, creating a firm basis for this collaboration. Various systems theory based tools such as optimization and optimal control can be used for this task. The field being highly developed, only some of the recent innovative techniques are reviewed in this chapter. This includes uncertainty modeling techniques from finance (real options theory), efficient optimization technique to solve stochastic nonlinear programming problems (L-shaped BONUS algorithm), and stochastic maximum principle to solve optimal control problems in the presence of uncertainty. As a proof of concept for this approach, the chapter discusses mercury waste management in water using the case study example of Savannah river watershed. In accordance with the idea of sustainability, it presented management options at individual industry, industrial sector and ecological level. At the industry level, the option is to design and optimize efficient mercury treatment technologies to reduce emissions. At the inter-industry level, pollutant trading offers an avenue to achieve tighter regulations at lower overall cost. Optimization framework (deterministic and stochastic) is proposed to ensure optimal decisions in the face of increased complexities. At the ecology level, the effect of emitted mercury is minimized through time dependent liming strategies for the water bodies to control bioaccumulation. The results discussed in the respective sections collectively point to the success of using a systematic approach in decision making. The approach enables quantification of the benefits and drawbacks of the decisions, and puts forth certain interesting results that are not intuitive.

The discussion in this chapter should however be looked at from a broader perspective, not restricted to the case of mercury waste. The intention is to make the reader aware of the scope of waste management problem and the handlers available to manage that problem. The approach is valid for any pollutant. One should incorporate specific aspects of the problem (such as technologies, fate and transportation, secondary pollution etc.) in this general framework. This should then pave the way to achieve sustainable industrial waste management.

References

Ambrose, J., Wool, T. & Martin, J. (1993). WASP5, A hydrodynamic and water quality model - model theory, user's manual and programmer's guide. Technical report: EPA/600/3-87/039, U.S. Environmental Protection Agency. Environmental Research Laboratory, Athens, GA.

Atkinson, S. & Tietenberg, T. (1982). The empirical properties of two classes of designs for transferable discharge permit markets. *Journal of Environmental Economics and Management*, **9**, 101–121.

Bertilsson, L. & Neujahr, H. (1971). Methylation of mercury compounds by methylcobalamin. *Biochemistry*, **10**(14).

Birge, J. & Louveaux, F. (1997). *Introduction to Stochastic Programming*. Springer series in Operations Research.

Bohi, D. & Burtraw, D. (1998). SO_2 allowance trading: How experience and expectations measure up. *The Electricity Journal*, **10**(7), 67–75.

Boumol, W. & Oates, W. (1988). *The Theory of Environmental Policy*. 2nd edn. New York: Cambridge University Press.

Burtraw, D., Evans, D., Krupnick, A., Palmer, K. & Toth, R. (2005). Economics of pollution trading for SO_2 and NO_x. *Annual Review and Environment and Resources*, **30**, 253–289.

Cason, T. & Gangadharan, L. (2003). Transaction costs in tradable permit markets: An experimental study of pollution market design. *Journal of Regulatory Economics*, **23**(2), 145–165.

Charnes, A. & Cooper, W. (1959). Chance-constrained programming. *Management Science*, **5**, 73–79.

Chukwu, E. (2001). *Stability and time-optimal control of hereditary systems - with applications to the economic dynamics of the U.S.*, vol. 60 of *Series on advances in mathematics for applied science*. World Scientific.

Crocker, T. (1966). *The economics of air pollution*, chap. The structuring of air pollution control systems. New York: W.W. Norton.

Dales, J. (1968). *Pollution, property and prices*. Toronto: University of Toronto Press.

Dantzig, G. & Glynn, P. (1990). Parallel processors for planning under uncertainty. *Annals of Operations Research*, **22**, 1–21.

Dantzig, G. & Infanger, G. (1992). *Computational and Applied Mathematics I*. Elsevier Science Publishers, B.V. (North Holland).

DeSimone, R., Penley, M., Charbonneau, L., Smith, G., Wood, J., Hill, H., Pratt, J., Ridsdale, S. & Williams, R. (1973). The kinetics and mechanism of cobalamine-dependent mthyl and ethyl transfer to mercuric ion. *Biochimica et Biophysica Acta*, **304**, 851–863.

Diwekar, U. (1996). *Batch Distillation: Simulation, Optimal Design and Control*. Taylor and Francis, Washington, DC.

Diwekar, U. (2003a). Greener by design. *Environmental Science & Technology*, **37**, 5432–5444.

Diwekar, U. (2003b). *Introduction to Applied Optimization*. Kluwer Academic Publishers, Dordrecht.

Diwekar, U. (2003c). Optimization under uncertainty in chemical engineering. *Proceedings of Indian National Science Academy*, **69,A**(3 & 4), 267–284.

Diwekar, U. (2005). Green process design, industrial ecology, and sustainability: A systems analysis perspective. *Resources, Conservation and Recycling*, **44**(3), 215–235.

Dixit, A. & Pindyck, R. (1994). *Investment under uncertainty*. Princeton University Press, Princeton, New Jersey.

Driscoll, C., Blette, V., Yan, C., Schofield, C., Munson, R. & Holsapple, J. (1995). The role of dissolved organic carbon in the chemistry and bioavailability of mercury in remote adirondack lakes. *Water, Air and Soild Pollution*, **80**, 499–508.

Eisler, R. (1987). Mercury hazards to fish, wildlife, and invertebrates: A synoptic review. Publication no. 85 (1.10), U.S. Fish and Wildlife Service, Department of the Interior, Washington, DC.

Elmaghraby, S. (1960). Allocation under uncertainty when the demand has continuous D.F. *Management Science*, **10**, 270–294.

Erikkson, A. (1998). Are the effects of acidification postponed and aggravated by lake liming? - a laboratory study on a re-acidification of the limed Lake Gårdsjön, SW Sweden. Technical report, Department of Physical Geography, Göteborg.

Farrell, A. (2000). The NOx budget: A look at the first year. *The Electricity Journal*, **13**(2), 83–93.

Farrell, A., Carter, R. & Raufer, R. (1999). The NOx Budget: market based control of tropospheric ozone in the northeastern United States. *Resources and Energy Economics*, **21**, 103–124.

Findlay, D. (1996). The effect of incremental pH recovery on the Lake 223 phytoplankton community. *Canadian Journal of Fisheries and Aquatic Sciences*, **53**, 856–864.

Gangadharan, L. (2000). Transaction costs in pollution markets: An empirical study. *Land Economics*, **76**(4), 601–614.

Gilmour, C., Henry, E. & Mitchell, R. (1992). Sulfate stimulation of mercury methylation in freshwater sediments. *Environmental Science & Technology*, **26**, 2281–2287.

Hahn, R. & Hester, G. (1989). Marketable permits: Lessons from theory and practice. *Ecology Law Quarterly*, **16**(2), 361–406.

Håkanson, L. (2003a). Consequences and correctives related to lake acidification, liming and mercury in fish - A case-study for lake Huljesjön, Sweden, using the LakeWeb-model. *Environmental Modeling and Assessment*, **8**, 275–283.

Håkanson, L. (2003b). A general management model to optimize lake liming operations. *Lakes & Reservoirs: Research and Management*, **8**, 105–140.

Håkanson, L. & Boulion, V. (2002). *The lake foodweb*. Backhuys Publishers, Leiden.

Henrikson, L. & Brodin, Y. (1995). *Liming of acidified surface waters*. Springer, Berlin.

Hesterberg, T. (1995). Weighted average importance sampling and defensive mixture distribution. *Technometrics*, **37**, 185–194.

Higlc, J. & Sen, S. (1991). Stochastic decomposition: an algorithm for two stage linear programs with recourse. *Annals of Operations Research*, **16**, 650–669.

Hincks, S. & Mackie, G. (1997). Effects of pH, calcium, alkalinity, hardness, and chlorophyll on the survival, growth, and reproductive success of zebra mussel (Dreissena polymorpha) in Ontario lakes. *Canadian Journal of Fisheries and Aquatic Sciences*, **54**, 2049–2057.

Hindar, A., Kroglund, F., Lyderson, E., Skiple, A. & Høgberget, R. (1996). Liming of wetlands in the acidified Lake Røynelandsvatn catchment in southern Norway: effects on stream water chemistry. *Canadian Journal of Fisheries and Aquatic Sciences*, **53**, 985–993.

Horan, R. (2001). Differences in social and public risk perceptions and conflicting impacts on pont/nonpoint trading ratios. *American Journal of Agricultural Economics*, **83**(4), 934–941.

Horan, R., Shortle, J., Abler, D. & Ribaudo, M. (2001). The design and comparative economic performance of alternative second-best point/nonpoint trading markets. Staff paper, Department of agricultural economics, Michigan State University, East Lansing, Michigan 48824.

Hung, M.-F. & Shaw, D. (2005). A trading-ratio system for trading water pollution discharge permits. *Journal of Environmental Economics and Management*, **49**, 83–102.

Jarvie, M. & Solomon, B. (1998). Point-Nonpoint effluent trading in watershed: A review and critique. *Environmental Impact Assessment Review*, **18**, 135–157.

Jensen, S. & Jernelöv, A. (1969). Biological methylation of mercury in aquatic organisms. *Nature*, **223**, 753–754.

Jung, J., Blau, G., Penky, J., Reklaitis, G. & Eversdyk, D. (2004). A simulation based optimization approach to supply chain management under demand uncertainty. *Computers & Chemical Engineering*, **28**, 2087–2106.

Kalagnanam, J. & Diwekar, U. (1997). An efficient sampling technique for off-line quality control. *Technometrics*, **39**(3), 308–319.

Kampas, A. & White, B. (2003). Selecting permit allocation rules for agricultural pollution control: a bargaining solution. *Ecological Economics*, **47**, 135–147.

Kheawhom, S. & Hirao, M. (2004). Decision support tools for environmentally benign process design under uncertainty. *Computers & Chemical Engineering*, **28**, 1715–1723.

Kirk, D. (1970). *Optimal Control Theory: An introduction.* Prentice-Hall, Englewood Cliffs, NJ.

Kling, C. (1994). Environmental benefits from marketable discharge permits or an ecological vs. economical perspective on marketable permits. *Ecological Economics*, **11**, 57–64.

Kolosov, G. (1997). Size control of a population described by a stochastic logistic model. *Automation and Remote Control*, **58**(4), 678–686.

Lewis, F. (1986). *Optimal Control.* John Wiley & Sons.

Lin, X., Janak, S. & Floudas, C. (2004). A new robust optimization approach for scheduling under uncertainty: I. Bounded uncertainty. *Computers & Chemical Engineering*, **28**, 1069–1085.

Louveaux, F. (1978). Piecewise convex programs. *Mathematical Programming*, **15**, 53–62.

Louveaux, F. (2001). Nonlinear stochastic programming. Technical report, Faculté des Sciences, Département de Mathématique.

Ludwig, D., Carpenter, S. & Brock, W. (2003). Optimal phosphorous loading for a potentially eutrophic lake. *Ecological Applications*, **13**(4), 1135–1152.

Malik, A., Letson, D. & Crutchfield, S. (1993). Point/Nonpoint source trading of pollution abatement: choosing the right trading ratio. *American Journal of Agricultural Economics*, **75**, 959–967.

Marshall, C. (1999). Results of water-based trading simulations: Final report. Tech. rep., Phillip Services, Fort Washington, PA (generate for the U.S. Environmental Protection Agency).

McKitrick, R. (1999). A derivation of the marginal abatement cost curve. *Journal of Environmental Economics and Management*, **37**, 306–314.

Montgomery, D. (1972). Markets in licences and efficient pollution control programs. *Journal of Economic Theory*, **5**, 395–418.

Morabito, G. (2001). Six years (19921997) evolution of phytoplankton communities after recovery by liming in Lake Orta, northern Italy. *Lakes & Reservoirs: Research and Management*, **6**, 305–312.

Nagase, H., Ose, Y., Sato, T. & Ishikawa, T. (1982). Methylation of mercury by humic substances in an aquatic environment. *Science of Total Environment*, **32**, 147–156.

Newell, R. & Stavins, R. (2003). Cost heterogeneity and potential savings from market based policies. *Journal of Regulatory Economics*, **23**(1), 43–59.

Oates, W., Portney, P. & McGartland, A. (1989). The net benefit of incentive based regulation: A case study of environmental standard setting. *The American Economics Review*, **79**(5), 1233–1242.

Ottosson, F. & Håkanson, L. (1997). Presentation and analysis of a model simulating the pH response of lake liming. *Ecological Modelling*, **105**, 89–111.

Porcella, D., Schofield, C., DePinto, J., Driscoll, C., Bukaveckas, P., Gloss, S. & Young, T. (1989). *Acidic precipitation*, vol. Volume 4: Soils, aquatic processes, and lake acidification of *Advances in Environmental Science*, chap. Mitigation of acidic conditions in lakes and streams. Springer-Verlag, pages 159–186.

Prepas, E., Pinel-Alloul, B., Chambers, P., Murphy, T., Reedyk, S., Sandland, G. & Serediak, M. (2001). Lime treatment and its effects on the chemistry and biota of hardwater eutrophic lakes. *Freshwater Biology*, **46**, 1049–1060.

Rico-Ramirez, V. & Diwekar, U. (2004). Stochastic maximum principle for optimal control under uncertainty. *Computers & Chemical Engineering*, **28**, 2845–2845.

Rico-Ramirez, V., Diwekar, U. & Morel, B. (2003). Real option theory from finance to batch distillation. *Computers & Chemical Engineering*, **27**, 1867–1882.

Riely, P. & Rockland, D. (1988). Evaluation of liming operations though benefit-cost analysis. *Water, Air and Soil Pollution*, **41**, 293–228.

Rockafellar, R. & Wets, R.-B. (1986). A Lagrangian finite generation technique for solving linear-quadratic problems in stochastic programming. *Mathematical Programming Study*, **28**, 63–93.

Rockafellar, R. & Wets, R.-B. (1991). Scenario and policy aggregation in optimization under uncertainty. *Mathematics of Operations Research*, **16**, 119–147.

Ruszczyński, A. (1986). A regularized decomposition for minimizing a sum of polyhedral functions. *Mathematical Programming*, **35**, 309–333.

Rya, J.-H., Dua, V. & Pistikopoulos, E. (2004). A bilevel programming framework for enterprise-wide process networks under uncertainty. *Computers & Chemical Engineering*, **28**, 1121–1129.

Sahin, K. & Diwekar, U. (2004). Better Optimization of Nonlinear Uncertain Systems (BONUS): A new algorithm for stochastic programming using reweighting through kernel denisty estimation. *Annals of Operations Research*, **132**, 47–68.

Sahinidis, N. (2004). Optimization under uncertainty: state-of-the-art and opportunities. *Computers & Chemical Engineering*, **28**, 971–983.

Shastri, Y. & Diwekar, U. (2006a). An efficient algorithm for large scale stochastic nonlinear programming problems. *Computers & Chemical Engineering*, **30**, 864–877.

Shastri, Y. & Diwekar, U. (2006b). Sensor Placement in Water Networks: A Stochastic Programming Approach. *Journal of Water Resources Planning and Management*, **132**(3), 192–203.

Shastri, Y. & Diwekar, U. (2006c). Sustainable ecosystem management using optimal control theory: Part 1 (Deterministic systems). *Journal of Theoretical Biology*, **241**, 506–521.

Shastri, Y. & Diwekar, U. (2006d). Sustainable ecosystem management using optimal control theory: Part 2 (Stochastic systems). *Journal of Theoretical Biology*, **241**, 522–532.

Shastri, Y. & Diwekar, U. (2008a). L-shaped BONUS algorithm with application to water pollutant trading. *Industrial & Engineering Chemistry Research*, **47**, 9417–9425.

Shastri, Y. & Diwekar, U. (2008b). Optimal Control of Lake pH for Mercury Bioaccumulation Control. *Ecological Modelling*, **216**, 1–17.

Shortle, J., Horan, R. & Abler, D. (1998). Research Issues in Nonpoint Pollution Control. *Environmental and Resource Economics*, **11**(3-4).

Silvermann, B. (1986). *Density estimation for statistics and data analysis*. Chapman and Hall (CRC reprint 1998), Boca Raton, USA.

Solomon, B. (1998). Five years of interstate SO_2 trading: Geographic patterns and potential cost savings. *The Electricity Journal*, **11**(4), 58–70.

Sorensen, J., Glass, G., Schmidt, K., Huber, J. & Rapp, G. (1990). Airborne mercury deposition and watershed characteristics in relation to mercury concentrations in water, sediments, plankton and fish of eighty northern minnesota lakes. *Environmental Science & Technology*, **24**, 1716–1727.

Tietenberg, T. (1985). *Emissions trading an exercise in reforming pollution policy*. Resources for the Future ; Baltimore : Distributed by Johns Hopkins University Press, Washington, D.C.

Tietenberg, T. (1995). Tradeable permits for pollution control when emission location matters: what have we learned? *Environmental and Resource Economics*, **5**, 95–113.

Tietenberg, T. (2000). *Environmental and natural resource economics*. New York: Addison Wesley Longman.

Tomlinson, C. (1987). *Our Common Future*. World Commision on Environment and Development, Oxford University Press, Oxford.

Ulas, S. & Diwekar, U. (2004). Thermodynamics uncertainties in batch processing and optimal control. *Computers & Chemical Engineering*, **28**, 2245–2258.

USDOI (2001). Total plant costs: For contaminant fact sheets. Technical report, U.S. Departement of Interior, Bureau of Reclamation, Water treatement engineering and research group, Denver CO 80225.

USEPA (1991). Guidance for water quality-based decisions: The TMDL process. Tech. rep., Washington, DC: United States Environmental Protection Agency, Office of Water.

USEPA (1996). Draft Framework for Watershed-Based Trading. Tech. rep., EPA 800-R-96-001. Washington, DC: United States Environmental Protection Agency, Office of Water.

USEPA (1997a). Capsule report: Aqueous mercury treatment. Technical report: EPA/625/R-97/004, United States Environmental Protection Agency, Office of Research and Development, Washington DC 20460.

USEPA (1997b). Mercury study report to congress. Report to congress: EPA-452/R-97-003, United States Environmental Protection Agency.

USEPA (2000). Mercury research strategy. Technical report: EPA/600/R-00/073, United States Environmental Protection Agency, Office of Research and Development, Washington DC 20460.

USEPA (2001a). Total Maximum Daily Load (TMDL) for total mercury in fish tissue residue in the middle and lower Savannah river watershed. Report, United States Environmental Protection Agency, Region 4.

USEPA (2001b). Watershed Characterization System Users Manual. User's manual, U.S. Environmental Protection Agency, Region 4, Atlanta, Georgia.

USEPA (2003). Water Quality Trading Policy. Tech. rep., Office of Water, Washington, DC: United States Environmental Protection Agency.

Van Slyke, R. & Wets, R.-B. (1969). L-shaped Linear Programs with Application to Optimal Control and Stochastic Programming. *SIAM Journal on Applied Mathematics*, **17**, 638–663.

Wang, R., Diwekar, U. & Grégorie Padró, E. (2004). Efficient sampling techniques for uncertainties and risk analysis. *Environmental Progress*, **23**(2), 141–157.

Weber, M. (2001). Markets for water rights under environmental constraints. *Journal of Environmental Economics and Management*, **42**, 53–64.

Winfrey, M. R. & Rudd, J. W. M. (1990). Environmental factors affecting the formation of methylmercury in low pH lakes. *Environmental Toxicology and Chemistry*, **9**, 853–869.

Wood, J., Kennedy, F. & Rosen, C. (1968). Synthesis of methyl-mercury compounds by extracts of a methanogenic bacterium. *Nature*, **220**, 173–174.

Woodward, R. (August, 2001). The environmentally optimal trading ratio. In *Annual meeting of the American Agricultural Economics Assosiation*. Chicago, IL.

Xun, L., Campbell, N. & Rudd, J. (1987). Measurements of specific rates of net methyl mercury production in the water column and surface sediments of acidified and circumneutral lakes. *Canadian Journal of Fisheries and Aquatic Sciences*, **44**, 750–757.

Zhang, W. & Wang, X. (2002). Modeling for pointnon-point source effluent trading: perspective of non-point sources regulation in China. *The Science of the Total Environment*, **292**, 167–176.

Ziemba, W. (1970). Computational algorithms for convex stochastic programs with simple recourse. *Operations Research*, **18**, 414–431.

Ziemba, W. (1972). Solving nonlinear programming problems with stochastic objective functions. *The Journal of Financial and Quantitative Analysis*, **7**(3), 1809–1827.

In: Industrial Waste...
Editor: John P. Samuelson, pp. 205-222
ISBN: 978-1-60692-720-5

Chapter 6

POLLUTANT DETECTION AND ENVIRONMENTAL MONITORING USING CONDUCTOMETRIC MICROBIOSENSORS

Mouna Marrakchi [1,2,a], ***Nicole Jaffrezic-Renault*** [1,b]
Sergei V. Dzyadevych [3] ***and Florence Lagarde*** [1]
[1]Université de Lyon, Laboratory of Analytical Sciences, UMR CNRS 5180,
Université Claude Bernard Lyon 1, 43 Boulevard 11 Novembre 1918,
69622 Villeurbanne Cedex, France
[2]Laboratory of Microbial and Ecological Technology, National Institute of Applied Sciences and Technology, Centre Urbain Nord, B.P.676, 1080 Tunis Cedex, Tunisia.
[3]Laboratory of Biomolecular Electronics, Institute of Molecular Biology and Genetics, National Academy of Sciences of Ukraine,150 Zabolotnogo St., Kiev 03143, Ukraine

Abstract

Traditional systems of analysis are no longer adequate, owing to the increasing demand for real-time and in situ control of pollution (industrial wastes, heavy metals, quality of water, etc.). The appearance of electrochemical biosensors is probably one of the most promising ways to solve some problems concerning sensitive, fast and cheap analytical techniques. In this chapter, we focus on conductometric biosensors because of the advantages they offer over other types of transducers: they can be produced through inexpensive thin-film standard technology, no reference electrode is needed, and differential mode measurements allow cancellation of a lot of interferences. In the first part, the principles of conductometric measurements in ionic media are presented. These types of measurements are first applied to the monitoring of biocatalytic reactions. The use of conductometric microtransducers is then presented and detailed in the case of pollutant detection for environmental monitoring. The performances obtained for the detection of different pesticides, herbicides and heavy metal ions, based on enzyme inhibition, are presented as well as those obtained for the detection of

[a] E-mail address: mounamarrakchi@yahoo.fr; Tel. (+216) 98 66 68 37, Fax: (+216) 71 57 35 26. Author to whom correspondence should be addressed.
[b] E-mail address: nicole.jaffrezic@univ-lyon1.fr, Tel. (+33) 4 72 43 11 82, Fax: (+33) 4 72 43 12 06. Author to whom correspondence should be addressed.

formaldehyde, 4-chlorophenol, nitrate and proteins as markers of dissolved organic carbon based on enzymatic microbiosensors. Finally, whole-cell conductometric sensors using microalgea and bacteria as bioreceptors for heavy metals, selenite and pesticides detection are described.

Keywords: interdigitated microelectrodes, conductometric measurements, enzymatic microsensors, whole-cell microsensors, pesticides, herbicides, nitrate, DOC, heavy metals, toxicity assessment, selenite

1. Introduction

During the last 30 years, intensive use of chemicals (like pesticides) in agriculture has considerably increased the pollution of food, water and soils. From an analytical point of view, these toxic compounds can be determined in environmental matrices using two different kinds of methods. In the first one, pollutants are identified and quantified by classical chromatographic techniques such as gas chromatography (GC) or high performance liquid chromatography (HPLC) [1]. Nevertheless, these techniques are time-consuming (because of sample preparation and need for pre-concentration), expensive, and cannot be performed easily outside the laboratory. In addition, they are restricted to a limited set of substances. The selection of target compounds may completely fail to recognize the most harmful toxic constituents, coming from, for example, degradation processes. In the second type of method, the compounds are not clearly identified, but measurements allow the assessment of toxicity of the tested samples. These techniques are very useful for assessing the potential risk of contaminated water samples. A variety of toxicity measurement systems exist, including those based on bacteria and algae, animal cells, small mammals, fish fly, and zooplankton. Some of these systems, e.g., animals and fish larvae, are difficult to handle and do not provide a rapid response. In addition, the use of some of these systems may be ethically objectionable. Other systems, such as mammalian cells, are expensive, and results are not always consistent. Also, the response of single toxicity assay is an insufficient measure of adverse biological impact of a compound in a generally diverse receiving ecosystem. Different toxicants act differently, and not all life forms are equally susceptible. Consequently, several different assays need to be used simultaneously to assess the toxicity adequately [2].

The analysis of various biological and chemical pollutants in environmental matrices has entered a new phase during the last decade. Improvements in instrumentation, sampling, and sample preparation techniques have become essential to keep up with the requirements of detection at low levels, as well as to achieve a faster analysis. Biosensors offer considerable promises as future analytical tools. They are consequently expected to play an increasingly important role in the improvement of the quality of life [3]. Indeed, biosensors are versatile analytical tools applied more and more, not only for environmental pollution monitoring [4], but also in various other fields, such as medicine, food quality and safety control.

A biosensor is an analytical device incorporating a deliberate and intimate combination of a specific biological element (that creates a recognition event) and a physical element (that transduces the recognition event) [5]. Functionally, the device consists of two parts. The first one consists of a biological sensing element (bioreceptor) forming a recognition layer (enzymes, microorganisms, antibodies, receptors and organelles). The bioreceptor is intimately associated with the second part of a biosensor that is transducer, via entrapment, adsorption, cross-linking or formation of covalent bonding. This association is necessary to ensure an easy, fast and specific response of the transducer to the analyte. The transducer critically acts like a translator, recognizes the biological/chemical event from the biological component, and transforms it into another signal for interpretation by the processor that converts it into a readable/measurable output [6]. The transducer may be: a) electrochemical (detects electron transfer reactions) based on potentiometric, impedimetric, amperometric and conductometric measurements [7]; b) optical (detects changes in light transmission) [8]; c) piezoelectric (detects a variation of mass caused by selective binding of the analyte) [9]; or d) calorimetric (measures temperature rise due to protein-protein reaction or enzyme catalysed reactions) [10].

The creation of electrochemical biosensors is probably one of the most promising ways to solve some analytical problems using sensitive, fast and inexpensive techniques. However, compared to amperometric and potentiometric biosensors, few works have been devoted to conductometric enzyme [11-14] or whole cell biosensors [46, 55, 56, 60] based on thin-film interdigitated electrodes. The conductometric biosensors are based on the fact that almost all enzymatic reactions involve either consumption or production of charged species and, therefore, lead to a global change in the ionic composition of the tested sample [15]. Biosensors based on the conductometric principle present several advantages: a) thin-film electrodes are suitable for miniaturisation and large-scale production using inexpensive technology, b) they do not require any reference electrode, c) transducers are not light sensitive, d) the driving voltage can be sufficiently low to decrease significantly the power consumption, and e) a large spectrum of compounds of different nature can be determined on the basis of various reactions and mechanisms.

The liquids analysed are mostly considered to have significant background conductivity which is easily modified by different factors. Therefore, the selectivity of this method is presumed to be low and, consequently, its potential use for different applications is somewhat doubtful [16]. However, in the case of an integrated microbiosensor, most of these difficulties can be overcome using a differential measuring scheme which compensates for changes in background conductivity, the influence of temperature variations and of other factors [17, 18].

This chapter describes basic theoretical principles of conductometry in bioanalytical practice and the application of conductometric biosensors for environmental monitoring.

2. Conductometric Method of Measurement

The conductivity of liquids results from the dissociation of the dissolved substance—electrolyte—into ions, and the migration of the latter induced by an electrical field. When the potential difference is applied to the electrode, there is an electrical field within the electrolyte, so the chaotic ion movement is influenced by the ordered, oppositely directed movement of ions (those with negative charge more towards anodes, while positively charged ones move towards cathodes) (Figure 1). Thus, the current in the electrolyte is caused by the ion movement towards the electrodes where the ions are neutralized and isolated as neutral atoms (or molecules).

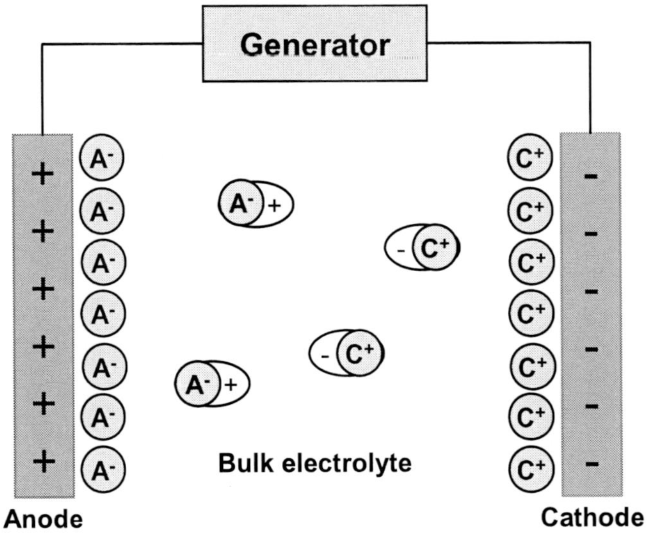

Figure 1. Ion migration in the solution volume and electrolyte conductivity.

The ion flux, i.e., the number of ions passing through a unit of electrolyte cross-section per unit of time (p_i) can be determined by the formula:

$$p_i = c_i\, v - k_i\, c_i \text{ grad } \mu_i - z_i\, v_i\, c_i\, F \text{ grad } \psi \tag{1}$$

where v – speed of solution flow due to natural or forced convection; c_i – ion concentration; k_i – diffusion coefficient; z_i – charge number; v_i – speed of ion movement caused by applied field; F – Faraday number; Ψ electrical potential; μ electrochemical potential..

Thus, the first member in formula (1) corresponds to the contribution of the convectional flow of ions at a concentration of c_i; the second – the contribution of their molecular diffusion; the third – that of the ion migration induced by the applied potential. Temperature, as a rule, is assumed to be constant (T = const, grad T = 0), so the ion's thermal diffusion can be ignored.

In reality all three processes usually coexist, but present an initial influence on each other. In any case some assumptions can be made so that only one of them is taken into account. Thus, in the case of homogenous immobile electrolyte the first and second members of equation (1) can be neglected, and only the ion migration caused by the electric field effect can be considered. Then

$$p_i = -z_i\, v_i\, c_i\, F \text{ grad } \psi = -c_i\, u_i \text{ grad } \psi \tag{2}$$

where v_i – speed of ion movement, c_i – ion concentration, $u_i = z_i\, v_i\, F$ – ion mobility which is a constant value for the given ion in the infinitely diluted solution.

The current density, i.e. the current per unit of the system cross-section is an algebraic sum of products of ion fluxes and ion charges:

$$j = F\, \Sigma\, z_i\, p_i = F \text{ grad } \psi\, \Sigma\, z_i\, c_i\, u_i \tag{3}$$

On the other hand, according to Ohm's law

$$j = S \text{ grad } \psi \tag{4}$$

where S – conductivity, i.e. reciprocal value to resistance.

Hence, from (3) and (4)

$$S = F \Sigma z_i c_i u_i \tag{5}$$

Thus, the conductivity of the electrolyte solution depends on the ion concentration and mobility.

The resistance of electrolyte solution is well-known to be in direct proportion to the distance "L" between the immersed electrodes and reciprocal to their area A, therefore

$$S = \chi \frac{A}{L} \tag{6}$$

where χ - specific conductivity.

This leads to the following conclusions: conductometric measurement commonly consists of determining the conductivity of a solution between two parallel electrodes; its value is a sum of all the ions within the solution tested.

3. Transducers for Conductometric Biosensors

The conductometric transducer is a miniature two-electrode device designed to measure the conductivity of the thin electrolyte layer adjacent to the electrode surface. Most authors agree that the best design for the development of conductometric electrodes is an interdigitated structure [19-25]. Thus, an enzyme reaction is followed by a change in the solution's conductivity in the vicinity of the electrode, which is recorded by the conductometric transducer.

An important aspect in the development of conductometric interdigital transducers is the proper selection of electrode material. Various materials have been tested: platinum [13, 23-30], gold [13, 19, 22, 26, 31-33], aluminium [19, 22, 34], nickel [19, 22, 34], copper [19, 22, 29], titanium [13, 19, 22], chromium [19, 22], Ta_2O_5 [20, 21], silver [30] and carbon [35]. In general, all these materials are suitable, especially when high-frequency ac is used. However, the electrodes made of precious metals have better characteristics. Titanium, chromium and aluminium electrodes have been revealed to be undesirable for operation with biological liquids since these electrodes have low sensitivity to changes in the ion strength of solution and reach conductivity saturation in a short time.

Concerning the dimensional characteristics of electrodes, the investigations of the authors and the data presented in [24, 26] show that transducer miniaturization does not require the number of electrode fingers to be increased by a corresponding decrease in their size, as was earlier thought. On the contrary, doing so resulted in lower transducer sensitivity.

Miniaturization should be performed by decreasing both the electrode working surface and its characteristic dimensions. On the other hand, an electrode with large fingers is not an appropriate design (though transducer sensitivity can be higher) because in this case the thickness of the biologically active membrane plays a significant role. The main parameter to determine sensor size is the correlation between membrane thickness, the electrode characteristic dimensions and its active area.

Conductometric transducers are mostly manufactured by microelectronics techniques – photolithography and vacuum spraying - whose advantages are described in detail in [36]. Some authors have used thick-film printing technology [30, 37], whose merits are reviewed in [36]. All authors, however, agree that it is much easier to produce conductometric transducers than electrochemical transducers of other type.

4. Conductometry in Enzyme Catalysis

The conductometric measuring method can be used in enzyme catalysis to determine substance concentration and enzyme activity, selectivity in this case being provided just by the enzymes which catalyze only defined reactions. As a matter of fact, the subject under consideration is not a biosensor as such but an application of this method in enzymology.

In 1961, one of the first research efforts in this field was published, showing how it might be possible to determine urea concentration in solutions [38]. This method is based on the difference between electric conductivity of urea solution and that of a solution of ammonium carbon formed as a result of urea hydrolysis by urease. In the experiments a bridge-measuring scheme was used. The urease activity was shown to decrease in the presence of heavy metal ions (Ag, Hg, etc.) in the solution. Such electrolytes as NaCl or KCl do not influence urease activity, but if their concentration in the solution is high it can lead to a wrong result, especially at low urea concentrations. In experiments carried at low electrolyte concentrations in the absence of buffering solution, the medium pH gradually changed from 7 to 9 during urea hydrolysis. However, this caused only an insignificant change in the urease activity while the solution conductivity during the reaction varied substantially. The urea concentration was determined within the 0.1 μM–2 mM range, at optimal pH 7. A comparison of the conductometric method with other methods of urea analysis carried out in that work has shown that the first is characterized by its high accuracy, speed and simplicity. Besides, in contrast to optical methods, the measurement accuracy of conductometry does not depend on solution colour.

In 1965 a paper was published on applying the conductometric method to the study of the kinetics of urea enzyme hydrolysis as well as to urease activity determination [39]. A differential measuring scheme was used. The system consisted of two pairs of platinum plates, each of them placed in its own measuring cell, one with the enzyme, another without. The difference between signals from both cells was registered, thus eliminating any error associated with variations of the parameters outside the cells (temperature, buffer concentration, etc.). The determination range of urea concentration was 1–75 mM, while that of urease activity was 0.04–2.5 activity units/ml. Comparing the data obtained with the results of classical photometric analysis showed that the conductometric method has all the merits of the classical one, and exceeds the latter in accuracy and speed.

At the same time, a research work [33] was presented demonstrating that the changes in conductivity during an enzyme reaction can be considered to be a universal characteristic of the substrate chemical transformation. Even if the conductivity of the products of the reaction and that of the substrate differs a little, the change in the solution viscosity and in the level of hydration of molecules and ions at the substrate transformation (especially in the presence of other current carriers in the solution) cause noticeable variation in the tested mixture. To prove experimentally the potential of conductometry, the authors chose reactions associated with different mechanisms of conductivity change. It was important to expose the character of conductivity change in the case of evident changes in the solution composition as well as when only solution viscosity and level of molecule hydration vary as a result of the reaction. The enzyme hydrolysis of acetylcholine and starch, on the one hand, and the enzyme depolymerization of gialuronate, on the other, were considered to be the reactions meeting these requirements.

Acetylcholine hydrolysis is accompanied by the rupture of complex ether bound and the formation of acetic acid dissociated into protons and CH_3COO^-. Proton does not participate in the total conductivity mechanism because the reaction takes place in a phosphate buffer, pH 7.8, while the appearance of CH_3COO^- anions in the solution results in an increase of the solution conductivity.

For starch hydrolysis, filtered human saliva was added to the starch solution to decrease the solution viscosity and increase the level of starch hydrolysis and so increase the solution conductivity. When gialuronidase was used, the conductivity increased as a result of reducing the viscosity of the solution due to the depolymerization of the gialuronate molecules.

The authors demonstrated that the conductometric method is preferable to the well-known routine methods of biochemical analysis because of its higher accuracy and lower labour consumption. Using a single device and a single method with no modifications, the kinetics of three enzyme processes with specific features were studied, while three devices based on different principles would be required for classical enzyme analysis.

In the next work the authors modified their conductometric device and used a differential mode to measure the activities of collagenase, trypsin, lactate dehydrogenase and pseudocholinesterase [41].

However, despite demonstrating the potential of the conductometric method to record enzyme processes, the character of all the works mentioned is that of a preliminary or feasibility study.

At the beginning of the 1980s, a detailed analysis concerning the potentials and limitations of the conductometric method of measurement was performed for the complex study of urease [42]. In the first part of the work, the effect on the conductivity of the medium pH, the urea, urease and salt concentrations of the studied solution was investigated in the absence of any enzyme reaction. It was shown that the changes in concentrations of the solution compounds cause no substantial variations of its conductivity which is why these changes can be neglected. The curve of conductivity-pH dependence is bell-shaped with a maximum at pH 8.0 for Tris-buffer and at pH 6.0 for a citrate buffer.

In the second part of the work the influence of the medium conditions mentioned on urea hydrolysis was investigated. Though pH 7.2 was shown to be optimal for urease in citrate buffer, the substantial dependence of the reaction speed on the solution ion strength was revealed at this value - the speed decreased with rising ion strength, while at pH 6.5 it hardly changed. The Michaelis constant for urease in citrate buffer, pH 4.5, was determined to be around 2.5 mM. The linear dependence of the speed of the enzyme reaction on the urease concentration in the solution was also obtained. The successful use of the conductometric method to study the activity of urease, described in this work, can be considered as a convincing argument for its potential due to high sensitivity and good agreement of the kinetic parameters obtained by the conductometric method compared with the results of classical biochemical analysis.

In [15] and [43] a 6-channel conductometer was used to study different enzymes. The investigation allowed the authors to formulate 5 factors that bring about, separately or in combination with each other, a change in conductivity, thus ensuring the potential of the conductometric method to record parameters during enzyme reactions (Table 1). In some reactions several factors work simultaneously, namely factors 3 and 5, in reactions with phosphorilase, factors 1, 3-5, in those with apirase. Factors 1 and 2 influence the conductivity in the most effective way and consequently are the most promising when using conductometry in enzymology. The decrease in size of charged particles does not cause considerable change in conductivity (phosphatase substrates).

Table 1. Factors resulting in conductivity changes

No	Source of changes in conductivity	Enzymes
1	Generation of ion groups	Amidases
2	Separation of different charges	Dehydrogenases and decarboxylases
3	Ion migration	Esterases
4	Change in level of ion particles association	Kinases
5	Change in size of charged groups	Phosphatases and sulfatases

In the reactions with proton migration some influence of the buffer type has to be taken into account as an additional factor - an anion buffer decreases conductivity while a cation one increases it due to protoning. Therefore, the choice of the type of buffer is very important. Experiments demonstrated a considerable increase in conductivity in Tris and imidasole buffers in reactions with lipase, and an insignificant change in conductivity of these enzymes in phosphate buffer. The conductometric method was also shown to be promising in all the reactions causing a change in pH, which is determined as a rule by pH-titration. The sensitivities of both methods are comparable, but conductometry is preferable due to its lower cost and simplicity of use. Besides, titration (unlike conductometry) can be performed only in the direction of pH change and requires more complicated manipulation at operation.

However, conductometric methods have some limitations. The ratio between the signal and noise level should not be higher than 2%. For this reason, the concentrations of buffer and some other ingredients, which can be added to the reaction mixture, are important. The method sensitivity drops in the presence of non-reacting ions in the solution. Buffers with low ion strength can be used, then, to measure low concentrations until the signal/noise ratio is enough. Another disadvantage of conductometry is its low specificity – it is incapable of distinguishing between reactions that can cause an artefact. The double layer capacity and the electrode polarisation during reaction can be also sources of method error. All the investigations reported have become a basis for further development of conductometric biosensors.

5. Conductometric Biosensors

The first conductometric biosensor for urea determination has been described in [13]. It was a device consisting of a silicium substance with a pair of gold interdigitated and serpentine electrodes. The experiments were carried out in both a laboratory and clinics; the biosensor response to urea was in the range of 0.1–10 mM in imidasole buffer, pH 7.5. The K_M of immobilised enzyme was higher than that in the solution; the authors explained it as a result of diffusion limitation. A comparison of the data obtained by the biosensor in the laboratory with the results of conventional clinical tests showed good agreement (the correlation coefficient was higher than 0.99).

Similar conductometric biosensor has also been used as a multisensory [31]. Urease was immobilised on the surface of the first electrode pair in the gel layer; on the second pair, L-asparaginase; on the third pair, a three-enzyme system "urease-creatinase-creatininase". This sensor was used for the determination of urea, L-asparagine and creatinine, respectively. The sensor was tested with every substrate and in a multi-substance medium, the kinetic and calibration curves were determined.

5.1. Conductometric Biosensors for Environmental Monitoring

In Table 2, the developments in the field of conductometric biosensors for environmental monitoring during the last decades are gathered.

Table 2. Developments of different conductometric biosensors for environmental monitoring

№	Substance	Bioreceptor	Information
1	Organophosphorous pesticides	Acetylcholinesterase, Butyrylcholinesterase Chlorella vulgaris microalgae (using acetylcholinesterase activity)	[11, 44, 46-48, 50-52, 54] [55]
2	Heavy metal ions	Urease *Chlorella vulgaris* microalgae (using alkaline phosphatase activity)	[45, 47-50, 54] [55,56]
3	Formaldehyde	Alcoholoxidase	[14, 47, 50]
4	4-Chlorphenol	Tyrosinase	[53, 54]
5	Triazine herbicides	Tyrosinase	[53, 54, 57]
6	Carbamate pesticides	Acetylcholinesterase	[54]
7	Nitrate	Nitrate reductase	[58,59]
8	Proteins as marker of DOC	Proteinase K	[60, 61]
9	Selenite (SeO_3^{2-})	*Herminiimonas arsenicoxydans* bacterium + glucose oxidase	[62]

A view of the design of the conductometric transducer (with a reference and a working electrode) and of the experimental set-up for conductometric measurements is presented in Figure 2.

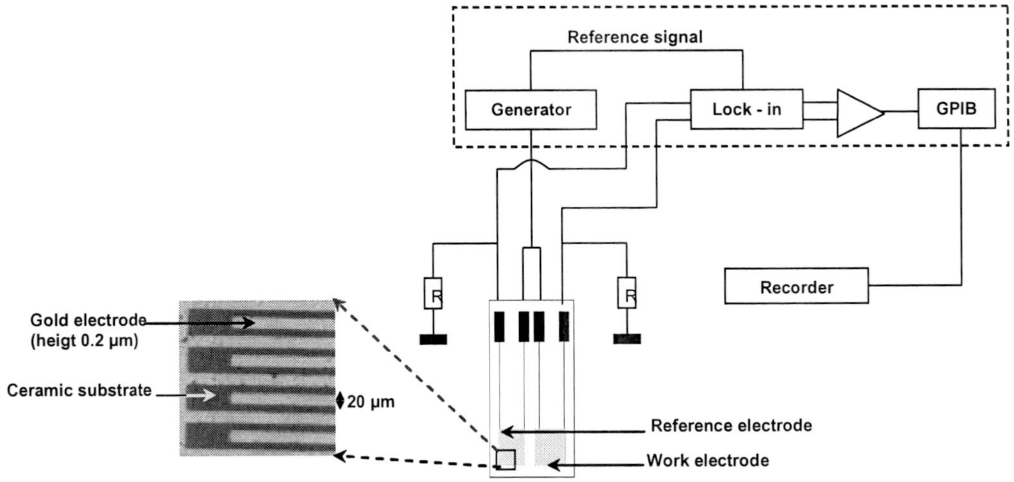

Figure 2. Design of the conductometric transducer (with two pairs of interdigitated thin film electrodes) and of the experimental set-up for conductometric measurements (differential mode)

The enzyme is immobilized on the working electrode mainly through co-reticulation with BSA (bovine serum albumin) by glutaraldehyde cross-linking. On the reference electrode only BSA is co-reticulated.

5.2. Conductometric Enzyme Biosensors for Environmental Monitoring

Conductometric biosensors for the detection of organophosphorous pesticides (based on enzyme inhibition)

The conductometric biosensor based on inhibition analysis, first described in [44], was intended for the determination of organophosphorous pesticides. As a sensitive element, the enzymes acetyl- and butyrylcholinesterase were used. The enzymatic reaction is the following one:

$$CH_3\text{-}(CH_2)_2\text{-}\overset{\overset{O}{\|}}{C}\text{-}O(CH_2)_2\text{-}\overset{+}{N}(CH_3)_3\ \overset{-}{Cl} \xrightarrow[H_2O]{BuChE} CH_3\text{-}(CH_2)_2\text{-}CO\overset{-}{O} + H^+ + HO\text{-}(CH_2)_2\text{-}\overset{+}{N}(CH_3)_3\ Cl^-$$

butyrylcholine chloride   butyric acid   choline

The organophosphorus compounds partly inhibit the biological activity of acylcholinesterase through phosphorylation of the serin group, according to the following reaction (example of the organophosphorus compound trichlorfon):

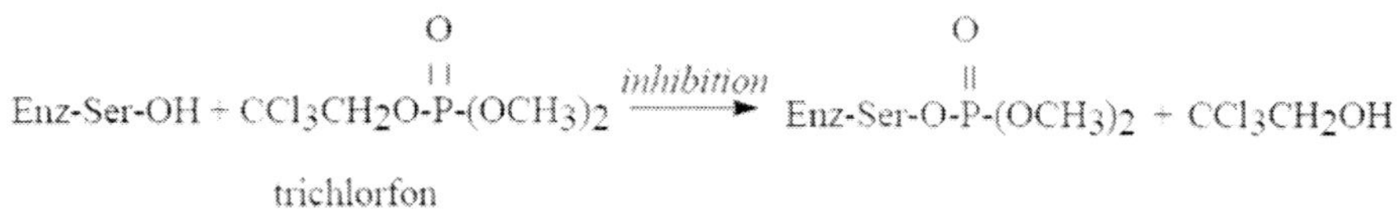

The sensor sensitivity to different pesticides (diisopropyl fluorophosphate, paraoxon-ethyl, paraoxon-methyl, trichlorfon) was investigated, the minimal detection limits for inhibitor concentrations were: 5×10^{-11} M for diisopropyl fluorophosphate, 10^{-8} M for paraoxon-ethyl, 5×10^{-7} M for paraoxon-methyl, and 5×10^{-7} M for trichlorfon. The dependence of biosensor response on how long the transducer was incubated in the pesticide solution was studied. The possibility of reacting the enzyme in the membrane by means of reactivator pyridine-2-aldoxime-methiodide, according to the following reaction, was shown.

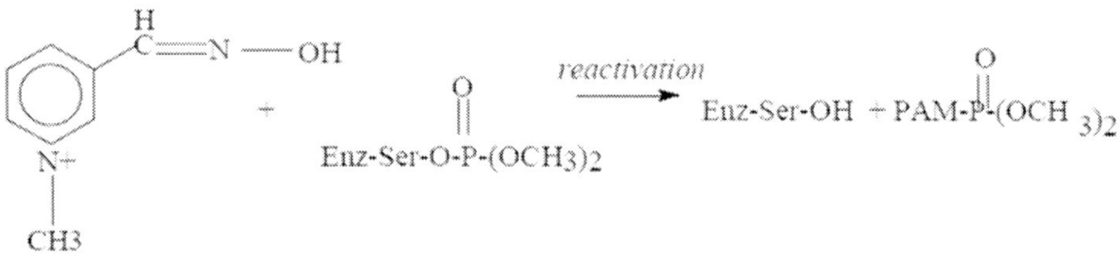

The conclusion was drawn that the described biosensors could be used for the analysis of organophosphorous pesticides in aqueous solutions.

Conductometric biosensors applied for analysing total solution toxicity at parathion-methyl photodegradation were presented in [51, 52]. The assessment of the toxicity of the photodegradation products was performed using the percentage of inhibition of cholinesterase. The results obtained were compared with the data from traditional high-sensitive method of HPLC and from the Lumistox device (Germany) for toxicity determination. It has been found that the strongest inhibitor of immobilised cholinesterases was the methyl paraoxon and a strong synergistic effect of methyl parathion and methyl paraoxon was occurred. The solution toxicity was shown to increase dramatically as pesticide photodegradation began. The toxicity remained once the parathion-methyl dissociation had been completed. However, the authors do not oppose the biosensor method to others, but consider it as an additional fast way for the early screening of numerous samples.

Conductometric biosensor for the detection of diuron and atrazine (based on tyrosine inhibition)

A conductometric tyrosinase biosensor for the detection of some toxic compounds including diuron, atrazine, and copper ions was developed in [53, 57]. Under the experimental conditions employed using 4-chlorophenol as enzyme substrate, and 30 min of contact with tyrosinase inhibitor, detection limits for diuron and atrazine were in the μg L^{-1} and dynamic range of 2.3–2330 μg L^{-1} and 2.15-2150 μg L^{-1} were obtained for diuron and atrazine respectively. A relative standard deviation of the output signal was estimated to be 5% and a slight drift of 1.5 μS/hour was observed. The 90% of the enzyme activity was still maintained after 23 days of storage in a buffer solution at 4°C. The different samples tested were solutions containing diuron, atrazine, copper, lead and zinc ions, mixtures of copper/atrazine or copper/diuron and real water samples coming from a Vietnamese river. In the last case, classical techniques such as GC-MS or atomic absorption spectrometry were used in order to estimate exact concentration of these species in real water samples. Results have shown that such a biosensor could be used as an early warning system for the detection of these pollutants, as no matrix effect coming from the real sample was observed and no synergetic or antagonist effects were found for the mixture of toxic compounds. In addition, results were coherent with the content of the tyrosinase inhibitors.

Conductometric biosensors for the detection of heavy metal ions (based on urease inhibition)

The potential of conductometric urease biosensors for the determination of heavy metal ions was demonstrated in [38]. The measurements of the urease residual activity have been carried out for a quantitative estimation of general water pollution with heavy-metal ions. The sensitivities of heavy metals towards urease varied as follows: $Hg^{2+} > Cu^{2+} > Cd^{2+} > Co^{2+} > Pb^{2+} > Sr^{2+}$. Reactivation of the inhibited enzyme with EDTA was shown to be possible.

Conductimetric biosensor for the detection of formaldehyde

A conductometric enzyme biosensor for determination of formaldehyde in aqueous solutions has been developed using interdigitated thin-films planar electrodes and immobilised alcohol oxidase from *Hansenula polymorpha* [6]. The enzymatic reaction was the following one:

$$CH_2O + H_2O \xrightarrow{\text{alcohol oxidase}} HCOOH + H_2O_2$$

$$HCOOH \rightarrow HCOO^- + H^+$$

The biosensor steady-state response was reached after about 1 min. Its dynamic range can vary from 0.05 mM to 500 mM formaldehyde and depends on the time of enzymatic membrane cross-linking by glutaraldehyde and on the buffer concentration used. The biosensor developed was not absolutely specific and selective. No response was observed for primary alcohols and some other substrates alone. Unfortunately, biosensor response towards formaldehyde was decreased in the presence of methanol. Biosensor response was found stable over more than 20 hours and the relative standard deviation appeared to be about 3%. Moreover, the storage stability was more than one month.

Conductimetric biosensor for the detection of nitrate

A highly sensitive, fast and stable conductometric enzyme biosensor for determination of nitrate in waters was described in [58, 59]. The conductometric electrodes were modified by methyl viologen mediator mixed with nitrate reductase from *Aspergillus niger* by cross-linking with glutaraldehyde in the presence of bovine serum albumin and Nafion® cation-exchange polymer, allowing retention of viologen mediator according to the following ion exchange reaction:

$$MV_{aq}^{2+} + 2(SO_3^-Na^+)_{film} \rightarrow [(SO_3^-)_2MV^{2+}]_{film} + 2Na_{aq}^+$$

The enzymatic reaction for the reduction of nitrate is the following one:

$$NO_3^- + 2MV^{\bullet +} + 2H^+ \rightarrow NO_2^- + 2MV^{2+} + H_2O$$

Linear calibration in the range of 0.02 and 0.25 mM with detection limits of 0.005 mM nitrate were obtained with a signal-to-noise ratio of 3. When stored in 5 mM phosphate buffer (pH 7.5) at 4℃, the sensor showed good stability over two weeks.

Conductometric biosensor for the detection of Dissolved Organic Carbon (DOC)

A conductometric proteinase K biosensor for organic matter monitoring in rivers has been developed, based on a conductometric transducer [60-61]. In fact, with approximately 30% of the total organic carbon, proteins were chosen to be used as indicators of urban pollution. Proteinase K hydrolyzes proteins into different ionic amino-acids which results in local conductivity changes. In this work, the biosensor response using bovine serum albumin (BSA) as standard protein was optimized. A stable biosensor with a constant repeatability and a detection limit about 0.5 μg/mL BSA were obtained. Then, response biosensor was tested with samples of rivers water. Good correlations between conductance changes and values given by standard methods (chemical oxygen demand and protein concentration evaluated by microBCA protein assay) have been shown. Correlation coefficients of 0.89 and 0.92 were respectively obtained.

5.2. Conductometric Whole-Cell Biosensors for Environmental Monitoring

Conductometric biosensor based on microalgea

A conductometric biosensor using immobilised *Chlorella vulgaris* microalgae as bioreceptors was used as a bi-enzymatic biosensor [55, 56]. Local conductivity variations caused by algae alkaline phosphatase and acetylcholinesterase activities could be detected. These two enzymes are known to be inhibited by distinct families of toxic compounds: heavy metals for alkaline phosphatase, carbamates and organophosphorous pesticides for acetylcholinesterase. The bi-enzymatic biosensors were tested to study the influence of heavy metal ions and pesticides on the corresponding enzyme. It has finally appeared that these biosensors are quite sensitive to Cd^{2+} and Zn^{2+} (limits of detection (LOD) = 10μg L^{-1} for a 30 minutes long exposure) while Pb^{2+} gives no significant inhibition as this ion seems to adsorb on albumin preferably. For pesticides, first experiments showed that paraoxon-methyl inhibits *Chlorella vulgaris* AChE contrary to parathion-methyl and carbofuran. Biosensors were then exposed to different mixtures (Cd^{2+}/Zn^{2+}, Cd^{2+}/MPx) but no synergetic or antagonist effect could be observed. A good repeatability could be achieved with biosensors since the relative standard deviation did not exceed 8% while response time was 5 to 7 minutes.

Conductometric biosensor based on heavy-metal resistant bacterium and glucose oxidase

A conductometric hybrid biosensor using immobilised bacterium *"Herminiimonas arsenicoxydans"* and glucose oxidase, for selenite (SeO_3^{2-}) determination was developed [62]. Indeed, selenium is an essential trace element which becomes toxic if taken in excess [63]. *"Herminiimonas arsenicoxydans"* is a bacterium with *lac Z* as reporter gene. In presence of selenium, the production of the enzyme β-galactosidase is induced and the presence of this enzyme can be detected after the addition of its substrate lactose: lactose is hydrolyzed by β-galactosidase and glucose and galactose are produced. The produced glucose is then oxidized by glucose oxidase (Figure 3) which leads to changes in conductivity of the enzymatic membrane measured by the conductometric electrode. The association of β-galactosidase and glucose oxidase for lactose determination was described in [64].

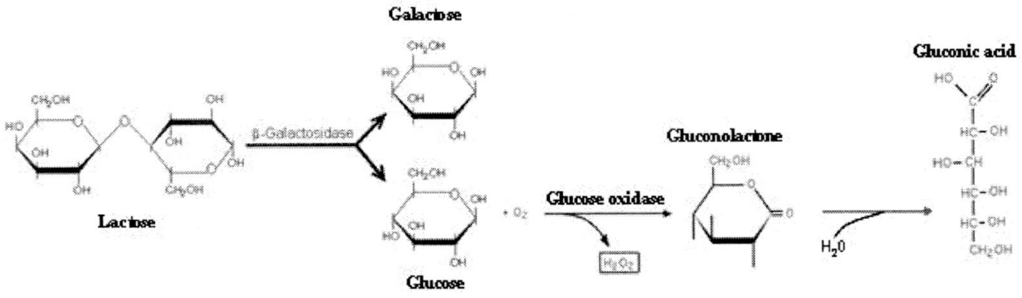

Figure 3. Enzymatic reactions catalyzed by β-galactosidase and glucose oxidase.

Hybrid biosensors were exposed to different selenite concentrations. First experiments show that this original system responds linearly in the concentration range between 0.1 and 0.5 mM. The evaluation of other performances (detection limits, selectivity, stability) are in progress.

6. Conclusion

Application of the conductometric measurement method to environmental monitoring is thoroughly examined and analysed regarding both standard conductometers and enzyme and whole-cell conductometric biosensors. As compared to conventional methods of analysis, the method considered is universal and features high accuracy and low labour costs.

Conductometric biosensors also have advantages over other types of transducers. First, they can be produced through inexpensive thin-film standard technology without any reference electrode. This, along with using an optimised method of immobilisation of biological material, results in considerable decrease in both the primary cost of devices and the total price of analyses. For integrated microbiosensors, it is easy to perform a differential measurement mode, thus compensating external effects and considerably increasing measurement accuracy.

The data is convincing evidence of the great potential of conductometric biosensors, especially for environmental monitoring. However, it is still a novel trend in the field of biosensors. The development of commercial devices has a promising future, especially for microbiosensor arrays.

References

[1] Patnaik, P. *Handbook of Environmental Analysis.* Chemical pollutants in air, water, soil, and solid wastes; CRC Press Inc.: Boca Raton, USA, 1997.

[2] Fernandez-Alba, A.R.; Guil, L.H.; Lopez, G.D.; Chisti, Y. Toxicity of pesticides in wastewater: a comparative assessment of rapid bioassays. *Anal. Chim. Acta* 2001, 426, 289-301.

[3] Castillo, J.; Gáspár, S.; Leth, S.; Niculescu, M.; Mortari, A.; Bontidean, I.; Soukharev, V.; S.A., Dorneanu; A.D., Ryabov; Csöregi, E. Biosensors for life quality Design, development and applications. *Sens. Actuators B* 2004, 102, 179–194.

[4] Rogers, K.R. Recent advances in biosensor techniques for environmental monitoring. *Anal. Chim. Acta* 2006, 568, 222–231.

[5] Fraser, D.M. Biosensors: Making Sense of Them. *Medical Device Technology* 1994, 5, 38-41.

[6] Anjum, V.; Pundir, C.S. Biosensors: Future analytical tools. *Sensors & Transducers* 2007, 76, 935-936;

[7] Thèvenot, D.R.; Toth, K.; Durst, R.A.; Wilson, G.S. Electrochemical biosensors: recommended definitions and classification, *Biosens. Bioelectron.* 2001, 16, 121-131.

[8] Mehrvar, M.; Bis, C.; Scharfer, J.M.; Moo-Young, M.; Luong, J.H. Fiber-optic biosensor-Trends and advances, *Anal. Sci.* 2000, 16, 677-692.

[9] Bunde, R.L.; Jarvi, E.J.; Rosentreter, J.J. Piezoelectric quartz crystal biosensors. *Talanta* 1998, 46, 1223–1236.

[10] Lal, R. Integrated biosensors: promises and problems. *Bioelectrochemistry & Bioenergetics* 1992, 27, 121-139;

[11] Dzydevich, S.V.; Shul'ga, A.A.; Soldatkin, A.P.; Nyamsi Hendji, A.M.; Jaffrezic-Renault, N.; Martelet, C. Application of conductometric for sensitive detection of pesticides biosensor based on the cholinesterases. *Electroanalysis* 1994, 6, 752-758.

[12] Shul'ga, A.A.; Soldatkin, A.P.; El'skaya, A.V.; Dzyadevich, S.V.; Strikha, V.I. Thin-film conductometric biosensor for glucose and urea determination. *Biosens. Bioelectron.* 1994, 9, 217-223.

[13] Watson, L.D.; Maynard, P.; Cullen, D.C.; Sethi, R.S.; Brettle, J.; Lowe, C.R. A microelectronic conductometric biosensor. *Biosensors* 1987/88, 3, 101-115.

[14] Dzyadevych, S.V.; Arkhypova, V.N.; Korpan, Y.I.; El'skaya, A.V.; Soldatkin, A.P.; Jaffrezic-Renault, N.; Martelet, C. Conductometric formaldehyde sensitive biosensor with specifically adapted analytical characteristics. *Anal. Chim. Acta* 2001, 445, 47-55.

[15] Lawrence, A.J.; Moores, G.R. Conductimetry in enzyme studies. *Eur. J. Biochem.* 1972, 24, 538-546.

[16] Dorokhova, E.N.; Prokhorova, G.V. Analytical chemistry. *Phisical-chemical methods of analysis;* Vysshaya shkola: Moscow, 1991.

[17] Gopel, W.; Jones, T.A.; Kleitz, M.; Lundstrom, J.; Seiyama, T. *Conductometry. In Sensors. A Comprehensive Survey*; Gopel, W.; Hesse, J.; Zemel, J. N., Eds; VCH Verlagsgesellschaft: Weinheim, 1991, Vol. 2, Part I, pp 314-337.

[18] Kell, D.B. The principles and potential of electrical admittance spectroscopy: an introduction. In *Biosensors: Fundamentals and Applications;* Turner, A. P. F.; Karube, I.; Wilson, G. S., Eds; Oxford University Press: Oxford, 1987, pp 427-468.

[19] Dzyadevich, S.V.; Shul'ga, A.A.; Patskovsky, S.V.; Arkhipova, V.N.; Soldatkin, A.P.; Strikha, V.I. Thin-film conductometric transducers for enzyme biosensors. *Rus. J. Electrochem.* 1994, 30, 887-891.

[20] Olthuis, W.; Volanschi, A.; Bomer, J.G.; Bergveld, P. A new probe for measuring electrolytic conductance. *Sens. Actuators B* 1993, 13-14, 230-233.

[21] Olthuis, W.; Smith, A.; Van der Zalm, R. A. J.; Bergveld, P. New operational modes for the Ta_2O_5-based electrolyte conductance cell. *Sens. Actuators B* 1994, 18-19, 65-68.

[22] Shul'ga, A. A.; Strikha, V. I.; Dzyadevich, S. V.; Patskovsky, S. V.; Gopel, W. In Eurosensors VII; Turner, A. P. F., Ed.; Elseiver Adv. Techn.: Oxford, 1993, 321.

[23] Weimar, U.; Gopel, W. A.C. measurements on tin oxide sensors to improve selectivities and sensitivities. *Sens. Actuators B* 1995, 26-27, 13-18.

[24] Sheppard, N. F.; Tucker, R. C.; Wu, C. Electrical conductivity measurements using microfabricated interdigitated electrodes. *Anal. Chem.*, 1993, 65, 1199-1202.

[25] Lee, W. Y.; Kim, S. R.; Kim, T. H.; Lee, K. S.; Shin, M. C.; Park, J. K. Sol-gel-derived thick-film conductometric biosensor for urea determination in serum. *Anal. Chim. Acta* 2000, 404, 195-203.

[26] Jacobs, P.; Suls, J.; Sansen, W. Performance of a planar differential-conductivity sensor for urea. *Sens. Actuators B* 1994, 20, 193-198.

[27] Hintsche, R.; Moller, B.; Dransfeld, I.; Wollenberger, U.; Scheller, F.; Hoffmann, B. Chip biosensors on thin-film metal electrodes. *Sens. Actuators B* 1991, 4, 287-291.

[28] Trebbe, U.; Niggemann, M.; Cammann, K.; Fiaccabrino, G. C.; Koudelka-Hep, M.; Dzyadevich, S.; Shulga, O. A new calcium sensor based on ion-selective conductometric microsensors - membranes and features. *Fres. J. Anal. Chem.* 2001, 371, 734-739.

[29] Mikkelsen, S. K.; Rechnitz, G. A. Conductometric transducers for enzyme-based biosensors. *Anal. Chem.* 1989, 61, 1737-1742.

[30] Bilitewski, U.; Drewes, W.; Schmid, R. D. Thick film biosensors for urea. *Sens. Actuators B* 1992, 7, 321-326.

[31] Cullen, D. C.; Sethi, R. S.; Lowe, C. R. Multi-analyte miniature conductance biosensor. *Anal. Chim. Acta* 1990, 231, 33-40.

[32] McNeil, C. J.; Athey, D.; Ball,M.; On Ho, W.; Krause, S.; Armstrong, R. D.; Wright, J. D.; Rawson, K. Electrochemical sensors based on impedance measurement of enzyme-catalyzed polymer dissolution: theory and application. *Anal. Chem.* 1995, 67, 3928-3935.

[33] Sergeeva, T. A.; Lavrik, N. V.; Rachkov, A. E.; Kazantseva, Z. I.; Piletsky, S. A.; El'skaya, A. V. Hydrogen peroxide – sensitive enzyme sensor based on phtalocyanine thin film. *Anal. Chim. Acta* 1999, 391, 289-297.

[34] Endres, H.-E.; Drost, S. Optimization of the geometry of gas-sensitive interdigital capasitors. *Sens. Actuators B* 1991, 4, 95-98.

[35] Besson, C.; Vessillier, S.; Gonzales, T.; Saulnier, J.; Wallach, J. Conductimetric assay of pyroglutamyl peptidase activity. *Anal. Chim. Acta* 1994, 294, 305-309.

[36] Dzyadevych, S.V.; Soldatkin, A.P.; Arkhypova, V.N.; Martelet, C.; Jaffrezic-Renault, N.; El'skaya, A.V. Electrochemical enzyme biosensors. IMBG Press: Kyiv, 2006.
[37] On Ho, W.; Krause, S.; McNeil, C. J.; Pritchard, J. A.; Armstrong, R. D.; Athey, D.; Rawson, K. Electrochemical sensor for measurement of urea and creatinine in serum based on ac impedance measurement of enzyme-catalyzed polymer transformation. *Anal. Chem.* 1999, 71, 1940-1946.
[38] Chin, W. T.; Kroontje, W. Conductivity method for determination of urea. Anal. Chem. 1961, 33, 1757-1760.
[39] Bourrelly, P.; Bourrelly-Durand, V. Methode d'etude par conductometric differentielle de la cinetique de l'hydrolise enzymatique de l'uree. *J. Chem. Phys.* 1965, 65, 673-677.
[40] Andreeev, V. S.; Rozengart, V. I.; Torubarov, V. A. Registration of kinetics of enzymatic reaction by high-frequency method. *Ukr. Biochem. J.* 1965, 37, 920-926.
[41] Andreev, V. S.; Bashtanov, A. V. Differential conductometric device fo registration of physical-chemical processes. *Zavodskaya laboratoriya* 1968, 34, 1546-1548.
[42] Hanss, M.; Rey, A. Application de la conductometrie a l'etude des reactions enzymatiques. Systeme uree-urease. Biochim. Biophys. Acta 1971, 227, 630-638.
[43] Lawrence, A. J. Conductimetric enzyme assays. *Eur. J. Biochem.* 1971, 18, 221-225.
[44] Dzyadevich, S. V.; Soldatkin, A. P.; Shul'ga, A. A.; Strikha, V .I.; El'skaya, A. V. Conductometric biosensor for organophosphorus pesticides determination. *J. Anal. Chem.* 1994, 49, 789-792.
[45] Zhylyak, G. A.; Dzyadevich, S. V.; Korpan, Y. I.; Soldatkin, A. P.; El'skaya, A. V. Application of urease conductometric biosensor for heavy-metal ion determination. *Sens. Actuators B* 1995, 24-25, 145-148.
[46] Dzyadevich, S. V.; Korpan, Y. I.; Soldatkin, A. P.; Shul'ga, A. A.; Strikha, V. I.; El'skaya, A. V. Conductometric biosensor for ethanol detection based on whole yeast cells. Ukr. Biochem. J. 1993, 65, 47-54.
[47] Soldatkin, A. P.; Dzyadevich, S. V.; Korpan, Y. I.; Arkhipova, V. N.; Zhylyak, G. A.; Piletsky, S. A.; Sergeeva, T. A.; Panasyuk, T. L.; El'skaya, A. V. Biosensors based on conductometric detection. *Biopolymers Cell* 1998, 14, 268-277.
[48] Arkhypova, V. N.; Dzyadevych, S. V.; Schuvaylo, O. N.; Soldatkin, A. P.; El'skaya, A. V.; Jaffrezic-Renault, N.; Martelet, C. Concept of multibiosensors for determination of different toxic compounds based on enzyme inhibitor analysis. *Biopolymers Cell* 2001, 17, 70-77.
[49] Arkhypova, V. N.; Dzyadevych, S. V.; Soldatkin, A. P.; El'skaya, A. V.; Jaffrezic-Renault, N.; Jaffrezic, H.; Martelet, C. Multibiosensor based on enzyme inhibition analysis for determination of different toxic substances. *Talanta* 2001, 55, 919-927.
[50] Dzyadevych, S. V.; Arkhypova, V. N.; El'skaya, A. V.; Jaffrezic-Renault, N.; Martelet, C.; Soldatkin, A. P. Conductometric enzyme biosensors for substrates or inhibitors analysis. *Cur. Topics Anal. Chem.,* 2001 2, 179-186.
[51] Dzyadevych, S. V.; Soldatkin, A. P.; Chovelon, J.-M. Assessment of the toxicity of parathion and its photodegradation products in water samples using conductometric enzyme biosensors. *Anal. Chim. Acta*, 2002 459, 33-41.
[52] Dzyadevych, S. V.; Chovelon, J.-M. A comparative photodegradation studies of methyl parathion by using Lumistox test and conductometric biosensor technique. *Mater. Sci. Engineer. C* 2002, 21, 55-60.

[53] Mai Anh, T.; Dzyadevych, S. V.; Chau Van, M.; Jaffrezic-Renault, N.; Duc Chien, N.; Chovelon, J.-M. Conductometric tyrosinase biosensor for the detection of diuron, atrazine and its main metabolites. *Talanta* 2004, 63, 365-370

[54] Dzyadevych, S. V.; Soldatkin, A. P.; Arkhypova, V. N.; El'skaya, A. V.; Chovelon, J.-M.; Georgiou, C.; Martelet, C.; Jaffrezic-Renault, N. Early-warning electrochemical biosensor system for the environmental monitoring based on enzyme inhibition effect. *Sens. Actuators B* 2005, 55, 81-87.

[55] Chouteau, C.; Dzyadevych, S.V.; Durrieu, C.; Chovelon, J.-M. A bi-enzymatic whole cell conductometric biosensor for heavy metal ions and pesticides detection in water samples. *Biosensors Bioelectron.* 2005, 21, 273-281.

[56] Chouteau, C.; Dzyadevych, S.V.; Durrieu, C.; Chovelon, J.-M. Development of novel conductometric biosensors based on immobilised whole cell Chlorella vulgaris microalgae. *Biosensors Bioelectron.* 2004, 19, 1089-1096.

[57] Mai Anh, T.; Dzyadevych, S.V.; Prieur, N.; Nguen Duc, C.; Pham, T.D.; Jaffrezic-Renault, N.; Chovelon, J.-M. Detection of toxic compounds in real water samples using a conductometric tyrosinase biosensor. *Mater. Sci. Engineer. C* 2006, 26, 453-456.

[58] Wang, X.; Dzyadevych, S.V.; Chovelon, J.-M.; Jaffrezic-Renault, N.; Ling, C.; Siqing, X. Development of conductometric nitrate biosensor based on Methyl viologen/Nafion® composite film. *Electrochem Comm.* 2006, 8, 201-205.

[59] Wang, X.; Dzyadevych, S.V.; Chovelon, J.-M.; Jaffrezic-Renault, N.; Ling, C.; Siqing, X. Conductometric nitrate biosensor based on Methyl viologen/Nafion®/Nitrate reductase interdigitated electrodes. *Talanta* 2006, 69, 450-455.

[60] Marrakchi, M.; Dzyadevych, S.V.; Namour, Ph.; Martelet, C.; Jaffrezic-Renault, N. A novel proteinase K biosensor based on interdigitated conductometric electrodes for proteins determination in rivers and sewers water. *Sens. Actuators B* 2005, 111-112, 390-395

[61] Marrakchi, M.; Dzyadevych, S.V.; Namour, Ph.; Martelet, C.; Jaffrezic-Renault, N. An enzyme biosensor based on gold interdigitated thin film electrodes for water quality control. *Anal. Lett.* 2007, 40, 1307-1316

[62] Marrakchi, M. Ph.D. thesis, Ecole Centrale de Lyon, Lyon, FRANCE, 2006.

[63] Agency for Toxic Substances and Disease Registry. *Toxicological profile for selenium,* URL http://www.atsdr.cdc.gov/toxprofiles/tp92.pdf.

[64] Marrakchi, M.; Dzyadevych, S.V.; Lagarde, F.; Martelet, C.; Jaffrezic-Renault, N. Conductometric biosensor based on glucose oxidase and beta-galactosidase for specific lactose determination in milk, *Mater. Sci. Engineer. C* 2008, 28, 872-875.

In: Industrial Waste…
Editor: John P. Samuelson, pp. 223-231
ISBN: 978-1-60692-720-5

Chapter 7

USING LEARNING MACHINES MORE INTELLIGENTLY AND RIGOROUSLY FOR THE GROUND-LEVEL OZONE PREDICTION

Wei-Zhen Lu[1] ***and Dong Wang***
Department of Building and Construction, City University of Hong Kong, Kowloon Tong, Kowloon, Hong Kong

Abstract

Multilayer perceptron (MLP) and support vector machine (SVM), as two popular learning machines, are used increasingly as alternatives to classical statistical models for the ground-level ozone prediction. However, without the sufficient awareness about their limitations employing learning machines can still lead to unsatisfactory results on modeling ozone evolving mechanism, especially during ozone episodes. With some spirit of review, this commentary discusses, in the ozone prediction context, the recently developed algorithms/technologies on treating the most prominent model-performance-degraded limitations which include, for MLP, the "black-box" property, i.e., hardly proving physical explanation for the trained model, overfitting and local minima problems, and for SVM, parameters identification and class imbalance problems. This commentary will impress the "would-be" environmental modelers that the underlying philosophy of using learning machines is by no means as trivial as simply fitting models to the data, because it has reflected on the difficulties, controversies or unresolved problems; meanwhile this commentary also serves as a reference point for further technical reading for experts in this field.

1. Introduction

Elevated ground-level ozone concentration poses risk to public health in various ways, which continuously engages environmental modelers in developing better forecasting models. From the last decade or more, multilayer perceptron (MLP) and support vector machine (SVM), as

[1] E-mail address: bcwzlu@cityu.edu.hk. Tel. 852 27844316, Fax. 852 27887612. (Corresponding author)

two popular learning machines, were used increasingly as alternatives to classical statistical models, e.g., Box-Jenkins and multivariate regression, for the ground-level ozone prediction. Such diversion is mainly due to that those learning machines are superior to classical statistical models in resolving ozone variation which is nonlinearly impacted by meteorology and primary emissions (Yi and Prybutok, 1996; Comrie, 1997; Gardner and Dorling, 1998; Wang and Lu, 2005). MLP and SVM, however, are never panaceas and if used indiscreetly, models' performance could also be degraded drastically. With some spirit of review, this commentary discusses, in the ozone prediction context, the recently developed algorithms/technologies on treating the most prominent model-performance-degraded limitations which include, for MLP, the "black-box" property, i.e., hardly proving physical explanation for the built model, overfitting and local minima problems, and for SVM, parameters identification and class imbalance problems. This commentary will impress the "would-be" environmental modelers that the underlying philosophy of using those two learning machines is by no means as trivial as simply fitting models to the data, because it has reflected on the difficulties, controversies or unresolved problems; meanwhile this commentary also serves as a reference point for further technical reading for experts in this field.

2. Ozone Prediction with MLP

2.1. Review

The magic of MLP for ozone prediction is empowered by its training algorithms. The training algorithms aim to discover the highly nonlinear relationship between ozone and its influential factors by searching the optimal weight matrix in its weight space. Such optimal weight matrix stores the information, which can largely represent such highly nonlinear relationship. Without giving specified remedies from the algorithm perspective, most of researchers stick to the gradient-based training algorithms developed in the maximum likelihood framework, which are the origin of the three notorious problems of MLP, i.e., "black-box" property, overfitting and local minima, although great achievements have been made for tackling these drawbacks in computer science in the past decade or more.

For a set of training examples $D = \{(x_i, O_i), i = 1,...,l\}$, training algorithms in the maximum likelihood framework can conceptually be expressed in the following formulas

$$O = f(x, w) + \xi \tag{1}$$

where ξ is the intrinsic data noise with a conditional probability density function $p(O \mid x; w) = q(O - f(x, w))$, O is the target data, f is the general function representing the MLP's mapping from input to output. l Is the number of training examples. With the normal distribution assumption for $p(O \mid x; w)$, all training algorithms developed in maximum likelihood framework are to minimize the error term or find "the lowest point in the error surface" in the following equation with respect to weight matrix w.

$$e(w) = \sum_{i=1}^{l} (O_t - f(x_i, w))^2 \tag{2}$$

As for the gradient-based training algorithms developed in the maximum likelihood framework, most researchers (Yi and Prybutok, 1996; Comrie, 1997; Nunnari et al., 1998; Spellman, 1999; Prybutok et al., 2000; Narasimhan et al., 2000; Abdul-Wahab and Al-Alawi, 2002; Balaguer Ballester et al., 2002; Viotti et al., 2002; Pastor-Barcenas et al., 2005; Gomez-Sanchisa et al., 2006) preferred to use MLP trained by back propagation (BP) learning algorithm and its variants (Rumelhart et al., 1986) to forecast ozone level. These models only include the first derivative of the error term $e(w)$, i.e., the gradient of $e(w)$. They adjust the weight matrix in the steepest descent direction, in which the error term $e(w)$ is decreasing most rapidly. However, associated with three notorious problems mentioned earlier, BP-trained MLP inclines to produce slow convergence rate, because the searching is always fluctuated wildly around the local optimal solution in the error surface at the later training stage.

Realizing the slow convergence to local minima property of BP family algorithms, some researchers begins to employ other gradient-based algorithms that include the first and second derivative of $e(w)$, i.e., the gradient and Hessian of $e(w)$. Gardner and Dorling (2000), Agirre-Basurko et al. (2006) and Lu et al. (2006) employed the scaled conjugate gradient (SCG) algorithm to train MLP for ozone forecast. Such algorithm searches along conjugate directions and it avoids a line search at each iteration and does not need compute Hessian matrix, which produces generally faster convergence rate than steepest descent directions (Moller, 1993). Chaloulakou et al. (2003) and Corani (2005) reported that using Levenberg-Marquardt (LM) to train MLP can obtain satisfied result comparing with other statistical models in their ozone prediction work. LM can give a good compromise between the speed of the Newton's method and the stability of the steepest descent method, and consequently it constitutes a good transition between these two methods. Schlink et al. (2003) carried out a remarkable inter-comparison of learning machines for ozone prediction and concluded that MLP trained by quasi-Newton techniques, a class of algorithms based on Newton's method, were among the favorite predictive models. The quasi-Newton techniques, similar with LM, are alternatives for fast convergence, since they does not need compute Hessian matrix either.

Although the modifications for the gradient-based algorithms did improve MLPs performance in ozone prediction literature to a degree, the inherent three notorious problems were not addressed from the algorithm perspective. One can find however that from the above-mentioned publications the early stopping, regularization, simple validation and cross-validation techniques are popular techniques for tackling overfitting problem, and reinitializing the weight matrix and re-training the MLP accordingly is a compromise way for solving local minima problem, while none effort has been made for "black-box" property problem so far.

2.2. Overfitting and "Black-Box" Problems Addressed by Bayesian MLP, and More

Different with MLP developed in maximum likelihood, Bayesian MLP attempts to maximize a posterior probability distribution $P(w \mid D)$ with respect to weight matrix w, and uses such distribution to estimate the possible distribution of MLP optimal weight w_{opt}. The form of $P(w \mid D)$ is expressed by Mackay (1995) in the following equations and can be shown that it approximates $e(w)$ in equation (2) when l goes to infinite (Neal, 1996),

$$P(w \mid D) = \frac{1}{Z_s} \exp(-S(w)) \tag{3}$$

$$S(w) = \frac{\beta}{2} \sum_{n=1}^{l} \{(f(x^n; w) - t^n)\}^2 + \frac{\alpha}{2} \sum_{i=1}^{V} w_i^2 \tag{4}$$

where w_i is the ith weight component in w, v is the number of weight component of w, $f(x^n; w)$ is the MLP output with respect to the nth input data , t^n is the nth target data, α and β are two hyperparameters that govern the magnitude of w and the variance of data noise respectively, and Z_s is a normalization factor independent of weight vector. Since the two hyperparameters can be automatically tuned with respect to data complexity in the evidence framework, Bayesian MLP is insensitive to MLP topology and thus resists the overfitting problem.

Wang and Lu (2006a) employed an extension form of Bayesian MLP, i.e., MLP with automatic relevance determination, MLP-ARD, for ozone prediction. In that model, the input-layer weights are grouped into N sub-weight groups, where N is the number of input variables. The weights in each sub-weight group are defined as the weights fanning out from each input variable, and the magnitude and the direction of weights in each sub-weight group are separately governed by each hyperparameter α_{w1j} ($j = 1,\ldots,N$). The trained MLP-ARD thus can rank the relative importance of jth input variable according to the magnitude of α_{w1j} and provide physical explanation for ozone influential factors which are MLP inputs. In addition, the authors stressed the interval prediction ability of MLP-ARD, and concluded that ozone peak were better predicted by the upper interval of model output, compared with the point prediction from MLP trained by LM algorithm. The possible defect of their work could be local minima, since in the training process of MLP-ARD, the SCG algorithm is used to maximize $P(w \mid D)$ with respect to w and other hyperparameters. Cross-validation technique or other population-search algorithms we will discuss in the following can be remedies.

2.3. Local Minima Problem Addressed by Particle Swarm Optimization Algorithm, and More

The gradient-based algorithms, whether they are within the maximum likelihood or Bayesian framework, for training MLPs in ozone prediction are the deterministic local optimization algorithms, e.g., LM and SCG, which are the individual-based searching techniques to find the minima in the error surface of $e(w)$. Although the magnitude of $e(w)$ is decreased rapidly at the early phase of training, this category of optimization algorithms has a notorious reputation of convergence to the local minima and easily overfitting (Lawrence and Giles, 2000). Hence, finding the optimal weight matrix may not be guaranteed. The other category is the stochastic global optimization algorithms, which are the population-based searching techniques, e.g., genetic algorithm and particle swarm optimization (PSO) algorithm (Gudise and Venayagamoorthy, 2003).

Recently, Lu et al. (2003) has firstly employed the standard PSO to train MLP (named as MLP-PSO) for forecasting several primary air pollutant levels (not including ozone), and also showed MLP-PSO outperformed MLP trained by the local optimization algorithms since local minima was avoided. Without applying such model directly to ozone prediction, Wang and Lu, (2006b) and Wang and Lu, (2006c) proposed two MLPs trained by two hybrid algorithms respectively based on the standard PSO. They argued that for ozone prediction, larger MLP weight space would be spanned compared with the one in the work of Lu et al. (2003), since more influential factors, accounting for ozone variation, would be involved as MLP inputs and thus the standard PSO would suffer the curse of dimension problem. It could inevitably lead to a situation, i.e., the larger the dimension of MLP weight space, the less the possibility of the solution converging to the global/appropriate minima. Or in other words, despite the fact that the standard PSO's global searching ability in the optimization, the 'bad' local minima still could not be avoided satisfactorily during the MLP training process.

Instead of initializing the weight matrices population randomly which is the common practice in the standard PSO for training MLP, Wang and Lu (2006b) employed the hybrid Monte Carlo (HMC) method to find an ergodic Markov chain that has $P(w \mid D)$ in the equation (3) as its equilibrium distribution, and then explore $P(w \mid D)$ to sample the initial weight matrix population, which in fact confines the search space to the area where the 'better' local minima exists. The standard PSO algorithm continues to train MLP with the weight matrices population initialized from HMC. The better prediction performance of the hybrid model was reported especially for the ozone peak, compared with one from MLP-PSO, which justified that the local minima also plagued the global optimization algorithms and can be avoided by hybriding with other algorithms accordingly.

Realizing the standard PSO does encounter the local minima during MLP training for ozone prediction, Wang and Lu (2006c) hybrided the standard PSO with LM algorithm, where the LM algorithm can be inserted automatically to replace the standard PSO when stagnation occurs and training MLP process continues with LM instead until convergence. The key in this work was that the authors indicates the local minima occurrence with stagnation which was at the later phase of the standard PSO training MLP, the difference of the interested error criterion between two successive PSO generations is becoming extremely small. The results manifested that combining the local and global optimization algorithms in

MLP training process can avoid local minima associated with the standard PSO, and MLP with such hybrid training algorithm outperformed MLP with the standard PSO in ozone prediction, especially during its episodes.

By developing MLP training algorithms, overfitting and local minima problems have been effectively addressed in ozone prediction literature respectively. However, it seems hardly to find or develop algorithms to overcome the two problems simultaneously in the MLP or even neural network domain (of course, other techniques like cross-validation, early stopping and regularization, are also effective to a degree but not from training algorithms perspective). Support vector machine, another category of learning machine, recently becomes a promise solution for this purpose.

3. Ozone Prediction with Support Vector Machine

Support vector machine was developed from the statistical learning theory (Vapnik, 1995). In this theory, all learning machines are supposed to minimize the expected value of the loss $R(w)$,

$$R(w) \leq R_{emp}(w) + Q(l, h\eta) \tag{5}$$

For neural networks, e.g., MLP models, implementing the empirical risk minimization principle, the aim is to minimize the first summand (in this case, $R(w) = R_{emp}(w) = e(w)$) in expression (5), which is in fact to minimize the training error and is the root of three problems mentioned before. While implementing the structural risk minimization principle, SVM minimizes both terms and in fact it minimizes the upper bound of generalization error, which endows it the strong ability of resisting overfitting. In addition, the solver of SVM usually comes down to quadratic programming or linear programming problem which can make it exempt from local minima, since only unique solution can be obtained. The structural risk minimization principle however is also implicitly associated other disadvantages which will be briefly mentioned in the following.

Compared with MLP models in ozone prediction application, the number of SVM application is rare, although SVM has the distinguished properties for tackling both overfitting and local minima problem simultaneously. Lu and Wang (2005) pioneered to use support vector regression (SVR) to predict several primary air pollutants level and found that performance was better from SVR, compared with the classical radial basis function network, a type of neural network. The authors also discussed SVR parameters identification, since parameters can considerably impact model's predictive performance. They used the computation-intensive trial-and-error method to indentify the appropriate SVR parameters, although better methods are available in SVM literature (Jordaan and Smits, 2002). Later, Lu and Wang (2008) firstly brought out the class imbalance problem for the standard support vector classifier (SVC) and SVR applications in ozone days classification/prediction. The authors stated that in most of collected field data, number of ozone polluted days is much smaller than that of non-polluted days, which will deteriorate any regressors/classifiers performance on minority class—i.e., ozone polluted days. Realizing SVC possessing the same highly overfiting resisting and local minima free properties like SVR, they modified SVC by proposing a cost-sensitive classification scheme, which can successfully tackle class imbalance problem and improves the prediction accuracy for ozone peak, compared with the standard SVC and SVR. To indentify the appropriate parameters for SVC and SVR, unfortunately, the authors used the grid search method which can have an exhaustive search in parameter space but is still computation-intensive. The extraordinary reward by using SVM in ozone prediction seems to be its good performance in ozone episodes. The most likely reason is unique trait for resisting overfitting and avoiding local minima.

4. Discussion and Conclusion

From the algorithm point of view, this commentary paper exemplifies solutions and remedies, although not exhaustively, for several model-performance-degraded limitations associated with MLP and SVM in ozone prediction application. For MLP models, they will continue to be the most popular models for ozone prediction, although the three most prominent notorious drawbacks presented in this paper could plague model's performance. Since these drawbacks are intertwined together and presented for every MLP used, e.g., Bayesian MLP is vulnerable for local minima problem although it is free from overfitting and black-box property, and MLP trained by PSO-based algorithms also need to tackle overfitting and suffer the black-box property problem although it can overcome local minima, to find an omnipotent model or training algorithm for all problems seems unrealistic. The compromise however is available, e.g., using Bayesian MLP with early stopping technique to tackle local minima, using MLP trained by PSO-based algorithms with cross-validation to resist overfitting. To circumvent the black-box property problem of MLP, sensitivity analysis for MLP inputs can be an alternative for Bayesian MLP. For SVM model, an increasing trend will be seen for its application in ozone or other air pollutants prediction (Wang, et al., 2008), since it possesses the peerless traits for effectively solving overfitting and local minima problems simultaneously. Beyond the parameters identification and class imbalance problems mentioned in this paper, some other challenges however lay ahead for modelers either, e.g., large dataset, which is usually true for the collected ozone and its influential factors data, could cause SVM break down because SVM uses a quadratic programming solver, and how can one choose an appropriate kernel in SVM for a specified dataset, because kernels have greatly impact on SVM performance.

The ground-level ozone prediction using machine learning models is by no means trivial work since its aim is to estimate the conditional mean of ozone target data. The whole paper therefore is sending the message that smart modelers should tackle the model-performance-degraded limitations more intelligently and rigorously which are rooted in the minimizing mechanism in the equation (2) for MLP and in expression (5) for SVM.

References

Abdul-Wahab, S.A., Al-Alawi, S.M., 2002. Assessment and prediction of tropospheric ozone concentration levels using artificial neural networks. *Environmental Modelling and Software* **17**, 219-228.

Agirre-Basurko, E., Ibarra-Berastegi, G., Madariaga., I., 2006. Regression and multilayer perceptron-based models to forecast hourly O_3 and NO_2 levels in the Bilbao area. *Environmental Modelling and Software* **21**, 430–446.

Balaguer Ballester, E., Camps, i., Valls, G., Carrasco-Rodriguez, J.L., Soria Olivas, E., del Valle-Tascon, S., 2002. Effective 1-day ahead prediction of hourly surface ozone concentrations in eastern Spain using linear models and neural networks. *Ecological Modelling* **156**, 27-41.

Chaloulakou, A., Saisana, M., Spyrellisa, N., 2003. Comparative assessment of neural networks and regression models for forecasting summertime ozone in Athens. *The Science of the Total Environment* **313**, 1–13.

Comrie, A.C., 1997. Comparing neural networks and regression models for ozone forecasting. *Journal of the Air andWaste Management Association*, **47**, 653-663.

Corani, G., 2005. Air quality prediction in Milan: feed-forward neural networks, pruned neural networks and lazy learning. *Ecological Modelling.* **185** 513–529.

Gardner, M.W., Dorling, S.R., 1998. Artificial neural networks (the multilayer perceptron) a review of applications in the atmospheric sciences. *Atmospheric Environment* **32**, 2627-2636.

Gardner, M.W., Dorling, S.R., 2000. Meteorologically adjusted trends in UK daily maximum surface ozone concentrations. *Atmospheric Environment* **34**, 171-176.

Gomez-Sanchis, J., Martın-Guerrero, J.D., Soria-Olivas, E., Vila-Frances, J., Carrasco, J.L., Valle-Tascon, D.S., 2006. Neural networks for analysing the relevance of input variables in the prediction of tropospheric ozone concentration. *Atmospheric Environment* **40**, 6173–6180.

Gudise, V.G., Venayagamoorthy, G.K., 2003. Comparison of particle swarm optimization and backpropagation as training algorithms for neural networks. Swarm Intelligence Symposium, *Proceedings of the IEEE* 110-117.

Jordaan, E.M., Smits, G.F., 2002. Estimation of the regularization parameter for support vector regression. Neural Networks, 2002. IJCNN '02. *Proceedings of the 2002 International Joint Conference on.*

Lawrence, S., Giles C.L., 2000. Overfitting and Neural Networks: Conjugate Gradient and Backpropagation. *Proceedings of the IEEE-INNS-ENNS International Joint Conference on Neural Networks*, Como, Italy. 114-119.

Lu, H.C., Hsieh, J.C., Chang, T.S., 2006. Prediction of daily maximum ozone concentrations from meteorological conditions using a two-stage neural network. *Atmospheric Research* **81**, 124– 139.

Lu, W.Z., Fan, H.Y., Lo, S.M., 2003. Application of evolutionary neural network method in predicting pollutant levels in downtown area of Hong Kong. *NeuroComputing* **51**, 387-400.

Lu, W.Z., Wang, D., 2008. Ground-level ozone prediction by support vector machine approach with a cost-sensitive classification scheme. *Science of the Total Environment* **395**, 109 -116.

Lu, W.Z., Wang, W.J., 2005. Potential assessment of the "support vector machine" method in forecasting ambient air pollutant trends. *Chemosphere* **59**, 693–701.

MacKay, D.J.C., 1995. Probable networks and plausible predictions - a review of practical Bayesian methods for supervised neural networks. Network: *Computation in Neural Systems* **6**,469-505.

Moller, M. F., 1993. A scaled conjugate gradient algorithm for fast supervised learning, *Neural Networks* **6**, 525-533.

Narasimhan, R, Keller, J., Subrarnaniam, G., Raasch, E., Croley, B., Duncan, K., and Potter, W.T. 2000. Ozone modeling using neural networks. *Journal of Applied Meteorology* **39**, 291-296.

Neal R.M., 1996. Bayesian Learning for Neural Networks, *Lecture Notes in Statistics* No **118**. Springer, New York, NY, USA.

Nunnari, G., Nucifora, A.F.M., and Randieri, C 1998. The application of neural techniques to the modelling of time-series of atmospheric pollution data. *Ecological Modelling* **11**, 187-205.

Pastor-Barcenas, O., Soria-Olivas, E., Martın-Guerrero, J.D., Camps-Valls, G., Carrasco-Rodrıguez, J.L,. Valle-Tascon, S. del. 2005. Unbiased sensitivity analysis and pruning techniques in neural networks for surface ozone modeling. *Ecological Modelling* **182**, 149-158.

Prybutok, V.R., Yi, J.S., Mitchell, D., 2000. Comparison of neural network models with ARIMA and regression models for prediction of Houston's daily maximum ozone concentrations. *European Journal of Operational Research* **122**, 31-40.

Rumelhart, D.E., McClelland, J.L., and the PDP Research Group, 1986. *Parallel Distributed Processing: Explorations in the Microstructure of Cognition*, vol. 2. M.I.T. Press, Cambridge, MA.

Schlink, U., Dorling, S., Pelikan, E., Nunnari, G., Cawley, G., Junninen, H., Greig, A., Foxall, R., Eben, K., Chatterton, T., Vondracek, J., Richter, M., Dostac, M., Bertucco, L., Kolehmainen, M., Doyle, M., 2003. A rigorous inter-comparison of ground-level ozone predictions, *Atmospheric Environment* **37**, 3237-3253.

Spellman, G., 1999. An application of artificial neural networks to the prediction of surface ozone concentrations in the United Kingdom. *Applied Geography* **19**, 123- 136.

Vapnik, V., The nature of statistical learning theory. New York, USA: Wiley; 1995.

Viotti, P., Liuti, G., and Di Genova, P., 2002. Atmospheric urban pollution: applications of an artificial neural network (ANN) to the city of Perugia. *Ecological Modelling* **148**, 27-46.

Wang, D., Lu, W.Z., 2006a. Interval estimation of urban ozone level and selection of influential factors by employing automatic relevance determination model. *Chemosphere* **62**, 1600-1611.

Wang, D., Lu, W.Z., 2006b. Ground-level ozone prediction using multilayer perceptron trained with an innovative hybrid approach. *Ecological Modelling* **198**, 332-340.

Wang, D., Lu, W.Z., 2006c. Forecasting of ozone level in time series using MLP model with a novel hybrid training algorithm. *Atmospheric Environment* **40**, 913-924.

Wang, W.J., Lu, W.Z., Wang, X.K., Leung, A.Y.T., 2003. Prediction of maximum daily ozone level using combined neural network and statistical characteristics. *Environment International* **29** 555– 562.

Wang, W.J., Men, C.Q., Lu, W.Z., 2008. Online prediction model based on support vector machine. *NeuroComputing* **71**, 550–558.

Yi, J., Prybutok, V.R., 1996. A neural network model forecasting for prediction of daily maximum ozone concentration in an industrialized urban area. *Environmental Pollution* **92**, 349-357.

In: Industrial Waste...
Editor: John P. Samuelson, pp. 233-246
ISBN: 978-1-60692-720-5

Chapter 8

PROFILING THE IMPLEMENTING OF PRODUCT STEWARDSHIP: A CASE STUDY OF SOME CHEMICAL INDUSTRIES IN GERMANY

Dieudonne Alemagi [1]
Department of Industrial Sustainability, Brandenburg University of Technology, Cottbus; Postfach 101344, 03013 Cottbus, Germany

Abstract

This paper had the fundamental objective of assessing product stewardship initiatives designed by a plethora of chemical industries in Germany. The paper justifies the need for product stewardship in Germany's chemical industry's and presents the results of a survey of some chemical industries in Germany in June 2003. A total of 122 questionnaires were administered to environmental managers in the selected industries using systematic random sampling. The return rate was 51% corresponding to 66 industries. Survey results revealed that despite the inadequacy in the implementation of some product stewardship tools like emission reduction (particularly solid waste emission), implementation of product stewardship tools by the sampled industries was advanced. It was concluded that most of the attempts made thus far by industries to address product stewardship have been major successes. However, it was emphasized that further improvement could be achieved if incentives for product stewardship implementation are promoted and the merits of strategic research in the area of reduction in solid waste emission is further developed and encouraged.

Keywords: Germany, Product Stewardship, Product Stewardship Tools, Chemical Industries

[1] E-mail address: dalemagi@yahoo.co.uk, Telephone: (0049) 0355 694 653, Fax: (0049)0355 694 700. Correspondence to the author can be made by post, email or telephone at: Dieudonne Alemagi, Department of Industrial Sustainability, Brandenburg University of Technology, Cottbus; Postfach 101344, 03013 Cottbus, Germany.

1. Introduction

The global production of chemicals increased from 1 million tonnes in 1930 to 400 million tonnes in 2001. The world production of chemicals in 2001 was estimated at 1,632 billion Euros. The European Union chemical industry accounted for 32% of this total amount, which generated a trade surplus of 41 Billion Euros (CEFIC 2002). The Federal Republic of Germany is the largest chemical producer in Europe and most of Germany's largest chemical industries are located in North – Rhine Westphalia State of (Figure 1).

The chemical industry is one of the key branches of the German economy and the growth engine for innovations and technological advance. The production and industrial uses of chemicals by Germany's chemical industry makes a vital contribution to the economic and social well being of citizens particularly in terms of trade and employment. Chemicals, their production and uses bring about benefits on which modern society is entirely dependent (table 1).

Figure 1. Location of major chemical industries in the North – Rhine Westphalia State of Germany (Source: Lemke, K.W, 2002).

Table 1. Industrial uses of some chemicals (Source: Clarkson R.J, 1998)

Chemical	Uses
Aluminium	Cooking foil and food containers. In alloys for making aircraft, Electric cables
Ammonia	Manufacture of nitric acid and ammonium salts
Chlorine	Batteries
Nitrate	Fertilizers/Explosives
Sulphate	Fertilizers
Chlorine	As a bleach, for cotton and paper, in making PVC
Copper	Electric cables
Silver Compounds	Photography

Table 1. Continued

Chemical	Uses
Zinc	Galvanizing iron, making brass
Sulphuric Acid	Phosphate and ammonium fertilizers, detergent pains
Sodium Hydroxide	For making soap
Polymers (in general)	Electrical insulators
Steel (Stainless)	Large- scale chemical reaction vessels and Cutlery production
Steel (Mild)	Car bodies
Calcium Carbonate	Manufacturer of iron and cement

This, notwithstanding, chemicals, their production and uses by chemical industries have caused serious damage to human health resulting in suffering and premature death as well as damage to the environment in general. Well-known examples include asbestos that is known to cause lungs cancer and benzene which leads to leukemia. Abundant use of DDT resulted to reproductive disorder in birds (Commission of the European Communities, 2001). The incidence of some disease, e.g. testicular cancer in young men and allergies, has increased significantly over the last decades. While the underlying reason for this has not been identified, there is justified concern that certain chemicals play a causative role for allergies. Corporate initiatives have been formulated by chemical industries in an attempt to mitigating the adverse impacts of chemicals on the environment. In Germany, the chemical industry has made key strides in the arena of Product Stewardship as a rigorous strategy to attain excellence in their pursuit for environmental protection.

The European Chemical Industry Council (CEFIC) defines product stewardship or Extended Product Responsibility (EPR) as the responsible and ethical management of the safety, health and environmental aspects of a product throughout its total life cycle. In other words, products must be managed and used safely every step of the way, through development, manufacture, packaging, distribution, use and ultimate disposal (CEFIC, 2002). Product stewardship is an essential part of the Responsible Care Program within the chemical industry and as Cramer *et al* (1999) opine, this branch of the industry is setting a trend for other branches. However, effluents emanating from chemical industries are still a nuisance to safety, health and the environment. The purpose of this paper is therefore to evaluate initiatives taken by a plethora of chemical industries in Germany towards the implementation of product stewardship and to prescribe robust recommendations for improvement.

The paper is organized as follows. In the following section I begin by providing a background to the paper. This sector of the paper presents the importance of Germany's Chemical industries from a socio-economic standpoint. In the penultimate part of this section, tool for product stewardship are outlined. This section terminates with a justification for product stewardship in Germany's chemical industry. The research methodology is outlined in the next section of the paper. In the section that follows, the research results are presented and reflected upon. Finally, the paper is concluded with a way forward with regards to product stewardship enhancement in Germany's chemical industry.

2. Background

2.1. The Socio- Economic Importance of Germany ´S Chemical Industry

The German chemical industry does not only consist of large international companies. In the year 2000, the share of small and medium sized enterprises (< 500 employees) was quite considerate as well. Small and Medium Size companies (SMEs) provided 35 percent of all jobs in Germany's chemical industry and realize 36% of total chemical sales in 2001. The German chemical industry is the third biggest producer of chemical products worldwide. In terms of employment, this industry employs more people than any other European country or Japan (Table 2). As detailed in table 2, Germany was the second largest exporter of chemicals in 2001 (largest exporter: USA), with a world market share of nearly 12.2 percent (VCI, 2002).

From a European perspective, Germany is Europe's largest producer of chemicals. It realized 26% of the total European Union (EU) sales in 2001 (Table 3). Indeed, the chemical industry in Germany offers almost every third job within the European chemical industry.

Table 2. Indicators of chemical industry in the world main chemical producer nations (Source: VCI, 2002)

2001	USA	Japan	Germany	France	Italy	UK
Sales (in bill. Euro)	476.0	237.9	133.9	85.0	72.9	56.9
Employees (in 1000)	1033	366	467	239	207	235
Exports (in bill. Euro)	87.6	37.6	81.4	50.2	25.5	46.0
Imports (in bill. Euro)	86.2	25.8	54.8	40.2	33.7	39.0

Table 3. Germany's Chemical Industry and its contribution to the EU economy (Source: VCI, 2002)

Major Contribution	Amount (in billion Euro)	Share of total EU	Rank within the EU
Sales	133.9	26.0%	1
Investment	7.0	27.2%	1
Imports of chemical products	54.8	18.7%	1
intra EU imports	42.2	-	1
Exports of chemical products	81.4	22.2%	1
Intra EU exports	43.4	-	1

On a national scale, the industry constitutes a key stone that underpins the German economy. In terms of sales, it is about 10 percent of Germany's manufacturing industry and as of the year 2001, it employed 467,000 people (7.4 percent of total manufacturing industry) and ranked 5th in terms of employment (VCI, 2002). New products and processes are the driving force of the international competitiveness of the chemical industry in Germany. A high trade surplus was realized by this industry in 2001 and together with the automobile industry, the industry contributed decisively to Germany's trade balance surplus in 2001. Despite the aforementioned importance of Germany's chemical industry, it is important to underscore the fact that this industry has inflicted considerable damages to safety, health and the environment. There is therefore a great need to balance the socio-economic value of this industry with the ecological needs. Thus, the next part of this section of the paper is dedicated to product stewardship tools and the fundamental need for product stewardship in Germany's chemical industry.

2.2. Product Stewardship: An Overview of Basic Tools

Germany's chemical industry has a long tradition of being concerned with environmental issues as an umbilical commitment to decrease the negative environmental impacts caused by their organization and activities and to further the process of sustainable development. Product stewardship is one of those tools that has been employed by Germany's chemical industry to sustain this commitment. Within the chemical industrial sector, the first guidance document on product stewardship was issued in 1992 by the US Chemical Manufacturers Association (CMA), soon followed by various other chemical sector organizations (E.g. the European Chemical Industry Council (CEFIC) and the Canadian Chemical Producers Association (CCPA)). Ultimately, product stewardship is all about manufacturers taking on more responsibility for reducing the health and environmental impacts of their products and packaging. It's an incentive to redesign products with fewer hazardous materials, so that they are more durable, reusable and recyclable. Lindhqvist (1992) defines extended product responsibility (EPR) as an environmental protection strategy to reach an environmental objective of a decreased environmental impact of a product, by making the manufacturer of a product responsible for the entire life cycle of a product and especially for the take-back, recycling and final disposal of the product. Lindhqvist argues that while the manufacturer, distributor, user, recycler and final disposal can all influence the environmental impacts of a product, primary responsibility rests with the manufacturer. According to Lindhqvist (1992), this is because the producer is responsible for product development and therefore has the unique opportunity to prevent environmental impacts through the adoption of environmentally conscious product development (or eco-design). Indeed, as Cramer *et al* (1999) further purport, the main purpose of product stewardship is to ensure that an industry systematically structures its safety health and environmental policy at the product level. In order to achieve this goal, a set of tools referred to as Product Stewardship Tools (PST) are necessary for product manufacturers (PPRC, 1999). These tools have been summarized by the United States of America Pacific Northwest Pollution Prevention Resource Centre as follows:

a. Design for the Environment

Design for the Environment (DFE) considers the potential environmental impacts of a product and the process used to make that product, including components and raw materials. It could be considered one facet of life cycle management. DFE principles evaluate facility and local impacts, as well as global impacts and include habitat disturbance, emissions and effluents, chemical releases, inefficient use of water and energy, solid waste, and much more. Application of DFE also considers recovery of the product at the end of its useful life, through Design for Disassembly, Design for Remanufacturing, and Design for Recycling.

b. Eco-Labeling

Eco-label programs are certification programs in which a product, process, or management system is certified to meet specific environmental criteria as established by an assessment body. Manufacturers wishing to achieve certification to an eco-label program may enjoy marketing advantages because the certification positively differentiates their products as environmentally sound. Typically, the assessment organization will certify the product/process and upon approval, allow the applicant to use a logo on the product label, packaging, and/or advertising. Typical examples include Germany's Blue Angel. Independently, a manufacturer can also label their products with non-certified, specific, environmental information, which is another form of eco-labeling. There are many international eco-labeling entities that certify products for their country, and for manufacturers who sell products in their country. Additional programs certify performance-specific features such as the energy efficiency or indoor air quality emissions of various products or processes. Eco-labeling programs have influenced product design in the areas of life-cycle impact, durability, hazards, recyclability, and more. The criteria differ for each program and can cover the product itself, and in addition, the processes used to manufacture the product, or other life-cycle impacts. Examples include efficient use of fossil fuels, reduction of greenhouse gas emission, conservation of resources, concentration restrictions on substances such as heavy metals and toxins.

c. Green Purchasing

These are procurement policies, measures, or preferences that specify environmentally preferable products and services. Businesses and consumers can use their purchasing power to leverage supplier practices, products, and decisions, for better environmental protection. By specifying purchasing preferences for products that have less environmental impact, organizations can spur manufacturers to: use recycled-content, design for recycling/reuse, consume less energy or water, minimize waste, or create less toxic products. Procurement preferences can provide an immediate market for environmentally and socially responsible products.

d. Take Back Programs

These programs are a fundamental tool for product stewardship. They provide manufacturers the physical responsibility for products and/or packaging at the end of their useful lives.

Indeed, by accepting used products, manufacturers can acquire low-cost feedstock for new manufacture or remanufacture, and offer a value-added service to the buyer. This play a vital role in limiting the adverse impacts of a product during the cradle to crave process.

2.3. Why the Need For Product Stewardship in Germany's Chemical Industry?

It is my respectful submission that of all the German industries, however, the one in greatest need of sustainable development is the chemical industry. The rationale behind this establishment is that this industry is susceptible to environmental damages on several fronts. Indeed, this industry has a history of ecological damages. More than 50% of all the heavy metals and organic pollutants in the river Rhine are of diffuse origin while the rest emanate from point sources including Germany's chemical Industries located along the Rhine catchment (ICP, 2001).

The river Rhine is an important natural resource in Western Europe and flows through several nations including Germany. Pollution of the Rhine from point sources ends up in the Netherlands at the port of Rotterdam, which is found downstream region of the Rhine. The Concentration of most hazardous substances in the Rhine are slowly decreasing or have stagnated at a low level during the year 1971 to 1998 which illustrate the results of three monitoring stations along the Rhine. Nevertheless, the International Commission for the Protection of the Rhine (ICPR) target values for heavy metals in the suspended matter has not yet been delivered. The last valuations in 2001 shows that the target values for Mercury, Cadmium, Zinc and Copper have not been reached. Elevated heavy metals (Mercury, Cadmium, Zinc and Copper) concentrations are associated with degraded water and sediment quality and can result in potential impacts on human health and the environment. In the eastern harbors of the port of Rotterdam, mainly influenced by inputs from the Rhine, Zinc, Cadmium and Copper frequently exceeds the sea/slufter values in the sedimental dredged material. After the year 2000, it is likely that some 30 million meters of sedimental dredged material will still be contaminated by heavy metals to the extent that it has to be stored in the slufter (Rotterdam Municipal Port Management, 2000). A National Academy of Sciences study in the United States of America suggested that one out of four developmental and behavioral problems in children may be linked to genetic and environmental factors, including exposure to heavy metals like lead and mercury (SFDTRA, 2000). Itai-itai disease is well known as a health hazard induced by Cadmium originating from chemical industries in cadmium-polluted areas. According to Keiko *et al* (1999), the main clinical features of Itai-itai disease are: Multiple proximal renal tubular dysfunctions, hormonal imbalance, and excruciating pain in the bones and all over the body.

At the BASF (One of Germany's largest chemical industry) "Aktiengesellschaft's" site in Ludwigshafen, Germany, approximately 3 Million metric tons of hazardous goods left the site in some 118,000 shipments by road, rail, inland waterways, sea and air in 2001. Four accidents pertaining to the shipment of these hazardous goods occurred (BASF 2001). During the year 1999, many operational disturbances and scenario accidents were recorded in BASF AG. BASF is located in the catchment of the river Rhine and most of the accidents recorded at BASF could pose severe threats to the river Rhine and the environment in general. In January 27, 1999, during the course of cleaning work, a dye was accidentally released into the environment through the wastewater treatment plant. Furthermore, a similar case of industrial contamination was recorded on November 15 of the same year, when a leaky valve caused tertiary butanol to be released accidentally into the environment through an outlet of wastewater treatment plant. These tacit and glaring cases of ecological contamination caused by BASF industrial activities indicate that irrespective of the cumulative efforts to check environmental hazards, plant disturbances and anomalies at the corporate level can never be completely ruled out.

In 1998, a consistent increase in dioxin concentrations in cow milk was observed by the German regional authorities (Malisch, 1998). Contamination was eventually traced back to citrus pulp pellets (CPP) that were contaminated with dioxins. Citrus pulps are used as animal feed. The European Union concluded that lime milk used in the production of CPP was the cause of contamination (EU, 1999). The characteristic of all dioxins is their high melting point, low vapor pressure and good solubility and affinity for non-polar conditions. Moreover, they accumulate and biomagnify in the food chain due to their fat solubility and pronounced resistance to metabolic degradation. As Irwin *et al* (1997) argue, dioxin is lethal and some of its effects to human beings include *inter alia* changes in the human liver, harmful effects on the human kidney, heart, lungs and nervous system, increased rates of human cancer, immunological, reproductive, fetotoxic, genotoxic effects.

3. Research Methodology

In June 2003, a survey was conducted among a series of chemical industries in Germany. A brief questionnaire was prepared and addressed to environmental managers of 122 randomly chosen large, small and medium sized chemical industries from all parts of Germany. This questionnaire based survey was undertaken in order to ascertain the extent to which these industries were implementing Product Stewardship Tools (PST). The raison d′etre for using this method was due to the large sample size and because a questionnaire based survey will undoubtedly enable detailed information about corporate product stewardship to be collected within the shortest possible time frame. All questionnaires were sent by email during the first week of June 2003 and by the end of that month, 63 questionnaires had been received corresponding to 51%. Since June 2003 when the survey was conducted, the sampled industries have undoubtedly accrued experience in the arena of product stewardship and it is therefore possible that if the same questions were posed to them today, their answers will surely be very different. Thus, it is my contention that the results of this study are only valid for the industries sampled during the month of June 2003. The following results were obtained.

4. Survey Results

In this section of the paper, questions raised in the questionnaire are presented and a summary of corresponding results obtained are outlined.

4.1. Have You Adopted Product Stewardship in Your Company?

This first question was formulated to elicit a simple answer (yes or no) to a very crucial question about the adoption of product stewardship as a tool for corporate sustainability. To ensure clarity, the CEFIC definition of product stewardship was given. A total of sixty one (about 97%) industries that responded to the survey had adopted product stewardship. Only two industries (about 3%) answered "no". Industries that answered "no" expressed that they had not fully adopted product stewardship.

4.2. When Was Product Stewardship Initiated in Your Company?

Any assessment of product stewardship needs to begin with the history and evolution of product stewardship and the possible trends for the future. Results obtained indicate that the first chemical industry initiated product stewardship in 1954 (Figure 2). There was a considerable increase in initiation in 1991. A peak was obtained in 1998 (Figure 2).

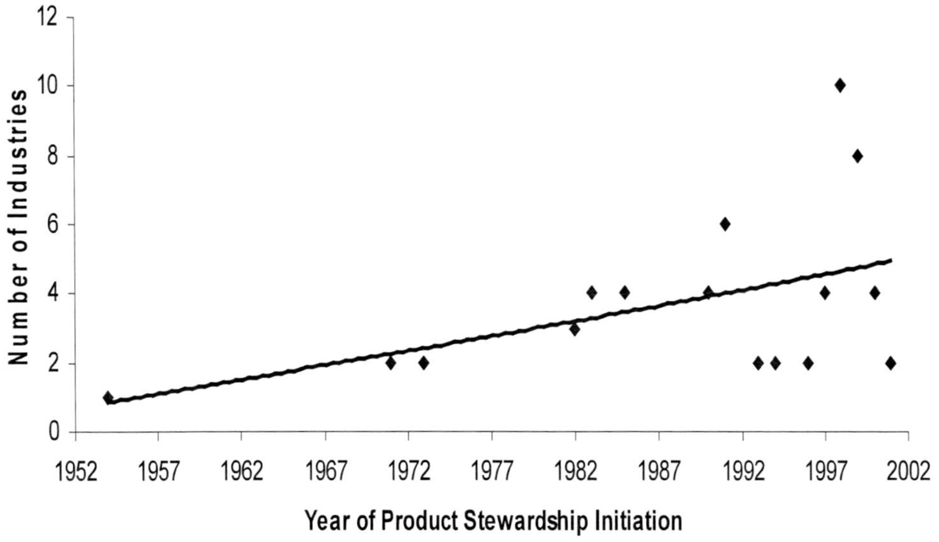

Figure 2. Polynomial trend of product stewardship initiation in some German chemical industry.

4.3. Do You Purchase Environmentally Friendly Raw Material to Be Used in the Design of Your Products? If Yes Please Give Examples

Survey respondents were asked to say whether they were committed to green purchasing. Forty –eight (about 76%) of all the respondents answered affirmatively (Figure 3). Examples of green purchasing were varied but some of the key examples that were mentioned included:

- Soya bean oil (it is used in the manufacture of margarine) in the place of synthetic oil which is less environmentally friendly;

- Formaldehyde free stabilizers (Stabilizers serve as wetting agents , preserve dyes, inhibit bacteria growth and harden films in the film industry);
- Bleaches with biodegradable chelates;
- Non halogenated solvents;
- Bee wax, chalk and dextrin (used in the manufacture of paints and colorants);
- Natural rubber (used in the production of tyres, belts, gaskets etc) instead of synthetic rubber that is synthesized from petroleum;
- A "respimat". This is a novel, reusable, propellant-free (no CFC) multi dose soft mist inhaler that is used in pharmaceutical industries for drug testing. It is an alternative to the conventional metered dose inhaler.

Ten (16%) respondents, however, declared tacitly that they were not committed to green purchasing (Figure 3). The rest did not provide any answers.

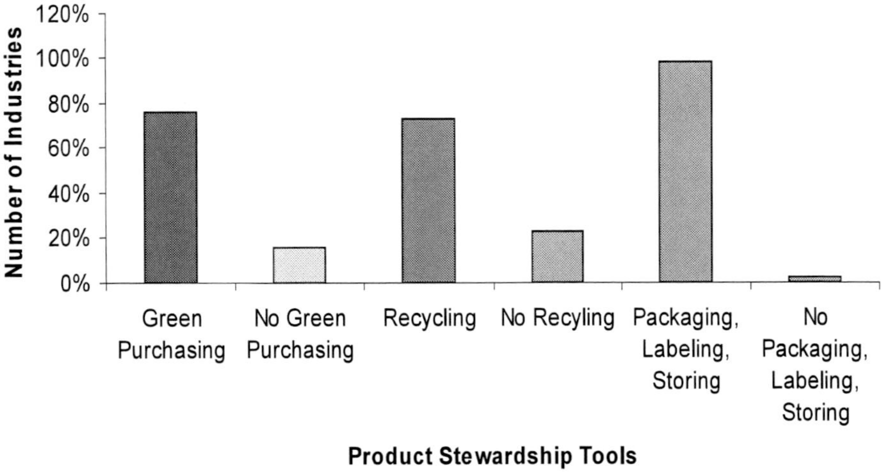

Figure 3. Level of implementation of green purchasing, recycling, packaging, labeling and storing of products in some chemical industries in Germany.

4.4. Do You Recycle and Reprocess Used Products?

This question was aimed at documenting recycling and take back programs as one of the fundamental tools of product stewardship. Of the sixty three respondents that responded to the survey, forty six (73%) maintained that they were engaged in take back and recycling programs while seventeen respondents (23%) declared the contrary (Figure 3). In answering affirmatively, major examples of recycled products provided by the respondents included:

- Polyvinyl chloride or PVC (a modern synthetic polymer that is used in the manufacture of plastics , electric cables , pipes etc);
- Sulfur;
- Solvents like Chlorine, butanol and Iso- butanol;
- Polyethylene;

- Styrofoam;
- Glycol;
- Used coolants;

4.5. Are Your Products Packaged, Labeled and Stored According to the Relevant EU Legislation?

Sixty two (about 98%) industries admitted that packaging, storing and labeling of their products was done according to the relevant EU legislation (Figure 3). Only one respondent (about 2%) submitted that packaging, labeling and storing of products as per the requisite EU legislation was not part of their services. This respondent did not state whether this was handled by a third party or not.

4.6. In What Areas Have You Predominantly Reduced Emissions?

In figure five, key arenas where emission reduction was achieved by the sampled industries is presented. Major reduction in emission to water and air were received as some of the answers to this question. Fifty-five (87%) and Fifty-two (82%) of the companies revealed that there had been a considerable reduction in their emission to water (example include reduction in Chemical Oxygen Demand and Biological Oxygen Demand) and air respectively. These respondents sited C0, $C0_2$, and $S0_2$ as key examples of major pollutants whose emission to air had been greatly reduced.

With regards to hazardous wastes, a total of forty eight (76%) respondents expressed significant reduction in hazardous wastes. The most commonly listed examples of these wastes were: Chlorine, butanol, and polyethylene. Solid waste was sited as an arena where key strides had been made as thirty three (52%) of the responded asserted the fact that there had been a reduction in the emission of solid waste (like glass, and scrap metals) from their operations. Thirty (48%) industries made a contrary declaration.

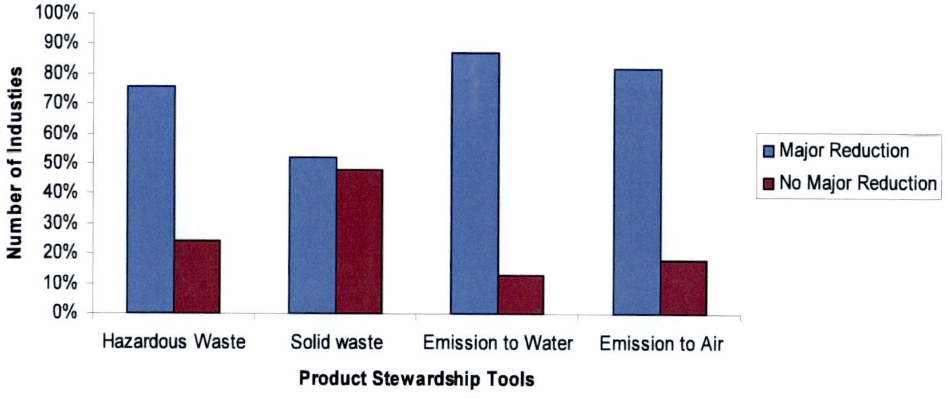

Figure 4. Level of implementation of some product stewardship tools (emission reduction in hazardous and solid waste, emission reduction to air and water) in some chemical industries in Germany.

5. Survey Reflections

The survey demonstrated that initiation of product stewardship in the sampled industries reached its peak in 1998 (Figure 2). A plausible explanation for this trend could be that most industries in Germany might have wanted to boost up their ecological credentials (achieve a "green" image) as a result of political motivation from Germany's coalition government that was formed in 1998 between the Green party and the Social Democratic Party (SPD).

The survey also revealed that almost all the industries that had adopted PST were large and medium sized enterprises. This is consistent with studies obtained by Alemagi *et al* (2006) in a study of environmental management systems in industries along the Atlantic coast of Cameroon. As Alemagi *et al* (2006) argue that larger industries have the requisite knowledge, financial and human resources to design and implement an environmental management system which like product stewardship is a tool for corporate ecological sustainability.

The survey found little differences in the number of industries that declared that they had predominantly reduced emission of solid waste with those that rejected this declaration (figure 4). A possible reason to underpin this observation could be that these industries with little reduction in solid waste may be generating household waste (like used paper) which in most instances could be very difficult to control. Indeed, a closer look at the annual environmental report of some of these industries revealed that solid waste generated during their operation was handled by a licensed third party which in most instances could be the local authority.

It is glaring from the survey responses that most survey respondents were steadfastly committed to the implementation of PST like recycling, green purchasing, packaging, labeling and storing of products (figure 3). This result is consistent with Lewis' (2005) assertion. This author maintains that the most essential element in EPR is the extension of producer responsibility to the post-consumer or end-of-life stage. He argues that this was first mandated in Germany in 1991 for packaging, and is now being applied to packaging and other product sectors in many of the world's industrialized economies. However, comparing our results with the prevailing situation in the United States, it would be fair to make a reasonable inference as to the fact that implementation of product stewardship could be far advanced in Germany's chemical industry as compared to the United States chemical industry. In support of this postulation, a survey of California local government electronic recycling programs revealed that take back programs are still at a premature stage (Boisson and associates 2002). The rationale behind this observation could be that implementation of EPR by Germany's chemical industry is generally regulated by legal obligations which are authorized and enforced by government. In the United States, legal enforcement has generally been avoided in favor of voluntary approaches with shared responsibilities and minimal or no regulation (Ehrenfeld, 2002).

6. The Way Forward

This thesis has provided an insight into the current product stewardship initiatives formulated by some chemical industries in Germany. It submitted that, the efforts made by these industries towards implementing product stewardship and its tools show a right tract to follow but needs to be strengthened. Indeed, further improvements in the arena of product stewardship can be made if these industries become more active in promoting the implementation of all tools for product stewardship particularly in the realm of solid waste reduction. Specifically, the German government needs to investigate various incentives for the implementation of PST. The merits of fundamental and strategic research in the area of solid waste emission could be further investigated and encouraged as this could serve to benefit both the environment and the chemical industry. Furthermore, continued research into PST like recycling and reuse technology with pilot projects in these areas could be given more emphasis. Finally, it is essential that the general public and different stakeholders are enabled and encouraged to become more actively involved in the development of practical implementation of policies and legislation relevant to product stewardship at the corporate level. This means that individuals be given all the relevant information necessary to enable them to exercise informed choices as stakeholders; they must also have access to data related to PST to enable them monitor the performance of industries.

This paper has revealed that, to a large degree, the principle of product stewardship has already been advanced in some Germany's chemical industry through the proactive implementation of PST. The overall conclusion to be drawn from this study is that most of the product stewardship initiatives in place thus far addressed key concept and tools of product stewardship, but in certain cases, particularly in the realm of reduction in emission of solid waste tend to be inadequate. However, important ground work for PST has been laid. Thus it is crucial that both the government and these industries became more proactive in promoting the aforementioned recommendations as this will undoubtedly guarantee that the industries continue along a path of improved corporate social responsibility and environmental management.

References

Alemagi D, Oben PM, Ertel J. (2006). A Survey of Environmental Management System Initiatives in Industries along the Atlantic Coast of Cameroon. *Corporate Environmental Strategy: International Journal for Sustainable Business* (In press).

BASF, (2001). Environment, Safety, Health Report, BASF Group Worldwide.

Boisson and Associates. (2002). Meeting California's E-waste challenge: A survey of California local government Electronic Recycling Programs. Prepared for the California Integrated Waste Management board, California.

CEFIC, (2002). *The European chemical industry in a world wide perspective*. Brussels, Belgium.

Clarkson, R .J. (1998). *Industrial uses of chemicals*. Available online at: http://www.rjclarkson.demon.co.uk/middle/middle8.htm

Commission of the European Communities. (2001). White Paper on Strategy for a Future Chemical Policy. Brussels, Belgium 2001.

Cramer, J, and Alders G. (1999). The design of a product stewardship management system within Akzo Nobel. *Eco-Management and Auditing* **6**: 135 – 139.

Ehrenfeld, J .R. (2002). Sustainability and enterprise: an inside view of the corporation. In: *Extended Producer Responsibility: A Materials Policy for the 21st Century,* Fishbein, B., Ehrenfeld, J., Young, J. (Eds.). Inform, New York.

EU, (1999). Organization of official inspections in the field of animal nutrition: Dioxin contamination of citrus pulp pellets. *EU Report* XXIV/1005/99-MR Final.

ICPR, (2001). *Conference of the Rhine Ministers*: The state of the Rhine Yesterday – Tomorrow. Strasbourg, France.

Irwin, R.J, Van Mouwerik M, Stevens L, Seese M.D and Basham W. (1997). *Environmental contaminants Encyclopaedia.* United States of America National Park Service, Water Resources Division, Ford Collins, Colorado, U. S. A.

Lemke, K .W. (2002). *Chemistry in North-Rhine Westphalia.* Zentrum Für Innovation und Technik in NRW, Germany.

Lewis, H. (2005). Defining product stewardship and sustainability in the Australian packaging industry. *Environmental science and policy* **8:** 45 – 55.

Lindhqvist, T. (1992). Extended producer responsibility. Proceedings of the Conference on Extended Producer *Responsibility as a Strategy to Promote Cleaner Production,* Trolleholm Castle, Sweden.

Malisch, R. (1998). Increase of PCDD/F contamination of milk and butter by use of contaminated citrus pulp as component in feedstuff. *Organohalogen Compounds* **35**: 131–134.

PPRC. (1999). Pacific Northwest Pollution Prevention Resource Centre, U.S.A. Available online at: http://www.pprc.org/pprc/howto.html

Rotterdam Municipal Port management, (2000). Dredged Material in the port of Rotterdam: Interface between the Rhine Catchments Area and the North Sea. GKSS Research Centre, Germany.

SFDTRA, (2000). *Scientific Frontiers in Developmental Toxicology and Risk Assessment.* National Academy of Sciences, National Academy Press, U.S.A.

VCI, (2002). *Branch portrait of the German chemical industry*, Frankfurt, Germany.

In: Industrial Waste...
Editor: John P. Samuelson, pp. 247-270
ISBN: 978-1-60692-720-5

Chapter 9

BIOACTIVE SUBSTANCES DERIVED FROM FISH PROCESSING WASTES AND THEIR BIOTECHNOLOGICAL APPLICATIONS

Se-Kwon Kim[a,b,*] ***and Y. Dominic Ravichandran***[a,c]
[a]Marine Bioprocess Research Center, Pukyong National University, Busan 608-737, Republic of Korea
[b]Department of Chemistry, Pukyong National University, Busan 608-737, Republic of Korea
[c]Department of Science and Humanities, Sri Krishna College of Engineering and Technology, Coimbatore-641 008, India.

Abstract

Approximately one billion tones of fish and shellfish are produced worldwide every year, but only 50–60% is used for human consumption. Fish processing wastes are defined as fish processing by-products, such as viscera, heads, bones and skins, that are not utilized in fishery products. These by-products create multiple environmental problems. Recently, great efforts have been made to utilize fish processing wastes by producing high-value compounds such as pharmaceuticals, nutraceuticals and cosmetics, which has resulted in isolation of some biologically active compounds such as functional enzymes, proteins, lipids, chitins and minerals. Bioactive substances isolated from fish processing by-products are commercially important and are used for development in biotechnological applications. Herein we discuss recent progress that has focused on producing bioactive compounds from fish processing wastes and also review their biotechnological applications.

1. Introduction

Global production of fish, shellfish and crustaceans has shown an increasing trend over the last decade, and this trend is expected to continue. Marine capture fisheries contribute more

* E-mail address: sknkim@pknu.ac.kr. Tel.: +82-51-620-6753; fax +82-51-621-0047. (Corresponding author.)

than 50% of the total fish production in the world, and more than 70% of this production is utilized for processing. As a result, a considerable amount of total catch is discarded as processing leftovers. This includes trimmings, fins, frames, heads, skin and viscera. In addition to fish processing, shells of crustaceans and shellfish from marine bioprocessing plants are accumulated as processing by-products. Approximately 32 million tonnes of fish processing waste is produced annually [1]. Fish processing wastes are an important source of environmental contamination. Currently, most of the wastes from the fish processing industries are used in the production of fish oil, fishmeal, fertilizer, animal food and fish silage [2, 3]. There are number of reports on the methods to convert these wastes into useful products [4]. Hence, there is a great potential in the marine bioprocess industry to convert and utilize these wastes as valuable products [5].

Fishery waste is a mixture of many biologically active components and can be extracted and purified with various technologies. Recently, a number of bioactive compounds from the remains of fish muscle proteins, fish skin, fish oil, fish bone, internal organs, shellfish and crustacean shells have been identified [5-12]. These compounds include bioactive peptides, oligosaccharides, fatty acids, enzymes, water-soluble minerals and biopolymers with biotechnological and pharmaceutical applications. Some of these bioactive compounds have also been identified as nutraceuticals, which are beneficial in human health promotion [13]. Therefore, development of new technologies in the search of novel bioactive compounds will transform fish processing wastes into value-added and profitable materials. This will not only reduce the waste disposal from the fish processing industry but also protect the environment from contamination.

2. Bioactive Substances Derived from Fish Muscle

The fish frames and cut-offs obtained from mechanically deboned fish contain considerable amounts of muscle proteins. These muscle proteins are nutritionally valuable, as they are easily digestible with well-balanced amino acid composition [14]. Hence, fish proteins derived from processing wastes can be hydrolyzed enzymatically to recover protein biomass otherwise discarded as processing waste. Protein hydrolysates from the fish muscle of the processing wastes have been analyzed for their nutritional and functional properties [15]. Moreover, the increasing knowledge of functional properties of fish protein hydrolysates has created a number of avenues for research on the developments of fish processing wastes into value-based products, which include functional foods and nutraceuticals.

2.1. Peptides

Peptides isolated from the protein hydrolysates by enzymatic digestion have exhibited different physicochemical properties and biological activities based on their molecular weights and amino acid sequences. Therefore, it has become imperative to develop methods to separate peptides with different molecular weights. Peptides of desired molecular weights can be produced from the fish hydrosylates using an ultrafiltration membrane system equipped with the appropriate molecular weight cutoff [16]. Several enzymes derived from varied sources such as microbes, plants and animals are employed for the hydrolysis of fish processing byproducts [17]. Furthermore, functionally active peptides of different molecular weights can be obtained by a series of enzymatic digestions in a system using a multi-step recycling membrane reactor combined with an ultrafiltration membrane, as shown in Figure 1 [18].

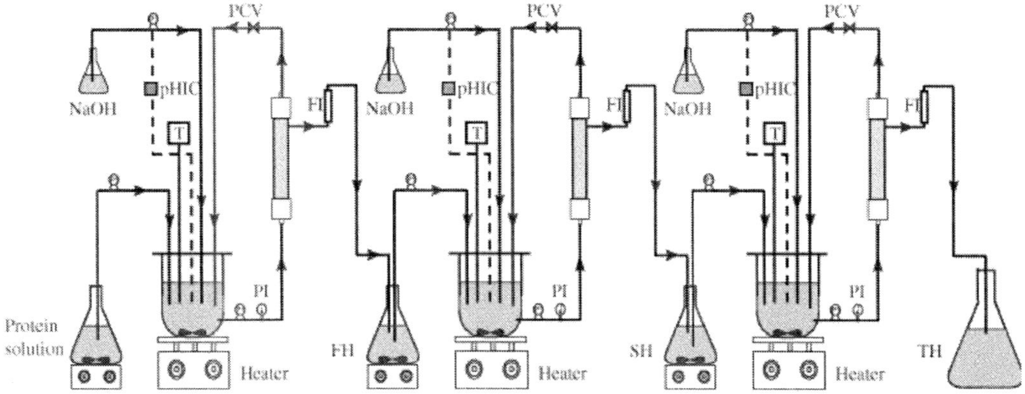

Figure 1. Schematic diagram of the three-step recycling membrane reactor for production and separation of enzymatic hydrolysates. TI, temperature indicator; PI, pressure indicator; FI, flow indicator; pHIC, pH indicator controller; FH, first hydrolysates; SH, second hydrolysates; TH, third hydrolysate.

Peptides isolated from various fish protein hydrolysates were found to exhibit a number of biological activities such as antioxidative, antihypertensive, antithrombotic and immunomodulatory activities.

2.1.1. Antioxidants

Peptides derived from fish proteins have shown potent antioxidative activities in different oxidative systems [19-21].

Protein hydrolysates from muscle proteins derived from fish processing wastes have been analyzed for their antioxidant activity. Peptides Leu-Pro-His-Ser-Gly-Tyr (672 Da) from the Alaska pollack frame protein [10] and Glu-Ser-Thr-Val-Pro-Glu-Arg-Thr-His-Pro-Ala-Cys-Pro-Asp-Phe-Asn (1801 Da) from hoki frame protein [7] have been isolated and were found to show an inhibitory effect on lipid oxidation similar to that of α-tocopherol. Furthermore, these peptides were also observed to exhibit radical scavenging activity.

2.1.2. ACE Inhibitors

The peptides derived from fish muscle also exhibit antihypertensive activity. The peptides Trp-Pro-Glu-Ala-Ala-Glu-Leu-Met-Met-Glu-Val-Asp-Pro (1.581 kDa) from the pepsin hydrolysate of *Thunnus obesus* dark muscle [22], Met-Ile-Phe-Pro-Gly-Ala-Gly-Gly-Pro-Glu-Leu (1.3 kDa) from the frame protein hydrolysates of yellow fin sole *Limanda aspera* [23], and Phe-Gly-Ala-Ser-Thr-Arg-Gly-Ala from the frame protein hydrolysates of Alaska Pollack *Theragra chalcogramma* [24] have been isolated and were found to act as non-competitive inhibitors against ACE I with IC_{50} values of 0.95 μM, 21.6 μM, 28.7 μM and

14.7 μM, respectively. Furthermore, these peptides were also found to exhibit in vitro activities by lowering blood pressure in spontaneously hypertensive rats.

2.1.3. Anticoagulants

Fish proteins enzymatically extracted from the marine fishs were found to show anticoagulant activity. The enzymatically hydrolyzed proteins from the muscle of the yellow fin sole [25] and from the muscle of granulated ark *Tegillarca granosa* [26] were found to show anticoagulant and antiplatelet properties in vitro. These results suggested the capability of fish peptides to inhibit coagulation factors in the intrinsic pathway.

2.1.4. Calcium Absorption Enhancers

Fish protein hydrolysates have been identified to possess hormone-like peptides and growth factors to accelerate calcium absorption [27]. Methods are developed to increase the solubility of calcium compounds, which can significantly improve the calcium availability in biological systems. The fish peptides capable of accelerating calcium absorption can decrease the number of osteoclasts by binding to cell surface receptors in the calcium metabolism and therefore could be used in the treatment of osteoporosis and Paget's disease [28, 29].

2.1.5. Immunostimulants

The acidic peptide fractions from Atlantic cod hydrolysate have shown strong immunostimulatory effects and these peptides have stimulated the oxidative burst of Atlantic salmon leucocytes [30]. Basically immunomodulators enhance the production of oxygen metabolites in macrophages that are responsible in determining the oxidative burst. Oxidative burst reactions are of major importance for the bactericidal power of phagocytes.

3. Bioactive Substances Derived from Fish Skin

3.1. Proteins (Collagen and Gelatin)

Fish skin wastes are potential source to isolate collagen and gelatin. Collagen is structurally formed as a triple helix by three extended protein chains that wrap around one another. Collagen and gelatin are different forms of the same macromolecule and gelatin is the partially hydrolyzed form of collagen. Heat denaturation converts collagen into gelatin. Collagen and gelatin are unique proteins compared to fish muscle proteins. This uniqueness is due to the rich non-polar amino acids (above 80%) such as Gly, Ala, Val and Pro. Collagen and gelatin are currently used in diverse fields including food, cosmetic, and biomedical industries. The main industrial sources of collagen and gelatin are bovine and porcine skin, but there are many reports on the extraction of collagen and gelatin from fish skin [31, 32], projecting it as an alternative source.

Collagen is generally extracted by acid treatment and solubilized without altering its triple helix. However, thermal treatment of collagen transforms its helical conformation into coiled conformation resulting in the formation of gelatin, by the cleavage of hydrogen and covalent bonds that stabilize the triple helix configuration in collagen [33]. Therefore hot water treatment is used to solubilize collagen in skin and extracted as gelatin. The properties of extracted gelatin may vary with the extraction conditions such as temperature and pH. Optimum conditions for the extraction depend on the fish species and the type of collagen [32].

3.1.1. Biomedical and Nutraceutical Applications

Collagen can be processed into a number of forms such as sheets, tubes, sponges, powders, fleeces, injectable solutions and dispersions, all of them have found use in medical practice 155. The aptness of collagen as a biomaterial is mainly because its nature and low immunogenicity and is therefore treated by the body as a normal constituent rather than a foreign matter [34]. Furthermore, attempts have been made to apply these systems for drug delivery in a variety of applications such as ophthalmology, wound and burn dressing, tumor treatment and tissue engineering [35,36]. The retention of drug for a long time and controlled release at the target sites promote the utilization of collagen as efficient drug carriers.

Collagen and gelatine have many biomedical applications. Microfibrous collagen sheets are used as promising drug carrier for the treatment of cancer [37] and collagen film/matrix are used as gene delivery agents promoting bone and cartilage formation [38]. Collagen has also been reported to play a role in formation of tissues, organs and functional expression of cells. Clinical investigations suggest that ingestion of collagen/gelatin hydrolysates reduces pain in patients suffering from osteoarthritis and the hydrolyzed collagen have been involved in cartilage matrix synthesis [39]. Moreover, collagen/gelatin has been used in the treatment for brittle nails and for the nourishment of scalp hair and currently is marketed as a supplement for the maintenance of normal bone integrity.

3.2. Peptides

Enzymatically hydrolyzed peptides from the fish skin were found to be better antioxidants and antihypertensive agents compared to the peptides derived from fish muscle protein [12, 18, 40]. This has been attributed to the unique Gly–Pro– amino acid sequence in the peptides derived from the fish skin.

3.2.1. Antioxidants

The peptides isolated from the skin of various fishes have been tested for their antioxidant activity in number of studies. Two peptides with strong antioxidative activity were isolated from Alaska Pollack skin with 13 and 16 amino acid residues respectively and both peptides contained a glycine residue at the C-terminus and the repeating motif of Gly-Pro-Hyp. Among these, the peptide with 16 amino acid with the sequence of Gly-Glu-Hyp-Gly-Pro-Hyp-Gly-Pro-Hyp-Gly-Pro-Hyp-Gly-Pro-Hyp-Gly was more effective in inhibiting the formation of TBARS in linolenic acid compared to the other peptide with 13 amino acid with the sequence of Gly-Pro-Hyp-Gly-Pro-Hyp-Gly-Pro-Hyp-Gly-Pro-Hyp-Gly [12].

Peptides Leu-Glu-Glu-Leu-Glu-Glu-Glu-Leu-Glu-Gly-Cys-Glu (1487 Da) from the skin protein of bullfrog *Rana catesbeiana Shaw* [6], His-Gly-Pro-Leu-Gly-Pro-Leu (797 Da) from tryptic hydrolysate of the Hoki *Johnius belengerii* skin gelatine [41], Phe–Asp–Ser–Gly–Pro–Ala–Gly–Val–Leu (880.18 Da) and Asn–Gly–Pro–Leu–Gln–Ala–Gly–Gln–Pro–Gly–Glu–Arg (1241.59 Da) from tryptic hydrolysate of jumbo squid *Dosidicus gigas* skin gelatine[42] have been isolated and were found to exhibit strong inhibition against lipid peroxidation much higher than that of natural antioxidant, α-tocopherol. These peptides were also found to scavenge DPPH, hydroxyl, superoxide and peroxyl radicals.

3.2.2. ACE Inhibitors

Peptides derived from fish skin were also found to show (ACE) inhibitory activity. Two peptides composed of Gly-Pro-Leu and Gly-Pro-Met with high ACE inhibitory activity have been isolated from the gelatin extract of Alaska Pollack *Theragra chalcogramma* skin. These peptides composed of Gly-Pro-Leu and Gly-Pro-Met were found to have IC_{50} values of 2.6 μM and 17.13 μM respectively [18].

3.3. Carbohydrates

The mucus coat of fish skin contains a variety of secretions from epidermal goblet cells and epithelial cells. These secretions are made of carbohydrates and have many important biological functions [43].

3.3.1. Anti Coagulants

Glycosaminoglycans have been reprted to possess anticoagulant activity. Heparin a highly-sulfated glycosaminoglycan with anticoagulant activity has been isolated from the tuna fish skin [44]. Furthermore a major glycosaminoglycan, dermatan sulphate with low anti-IIa activity has also been isolated from the skin of eel *Anguilla japonica* by actinase and endonuclease digestions, followed by β-elimination reaction and purification by DEAE-Sephacel chromatography [45].

4. Bioactive Substances Derived from Fish Bone

Fish bones represent a significant part of the cut offs from the fish processing industry and a better utilization of these materials for various applications is of great scientific interest. Fish bone consists of 60–70% of inorganic substances such as calcium phosphate and hydroxyapatite and 30% of organic component mainly collagen [46]. Therefore, fish bone is considered as a source to isolate collagen and gelatin in addition to fish skin [47]. Bones from whole fish and cut offs could be used as a raw material for health products and as ingredient in feeds for aquaculture [48].

4.1. Bioactive Calcium

Fish bone is considered as a potential source for calcium, which is an essential element for numerous functions in our body like strengthening of teeth and bones, nerve functions and enzymatic reactions. However, very few study have been carried out to identify bioavailability of calcium from fish bone and its potential applications. Pepsin hydrolysis has been found to be an effective method for the recovery of the bioavailable calcium from Alaska pollack fish backbones [49]. Generally, calcium is obtained from regular diet but it is severely deficient in most of the regular diets. Currently, there are several calcium-fortified products to improve calcium intake but it doesn't match the demand and hence there is always a lookout for a new source. Consumption of whole small fish is nutritionally beneficial providing a good amount of calcium. Calcium in fish could be absorbed by the body as tested *in vivo* [50]. However, there is no information available on the beneficial effects of the bones from the larger fishes. Few attempts have been made to find the benefits of these on human health. The bone material derived from processing of large fish is a useful source for calcium. Fish bone should be converted into soft and edible form by softening its structure in order to be incorporated into calcium-fortified food. There are different methods for converting the calcium obtained from the fish bones into edible forms. Some of them are hot water treatment, hot acetic acid treatment and superheated steam treatment [51]. The superheated steam reduces the loss of soluble components from fish tissue and that enabled better recovery of calcium within a short period.

4.2. Biomedical Applications of the Inorganic Compounds

Fish bone material has many important biomedical applications due to the presence of hydroxyapatite [$Ca_{10}(PO_4)_6(OH)_2$]. It is an inorganic material derived from coral and fish bone [52]. High strength hydroxyapatite has been isolated from the bones of the fish processing wastes by heat treatment. The heating of the bones to a temperature of 1300 °C gives a higher strength to hydroxyapatite structure [53, 54] and yield an excellent biocompatible inorganic substance [55-57]. The nanocrystalline calcium hydroxyapatite, from fish bone wastes can also be extractd by a hydrothermal route [58]. Unlike other calcium phosphates, hydroxyapatite does not break under physiological conditions. It is also thermodynamically stable at physiological pH and actively takes part in bone bonding. The property to support bone ingrowth and osseointegration has made it as a bone graft material in orthopaedic, dental and maxillofacial applications. Recently, there is a growing need for bone regeneration due to various clinical bone diseases such as bone infections, bone tumors and bone loss by trauma [59].

5. Bioactive Substances from the Internal Organs of Fish

5.1. Enzymes

Enzymes are biological catalysts which can be used for enhancing the speed of the reaction. The internal organs of fish are rich in enzymes. There is great potential for the recovery and use of digestive proteases from fish processing wastes. The most commonly studied fish digestive proteolytic enzymes are pepsin, trypsin, chymotrypsin, gastricsin and elastase. Proteases represent an important class of industrial enzymes, accounting for about 50% of the total sale of enzymes in the world. There are many studies on new sources of proteolytic enzymes, and proteases from fish processing wastes. Enzymes from cold adapted ectothermic organisms differ from homologous proteases of warm-blooded animals because of their living temperature and other characteristics and hence have many applications in the food industry [60, 61]

Currently fish processing byproducts are used for enzyme extraction because of the specific characteristics of these enzymes. A range of proteolytic enzymes including pepsin, trypsin, chymotrypsin and collagenases have been commercially extracted from the viscera of marine fish in a large scale [62-64]. Two trypsin isoforms (GT-A and GT-B) with intermediate thermal stablities have also been isolated and purified from the intestine of grass carp *Ctenopharyngodon idellus* [65]. The partial purification of trypsin-like enzyme from the intestine and pyloric caeca of spotted goatfish (*Pseudupeneus maculatus*) by a simple three steps procedure (heat treatment, ammonium sulphate precipitation and Sephadex G-75 filtration) has provided the basis for the proposition to prepare protease from the tons of intestine and pyloric caeca, that are usually discarded as waste [66].

Fish proteinases possess better properties such as higher catalytic efficiency at low temperatures, lower sensitivity to substrate concentrations and greater stability at broader pH range. Therefore these enzymes find varied applications in many food-processing operations [67].

Even though enzymes derived from marine fishes do not have direct applications in the field of functional foods or nutraceuticals they can be used to produce bioactive components in a large scale. Recently, several attempts have been made to extract crude enzyme mixtures from internal organs of some fish species and to utilize them in isolation of bioactive components from fish protein hydrolysates. It has also been reported that chitinases as well as chitosanases can be isolated from digestive tract and other organs of some marine fish species. These enzymes promote the recovery of chitin and chitosan from marine by-products necessary for a wide array of biomedical applications. Furthermore, these enzymes would be more economical in the process of isolating bioactive compounds from marine shell wastes [5].

5.2. Peptides

The hydrolysate prepared from visceral waste has a balanced amino acid composition and hence has the potential to be an ingredient in balanced fish diets [68], in aquaculture/animal feeds and effective nitrogenous source in microbiological growth media [69]. Certain peptides derived from cod processing wastes have been reported to show stronger antihypertensive activity inhibiting the action of angiotensin I converting enzyme (ACE) than that of many other natural peptides [70].

6. Shells of Crustacean and Shellfish

Shells of crustacean and shellfish are another important category of wastes from fish processing industry. The increased accumulation of these wastes from fish processing industries and the slow natural degradation of these materials have created major environmental problems and the efficient utilization of these shells has become an environmental priority. However, these shells are the potential source for the isolation of chitin.

6.1. Chitin, Chitosan and Their Oligomers

The shells of crustacean and shellfish are used for the commercial production of chitin, chitosan and their oligomers (Fig. 2). Chitin is one of the major structural components of these shells and has been identified as a biologically active polysaccharide made up of N-acetyl-D-glucosamine residue connected by β-(1-4) linkage. It has many valuable applications. It is the second abundant biopolymer in nature next only to cellulose and differs structurally from cellulose by the substitution of acetamide groups at the C-2 position in place of hydroxyl groups [71, 72]. Chitin can easily be converted into many other bioactive derivatives and among them chitosan is the most common form. Chitosan is a partially deacetylated chitin molecule and is a mixture of both D-glucosamine and N-acetyl-D-glucosamine monomers connected by β-(1-4) linkages. The degrees of N-acetylation of these large, cationic copolymers range from 5 to 30% [73, 74]. Generally, chitin and chitosan oligomers are produced by chemical or microbiological treatments. However, thermochemical treatments like deproteination, demineralization and deacetylation of the starting materials have been employed in the commercial preparations of chitin and chitosan [75]. The molecular weight and the degree of deacetylation of chitosan depend greatly on the conditions employed during the production process.

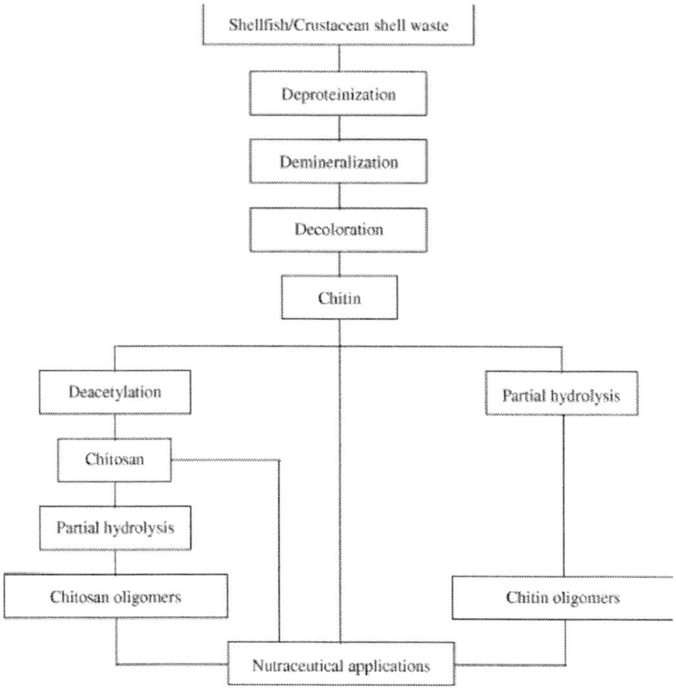

Figure 2. Schematic diagram for preparation of chitin, chitosan and their oligomers.

Recently much attention has been paid to chitin, chitosan and their oligomers as natural bioactive materials due to their non-toxic, biocompatible and biodegradable nature [76]. These materials prove to be very attractive with many applications in diversified fields due to their structural and functional properties. Among these natural bioactive materials chitosan was found to have many applications in biomedical and food industries [77]. Chitosan, the N-deacetylated derivative of chitin, has drawn increasing attention as a drug, gene and drug carrier. The reactive amino groups in the backbone of chitosan have made it possible to chemically conjugate with various biological molecules such as ligands and antibodies. This property has improved the targeting efficiency of the drug to the site of action [78]. However, the high viscosity and low solubility of these materials at neutral pH limit the applications of these materials in functional food and pharmaceuticals. This has created a growing need for the conversion of chitin and chitosan into their oligomers with better functional properties and improved absorption in the human digestive track. Production of chitin and chitosan oligomers can be achieved either by chemical or enzymatic hydrolysis. The enzymatic hydrolysis is more preferred to chemical hydrolysis as it yields higher amount of oligomers with higher degree of polymerization [79, 80].

Some of the practical applications of these materials include cationic agents for polluted waste-water treatment, agricultural materials, food and feed additives, hypocholesterolemic agents, biomedical and pharmaceutical materials, wound-healing materials, blood anticoagulant, antithrombogenic and hemostatic materials, cosmetic ingredients, textile, paper, film and sponge sheet materials, chromatographic and immobilizing media, and analytical reagents [75]. In addition to their array of direct biomedical applications chitin and their derivatives have also been reported to play a role in drug delivery systems to obtain the controlled release [81, 82].

6.1.1. Antioxidants

Chitosan and its oligomers are very good antioxidants. The antioxidant activity of these materials is by the scavenging of oxygen radicals and it depends on their molecular weights as well as the degree of deacetylation [83]. The lower molecular weight chitosan oligomers were found to show better activity than the higher molecular weight oligomers. Highly deacetylated (90%) chitosan oligomers have been reported to exhibit radical scavenging activity in case of hydroxyl, superoxide, alkyl and highly stable DPPH radicals when tested *in vitro* [84, 85]. Chitosan and their derivatives have also been reported as hydrogen donors to prevent the oxidative sequence [86]. However no detailed mechanistic study on the antioxidant and radical scavenging activities of chitin and chitosan derivatives has been reported.

6.1.2. ACE Inhibitors

Chitin and their derivatives have also been reported as ACE inhibitors [87, 88]. The natural substrates of ACE are peptides and the mechanism of inhibition of ACE by chitosan oligomers is not well understood but the structural modification of chitooligosaccharides (COS) was found to promote the action as ACE inhibitors. Carboxylation of COS with $-COCH_2CH_2COO^-$ groups gave structural features similar to Captopril ® and their action as ACE inhibitor was found to increase with increase in the degree of substitution [89]. The chemical modification of COS with a molecular weight of 800–3000 Da (90% deacetylation), by grafting with 2-chloroethylamino hydrochloride at C-6 position was found to promote the ACE inhibitory activity of COS [90].

6.1.3. Immunostimulants

Carbohydrates and their derivatives were found to possess immunostimulting properties [91]. The non-specific immune system in animals is responsible for initiating rapid and general responses against invasion of microorganisms. Immunostimulants are generally identified as compounds that stimulate non-specific immune system by enhancing the capability of defending activity of phagocytes (macrophages and neutrophils). Most immunostimulants can bind specifically with the cell surface receptor proteins of phagocytes or lymphocytes, and it stimulates the production of immune response compounds such as interferon, interleukin and complement proteins that activate the immune system. In addition, some immunostimulants are capable of competing with specific receptor molecules in the target cells of infectious organisms [92].

Oligomers of chitin and chitosan are effective in enhancing migratory activity of macrophages. Moreover, this activity can be explained by their chemotatic effects on macrophages[93] but chitin and chitosan are reported to reduce the migration of macrophages and it suggested that chitin and chitosan might absorb some substances in culture medium that are involved in migration of macrophages [94]. Further, water-soluble chitosan can activate the production of cytokines such as interleukin-1 (IL-1β), tumor necrosis factor-alpha (TNF-α), and reactive oxygen intermediates to promote the defense system against microbial infections [95, 96].

6.1.4. Hypocholesterolemic Agents

The biological activities of chitosan and its oligomers are mainly attributed to the positively charged nature of these compounds. Chitosan and its oligomers have been effective in reducing LDL-cholesterol level in liver and blood [97]. Even though the exact mechanism has not been clearly explained, it has been reported that these molecules remove fat and cholesterol via excretion by their action as fat scavengers in the digestive tract [98]. It has also been reported that almost all the oligomers of chitosan with varied molecular weights could prevent the increase in blood cholesterol when tested with 5% dietary level [99], but there was an argument stating that the relatively low-molecular weight chitosans were more beneficial in fat lowering and helpful for the prevention of cardiovascular diseases[100]. Furthermore, it has also been reported that chitosan hydrolysate having a molecular weight of 10 kDa could significantly enhance fecal excretion of neutral steroids [98].

6.1.5. Antitumor Agents

In addition to the bioactivities discussed above, chitin and its derivatives have been reported to possess antitumor activities when tested both in vitro and in vivo [11]. Among them, *N*-acetyl-D-glucosamine $(GlcNAc)_6$ isolated from the crude enzyme hydrolysates of chitin, colloidal chitin and water-soluble chitosan have been found to show higher antitumor activity. The hydrolysates of water-soluble chitosan have been more effective and were found to inhibit the growth of CT26 cells in both concentration and time dependent manner [101]. Furthermore, the water-soluble chitosan can activate the production of cytokines such as interleukin-1 (IL-1β), tumor necrosis factor-alpha (TNF-α), and reactive oxygen intermediates to promote the defense system against microbial infections [95, 96]. Studies on mice have revealed that the low-molecular weight chitosan has significant antimetastatic effects against

Lewis lung carcinoma. Chitin with carboxymethyl groups as well as partially deacetylated chitin has also been effective to demote tumor progression [102]. Immunostimulating effects of chitin and its carboxymethyl derivatives via stimulation of cytolytic T-lymphocytes were suggested as the possible mechanism. The activity in case of smaller molecules was found to be due to the immunostimulating effect that activate peritoneal macrophages and stimulate non-specific host resistance but in case of higher molecular weight oligomers the activity was due to immunostimulating effect via increased production of lymphokines by activated lymphocytes [103].

6.1.6. Antimicrobial Agents

Chitosan and its derivatives exhibit antimicrobial activity against several bacterial species and is considered as one of the most important properties linked directly to their possible biological applications. The antibacterial activity of these compounds were found to be influenced by a number of factors such as degree of polymerization[104-106], level of deacetylation [107,108], type of microorganism [109,104,80] and some other physico-chemical properties [92].

In addition to antibacterial activity chitosan and its derivatives exhibit antifungal activity on mold and yeast due to their polycationic nature. The inhibition mechanism of COS against fungi was found to be similar to that as against bacteria. The formation of polyelectrolyte complexes between COS and negatively charged groups on the cell surface directly interferes the growth and normal physiological functions of fungi, suggesting that the charge distribution of COS has a correlation with its antifungal activity [110]. Moreover the antifungal activity of these compounds appeared to be more pronounced at lower pH and lower temperatures.

It is interesting to note that these compounds can also exhibit antiviral activity in addition to other diversified applications. The mechanism of antiviral activity of these compounds has not been well established. There are also conflicting observations with regard to the inhibition mechanism of antiviral activity with respect to chitosan and COS from different studies. However, one possible explanation is that, cationic charges of amino groups of chitosan and COS may have additional functions to activate the immune and defense systems in plants and animals [92].

6.1.7. β-secretase Inhibitors

Beta-secretase enzyme plays a major role in the occurrence of Alzheimer's disease by the progressive accumulation of β-amyloid peptides in the brain forming fibrillary tangles and insoluble plaques resulting in severe memory loss and neuronal cell death. Chitosan oligosaccharides are the first reported carbohydrate β-secretase inhibitors but their mechanism of action has not been explained [5]. Chitosan and their oligomers are capable of inhibiting β-secretase activity which depends on their molecular weights and degree of deacetylation. The highest inhibition has been exhibited in the case of chitosan oligosaccharides having molecular weights below 5 kDa and 90% deacetylation. Moreover, it has been reported that the chitosan oligomers exhibit β-secretase activity in a non-competitive manner [111].

6.1.8. Wound-Healing Agents

Rapid healing of wounds is desirable for patients, especially those suffering from diabetes as they show an extremely slow rate of healing. The role of chitin and chitosan as possible wound-healing accelerators has been investigated for a long time. These materials can be applied safely to animals, as well as to humans. Various forms of chitin-based products are available for medical applications, including finely divided powder, nonwoven fabrics, porous beads, lyophilized soft fleeces or gels, gauges, laminated sheets, and transparent films. Water-soluble derivatives of chitin in isotonic saline can be administered intramuscularly or intravenously [112].

The wound-healing activity of chitin and chitosan was found to be associated with their immunostimulating property, which involves higher production of macrophages that release cytokines necessary for the healing process [93]. It was also reported that wound-healing property of chitosan oligomers is due to their ability to stimulate fibroblast production by affecting the fibroblast growth factor. Subsequent collagen production further facilitates the formation of connective tissues [113]. Furthermore the wound break strength and the collagenase activity of the chitosan group (D-glucosamine (GlcN), chito-oligosaccharide (COS), chitosan) was found to be higher than the chitin group (*N*-acetyl-D-glucosamine (GlcNAc), chiti-oligosaccharide (NACOS), chitin) and stronger the break strength with higher degree of deacetylation [114].

6.1.9. Artificial Skin

Chitosan could also be considered for developing a skin replacement due to its structural characteristics similar to that of glycosaminoglycans. The design for artificial skin applicable to long-term chronic use was proposed with the focus on a nonantigenic membrane that performs as a biodegradable template for neodermal tissue [115]. A commercial product composed of chitin nonwoven fabric Beschitin W has been developed as a new wound dressing material and it has been found to be beneficial in clinical practice [116]. Polyelectrolyte complexes of chitosan and sulfonated chitosan have also been developed in to a wound covering material. In addition to acceleration of wound healing the oligomers of chitosan have also been found to be effective in regenerating the skin tissue of the wound area. Moreover, an artifical skin (a chitosan-collagen composite) to enhance the recovery from surgical wounds or burns has also been developed by the pharmaceutical company Katakurachikarin, based in Hokkaido, Japan [112].

6.2. Proteins

The rapid expansion of fisheries is demanding an adequate supply of efficient, nutritious and inexpensive fish feed, because feed contributes highly to the cost of fish production. Shrimp head, a waste product from the shrimp export industry qualifies as an economical, abundant and good quality protein source for fish feeds [117].

7. Bioactive Substances Derived from Other Fish Processing Wastes

7.1. Scales

Fish scales are composed of connective tissue protein, collagen, covered with calcium salts. The amount of protein ranges from 41 to 84% and the remaining is calcium phosphate and calcium carbonate [118]. Fish scales are potential source for the extraction of collagen even though their bioactivities and functional properties have not been established [46]. The composition, structure, and properties of fish scales have been analyzed to find their optimum utilization [119,120]. A simple and inexpensive method to extract DNA from fresh and preserved fish scales has been developed. Boiling the scales in 5% Chelex 100, followed by digestion with proteinase K yielded genomic DNA that has been subsequently isolated by absorption on silica. This technique will allow analysis of genetic diversity, mating systems, and parentage in populations of endangered or ornamental fish with minimal experimental influence [121].

7.2. Eggs

Fish eggs resulted from fish processing can be easily processed in to caviar or fish bait. Fish eggs are rich sources of lectins which are naturally occurring glycoproteins that can bind with carbohydrates to form stable complexes. It has the potential to be an alternative for antibiotics by making lectin-pathogen complexes due to its ability to bind with carbohydrates. Recently it has been identified that lectins from the fish eggs has a great potential to be developed into nutraceuticals [5,122,123].

7.3. Fins

Collagen prepared from fish fin, indicate that these fish waste materials have potential in supplementing the skin of land vertebrates as a source of collagen [47]. It has also been reported that gelatin was extracted from the precooked tuna caudal fin [124].

7.4. Offal

Fishing is not encouraged for extracting their oil due to the depleting resources of marine fisheries. Therefore, the large amount of offal generated from fish processing, would be a potential source to produce good quality fish oil for human consumption. The composition of fish oil is different from other oils and mainly composed of two types of fatty acids, eicosapentaenoic acid (EPA) and docosahexaenoic acid (DHA). These polyunsaturated fatty acids are classified as omega-3 fatty acids and are predominantly found in many marine animals including cold-water fish species with a higher unsaturated fat content. Compared to saturated fats, polyunsaturated fatty acids in fish oil are readily digested for energy production and have been reported to have various bioactivities [5].

The quality of the fish oil mainly depends on the methods and conditions of extraction. Currently, several methods are employed to extract fish oil. Some of the methods are high-speed centrifugation [125], low temperature solvent extraction [126], supercritical fluid extraction [127] and wet and steam rendering methods [128]. Development of proper separation methods and advanced processing technologies could promote the production of higher quality fish oil from the fish processing wastes. Among the different constituents of fish oil omega-3 fatty acids is of major interest because of its various functions in health promotion. Fish oil because of omega-3 fatty acids may play an important role in the pharmaceutical and food industry in the near future.

7.4.1. Fish Oils and Their Health Benefits

Fish oil is an excellent source of omega-3 fatty acids has many health benefits with the ability to fight against numerous diseases. It has been found that eicosapentaenoic acid (EPA) and docosahexaenoic acid (DHA) are the main protective components of fish oil that act against different types of diseases. Antiathrogenic and antithrombotic effects of omega-3 fatty acids inhibit the growth of lipid rich atherosclerotic plaques and decrease the risk of thrombosis to stop the formation of thrombus [129]. It has been found that fish oil also exert their protective effect against heart diseases by decreasing serum triglyceride levels, improving vascular endothelial function, lowering blood pressure, and decreasing inflammation [130]. Furthermore, it has been found that dietary $n-3$ fatty acids act to prevent arrhythmias that can lead to sudden cardiac death [131].

The beneficial effects of fish oil also include its effects against diabetes mellitus. Fish oil was found to improve many of the metabolic sequelae of insulin resistance by lowering blood pressure and triacylglycerol concentrations [132]. Moreover, the dietary intake of fish oil has been reported to accelerate glucose uptake and maintain normal glucose metabolism [133].

The omega-3 fatty acids were also found to exhibit anti-inflammatory action. Clinical studies have revealed that polyunsaturated fish oil treatments relieve patients from rheumatoid arthritis and this was found to be due to the ability of EPA and DHA to be converted into natural anti-inflammatory substances called prostaglandins and leukotrienes that help to decrease inflammation and pain [134]. Moreover, it has also been found that fish oils were used for the treatment of other inflammatory diseases such as Crohn's disease [135] and kidney diseases [136] due to their anti-inflammatory character.

Omega-3 polyunsaturated fatty acids were also found to have a significant immunomodulating effect and to prolong the survival of malnourished patients with generalized malignancy against various human carcinomas that include breast, colon, skin, pancreatic, prostate, lung and larynx cancer [137].

8. Conclusion

Fish processing wastes are one of the major contaminations that have caused environmental concern in recent times. Even though fishery wastes are mixtures of many biologically active components, the extraction and purification has not been focused, as it involves modern technologies and manpower with additional cost. Recently, the development of technologies for the extraction and purification of bioactive substances from the fish processing wastes and their biotechnological and biomedical applications have attracted the attention of industries. Until now, only a limited number of bioactive substances have been isolated from fish processing wastes, and further research is required to explore new and bioactive compounds with diverse applications.

Acknowledgements

This research was supported by a grant from the marine bioprocess research center of the marine bio 21 center funded by the Ministry of Land, Transport and Maritime, Republic of Korea.

References

[1] Kristinsson, H.G. & Rasco, B. A. (2000). Fish protein hydrolysates: production, biochemical, and functional properties. *Crit. Rev. Food Sci. Nutr,* **40** , 43–81.

[2] Choudhury, G. S. Bublitz, C. G. (1996) Computer-based controls in fish processing industry. In: G.S. Mittal, Editor, *Computerized control systems in the food industry*, Marcel Dekker Inc., New York, 513–538.

[3] Choudhury, G.S. & Gogoi, B.K. (1995). Extrusion processing of fish muscle. *Journal of Aquatic Food Product Technology*, **4**, 37–67.

[4] Martone, C.B. Borla, O.P.& Sánchez, J.J. (2005). Fishery by-product as a nutrient source for bacteria and archaea growth media. *Bioresource Technology*, **96**,383-387.

[5] Kim, S.K. & Mendis, E. (2006). Bioactive compounds from marine processing by-products - a review, *Food Research International,* **39,** 383–393.

[6] Qian, Z.J. Jung, W.K. &Kim, S.K. (2008). Free radical scavenging activity of a novel antioxidative peptide purified from hydrolysate of bullfrog skin, *Rana catesbeiana Shaw. Bioresource Technology*, **99**, 1690-1698.

[7] Kim, S.Y. Je, J.Y. & Kim, S.K. (2007). Purification and characterization of antioxidant peptide from hoki (*Johnius belengerii*) frame protein by gastrointestinal digestion. *The Journal of Nutritional Biochemistry*, **18**, 31-38.

[8] Qian, Z.J. Jung, W.K. Byun, H.G. & Kim, S.K. (2007). Protective effect of an antioxidative peptide purified from gastrointestinal digests of oyster, *Crassostrea gigas* against free radical induced DNA damage. *Bioresource Technology*, **99**, 3365-337

[9] Je, J.Y. Qian, Z.J. & Kim, S.K. (2007). Antioxidant Peptide Isolated from Muscle Protein of Bullfrog, *Rana catesbeiana* Shaw. *Journal of Medicinal Food*, **10**, 401-407.

[10] Je, J.Y. Park, P.J. & Kim, S.K. (2005). Antioxidant activity of a peptide isolated from Alaska pollack (*Theragra chalcogramma*) frame protein hydrolysate. *Food Research International*, **38**, 45-50.

[11] Jeon, Y. J. & Kim, S. K. (2002). Antitumor activity of chitosan oligosaccharides produced in an ultrafiltration membrane reactor system, *J. Microbiol. Biotechnol.*, **12**, 503-507

[12] Kim, S. K. Kim, Y. T. Byun, H. G. Nam, K. S. Joo, D. S. & Shahidi, F. (2001). Isolation and characterization of antioxidative peptides from gelatin hydrolysate of alaska pollack skin. *J. Agric. Food Chem*, **49**, 1984-1989.

[13] Defelice, S. L. (1995). The nutritional revolution: Its impact on food industry R&D. *Trends in Food Science and Technology,* **6**, 59–61.

[14] Venugopal, V. Chawla, S.P. & Nair, P.M. (1996). Spray dried protein powder from threadfin beam: Preparation, properties and comparison with FPC type B, *Journal of Muscle Foods,* **7**, 55–58

[15] Benkajul, S. & Morrissey, M.T. (1997). Protein hydrolysates from Pacific whiting solid wastes. *Journal of Agricultural and Food Chemistry,***45**, pp. 3423–3430.

[16] Jeon, Y.J. Byun H.G. & Kim, S.K. (2000). Improvement of functional properties of cod frame protein hydrolysates using ultrafiltration membranes. *Process Biochemistry,* **35**, 471–478.

[17] Simpson, B.K. Nayeri, G. Yaylayan, V. & Ashie, I.N.A. (1998). Enzymatic hydrolysis of shrimp meat. *Food Chemistry* , **61**, 131–138.

[18] Byun, H.G. & Kim, S.K. (2001). Purification and characterization of angiotensin I converting enzyme (ACE) inhibitory peptides from Allaska Pollack *Theragra chalcogramma* skin. *Process Biochemistry,* **36**, 1155–1162.

[19] Jun, S.H. Park, P.J. Jung, W.K. & Kim, S.K. (2004). Purification and characterization of an antioxidative peptide from enzymatic hydrolysate of yellowfin sole *Limanda aspera* frame protein. *European Food Research Technology,* **219**, 20–26.

[20] Kim, S. K. Choi, Y. R. Park, P. J. Choi, J. H. & Moon, S. H. (2000). Purification and characterization of antioxidative peptides from enzymatic hydrolysate of cod teiset protein, *Korean Fisher. Sci.,* **33**, 198- 204.

[21] Rajapakse, N. Mendis, E. Byun, H.G. & Kim, S.K. (2005). Purification and in vitro antioxidative effects of giant squid muscle peptides on free radical-mediated oxidative systems. *Journal of Nutritional Biochemistry,* **16**, 562–569.

[22] Qian,Z.J. Je,J.Y. & Kim.S.K. (2007). Antihypertensive Effect of Angiotensin 1 Converting Enzyme-Inhibitory Peptide from Hydrolysates of Bigeye Tuna Dark Muscle, Thunnys obesus. *Journal of Agricultural and Food Chemistry,* **55**, 8398-8403.

[23] Jung,W.K. Mendis, E. Je, J.Y. Park,P.J. Son, B.W. Kim,H.C. Choi,Y.K. & Kim, S.K.(2005). Angiotensin I-converting enzyme inhibitory peptide from yellowfin sole (*Limanda aspera*) frame protein and its antihypertensive effect in spontaneously hypertensive rats. *Food Chemistry*. **94**, 26-32.

[24] Je, J.Y. Park, P.J. Kwon, J.Y. & Kim, S.K. (2004). A novel angiotensin I converting enzyme (ACE) inhibitory peptide from Alaska pollack *Theragra chalcogramma* frame protein hydrolysate. *J. Agric. Food Chem*., **52**, 7842-7845.

[25] Rajapakse, N. Jung, W.K. Mendis, E. Moon, S.H. & Kim, S.K. (2005). A novel anticoagulant purified from fish protein hydrolysate inhibits factor XIIa and platelet aggregation. *Life Sciences,* **76**, 2607–2619.

[26] Jung, W.K. Jo, H.Y.Quian, Z.J. Jeong, Y.J. Park, S.G. Choi, I.W. & Kim, S.K. (2007). A novel anticoagulant protein with high affinity to blood coagulation factor Va from *Tegillarca granosa. Journal of Biochemistry and Molecular Biology,* **40**, 832-838

[27] Fouchereau-Peron, M. Duvail, L. Michel, C. Gildberg, A. Batista, I. & Gal, Y.I. (1999). Isolation of an acid fraction from a fish protein hydrolysate with a calcitonin–gene-related-peptide-like biological activity. *Biotechnology and Applied Biochemistry,* **29**, 87–92

[28] Kim, S.K. Jeon, Y.J. Byun, H.G. Park, P.J. Kim, G.H. & Choi, Y.R. *et al.* (1999). Calcium absorption acceleration effect on phosphorylated and non-phosphorylated

peptides from hoki *Johnius belengeri* frame. *Journal of the Korean Fisheries Society,* **32**, 713–717.

[29] Jung, W.K. Park, P.J. Byun, H.G. Moon, S.H. & Kim, S.K. (2005) Preparation of hoki *Johnius belengerii* boneoligophosphopeptide with a high affinity to calcium by carnivorous intestine crude proteinase. *Food Chemistry,* **91**, 33–340.

[30] Gildberg, A. Bogwald, J. Johansen, A & Stenberg, E. (1996) Isolation of acid peptide fractions from a fish protein hydrolysate with strong stimulatory effect on Atlantic salmon (*Salmo salar*) head kidney leucocytes. *Comparative Biochemistry and Physiology,* **114**, 97–101.

[31] Gomez-Gillen, M.C. Turnay, J. Fernandez-Diaz, M.D. Ulmo, N. Lizarbe, M.A. & Montero, P. (2002). Structural and physical properties of gelatin extracted from different marine species: A comparative study. *Food Hydrocolloids,* **16**, 25–34.

[32] Kim, S.K. Byun, H.G. & Lee, E.H. (1994). Optimum extraction conditions of gelatin from fish skins and its physical properties. *Journal of Korean Industrial and Engineering Chemistry,* **5**, 547–559.

[33] Djabourov, M. Lechaire, J.P. & Gaill, F. (1993). Structure and rheology of gelatin and collagen gels. *Biorheology,* **30**, 191–205

[34] Piez, K.A. (1984). Molecular and aggregate structures of the collagens, In: K.A. Piez, A.H. Reddi (Eds.), Extracellular Matrix Biochemistry, Elsevier, New York, 1–40.

[35] Friess,W. (1998). Collagen – biomaterial for drug delivery. *European Journal of Pharmaceutics and Biopharmaceutics*, **45**, 113-136.

[36] Lee, H.C. Singla, A. & Lee, Y. (2001). Biomedical applications of collagen. *International Journal of Pharmaceutics,* **221**, 1–22

[37] 36. Sato, H. Kitazawa, H. Adachi, I. & Horikoshi, I. (1996). Microdialysis assessment of microfibrous collagen containing a *p*-glycoprotein-mediated transport inhibitor, cyclosporine A, for local delivery of etopocide. *Pharmacological Research,* **13**, 1565–1569.

[38] Nakagawa, T. & Tagawa, T. (2000). Ultrastructural study of direct bone formation induced by BMPs-collagen complex implanted into an ectopic site. *Oral Diseases,* **6**, 172–179.

[39] Moskowitz, R.W. (2000). Role of collagen hydrolysate in bone and joint disease. *Semin Arthritis Rheum,* **30**, 87–99.

[40] Mendis, E. Rajapakse, N. & Kim, S.K. (2005). Antioxidant properties of a radical-scavenging peptide purified from enzymatically prepared fish skin gelatin hydrolysate. *Journal of Agricultural and Food Chemistry,* **53**, 581–587.

[41] Mendis, E. Rajapakse, N. & Kim, S.K. (2005). Antioxidant properties of a radical scavenging peptide purified from enzymatically prepared fish skin gelatin hydrolysate. *J. Agri. Food Chem.*, **253**, 581-587.

[42] Mendis, E. Rajapakse,N. Byun, H.G. & Kim, S.K. (2005). Investigation of jumbo squid (*Dosidicus gigas*) skin gelatin peptides for their in vitro antioxidant effects. *Life sciences*, **77**, 2166-2178

[43] Kimura, M. Hama, Y. Sumi, T. Asakawa, M. Narasinga Rao, B.N. Horne A.P. Li,S.C.Li,Y.T.& Nakagawa,H. (1994). Characterization of a deaminated neuraminic acid containing glycoprotein from the skin mucus of the Loach, Misgurnus anguillicaudatus. *The Journal of Biological Chemistry,* **269**, 32138-32143.

[44] Jeske,W.P. McDonald, M.K. Hoppensteadt,D.A. Bau, E.C. Mendes, A. Dietrich,C.P. Jeanine, M. Walenga, J.M., Coyne, E. (2007). Isolation and Characterization of Heparin From Tuna Skins. *Clinical and Applied Thrombosis/Hemostasis*, **13**, 137-145

[45] Sakai, S. Kim, W.S. Lee, I.S. Kim, Y.S. Nakamura, Toida, A.T. & Imanari, T. (2003). Purification and characterization of dermatan sulfate from the skin of the eel, *Anguilla japonica. Carbohydrate Research,* **338**, 263-269

[46] Nagai, T. Izumi, M. & Ishii, M. (2004). Fish scale collagen. Preparation and partial characterization. *International Journal of Food Science and Technology*, **39**, 239–244.

[47] Nagai, T. & Suzuki, N. (2000). Isolation of collagen from fish waste material-skin, bone and fins. *Food Chemistry,* **68**, 277–281.

[48] Toppe, J. Aksnes, A. Hope, B. & Albrektsen, S. (2006). Inclusion of fish bone and crab by-products in diets for Atlantic cod, *Gadus morhua. Aquaculture,* **253**, 636–645.

[49] Karawita, R. Heo, S.J. Lee, B.J. Kim, S.K. Song, C.B. & Jeon, Y.J. (2006). Recovery of Bioavailable Calcium from Alaska Pollack *Theragra chalcogramma* Fish Backbone By-products by Pepsinolytic Hydrolysis. *Journal of Food Science and Nutrition*, **11**, 120-126.

[50] Larsen, T. Thilsted, S.H. Konsbak, K. & Hansen, M. (2000). Whole small fish as a rich calcium source. *British Journal of Nutrition,* **83**, 191–196.

[51] Ishikawa, M. Kato, M. Mihori, T. Watanabe, H. & Sakai, Y. (1990). Effect of vapor pressure on the rate of softening of fish bone by super-heated steam cooking. *Nippon Suisan Gakkaishi*, **56**, 1687–1691.

[52] Jensen, S.S. Aaboe, M. Pinhold, E.M. Hansen, Z. H. Melsen, F. & Ruyter, I.E. (1996). Tissue reaction and material characteristics of four bone substitutes. *International Journal of Oral & Maxillofacial Implants,* **11**, 55–66.

[53] Choi, J.S. Lee, C.K. Jeon, Y.J. Byun, H.G. & Kim, S.K. (1999). Properties of the ceramic composites and glass ceramics prepared by using the natural hydroxyapatite derived from tuna bone. *Journal of Korean Industrial and Engineering Chemistry,* **10**, 394–399.

[54] Ozawa, M. & Suguru, S. (2002). Microstructural development of natural hydroxyapatite originated from fish-bone waste through heat treatment. *Journal of the American Ceramists Society,* **85**, 1315–1317.

[55] Kim, S.K. Choi, J.S. Lee, C.K. Byun, H.G. Jeon, Y.J. & Lee, E.H. (1998). Synthesis and biocompatibility of the hydroxyapatite ceramic composites from tuna bone (III)—SEM photographs of bonding properties hydroxyapatite ceramic composites in the simulated body fluid. *Journal of Korean Industrial and Engineering Chemistry,* **9**, 322–329.

[56] Kim, S. K. & Park, P. J. (2000). Evaluation of mucous membrane irritation by hydroxyapatite sinter produced from tuna bone in syrin hamsters. *Korean J. Life Sci.,* **10**, 605-609.

[57] Kim, S.K. Park, P.J. & Kim, Y.T. (2001). Study on acute subcutaneous toxicity of hydroxyapatite sinter produced from tuna bone in Sprague–Dawly rats. *Korean Journal of Life Science,* **11**, 97–102.

[58] Parhi, P. Ramanan, A. & Ray, A. R. (2006). Hydrothermal Synthesis of nanocrystalline powders of alkaline-earth hydroxyapatites, $A_{10}(PO_4)_6(OH)_2$ (A = Ca, Sr and Ba). *Journal of Material Science,* **41**, 1455-1458.

[59] Braddock, M. Houston, P. Campbell, C. & Ashcroft, P. (2001). Born again bone: tissue engineering for bone repair. *News Physiol Sci,* **16**, 208–213.

[60] Vecchi, S. D. & Coppes, Z. (1996). Marine fish digestive proteases – relevance to food industry and the south-west Atlantic region – a review. *Journal of Food Biochemistry,* **20**, 193–214.

[61] Haard, N.F. (1992). A review of proteolytic enzymes from marine organisms and their application in the food industry. *Journal of Aquatic Food Product Technology,* **1** , 17–35.

[62] Byun, H.G. Park, P.J. Sung, N.J. & Kim, S.K. (2003). Purification and characterization of a serine proteinase from the tuna pyloric caeca. *Journal of Food Biochemistry,* **26**, 479–494.

[63] Kim, S.K. Park, P.J. Kim, J.B. & Shahidi, F. (2002). Purification and characterization of a collagenolytic protease from the filefish, Novoden Modestrus, *Journal of Biochemistry and Molecular Biology,* **35**, 165–171.

[64] Park, P.J. Lee, S.H. Byun, H.G. Kim, S.H. & Kim, S.K. (2002). Purification and characterization of a collagenase from the mackerel, *Scomber japonicas*. *Journal of Biochemistry and Molecular Biology,* **35**, 576–582.

[65] Liu, Z. Y. Wang, Z. Xu, S.Y. Xu, L.N. (2007). Two trypsin isoforms from the intestine of the grass carp (Ctenopharyngodon idellus). *Journal of Comparative Physiology B,* **177**, 655-666

[66] Souza, A.G. Amaral, I. P.G. Santo, A. R. E. Carvalho, L. B. & Bezerra, R. S. (2007). Trypsin-like enzyme from intestine and pyloric caeca of spotted goatfish *Pseudupeneus maculatus*. *Food Chemistry,* **100**, 1429-1434.

[67] Haard, N.F. & Simpson. B.K. (1994). In: A.M. Martin, Editor, *Fisheries processing: Biotechnological application*, Champan & Hall, London. 132–153.

[68] Bhaskar, N. Mahendrakar, N.S. (2008). Protein Hydrolysate from visceral waste proteins of Catla (Catla catla): Optimization of hydrolysis conditions for a commercial neutral protease. *Bioresource Technology*, **99**, 4105-4111.

[69] Bhaskar, N. Benila, T. Radha, C. Lalitha, R.G. (2008). Optimization of enzymatic hydrolysis of visceral waste proteins of Catla (Catla catla) for preparing protein hydrolysate using a commercial protease. *Bioresource Technology*, **99**, 335-343.

[70] Choi, Y. R. Park, P. J. Choi, J. H. Byun, H. G. Jeong, I. C. Moon, S. H. & Kim, S. K. (2000). Purification and characterization of angiotensin I converting enzyme inhibitory peptides from enzymatic hydrolysate of cod liver protein. *Korean J. Life Sci.*, **10**, 140-149

[71] Synowiecki, J. & Al-Khateeb, N. A. (2003). Production, Properties, and Some New Applications of Chitin and Its Derivatives. *Rev. Food Sci. Nutr*, , **43**, 145-171.

[72] Shahidi, F. & Abuzaytoun, R. (2005). Chitin chitosan and coproducts: chemistry production applications and health effects. *Adv. Food Nutr. Res.* **49**, 93-135.

[73] Kurita, K. (2006). Chitin and Chitosan: functional biopolymers from crustaceans, *Mar. Biotechnol.*, **8**, 203-226.

[74] Ylitalo, R. Lehtinen, S. Wuolijoki, E. Ylitalo, P. & Lehtimaki, T. (2002).Cholesterol-lowering properties and safety of chitosan. *Arzneimittel-Forschung,* **52**, 1-7.

[75] Hirano, S. (1996). Chitin biotechnology applications. *Biotechnol. Annu. Rev.*,**2**, 237–258.

[76] Kim, S.K. Park, P.J. Yang, H.P. & Hanm, S.S. (2001). Subacute toxicity of chitosan oligosaccharide in Sprague–Dawly rats. Antibiotics. *Arzneim-Forsch. Drug Research,* **51**, 769–774.

[77] Jeon, Y.J. Shahdi, F. & Kim, S.K. (2000). Preparation of chitin and chitosan oligomers and their application in physiological functional foods. *Food Reviews International,* **16**, 159–176

[78] Park, J.H. Kwon, S. Lee, M. Chung, H. Kim, J.H. Kim, Y.S. Park, R.W. Kim, I.S. Seo, S. B. Kwon, I. C. & Jeong, S.Y. (2006) Self-assembled nanoparticles based on glycol chitosan bearing hydrophobic moieties as carriers for doxorubicin: In vivo biodistribution and anti-tumor activity. *Biomaterials,* **27**, 119-126.

[79] Jeon, Y.J. & Kim, S.K. (2000). Continuous production of chitooligosaccharides using a dual reactor system, *Process Biochemistry,* **35**, 623–632.

[80] Uchida, Y. Izume, M. & Ohtakara, A. (1989). Preparation of chitosan oligomers with purified chitosanase and its application. In: G. Braek, T. Anthonsen and P. Sandford, Editors, *Chitin and chitosan: Sources, chemistry, biochemistry, physical properties and applications*, Elsevier Applied Science, London, United Kingdom, 373–382

[81] Miyazaki, S. Yamaguchi, H. Takada, M. Hou, W.M. Takeichi, Y. & Yasubuchi, H. (1990). Pharmaceutical application of biomedical polymers. XXIX. Preliminary study on film dosage form prepared from chitosan for oral drug delivery. *Acta Pharmacology Nord* , **12**, 401–406

[82] Vasudev, S.C. Chandy, T. & Sharma, C.P. (1997). Development of chitosan/ polyethylene vinyl acetate co-matrix: Controlled release of aspirin–heparin for preventing cardiovascular thrombosis. *Biomaterials,* **18**, 375–381

[83] Park, P.J. Je, J.Y. & Kim, S.K. (2003).Free radical scavenging activity of chitooligosaccharides by electron spin resonance spectrometry. *Journal of Agricultural and Food Chemistry,* **51**, 4624–4627

[84] Je,J.Y. Park, P.J.& Kim, S.K. (2004). Free radical scavenging properties of hetero chitooligosaccharides using an ESR spectroscopy. *Food and Chemical Toxicology,* **42**, 381–387

[85] Je, J.Y. Park, P.J. & Kim, S.K. (2004). Radical scavenging activity of hetero chitooligosaccharides. *European Food Research and Technology* **219**, 60–65.

[86] Sun, T. Xie, W. & Xu, P. (2003). Antioxidant activity of graft chitosan derivatives, *Macromolecular Bioscience,* **3**, 320–323.

[87] Hong, S.P. Kim, M.H. Oh, S.W. Han, C.H. & Kim, Y.H. (1998). ACE inhibitory and antihypertensive effect of chitosan oligosaccharides in SHR. *Korean Journal of Food Science and Technology,* **30**, 1476–1479

[88] Park, P.J. Je, J.Y. & Kim, S.K. (2003). Angiotensin I converting (ACE) inhibitory activity of hetero-chitooligosacchatides prepared from partially different deacetylated chitosans. *Journal of Agricultural and Food Chemistry,* **51**, 4930–4934.

[89] Huang, R. Mendis, E. & Kim, S.K. (2005). Improvement of ACE inhibitory activity of chitooligosaccharides (COS) by carboxyl modification. *Bioorganic & Medicinal Chemistry*, **13**, 3649-3655.

[90] Ngo, D.N. Qian, Z.J. Je, J.Y. Kim, M.M. & Kim, S.K. (2008). Aminoethyl chitooligosaccharides inhibit the activity of angiotensin converting enzyme. *Process Biochemistry,* **43**,119-123.

[91] Matsuo, K. & Miyazono, I. (1993). The influence of long-term administration of peptidoglycan on disease resistance and growth of juvenile rainbow trout.*Nippon Suisan Gakkaishi,* **59**, 1377–1379.

[92] Kim, S.K. & Rajapakse, N. (2005). Enzymatic production an biological activities of chitosan oligosaccharides (COS): A review. *Carbohydrate Polymers,* **62**, 357-368.

[93] Okamoto, Y. Inooue, A. Miyatake, K. Ogihara, K. Shigemasa, Y. & Minami, S. (2003). Effects of chitin/chitosan and their oligomers/monomers on migrations of macrophages. *Macromolecular Bioscience,* **3**, 587–590.

[94] Okamoto, Y. Kawakami, K. Miyatake, K. Morimoto, M. Shigemasa, Y. & Minami, S. (2002). Analgesic effects of chitin and chitosan, *Carbohydrate Polymers.***49**, 249–252.

[95] Feng, J. Zhao, L. & Yu, Q. (2004). Receptor-mediated stimulatory effect of oligochitosan in macrophages. *Biochemical and Biophysical Research Communications,* **317**, 414–420.

[96] Shibata, Y. Metzger, W.J. & Myrvik, Q.N. (1997). Chitin particle-induced cell-mediated immunity is inhibited by soluble mannan: Mannose receptor-mediated phagocytosis initiates IL-12 production. *Journal of Immunology,* **18**, 947–951.

[97] Kanauchi, O. Deuchi, K. Imasato, Y. Shizukuishi, M. & Kobayashi, E. (1995). Mechanism for the inhibition of fat digestion by chitosan and for the synergistic effect of ascorbate. *Bioscience Biotechnology and Biochemistry,* **59**, 786–790.

[98] Ikeda, I. Sugano, M. Yoshida, K. Sasaki, E. Iwamoto, Y. & Hatano, K. (1993). Effects of chitosan hydrolysates on lipid absorption and on serum and liver lipid concentration in rats. *Journal of Agricultural and Food Chemistry,* **41**, 431–435.

[99] Sugano, M. Yoshida, K. Hashimoto, M. Enomoto, K. & Hirano, S. (1992). Hypocholesterolemic activity of partially hydrolyzed chitosans in rats. In: C. Brine, J.P. Zikakis and P. Sandford, Editors, *Advances in chitin and chitosan*, Elsevier, Amsterdam, 472–478.

[100] Enomoto, M. Hashimoto, M. Kuramae, T. & Kanno, M. (1992). Low molecular weight chitosan as anticholesterolemic. *Japanese Kokai. Tokkyo Koho*. JP. 04,108,734. CA. 117:104268.

[101] Liang, T.W. Chen, Y.J. Yen, Y. H. & Wang, S.L. (2007). The antitumor activity of the hydrolysates of chitinous materials hydrolyzed by crude enzyme from *Bacillus amyloliquefaciens. Process Biochemistry,* **42**, 527-534.

[102] Tsukada, K. Matsumoto, T. Aizawa, K. Tokoro, A. Naruse, R. & Suzuki, S. *et al.* (1990). Antimetastatic and growth inhibitory effects of *N*-acetyle chitohexaose in mice bearing Lewis lung carcinoma. *Japanese Journal of Cancer Research,* **81**, 259–265.

[103] Suzuki, K. Mikami, T. Okawa, Y. Tokoro, A. Suzuki, S. & Suzuki, M. (1986). Antitumor effect of hexa-*N*-aetylchitohexaose and chitohexaose. *Carbohydrate Research,* **151**, 403–408.

[104] Park, P.J. Je, J.Y. Byun, H.G. Moon, S.H. & Kim, S.K. (2004). Antimicrobial activity of hetero-chitosans and their oligosaccharides with different molecular weights. *Journal of Microbiology and Biotechnology,* **14**, 317–323.

[105] Park, P.J. Kim, S.K. & Lee, H.K. (2002). Antimicrobial activity of chitooligosaccharides on *Vibrio parahaemolyticus. Journal of Chitin and Chitosan,* **7**, 225–230.

[106] Yun, Y.S. Kim, S.K. & Lee, Y.N. (1999). Antibacterial and antifungal effect of chitosan. *Journal of Chitin and Chitosan,* **4**, 8–14.

[107] Chung, Y.C. Su, Y.P. Chen, C.C. Jia, G. Wang, H.L. & Wu, J.C. *et al.* (2004). Relationship between antibacterial activity of chitosan and surface characteristics of cell wall. *Acta Pharmacologica Sinica*, **25**, 932–936.

[108] Tsai, G.J. Su, W.H. Chen, H.C. & Pan, C.L. (2002). Antimicrobial activity of shrimp chitin and chitosan from different treatments and applications of fish preservation. *Fisheries Science,* **68**, 170–177.

[109] Gerasimenko, D.V. Avdienko, I.D. Bannikova, G.E. Zueva, O.Y. & Varlamov, V.P. (2004). Antibacterial effects of water-soluble low-molecular-weight chitosans on different microorganisms. *Applied Biochemistry and Microbiology,* **40**, 253–257.

[110] Hirano, S. & Nagao, N. (1989). Effects of chitosan, pectic acid, lysozyme and chitinase on the growth of several phytopathogens. *Agricultural and Biological Chemistry,* **53**, 3065–3066.

[111] Byun, H.G. Kim, Y.T. Park, P.J. Lin, X. & Kim, S.K. (2005). Chitooligosaccharides as a novel ß-secretase inhibitor. *Carbohydrate Polymers,* **61**, 198–202.

[112] Singh, D. K. & Ray, A. R. (2000). Biomedical Applications of Chitin, Chitosan, and Their Derivatives. *Polymer reviews,* **40**, 69-83.

[113] Howling, G.I. Dettmar, P.W. Goddard, P.A. Hampson, F.C. Dornish, M. &Wood, E.J. (2002). The effect of chitin and chitosan on fibroblast-populated collagen lattice contraction. *Biotechnology and Applied Biochemistry,* **36**, 247–253.

[114] Minagawa, T. Okamura, Y. Shigemasa, Y. Minami, S. & Okamoto, Y. (2007). Effects of molecular weight and deacetylation degree of chitin/chitosan on wound healing. *Carbohydrate Polymers,* **67**, 640-644

[115] Yannas, I. V. & Burke, J. F. (1980). Design of an artificial skin. I. Basic design principles. *Journal of Biomedical Materials Research,* **14**, 65 -81.

[116] Kifune, K. Yamaguchi, Y. & Kishimoto, S. (1988) Wound healing effect of chitin surgical dressing. *Trans Soc Biomater*; *XI*, 216.

[117] Cavalheiro, J. M. O. Souza, E. O. D. & Bora, P. S. (2007). Utilization of shrimp industry waste in the formulation of tilapia *Oreochromis niloticus* Linnaeus feed. *Bioresource Technology,* **98**, 602-606.

[118] Sankar, S. Sekar, S. Mohan, R. Rani, S. Sundaraseelan, J. & Sastry, T.P.(2008). Preparation and partial characterization of collagen sheet from fish (Lates calcarifer) scales. *International Journal of Biological Macromaolecules,* **42**, 6-9.

[119] Ikoma, T. Kobayashi, H. Tanaka, J. Walsh, D. & Mann, S. (2003). Microstructure, mechanical, and biomimetic properties of fish scales from *Pagrus major*, *Journal of Structural Biology,* **142**, 327–333.

[120] Jeon, Y. J. Kim, Y. T. & Kim, S. K. (1998). Analysis of compositions for effective utilization of fish scales. *Korean J. Life Sci.,* **8**, 589-597.

[121] Yue, G.H. Orban, L. (2001). Rapid isolation of DNA from fresh and preserved fish scales for polymerase chain reaction. *Marine Biotechnology*, **3**, 199-204.

[122] Bazil, V. & Entlicher, G. (1999). Complexity of lectins from the hard roe of perch (*Perca fluviatilis* L.). *International Journal of Biochemistry and Cell Biology,* **31**, 431–442.

[123] Jung, W.K. Park, P.J. & Kim, S.K. (2003). Purification and characterization of a new lectin from the hard roe of skipjack tuna, *Katsuwonus pelamis*. *International Journal of Biochemistry and Cell Biology,* **35**, 255–265.

[124] Aewsiri, T. Benjakul, S. Visessanguan, W.& Tanaka, M.(2008). Chemical compositions and functional properties of gelatin from pre-cooked tuna fin. *International Journal of Food Science & Technology,* **43**, 685-693.

[125] Hirata, H. Saeki, H. Nonaka, M. Kawasaki, K. Ooizumi, T. & Motoe, K. (1993). Recovery of fish oil from the manufacturing process of highly nutritional fish meat for foodstuffs from sardine. *Nippon Suisan Gakkaishi,* **59**, 111–116.

[126] Moffat, C.F. McGill, A.S. Hardy, R. & Anderson, R.S. (1993).The production of fish oils enriched in polyunsaturated fatty acids containing triglycerides. *Journal of the American Oil Chemists Society,* **70**, 133–138.

[127] Dunford, N.T. Temmeli, F. & LeBlanc, E. (1997). Supercritical CO_2 extraction of oil and residual proteins from Atlantic mackerel (*Scomber scombrus*) as affected by moisture content. *Journal of Food Science,* **62**, 289–294.

[128] Chantachum, S. Benjakul, S. & Sriwirat, N. (2000). Separation and quality of fish oil from precooked and non-precooked tuna heads, *Food Chemistry,* **69**, 289–294.

[129] Schacky, C.V. (2000). $n-3$ fatty acids and the prevention of coronary atherosclerosis. *American Journal of Clinical Nutrition,* **71**, 224–227.

[130] Kris-Etherton, P.M. Harris, W.S. & Appel, L.J. (2003). Omega-3 fatty acids and cardiovascular disease: New recommendations from the American Heart Association. *Arteriosclerosis Thrombosis and Vascular Biology,* **23**, 151–152.

[131] Christensen, J.H. Korup, E. Aaroe, J. Toft, E. Moller, J. & Rasmussen, K. *et al.*, (1997). Fish consumption, $n-3$ fatty acids in cell membranes, and heart rate variability in survivors of myocardial infraction with left ventricular dysfunction. *American Journal of Cardiology,* **79**, 1670–1673.

[132] Sheehan, J.P. Wei, I.W. Ulchaker, M & Tserng, K.Y. (1997). Effect of high fiber intake in fish oil-treated patients with non-insulin-dependent diabetes mellitus. *American Journal of Clinical Nutrition,* **66**,. 1183–1187.

[133] Berry, E.M. (1997). Dietary fatty acids in the management of diabetes mellitus. *American Journal of Clinical Nutrition,* **66**, 991–997.

[134] Belch, J.J.F. & Muir, A. (1998). $n-6$ and $n-3$ Essential fatty acids in rheumatoid arthritis and other rheumatic conditions, *Proceedings of the Nutrition Society,* **57**, 563–569.

[135] Belluzzi, A. Brignola, Campieri, C. M. Pera, A. Boschi, S. & Miglioli, M. (1996). Effect of an enteric-coated fish-oil preparation on relapses in Crohn's disease. *The New England Journal of Medicine,* **334**, 1557–1560.

[136] Donadio Jr., J.V. Bergstrahl, E.J. & Offord, K.P. (1994). A controlled trial of fish oil in IgA nephropathy. *New England Journal of Medicine,* **331**, 1194–1199.

[137] Gogos, C.A. Ginopoulos, P. Salsa, B. Apostolidou, E. Zoumbos, N.C. & Kalfarentzos, F. (1998). Dietary omega-3 polyunsaturated fatty acids plus vitamin E restore immunodeficiency and prolong survival for severely ill patients with generalized malignancy. *Cancer,* **82**, 395–402.

In: Industrial Waste...
Editor: John P. Samuelson, pp. 271-287

ISBN: 978-1-60692-720-5

Chapter 10

INDUSTRIAL WASTEWATER TREATMENT USING ADVANCED CHEMICAL OXIDATION WITH FENTON REAGENT

Carmen S.D. Rodrigues[1,2], Rui B.A. Pinheiro[3] and Nídia S. Caetano[2,3,4]*

[1]LSRE – Laboratory of Separation and Reaction Engineering,
[2] Department of Chemical Engineering, Faculty of Engineering at Porto University, R. Dr. Roberto Frias, s/n, 4200-465 Porto, Portugal.
[3]Department of Chemical Engineering, School of Engineering, Polytechnic Institute of Oporto, Rua Dr. António Bernardino de Almeida, 4200-072 Porto
[4]LEPAE – Laboratory for Process, Environmental and Energy Engineering

Abstract

Advanced oxidation with Fenton's reagent is a promising technology for removal of toxic and non biodegradable organic compounds present in wastewaters produced in semiconductor and wood production industries.

The objective of this work was to optimize COD removal from the industrial wastewaters, by applying advanced oxidation processes with Fenton's reagent.

In this study, the effect of three variables (reaction time, initial hydrogen peroxide and ferrous ion concentration) on COD removal from two different industrial wastewaters was evaluated.

For this purpose, an experimental design with two levels and three variables was used leading to the conclusion that reaction time is the only variable that has a statistical meaning for semiconductor industry wastewater treatment. The best %COD removal that could be attained for Fenton oxidation of this wastewater (with an initial COD of 21-32 g O_2/L) was 75%, when reaction started at pH = 3, adding hydrogen peroxide at 4.2M concentration and ferrous ions at a $FeSO_4$: H_2O_2 mass ratio of 1:5 for a 4h reaction time.

When optimizing the treatment of wastewater from wood production (with an initial COD of 25 g O_2/L) with Fenton's reagent, also an experimental design with two levels and three variables (reaction time, initial hydrogen peroxide and ferrous ion concentration) was

* E-mail address: nsc@isep.ipp.pt. Tel. +351-228340500; Fax: +351-228321159; (Corresponding author)

used. For this wastewater, all of the three variables have shown statistical meaning, as well as the interaction between each other. The conditions that conducted to the best results (%COD removal = 82%) were: initial pH = 3, $[FeSO_4]$ = 50 mM, $[H_2O_2]$ = 4.2 M and reaction time = 60 min.

Keywords: Advanced Oxidation Processes, Fenton Reaction, Industrial Wastewater Treatment

1. Introduction

Wastewater is produced in all human activities as a result of the use of water in several different operations. Concern with the effect of industrial wastewater discharges on health and the environment has grown and modern industries apply different treatment technologies to reduce physical, chemical and biological composition to the values required by regulations, before the treated wastewater can be discharged in the environment.

Elimination of toxics or non biodegradable soluble organic compounds contained in the wastewater constitutes a difficult problem. For the industrials, the problem can be resumed in what to make to obtain an effective degradation of the organic load and a reduction of the wastewater toxicity, without collateral effects, associating low cost and high efficiency of organic pollutant removal.

Most organic compounds are resistant to bacterial activity or toxic to bacteria, and direct biological treatment are not effective. Consequently, alternative technology must be developed to help solve the problem. Advanced oxidation processes (AOPs) have been developed for the destruction of certain pollutants present in wastewater, by conversion into water and carbon dioxide [1].

Use of advanced oxidation processes (AOPs) has been widely proposed, as they are run at temperature near ambient and atmospheric pressure. These processes involve the generation of highly reactive radicals (especially hydroxyl radicals) in quantity enough to effect water purification, and their use is justified by the high organic content of the wastewaters to be treated, besides the low reaction temperatures, thus requiring the presence of very active oxidation agents.

The most frequently used AOPs include: heterogeneous photocatalytic oxidation [2,3], treatment with ozone (often combined with H_2O_2, UVA or both) [4,5], H_2O_2/UV systems [6,7] and combined H_2O_2/Fe^{2+} (Fenton reagent).

In particular, the oxidation using Fenton's reagent has proved to be a promising and attractive treatment method for the effective destruction of a large number of hazardous and organic pollutants [8]. The generated $^{\bullet}OH$ radicals are highly oxidative, non-selective, and able to decompose many organic compounds.

Several studies can be found in the open literature focused on use the Fenton's process for wastewater treatment: tannery effluents [9], textile [10] and carpet dyeing wastewaters [11], alkyd resin wastewaters [12], cosmetic effluents [13], olive oil processing effluent [14] and combined the leachate effluent with two organic wastes (wastewater sludge and organic fraction of municipal solid wastes) [15].

This technique, is based on ferrous ion and hydrogen peroxide, and exploits the very high reactivity of the hydroxyl radical produced in acidic solution by the catalytic decomposition of H_2O_2 [16]:

$$Fe\ 2+ + H2O2 \rightarrow Fe\ 3+ + OH- + \bullet OH \quad (1)$$

The organic compounds (represented as RH) present in wastewater may react with hydroxyl radicals, starting a chain reaction according to the redox reactions that we show below [17-19]:

$$OH + RH \rightarrow H_2O + R\bullet \quad (2)$$

$$R\bullet + Fe^{3+} \rightarrow R^+ + Fe^{2+} \quad (3)$$

$$R^+ + H_2O \rightarrow ROH + H^+ \quad (4)$$

Fenton oxidation depends on various factors: pH, reaction time, H_2O_2 concentration ($[H_2O_2]$), temperature and amount of catalyst (usually expressed as the Fe^{2+}:H_2O_2 ratio).

Although many researchers have usually only focused on the single-factor-at-a-time approach, studying the effect of each experimental parameter on the process performance while keeping all other conditions constant, this type of approach does not take into account cross-effects from the factors considered and leads to a poor optimization result [20].

The experimental design methodology is a technique that uses mathematical and statistical tools to analyse the influence of independent variables on a specific dependent variable (response). By using experimental design methodology it is also possible to estimate linear interaction effects of the factors (variables) and to make a prediction model for the response.

In this work we used an experimental design of the type 2^3, with two levels (-1 and +1) and tree variables, corresponding to 16 assays plus 2 or 3 assays in the central point (level 0).

2. Materials and Methods

In this work, the degradation of organic compounds in semiconductor and wood production wastewaters using Fenton's reagent was studied. The wastewaters were provided at the industries site in the North of Portugal.

Chemical oxidation of the wastewaters with Fenton's reagent was carried out in closed glass flasks (1 L capacity). The reactor vessel is provided with constant stirring, accomplished through a magnetic bar and a magnetic stirrer (Framo Gerätetechnik, model M20/1).

Chemical oxidation occurred at pH = 3, because preliminary studies demonstrated that better efficiency of chemical oxygen demand (COD) removal was obtained at this pH – also suggested by Rivas et *al.*, 2004 [21]; Waite, 2002 [22]. Initial pH was adjusted to 3 with sulphuric acid at 98% (Fluka, p.a.) or NaOH solution (6 M, prepared with NaOH p.a. from Panreac). Monitoring of the pH of the wastewaters was done with a pH meter (Hanna Instruments, HI1333) with an electrode (Hanna Instruments, HI8417).

The hydroxyl radical was produced *in situ*, by addition of iron sulphate solution (100 mg/L, prepared with $FeSO_4.7H_2O$ p.a., Panreac) and hydrogen peroxide at 35% (commercial, José M. Vaz Pereira) into the wastewater.

Soluble COD determination was carried out according to the Standard Methods for the Examination of Water and Wastewater (Method 5220D) [23], after stopping the reaction by rising the treated wastewater pH to 12 with NaOH solution (6M) and heating in thermostatic bath to 80 °C for 10 minutes [24,25]. After this, the sample was left to settle for sedimentation of the resulting sludge and COD of the clarified was then measured.

A design of experiments approach was used to model and optimize the process performance. The model considered to describe our data was a first-order polynomial and the coefficients were calculated using MS-Excel.

3. Experimental Design

The experimental study of a three parameter treatment constitutes a difficult and long lasting problem, because there are a high number of variables and the interaction between each might influence the results.

In conventional optimization studies, the effect of each of the single variables can be evaluated in the range values considered. This procedure can lead to a quantitative relation between each variable, the corresponding answer and the mathematical model for each of the variables as isolated. However, this procedure, besides being inefficient (because it requires separate observations for each variable), is incapable to detect interactions between the variables and their effect in the final response.

Thus there are several statistical analysis showing how data can be manipulated, using all available information, in order to obtain satisfactory results.

The main objective of the experimental design (or DoE) is to determine, with a minimum effort (i.e., with a reduced number of experiments), the effects of each factor and of their interactions in the process response(s), within the range of the studied variables. Besides, the information obtained allows to decide how many and which factors and/or interactions are statistically significant, and if it is or not necessary to consider a more complex model in order to describe conveniently the phenomenon under study. For that, statistical techniques (e.g. analysis of variance – ANOVA) need to be used [26]. For the quantification of variance of the response it is necessary to run experiments in dissipated points in the domain of the problem, so that variance can be obtained with maximum possible information [27].

The statistical analysis proceeds with an ANOVA, which evaluates the adequacy of the model fitting. ANOVA subdivides the total variance in two components: the variance associated with the variable and the one associated with the experimental error [28]. If the value of variance associated by variable is large when compared with the experimental error, it means that the changes in the response can be attributed to variable only. This comparison is given by F_i, wich is computed using the equation below [28]:

$$F_i = \frac{V_i}{V_e} \quad (5)$$

$$V_i = \frac{E_i}{N * R} \quad (6)$$

$$E_i = \Sigma(r - x_i) \quad (7)$$

$$V_e = \frac{\sum (r - r_{rep})^2}{2 * (N * (R - 1) * (n - 1))} \tag{8}$$

where V_i is the variance of variable i, V_e is the variance of the error, N is the number of run, R is the number of replicates, E_i is the effect, r is the response of the system, r_{rep} is the response of the system obtained in replicated run, x_i is codified variable and n is the number of central runs.

The objective for using the DOE procedure in this work, was to collect the information that permitted to determine the mathematical model representing the response to the experimental conditions applied (eq.9):

$$\boldsymbol{r} = a_0 + \sum_i a_i X_i + \sum_j \sum_i a_{ij} X_i X_j + ... + \xi \tag{9}$$

where ξ includes the random experimental error and the factors that are not considered in the model, X_i represent the variables considered, a is the coefficient and $\boldsymbol{r}$ is the response of the system.

The coefficients are determined using the following equations:

$$a_0 = \frac{\sum r}{(N * R) + n} \tag{10}$$

$$a_i = \frac{E_i}{N * R} \tag{11}$$

For the statistical calculations, the natural variables (denoted as X_i) are converted into dimensionless codified values (x_i), in order to allow for the comparison of factors of different nature with different units and to decrease the error in the polynomial fit [29]. This is done using the following relationship:

$$x_i = \frac{X_i - X_o}{\Delta X} \tag{12}$$

where X_0 refers to the value of the variable i in the centre of the domain (i.e., it corresponds to $x_i = 0$) and ΔX corresponds to the difference of that variable between the maximum and minimum levels (variation of the coded variable equal to +1).

To determine the variables and/or interactions that are significant to the system under study an algorithm of Yates is used [26]. Basically this provides a table with a standard disposition of the runs to perform in the factorial design.

The obtained set of experiments resulting from the 2^3 type factorial design adopted (two levels, three variables) is shown in tables 3 (for the semiconductor industry wastewater) and 7 (for the wood production wastewater). The negative and positive signs refer to levels -1 and

+1, respectively, of the experimental plan at two levels, being -1/+1 the minimum/maximum value for each coded variable in the studied.

In the algorithm of Yates, the columns of interactions are obtained by the algebraic product of values inscribed in corresponding factor column. Each line corresponds to a different treatment condition.

4. Results and Discussion

4.1. Semiconductor Industry Wastewater

Wastewater used in the experiments was obtained from a semiconductor manufacturing plant in the North of Portugal and had a light green colour (Figure 1-a). Characteristics of the wastewater are summarized in Table 1.

The industrial wastewater was pre-treated by adjusting pH to a value of 2 with sulphuric acid, producing a blue sludge – as seen in Figure 1-b) – that could be removed by filtration using a glass funnel and filtering paper (18.5 cm, Whatman N°4). The wastewater obtained after filtration had a yellow colour – as seen in Figure 1-c). By pre-treatment application, a COD removal of ~19% was obtained.

Table 1. Main characteristics of the semiconductor industry wastewater

Parameter	Value
Soluble COD	21-32 g O2/L
BOD	10 mg O2/L
Ammonia	42 mg NH3/L
NTK	3472 mg NH3/L
Nickel	< 0.08 mg/L
Copper	0.78 mg/L
Zinc	0.06 mg/L
pH	10.5 – 11.0

Figure 1. a) Semiconductor raw wastewater. The pre-treatment application effect on semiconductor wastewater: b) wastewater after pH adjustment to 2, c) wastewater after filtration of the resulting sludge.

4.1.1. Preliminary Runs

For the oxidation reaction with the Fenton's reagent, three operating variables had to be considered for a given parent compound concentration, the concentration of hydrogen peroxide, the reaction time and $FeSO_4$:H_2O_2 mass ratio. A high number of experiments would be required if these three variables are considered in the experimental design. For that reason some preliminary runs were carried out, to find the levels for each considered variable in the experimental design (see Table 3).

4.1.1.1. Effect of Feso$_4$: H_2O_2 Mass Ratio

For the evaluation of the $FeSO_4$: H_2O_2 mass ratio that conducted to the maximization of the COD removal, a set of 8 experimental runs was analysed. Experimental conditions and corresponding results can be found in Table 2.

Table 2. Experimental conditions in Fenton oxidation and the COD removal (%) obtained

Run	pH	Reaction time - t (h)	$[H_2O_2]$ (M)	$FeSO_4$: H_2O_2 (w/w)	COD removal (%)
1	3	4	0.1	3:4	7.2
2				3:2	3.4
3			0.6	1:7	38.1
4				1:3	3.4
5			2.1	1:9	59.1
6				1:5	61.1
7			3.5	1:10	50.3
8				1:5	53.2

The analysis of the results lead to the conclusion that maximum COD removal (61%) would be obtained when the operational conditions were a $FeSO_4$:H_2O_2 mass ratio of 1:5 and $[H_2O_2] = 2.1M$, at a pH = 3 and with a reaction time of 4 h.

4.1.1.2. Effect of H_2O_2 Concentration

In figure 2, the effect of initial hydrogen peroxide concentration (2.1 to 5.9 M) on COD removal is shown, when initial pH = 3, reaction time, t = 4h and $FeSO_4$: H_2O_2 mass ratio = 1:5. The results show that COD removal increases with hydrogen peroxide concentration up to a hydrogen peroxide concentration of 4.2 M and above this concentration the mineralization of organic compounds decreases again. This small decrease is due to the fact that at higher H_2O_2 concentration scavenging of $^{\bullet}OH$ radicals will occur; which can be expressed by eqs. (13) and (14) [30,31]:

$$OH + H_2O_2 \rightarrow H_2O + HO_2^{\bullet} \quad (13)$$

$$HO_2^{\bullet} + {}^{\bullet}OH \rightarrow H_2O + O_2 \quad (14)$$

4.1.1.3. Reaction Time

To determine the optimum reaction time an experimental was run for 5 hours, along which several samples were taken. The experiment started at a pH = 3, with hydrogen peroxide addition at a concentration = 3.1M and $FeSO_4$: H_2O_2 mass ratio = 1:5. Figure 3 shows COD removal obtained for several reaction times, showing that the best COD removal was obtained for 3 hours reaction time (69.9% COD removal) after which COD removal remained approximately constant.

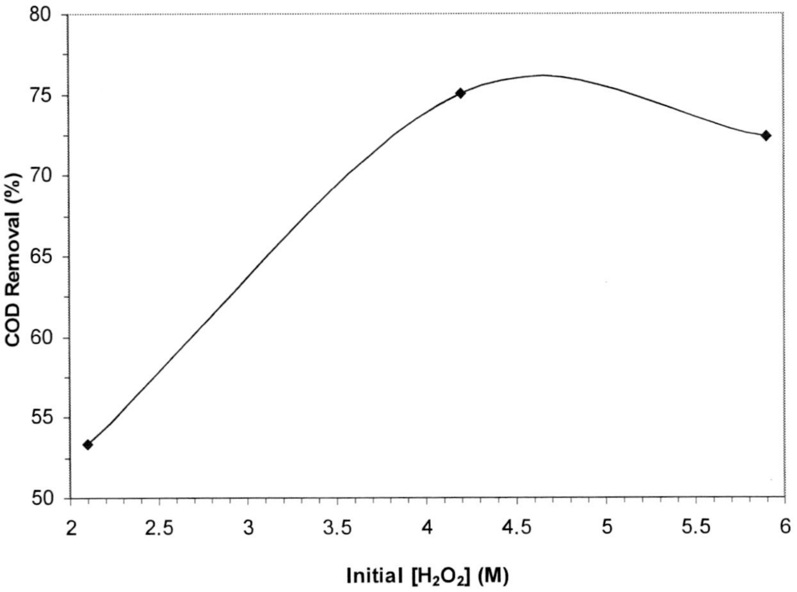

Figure 2. Effect of the initial hydrogen peroxide concentration on COD removal (initial pH = 3, reaction time, t = 4h and FeSO4 : H2O2 mass ratio = 1:5).

Mineralization of organic compounds is much faster in the first minutes of the reaction (0.5 hours), and the main reason for this well known behaviour is that ferrous ions react very quickly with H_2O_2 (rate constant is 53 $mol^{-1}dm^3s^{-1}$) to produce large amounts of hydroxyl radicals, as shown in eq. (1), that can then react rapidly with organic compounds. Ferric ions produced can react with H_2O_2 to produce hydroperoxyl radicals ($HO_2^{\bullet}$) and restore ferrous ions through the following reaction scheme [20]:

$$Fe^{3+} + H_2O_2 \leftrightarrow Fe\text{-}OOH^{2+} + H^+ \quad (15)$$

$$Fe\text{-}OOH^{2+} \rightarrow HO_2^{\bullet} + Fe^{2+} \quad (16)$$

However, the reaction rate for iron regeneration is much slower than that in eq.(1) (rate constant is now 0.02 $mol^{-1}dm^3s^{-1}$) [20]. Consequently, the rate of oxidation in this stage is slower than in the first one due to the slow production of Fe^{2+} from Fe^{3+}. We can thus say that reaction rate decrease on the second stage of the Fenton oxidation is essentially due to the fact that ferrous ions are consumed quickly, but reproduce slowly. Consequently, the oxidation rate of organic compounds is fast when large amounts of ferrous ions are present because large amounts of hydroxyl radicals are produced.

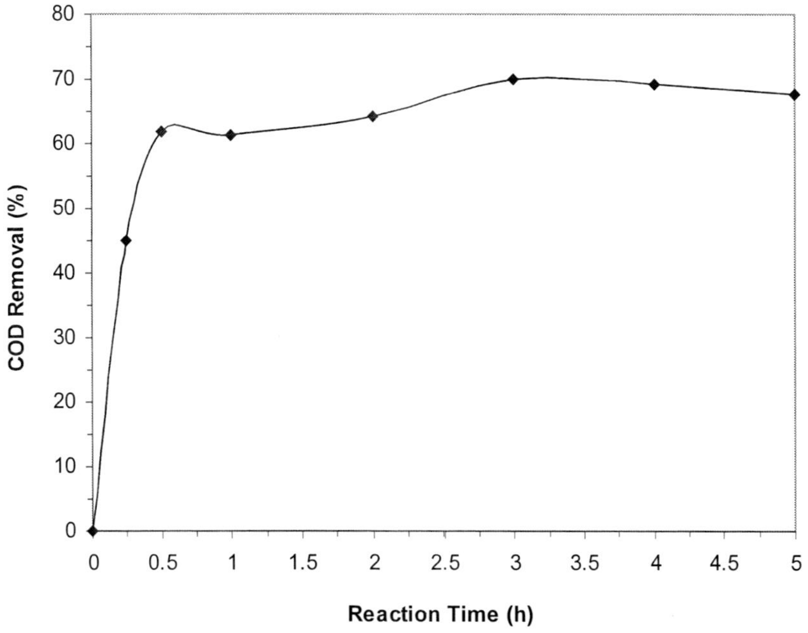

Figure 3. COD removal for different reaction times (Initial pH = 3, hydrogen peroxide concentration = 3.1M and FeSO4 : H2O2 mass ratio = 1:5).

4.1.2. Experimental Design

After the exploratory runs, wherein the effects of the process variables (concentration of hydrogen peroxide – x_1 and $FeSO_4$:H_2O_2 mass ratio – x_2 and reaction time – x_3) were briefly analysed, the experimental plan was drawn. In Table 3 values to be used in the experimental design were established and the corresponding response (r – % COD removal) is shown.

Table 3. Planning of the assays corresponding to the factorial drawing and % COD removal (r)

		Codified value			Experimental value		
Run	r (%)	x_1	x_2	x_3	x_1	x_2	x_3
1	36.30	-1	-1	-1	4.2	1:10	1
2	31.98	1	-1	-1	6.3	1:10	1
3	30.28	-1	1	-1	4.2	1:5	1
4	14.87	1	1	-1	6.3	1:5	1
5	11.07	-1	-1	1	4.2	1:10	3
6	10.10	1	-1	1	6.3	1:10	3
7	11.14	-1	1	1	4.2	1:5	3

Table 3. Continued

		Codified value			Experimental value		
Run	r (%)	x_1	x_2	x_3	x_1	x_2	x_3
8	15.80	1	1	1	6.3	1:5	3
9	19.04	-1	-1	-1	4.2	1:10	1
10	14.40	1	-1	-1	6.3	1:10	1
11	10.97	-1	1	-1	4.2	1:5	1
12	9.21	1	1	-1	6.3	1:5	1
13	9.11	-1	-1	1	4.2	1:10	3
14	0.90	1	-1	1	6.3	1:10	3
15	0.90	-1	1	1	4.2	1:5	3
16	5.08	1	1	1	6.3	1:5	3
17	7.59	0	0	0	5.2	1:6.5	2
18	10.57	0	0	0	5.2	1:6.5	2
19	4.28	0	0	0	5.2	1:6.5	2

Note: Calculation of COD removal took in account the dilution promoted for the hydrogen peroxide and iron solution addition.

Table 4 shows the values of the effects of the parameters and corresponding interactions, coefficients, variance of the results and the statistical F for the three parameters under study and all the possible interactions between them.

Table 4. Statistical analysis of the results corresponding to the experimental planning

Term	Effect	Coefficient	Variance of the Results (Vi)	Variance of the error (Ve)	Statistical F (F_i)
Intercept	-	13.35	-	-	-
x_1	- 27.34	-1.71	46.71	66.85	0.76
x_2	-33.76	-2.11	71.25		1.07
x_3	-103.83	-6.49	673.75		10.08
x_1x_2	10.70	0.67	7.15		0.11
x_1x_3	24.91	1.56	38.79		0.58
x_2x_3	39.02	2.44	95.16		1.42
$x_1x_2x_3$	27.12	1.69	45.96		0.69

After the statistical analysis of the results, a Fisher test was applied in order to eliminate the effects and interactions with no meaning. Comparing the F_i to the values of $F\alpha1\alpha2$ that can be found in F distribution tables for the degrees of freedom α_1 (n° of degrees of freedom for the numerator = n° of levels -1 = 1) and α_2 (n° of degrees of freedom for the denominator = αe = 10), we obtained the results shown in table 5.

Comparing the values of Fisher F (Table 5), with the statistical F (Fi) for each variable and interaction in Table 4, we observe that the statistical F of the variable reaction time is the only one that exceeds Fisher F values. Thus, we conclude that, for the range of tested conditions, reaction time is the only variable that presents statistical meaning for the treatment of this wastewater, generated in the production of printed circuits, by chemical oxidation using Fenton's reagent.

Table 5. Values of Fisher F

Probability Level (%)	Fisher F
1	10.04
5	4.96

Thus the model that best represents %COD removal for this industrial wastewater is:

$$\%\text{COD removal} = 13.35 - 6.49\ (t - 2) \qquad (17)$$

The comparisons of the model prediction vs. the experimental responses are given in the parity plot of Figure 4. Although absolute deviations as high as 16.2% was observed, on the average they are only 6.6%.

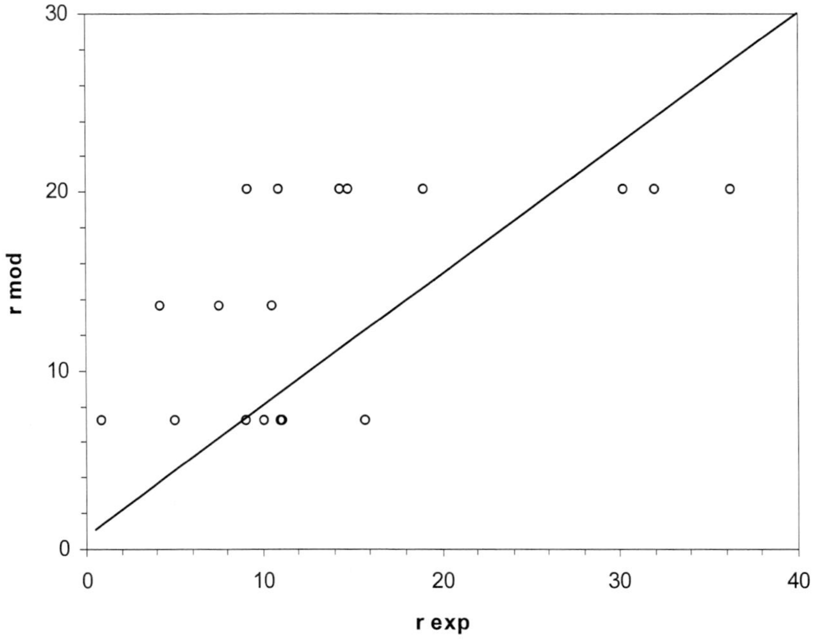

Figure 4. Parity plot comparing the experimental %COD removal (cf. Table 3) with the model predictions by Eq. (17).

4.2. Wood Production Wastewater

Wastewater tested in this part of the work was obtained from a wood manufacturing plant (in Oliveira do Hospital-North of Portugal). The wastewater had a COD of 25g O_2/L, pH was between 5 and 6 and had a white colour, as shown in Figure 5.

Figure 5. Raw wood production wastewater.

The wastewater was treated with Fenton's reagent. To optimize the experimental conditions for this kind of treatment an experimental design with two levels and three variables, was planned, as describe before.

The amounts of reagent and reaction time were fixed according to preliminary runs that permitted to choose the central point and the work amplitude (see table 6). The initial reaction pH was 3.

In Table 6 we present the algorithm of Yates and the respective COD removal obtained.

Table 6. Algorithmic of Yates and COD removal obtained

Run	r	Codified value			Experimental value		
		$[FeSO_4]$ (nM) x_1	$[H_2O_2]$ (M) x_2	t (h) x_3	$[FeSO_4]$ (nM) x_1	$[H_2O_2]$ (M) x_2	t (h) x_3
1	50.87	-1	-1	-1	17	2.5	1
2	72.21	1	-1	-1	50	2.5	1
3	59.94	-1	1	-1	17	4.2	1
4	81.33	1	1	-1	50	4.2	1
5	75.16	-1	-1	1	17	2.5	3
6	73.33	1	-1	1	50	2.5	3
7	72.36	-1	1	1	17	4.2	3
8	81.16	1	1	1	50	4.2	3
9	51.82	-1	-1	-1	17	2.5	1
10	72.67	1	-1	-1	50	2.5	1
11	59.00	-1	1	-1	17	4.2	1
12	82.12	1	1	-1	50	4.2	1

Table 6. Continued

Run	r	Codified value [FeSO$_4$] (nM) x_1	Codified value [H$_2$O$_2$] (M) x_2	Codified value t (h) x_3	Experimental value [FeSO$_4$] (nM) x_1	Experimental value [H$_2$O$_2$] (M) x_2	Experimental value t (h) x_3
13	71.23	-1	-1	1	17	2.5	3
14	70.99	1	-1	1	50	2.5	3
15	73.47	-1	1	1	17	4.2	3
16	82.40	1	1	1	50	4.2	3
17	80.30	0	0	0	33	3.4	2
18	80.88	0	0	0	33	3.4	2

In Table 7 we present the interactions and the effects values, coefficient, variance of the results and the statistical F for the three variables and the possible corresponding interactions between them.

Table 7. Statistical analysis of the results obtained with the experimental design

Term	Effect	Coefficient	Variance of the Results (Vi)	Variance of the error (Ve)	Statistical F (F_i)
Intercept	-	**71.74**	-	-	-
x_1	102.36	6.40	654.85	1.46	447.98
x_2	53.5	3.34	178.89		122.38
x_3	70.14	4.38	307.48		210.34
x_1x_2	22.12	1.38	30.58		20.92
x_1x_3	-71.04	-4.44	315.42		215.78
x_2x_3	-16.14	-1.01	16.28		11.14
$x_1x_2x_3$	17.48	1.09	19.10		13.06

When statistical F value is higher than F Fisher for probability level of 1 and 5% (see Table 8), we conclude that the variable or interaction has a statistical meaning and this variables or interactions have an influence on COD removal by oxidation with Fenton's reagent.

Table 8. F Fisher table

Probability level (%)	Fisher F
1	10,56
5	5,11

For the treatment of the wood production wastewater with Fenton oxidation, we conclude that all of the three variables and the interaction between each other have a statistical meaning and the mathematical model that represents this is:

$$\%COD\ removal = 71.36 + 6.40\frac{[FeSO_4]-33}{16} + 3.34\frac{[H_2O_2]-3.4}{0.9} + 4.38\frac{t-90}{30} + \\ +1.38\frac{[FeSO_4]-33}{16} * \frac{[H_2O_2]-3.4}{0.9} - 4.44\frac{[FeSO_4]-33}{16} * \frac{t-90}{30} - \\ -1.01\frac{[H_2O_2]-3.4}{0.9} * \frac{t-90}{30} + 1.09\frac{[FeSO_4]-33}{16} * \frac{[H_2O_2]-3.4}{0.9} * \frac{t-90}{30} \qquad (18)$$

In Figure 6 we can show the comparison of experimental %COD removal against %COD removal obtained from the model in eq. 18. An absolute deviation as high as 9.5% was observed, but on average it was only 1.9%.

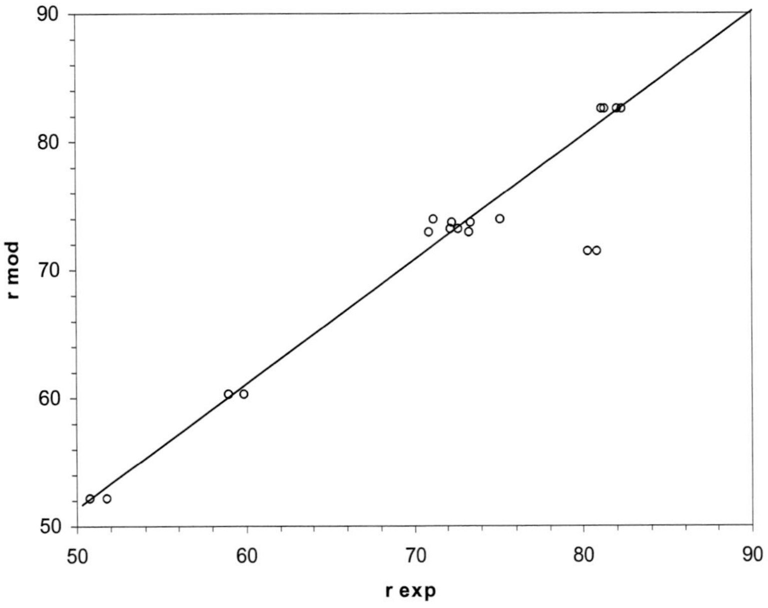

Figure 6. Parity plot comparing the experimental %COD removal (cf. Table 6) with the model predictions by Eq. (18).

5. Conclusions

This work shows that organic compounds present in semiconductor industry and wood production wastewaters can be mineralized using Fenton's reagent technology.

The effect: of hydrogen peroxide concentration, ferrous ion concentration and reaction time on the %COD removal when the industrial wastewaters were treated with Fenton's reagent was evaluated.

The experiments with industrial semiconductor wastewater started with preliminary runs, which evaluated the isolated effects of reaction time (0.25 to 5 hours), hydrogen peroxide concentration $[H_2O_2]$ (2.1 to 5.9 M) and $FeSO_4$: H_2O_2 mass ratio (1:5 to 3:2). Higher COD removal (75%) was obtained with the following conditions: pH = 3, reaction time of 4 hours, hydrogen peroxide concentration = 4.2 M and $FeSO_4$: H_2O_2 mass ratio of 1:5. As the wastewater was pre-treated, with a %COD removal of 19%, the global COD removal was 80%. From experimental design experiments it was possible to conclude that reaction time is the only variable that has a statistical meaning.

On the other hand, treatment of the wood production wastewater by chemical oxidation with Fenton's reagent was done without any preliminary runs. So the conditions that conducted to the best results (%COD removal = 82%) were: pH = 3, [$FeSO_4$] = 50 mM, [H_2O_2] = 4.2 M and reaction time = 60 min. In this study all of the variables had a statistical meaning in %COD removal by Fenton's reagent (reaction time, [H_2O_2] and [$FeSO_4$]) as well as the interaction between these variables.

References

[1] Martínez, E. C. and Lopez, G. D. (2001) Tratamiento químico de contaminantes orgánicos-El proceso Fenton, *Ingeniería Química (España)*, **375**, 149-151.

[2] Hsueh,C. L.; Huang, Y. H.; Wang, C. C.; Chen, C. Y. (2006). Photooxidation of azo dye Reactive Black 5 using a novel supported iron oxide: heterogeneous and homogeneous approach, *Water Science Technology*, **53**,195–201.

[3] So, C. M.; Cheng, M. Y.; Yu, J. C.; Wong, P. K. (2002). Degradation of azo dye Procion Red MX-5B by photocatalytic oxidation, *Chemosphere,* **46**, 905-912.

[4] Meric, S.; Selçuk, H.; Belgiorno, V. (2005). Acute toxicity removal in textile finishing wastewater by Fenton's oxidation, ozone and coagulation–flocculation processes, *Water Research.*, **39**, 1147–1153.

[5] Turan-Ertas, T.; Gurol, M. D. (2002). Oxidation of diethylene glycol with ozone and modified Fenton processes, *Chemosphere*, **47**, 293–301.

[6] Arslan,I.; Balcioglu, I. A.; Bahnemann, D.W. (2002). Advanced oxidation of a reactive dyebath efluent: comparison of O3, H2O2/UV-C and TiO2/UV-A processes, *Water Research.*, **36**, 1143–1154.

[7] Neamtu, M.; Simiceanu, I.; Yediler, A.; Kettrup, A. (2002). Kinetics of decolorization and mineralization of reactive azo dyes in aqueous solution by the UV/H_2O_2 oxidation, Dyes Pigments, **53**, 93–99.

[8] Oliveira, R.; Almeida, M. F.; Santos, L.; Madeira, L. M. (2006). Experimental Design of 2,4-Diclorophenol Oxidation by Fenton's Reaction. *Industrial and Engineering Chemistry Research*, **45**, 1266-1276.

[9] Schrank, S. G.; José, H. J.; Moreira R. F. P. M.; Schröder, H. Fr. (2005). Applicability of Fenton and H_2O_2/UV reactions in the treatment of tannery wastewaters. *Chemosphere*, **60**, 644-655.

[10] Papadopoulos, A. E.; Fatta, D.; Loizidou, M. (2007). Development and optimization of dark Fenton oxidation for the treatment of textile wastewaters with high organic load. *Journal of Hazardous Materials*, **146**, 558-563.

[11] Gulkaya I.; Suruc G. A.; Dilek F. B. (2006). Importance of H_2O_2/Fe^{2+} ratio in Fenton's treatment of a carpet dyeing wastewater. *Journal of Hazardous Materials B,* **136**, 763-769.

[12] Oliveira, I. S.; Viana, L.; Verona, C.; Fallavena, V. L. V.; Azevedo, C. M. N.; Pires, M. (2007). Alkydic resin wastewaters treatment by fenton and photo-fenton processes. *Journal of Hazardous Materials*, **146**, 564-568.

[13] Bautista, P.; Mohedano A. F.; Gilarranz M. A.; Casas, J. A.; Rodriguez J. J. (2007). Application of Fenton oxidation to cosmetic wastewaters treatment. *Journal of Hazardous Materials*, **143**, 128-134.

[14] Ahmadi, M.; Vahabzadeh, F., Bonakdarpour, B.; Mofarrah, E.; Mehranian M. (2005). Application of the central composite design and response surface methodology to the advanced treatment of olive oil processing wastewater using Fenton's peroxidation, *Journal of Hazardous Materials B,* **123**, 187-195.

[15] Turjillo D.; Font X.; Sánc hez A. (2006). Use of Fenton reaction for treatment of leachate from composting of different wastes. *Journal of Hazardous Materials B*, **138**, 201-204.

[16] Walling, C. (1975). Fenton's Reagent Revisited. *Accounts of Chemistry Research*, **8**, 125-131.

[17] Azbar, N.; Yonar, T.; Kestioglu, K. (2004). Comparison of Various Advanced Oxidation Processes and Chemical Treatment Methods for COD and Color Removal from a Polyesther and Acetate Fiber Dyeind Effluent. *Chemosphere*, **55**, 35 – 43.

[18] Kuo, W. G. (1992). Decolorizing Dye Wastewater with Fenton´s Reagent. *Water Research*, **26**, 881-886.

[19] Sanz, J.; Lombrana, J. I.; Luis, A. M. De Ortueta, M.; Varona, F. (2003). Microwave and Fenton´s Reagent Oxidation of Wastewater. *Environmental Chemistry Letters* **1**, 45-50.

[20] Ramirez, J. H.; Costa, C. A.; Madeira, L. M. (2005) Experimental Design to Optimize the Degradation of the Synthetic Dye Orange II Using Fenton´s Reagent. *Catalysis Today*, **107–108**, 68-76.

[21] Rivas, F. J., Navarrete, V., Beltrán, F. J. and Garcia–Araya, J. F. (2004) Simazine Fenton's Oxidation in a Continuous Reactor. Applied Catalysis B: *Environmental,* **48**(4), 249-258.

[22] Waite, T. D. (2002) Challenges and Opportunities in the Use of Iron in Water and Wastewater Treatment. *Reviews in Environmental Science and Biotechnology*, **1**(1), 9-15.

[23] APHA, AWWA, WEF (1995). "Standard Methods for the Examination of Water and Wastewater". 19th ed., American Public Health Association, American Water Works Association, Water Environment Federation, Washington, DC.

[24] Heredia, J. B.; Dominguez, J. R.; Lopez, R. (2004). Advanced Oxidation of Cork-Processing Wastewater Using Fenton´s Reagent: Kinetics and Stoichiometry. *Journal of Chemical Technology and Biotechnology,* **79**, 407-412.

[25] Martínez, N. S. S.; Fernández, J. F.; Segura, X. F.; Ferrer, A. S. (2003). Pre-oxidation of an Extremely Polluted Industrial Wastewater by the Fenton´s Reagent. , *Journal of Hazardous Materials B*, **101**, 315 – 322.

[26] Barker, T.B. (1994). *Quality by Experimental Design.* 2nd Edition, Marcel Dekker.

[27] Morgado, R., 1979. Experimentação Factorial. Internal Report, FEUP.

[28] Montgomery, D.C.; Runger,G.C.; (1999). *Applied Statistics and Probability for Engineers,* 2nd Edition, John Wiley and Sounds, New York.

[29] Montgomery, D.C. (2001), *Design and Analysis of Experiments*, 5th Edition, John Wiley and Sons, New York.

[30] Lin, S. H.; Lo, C. C. (1997). Fenton Process for Treatment of Desizing Wastewater. *Water Research*, **31**, 2050-2056.

[31] Lucas, M. S.; Peres, J. A. (2006) Decolorization of the Azo Dye Reactive Black 5 by Fenton and Photo-Fenton Oxidation. *Dyes and Pigments*, **71**, 236-244.

In: Industrial Waste...
Editor: John P. Samuelson, pp. 289-324
ISBN: 978-1-60692-720-5

Chapter 11

Coagulation-Flocculation Processes Applied in Water or Wastewater Treatment

A.I. Zouboulis[1], P.A. Moussas and N.D. Tzoupanos
Division of Chemical Technology, Department of Chemistry
Aristotle University of Thessaloniki, GR-54124, Greece

Abstract

The coagulation process is essential in a number of disciplines including biochemistry and most notably in water and wastewater treatment. The treatment of water using coagulation agents has been noted from ancient times, as early as 2000 BC in ancient Egypt. In modern water and wastewater treatment, coagulation process is an integrated step of the overall suite of treatment process. The continuous need for a thorough understanding of coagulation – flocculation phenomena is as important today as it has been in the past, taking into account the water scarcity as well as the fact that the requirements for water and wastewater quality has become increasingly strict. The aim of this paper is to provide a general description of coagulation process, focusing on the theoretical aspects and mechanisms via which coagulation proceeds. Moreover, the most commonly coagulation reagents, including metal coagulants and polymers, are presented, while the recent research trend and latest developments regarding the evolution of coagulation reagents are also discussed. Finally, the significance of coagulation – flocculation process in the area of water and wastewater treatment is reviewed and evaluated, emphasizing on the series of applications employed, including destabilization of colloids, removal of inorganic and organic matter (particulate and/or dissolved), removal of metals and anions (arsenic, phosphate etc) as well as removal of pathogen microorganisms.

Keywords: coagulation; flocculation; water and wastewater treatment; destabilization mechanisms.

---

[1] E-mail address: zoubouli@chem.auth.gr, Tel: +302310997794. (Corresponding author)

1. Introduction

Coagulation and flocculation are essential processes in water and wastewater treatment. In the case of potable water treatment, the application of coagulation reagents for the clarification of water has been first mentioned in Ancient Egypt as early as 2000BC, when Egyptians used crushed seeds (most notably, almonds seeds). Later, around 77 AC, Romans used alum as a coagulation reagent for removing the bitter flavor of the potable water. It is worth noting that the terms coagulation and flocculation derive from the Latin words "*coagular*", which means the union of two or more units and "*floccus*". For hundreds of years, people used these processes for removing pollutants from potable water, without realizing and understanding the true value of these techniques. Only in the middle of 20th century, scientists started to systematically gather information and experimental data from laboratory studies as well as from full scale treatment plants, thus establishing the theoretical base of coagulation and flocculation process. The disciplines of colloid chemistry and fluid dynamics were used for understanding and explaining the phenomena with scientific terms. However, due to the complex and sometimes crosscurrent physico-chemical interactions, experiment and field experience are still regarded as the most reliable method for developing the process and determining the optimum operational conditions.

In general, coagulation and flocculation process is a core environmental protection technology, which finds wide range of applications in water or wastewater treatment facilities. In particular, coagulation is regarded as a viable option for the treatment of surface waters to remove suspended solids (SS), colloidal particles, natural organic matter (NOM) and other soluble, mostly inorganic compounds, such as phosphate ions, as well as of heavy loaded, toxic or hazardous wastewaters, including landfill leachate. Several research efforts suggested that coagulation process can be utilized as a pre- or post-treatment method for landfill leachate treatment [1, 2]. The combination of coagulation with other appropriate physicochemical or biological treatment processes, such as ozonation, photo-oxidation or submerged biological filters, results in an enhancement of the efficiency, when treating such wastewaters [3-6]. The coagulation process by the addition of coagulation reagents can involve: (1) the destabilization of suspended and colloidal particulate matter, (2) the adsorption of portions of colloidal and dissolved organic matter onto the surface of the destabilized particles, and (3) the aggregation of the destabilized particles and the subsequent creation of flocs, which enmesh suspended, colloidal and dissolved material as they settle [7].

2. Coagulation – Flocculation Process

2.1. Types of Pollutants in Waters and Wastewaters

Drinking water of high quality is essential for human beings. Drinking water quality is evaluated by means of certain physical, chemical and biological parameters and consequently, limiting values has been set in accordance to national or international regulations. The natural surface waters are usually polluted from the air and ground by organic or inorganic impurities, which can be either in dissolved, or in particulate form. Inorganic particulates, consisting of clay, mineral oxides and silt, typically enter surface waters by natural erosion processes. On the other hand, organic constituents consisting from viruses, bacteria, algae, protozoan cysts etc can be found in particulate form. Additionally, surface waters usually contain colloidal and dissolved organic substances, such as humic or fulvic acids, arising from the decay of organic debris. Particulate and dissolved organic matter is often identified by the term Natural Organic Matter (NOM) [7]. As shown in Table 1 much of the suspended material found in waters is in microscopic to submicroscopic size range.

Most of the impurities found in natural waters, lie in the colloidal size range (i.e. <1 μm) and are negatively charged; hence, the colloidal dispersion is stable, due to electrostatic repulsive forces, which normally overwhelm the attractive van der Waals forces [9]. The removal of such contaminants is essential, as their presence deteriorates the water quality by: (a) reducing the clarity of water (turbidity, color), (b) causing infection, and (c) carrying toxic compounds, adsorbed on their surfaces. In addition, NOM is the main precursor to the formation of disinfection byproducts, when chlorine is applied as disinfection agent.

Table 1. Classification of particles according to size [8].

Particle size (mm) – estimation	Classification	Examples	Estimation of Total surface area (m^2/cm)	Time required to settle for a distance of 100mm, when specific gravity is 2.65 g/cm^3
10 1 10^{-1}	Coarse dispersion (visible to naked eye)	Gravel, coarse sand, mineral substances, precipitated and flocculated particles, silt macroplakton	6 x 10^{-4} 6 x 10^{-3} 6 x 10^{-2}	0.1 s 1 s 13 s
10^{-2} 10^{-3} 10^{-4}	Fine particulate dispersion (visible under microscope)	Mineral substances, precipitated and flocculated particles, silt, bacteria, plankton and other organisms	0.6 6 60	11 min 20 hours 80 days
10^{-5} 10^{-6}	Colloidal dispersion (submicroscopic)	Mineral substances, hydrolysis and precipitated products, macromolecules, biopolymers, viruses	600 6000	2 years 20 years
< 10^{-6}	Solution	Inorganic simple and complex ions, molecules and polymeric species, polyelectrolytes, organic molecule, undissociated solutes		

To comply with drinking water standards, water treatment plants remove these impurities by sedimentation and/or filtration, following the processing of water by coagulation/flocculation. Typically chemicals (coagulation reagents) used for water (and/or wastewater treatment) are predominantly inorganic salts of Al or Fe, such as $AlCl_3$, $Al_2(SO_4)_3$, $FeCl_3$, $Fe_2(SO_4)_3$ etc.

2.2. Chemistry of Al and Fe Salts

Once ferric and aluminium salts are added to the water, several parallel and sequential reactions occur. Fe(III) and Al(III) salts will dissociate, yielding Fe^{3+} and Al^{3+} ions.

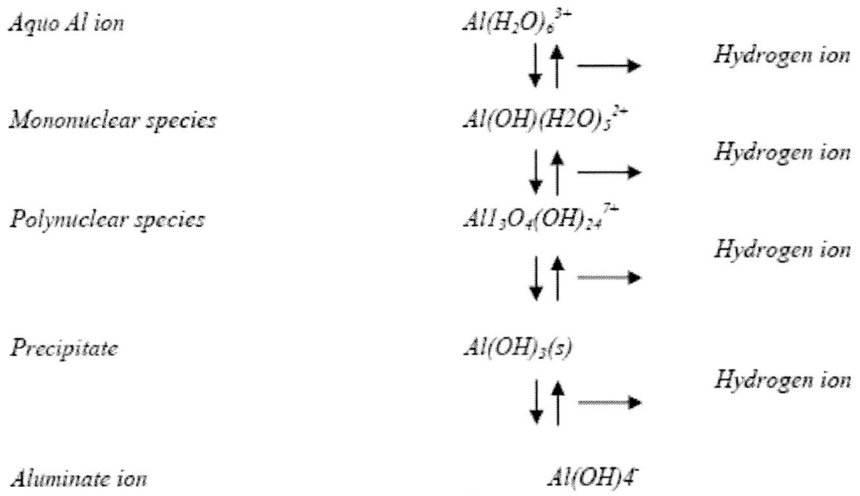

Figure 1. Aluminium Hydrolysis products [7].

Then the trivalent ions will hydrolyze resulting in the formation of aquo-metal complexes, i.e. $Me(H_2O)_6^{3+}$, which in turn will participate in several hydrolysis reaction yielding a wide variety of soluble mononuclear and/or polynuclear species, as illustrated in Figure 1 (in the case of aluminium). These reactions involve a displacement of a proton with hydroxyl ions, while the increase of pH favors these reactions. Figure 1 is a simplified description of the hydrolysis process. In fact, the hydrolysis is a complex process involving, apart from the hydrolysis reactions, parallel polymerization reactions which result in a great variety of polynuclear species [10]. The nature and extent of polymerization is not yet fully understood, and the efforts for explaining this phenomenon are based on specific products that have been experimentally determined and on observations from potentiometric titrations. Jia et al. [11] proposed a simplified scheme (Figure 2) about the possible hydrolysis and reaction pathways of aluminium.

The reactions 1-4 represent the hydrolysis products of Al^{3+} with increasing pH values, produced relatively quickly. The other reactions refer to possible polymerization reactions that lead to the formation of larger polymeric structures, which require more time to be formed, while the extent of their formation is not clear. Among the possible products, the dimmer ($Al_2(OH)_2^{4+}$), trimmer ($Al_3(OH)_4^{5+}$), tridecamer (Al_{13} - $AlO_4Al_{12}(OH)_{24}(H2O)_{12}^{6,7+}$) and more recently Al_{30} ($Al_{30}O_8(OH)_{56}(H2O)_{24}^{18+}$) have been determined through ^{27}Al-NMR and X-Ray Diffraction studies [12, 13]. It is suggested that a poly-nuclear species with up to 54 Al atoms ($Al_{54}(OH)_{144}^{18+}$) can be also formed [10]. The most important reaction is regarded to be the one leading to $Al(OH)^{4-}$ formation (reaction 4), because this aluminium anionic form is claimed to be the precursor for the polymeric Al_{13} creation [14]. Al_{13} with Keggin-like structure is claimed to be the most stable aluminium species in a partially neutralized Al - solution [13] and the improved coagulation properties of PAC are thought to be due to its existence, i.e. increased charge neutralization capability (coagulation) and increased molecular size and aggregation ability (flocculation).

Moreover, the decrease of mono-meric Al in favor of Al_{13} and of the other polymeric compounds eliminates the hydrolysis reactions and therefore results in a minor impact of the pH value after treatment. Similarly, The Fe(III) hydrolysis products are considered to be the following: five type of monomers (Fe^{3+}, $Fe(OH)^{2+}$, $Fe(OH)_3$ (molecule), $Fe(OH)^{4-}$), a dimmer ($Fe_2(OH)_2^{4+}$), a trimmer $Fe_3(OH)_4^{5+}$, and a solid precipitate $Fe(OH)_3$. Additionally, there also a wide range of dissolved polymeric species Fe(III) species with medium and high molecular weight. These polymeric species may well be described by the following general formulations: $Fex(OH)y^{(3x-y)+}$ or $[FexOy(OH)x+r]^{(2x-2y-r)+}$ [2, 15-17].

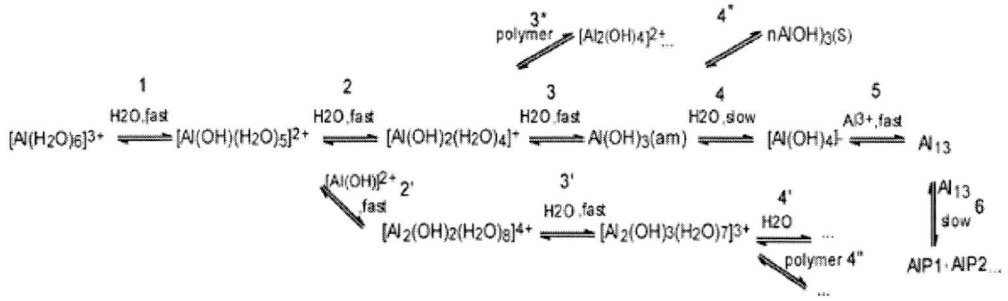

Figure 2. Hydrolysis and possible polymerization reactions of aluminium.

It is worth noting that these mononuclear and polynuclear species interact with the particles in water, promoting their coagulation and flocculation. However, the formation of such species is highly depending on the quality characteristics of the water to be treated (i.e. pH, temperature, alkalinity), which eventually affects the coagulation performance of the metal salts. As discussed later (in Section 3), this drawback led to the development of a new kind of coagulation reagents, the pre-hydrolyzed metal salt coagulants [9, 18].

2.3. Mechanisms of Coagulation

Generally, in an electrolyte the charge of a particle surface is balanced by an equivalent number of oppositely charged ions, which are distributed around the particle. This distribution is described by the Stern model according to which, the distribution of the counter-ions is divided into two zones (Figure 3). The inner zone (known as Stern layer) consists of a compact layer of counter-ions adsorbed onto the surface of particles, while in the outer zone (known as diffuse layer) the counter-ions are distributed according to the intensity of electric (attractive/repulsive) forces and random thermal motion (Brownian motion). Overall, the charge of the particle surface and the layer of counter-ions constitute an Electrical Double Layer (EDL).

When two similarly charged particles approach, the sheaths of counter-ions interact and repel each other. However, if the particles come sufficiently close, the electric repulsive forces become overwhelmed by the attractive forces (known as van der Waals forces) and the particles coalesce. Otherwise, the repulsive electric forces are dominant and, therefore, the particles cannot coalesce, thus remaining in a disperse state (normally termed as stable state). The combination of attractive van der Waals and electric repulsive forces is the basis of DLVO (named after the scientists that developed it; Derjaguin, Landau, Verwey, Overbeek) theory.

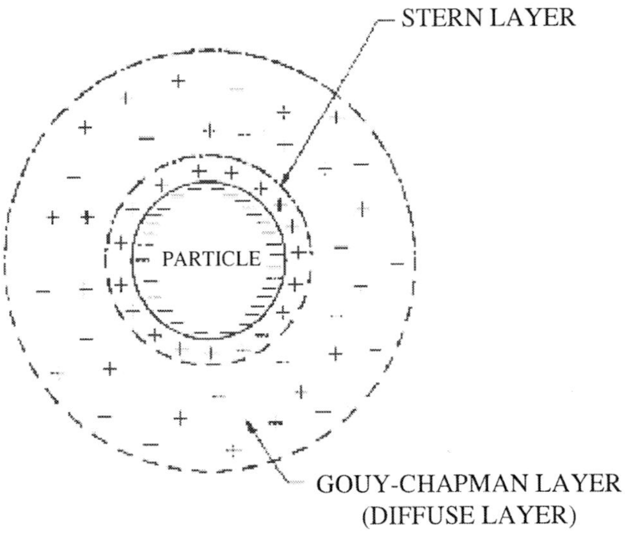

Figure 3. Electrical Double Layer [19].

2.2.1.DLVO Theory

As described earlier, the stability of particles is due to the repulsive electrical forces. Therefore, the net repulsion may be described as an *energy barrier* that must be overcome so as aggregation to occur. The intensity of the energy barrier depends on the initial charge of the particles and on the ionic strength of the solution. Therefore, for achieving an effective agglomeration of the particles, reduce of the energy barrier is required. The latter can be attained either by compressing the thickness of the electrical double layer or by reducing (neutralizing) the charge of the particles. Coagulation reagents, such as metal salts, pre-polymerized metal salts and polymeric coagulants are widely used for that purpose [19, 20].

2.2.2. Double Layer Compression

It is suggested that if the electrical double layer is compressed, the repulsive forces among the particles would be reduced accordingly, thus promoting the approach of the particles via the Brownian motion and the attractive van der Waals forces. Eventually, the energy barrier could be reduced up to zero provided that the ionic strength is sufficiently high. This condition is identified by a critical coagulant concentration (c.c.c.), which is independent from the colloids concentration [19]. Coagulation reagents that act in this manner are described as indifferent electrolytes due to the fact that they do not have any effect on the surface chemistry of the particles but they simply increase the ionic strength of the solution.

The effectiveness of the added coagulant in the destabilization of charged colloids is depended upon the charge of the ion, i.e. the trivalent aluminium ion Al^{3+}, is 1000 times more efficient than the mono-valent sodium ion, Na^{+}. Therefore, counter-ions with high charge are applied, such as Al^{3+} and Fe^{3+}. However, these coagulants are not regarded as indifferent electrolytes, as they significantly interact with the particles by being adsorbed onto the particles surface, thus altering (reducing) their surface charge and eventually lowering the energy barrier. Finally, it is worth noting that double layer compression is not practiced in water treatment due to the large amounts of indifferent electrolytes required. It only becomes usable in particular industrial applications, in which the ionic strength is high.

2.2.3. Charge Neutralization

As mentioned in section 2.3.2, particulates can be destabilized by adsorption of opposite charged ions and/or polymers. Provided that most of the particulates found in surface waters are negatively charged (around neutral pH), coagulation reagents including simple metal salts, pre-polymerized metal salts and cationic organic polymers are suitable for such a purpose. In particular, cationic coagulant species that enter the Stern layer and being held by strong forces are described as being adsorbed onto the particle surface and may affect the surface chemistry by reducing (neutralising) the surface charge. Consequently, the extent of electrical double layer repulsive interaction between adjacent particles is sufficiently reduced thus enabling them to approach each other as the van der Waals forces become dominant. Generally, a coagulant dose is regarded as optimum, when the particle surface is partially (- 50%) covered. The optimum dose is proportional to the surface area concentration of the particulates. When excess coagulant is added, the particle surface will attain a positive charge, resulting in the particles re-stabilisation [8, 9, 19].

2.3.4. Adsorption and Inter-Particle Bridging/Electrostatic Patch Mechanism

Pre–polymerized metal (with high concentration of polymeric species) and organic polymeric coagulants can also operate efficiently via another coagulation mechanism known as adsorption-inter-particle bridging. In particular, polymer chains adsorb on particle surface at one or more sites as a result of a) electric (charge- charge) interactions, b) dipole interaction, c) hydrogen bonding, and d) attractive van-der-Waals forces. When a polymer chains collides with a particle in suspension, a fragment of the polymer adsorbs onto the particles surface. The remaining of the polymer chain may be extended into the solution and adsorb on available surface sites of adjacent particles thus, forming a particle-polymer-particle complex, in which the polymer serves as a molecular bridge (Figure 4). Taking into account that higher molecular-weight polymers are able to be adsorbed on more than one particle, a complex structure of large flocs are formed. The stages of the aforementioned mechanism can be summarized as follows:

1) Dispersion of polymer in the suspension
2) Adsorption at the solid – liquid interface
3) Compression or settling down of the adsorbed polymer

4) Collision of adjacent polymer – coated particles to form bridges and consequently increasingly larger flocs.

However, in case that charged (cationic or anionic) polyelectrolytes are applied to dispersions with particle with opposite sign the adsorption–inter-particle bridging model is often inadequate to thoroughly describe the type of interactions occurring during the process. Such systems include a) anionic polyelectrolytes and positively charged colloidal dispersions (which are already destabilized by the introduction of metal coagulants) and b) cationic polyelectrolytes and negatively charged colloidal dispersions. For those systems, Gregory [18] proposed a different coagulation mechanism known as Electrostatic Patch mechanism (Figure 5). According to that mechanism, the polymer chain is completely adsorbed onto the particles surface and, consequently no inter-particle bridging can be attained, as no polymer fragment is extended within the suspension.

As several polymer chains are adsorbed onto different spots on the particles surface, alternating regions (patches) of positive and negative charges are formed. Although, the particle is electrical neutral in total, this type of charge distribution enables the opposite charged patches of adjacent particles to approach resulting in attachment. The electrostatic patch mechanism is dominant when high molecular weight polymers are employed.

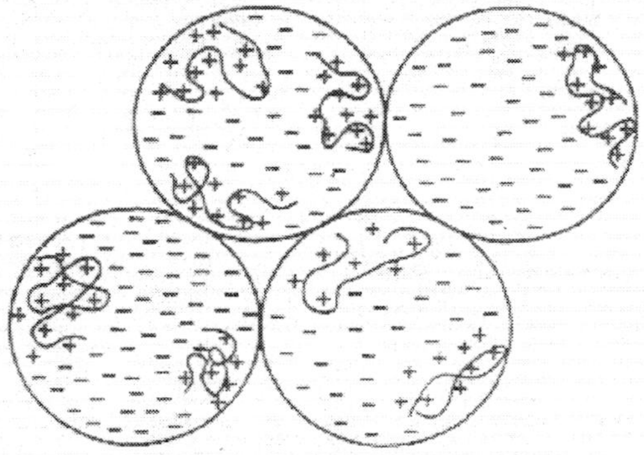

Figure 5. Electrostatic patch mechanism [19].

2.3.5. Precipitation and Enmeshment

When excessively high coagulation dosages are used, iron and aluminium may form insoluble precipitates (mainly metal hydroxides). As these precipitates are formed, particulate matter can be entrapped in the amorphous phase and removed from the solution as the precipitates settle down. This type of coagulation mechanism is known as precipitation and enmeshment or sweep flocculation. Common coagulation reagents for this kind of treatment are aluminium ($Al_2(SO_4)_3$) and ferric ($FeCl_3$) salts. This mechanism can be used in water treatment applications where pH values lie between pH 6 to 8 and aluminium and iron are used at excessively high concentrations, which exceed saturation in respect to the amorphous metal hydroxide solid that is formed. Metal hydroxide is an insoluble, gelatinous precipitate that is initially in the very small colloidal range. Then as these particles collide with each other, larger flocs are produced that enmesh the primary colloidal particles. In general, this mechanism does not depend on the type of particles to be removed, thus the required dosage for an effective coagulation is the same regardless of the type of particles that may be present (in the absence of NOM, as they may interact with the particles). However, the required dosage and, generally, the efficiency of the coagulants are highly dependent on the pH of the solution.

2.4. Evaluation of Coagulation Performance

Because of the random nature of the waters /wastewaters to be treated as well as of the different and (sometimes competing) reactions and coagulation mechanisms, the selection of coagulants (type, dosage) and conditions (mixing and settling times, mixing velocity) are empirically determined using jar tests and/or pilot plant studies. The most widely used bench scale experimental procedure is the "jar test" procedure. The aim of such a procedure is to simulate, to the extent possible, the conditions and optimize the operation of coagulation/flocculation facilities. Jar test procedure can be a quick, accurate, viable procedure for the design modification and operation of coagulation plants, as their scale up correlation with full scale plant is excellent [7]. A modern jar tests apparatus is shown on figure 3. As shown on figure 3 the apparatus consists of six batch reactors, each equipped with a paddle mixer. Square or circular beakers of various sizes can be used.

The jar test procedure consists of 3 phases: a) Rapid mixing phase: The tested suspension is mixed vigorously (time varies from few seconds to several minutes) while a certain coagulant dosage is added, b) slow mixing phase: the first phase is followed by a slow-mixing phase (duration up to several minutes), which simulates the flocculation process, allowing the formation of flocs, c) Sedimentation phase: flocs are allowed to settle and samples are taken from the supernatant for further analysis. The coagulation efficiency is assessed by measuring the following parameters: i) turbidity or suspended solids removal, ii) NOM removal as measured by DOC or as UV_{254} absorbance which is an alternative measure of dissolved NOM, iii) residual metal concentration, iv) pH, conductivity and alkalinity variation, v) zeta potential variation, vi) volume of settled flocs.

Overall, jar tests procedure enables the determination of optimum experimental conditions including the type and dosage of coagulation reagents, the pH and alkalinity of tested suspensions as well as the duration and intensity of the three phases for a certain purpose, i.e. turbidity or NOM removal.

References (For this Part)

[1] Amokrane A., Comel C., Veron J., Landfill leachates pretreatment by coagulation-flocculation, *Water Res.* **31** (1997) 2775-2782.

[2] Tatsi A.A., Zouboulis A. I., Matis K. A., Samaras P., Coagulation-flocculation pretreatment of sanitary landfill leachates, *Chemosphere* **53** (2003) 737-744.

[3] Ntampou X., Zouboulis A. I., Samaras P., Appropriate combination of physico-chemical methods (coagulation/ flocculation and ozonation) for the efficient treatment of landfill leachates, *Chemosphere* **62** (2006) 722-730.

[4] Perez A. G., Ramos A., Moreno, B., Zamorano Toro M., Coagulation-flocculation as a submerged biological filter pre-treatment with landfill leachate in *Waste Management and the Environment II* (2004) 213-222.

[5] Trebouet D., Schlumpf J. P., Jaouen P., Quemeneur F., Stabilized landfill leachate treatment by combined physicochemical-nanofiltration processes, *Water Research*, **35** (2001) 2935-2942.

[6] Wang Z. P., Zhang Z., Lin Y.J., Deng N.S., Tao T., Landfill leachate treatment by a coagulation-photooxidation process, *Journal of Hazardous Materials*, **95** (2002) 153-159.

[7] Crittenden, J.C., Trussel, R.R., Hand, D.W., Howe, K.J., Tchobanoglous, G. (eds.) "Coagulation, mixing and flocculation", in *Water Treatment: Principles and Design.* 2nd ed., J. Wiley and Sons, 2005, New Jersey, USA.

[8] Bratby, J., *Coagulation and Flocculation in Water and Wastewater Treatment*, second ed., IWA Publishing, London, 2006.

[9] Duan, J., Gregory, J., Coagulation by hydrolysing metal salts, *Advances in Colloid and Interface Science*, **100-102** (2003) 475-502.

[10] Bi S., Wang C., Cao Q., Zhang, C., Studies on the mechanism of hydrolysis and polymerization of aluminium salts in aqueous solution: correlations between the Core-links model and Cage-like keggin-Al13 model, *Coordination Chemistry Reviews*, **248** (2004), 441-455.

[11] Jia, Z., He, F. and Liu, Z., Synthesis of poly-aluminum chloride with a membrane reactor: operating parameter effects and reaction pathways, *Industrial Chemical Research*, **43** (2004), 12-17.

[12] Bertsch, P.M., Grant, W.T., Barnhisel, L., Characterization of hydroxyl-aluminum solutions by aluminium-27 Nuclear Magnetic Resonance Spectroscopy, Soil Science *Society of America Journal*, **50** (1986) 825-829.

[13] Wang, M. and Muhammed, M., Novel synthesis of Al13-cluster based alumina materials, *NanoStructured Materials*, **11** (1999), 1219-1229.

[14] Parker, D.R. and Bertsch, P.M., Formation of the Al13 tridecameric poly-cation under diverse synthesis conditions, *Environmental Science and Technology*, **26** (1992) 914-921.

[15] Flynn C.M., Hydrolysis of Inorganic Iron (III) salts, *Chemical Reviews*, **84** (1984) 31-41.

[16] Tang H.X., Stumm W., The coagulating behaviors of Fe(III) polymeric species, *Water Research*, **21** (1987) 115-121.

[17] Jiang J., Development of coagulation theory and pre-polymerised coagulants for water treatment, *Separation and Purification Methods*, **30**, (1) (2001) 127-141.

[18] Gregory R., Rates of flocculation of latex particles by cationic polymers, *Journal of Colloid and Interface Science,* **42**, (1973) 448- 456.

[19] Kim Y.H., Coagulants and flocculants, *Theory and Practice*, Littleton, CO: Tall Oaks Publishing., 1995.

[20] D.H. Bache, R. *Gregory, Flocs in Water Treatment*, IWA Publishing, London, 2007.

3. Coagulation Reagents

The main purpose of the whole process of coagulation-flocculation is the formation of relatively larger particles (flocs) through destabilization and agglomeration than the suspensions ingredients, which can be separated easier usually through gravity settling. Destabilization is achieved with the use of appropriate chemicals, the so-called coagulant agents or coagulants. The coagulants used can be divided into two main categories; the inorganic metal salts and the organic polymers.

The commonly used metal coagulants include aluminium or iron salts, and fall into two sub-categories: the simple metal salts (conventional coagulants) and the pre-polymerized metal salts. After the addition of a metal salt into a sample to be treated, various reactions occur (chapter 2.2). The nature and extent of these reactions depend mainly on the nature of the metal salt, i.e. conventional or pre-polymerized and is strongly correlated with their respective mechanism of action. One or a combination of two or more of the aforementioned mechanisms in chapter 2.3 can be responsible for the destabilization of the colloidal matter. Which mechanism predominates, depends mainly on the nature and the concentration of the metal salt used, i.e the action of pre-polymerized metal salts relies more to charge neutralization and to enhanced particles bridging, whereas in the case of conventional metal salts, enmeshment in precipitate is usually the main mechanism. Excess coagulant addition is required in this case and the process is usually called sweep flocculation, where charge neutralization has a minor contribution to the particles destabilization.

Organic polymers include synthetic or natural (bio-) polymers. They are used in specific cases as primary coagulants, but their main application is for the enhancement of particles aggregation, i.e. as flocculant aids.

In this chapter, the most commonly coagulation reagents are presented (inorganic or organic); their properties are discussed, as well as the latest innovations regarding the development of new coagulant agents are described in detail.

3.1. Inorganic Metal Salts

3.1.1. Simple Metal Salts (Conventional Coagulants)

The commonly used metal coagulants fall into two general categories: those based on aluminium and those based on iron. The most widely used metal coagulant is probably the aluminium sulfate ("alum"), which has been used for water treatment since ancient times and was first mentioned by Pliny (approx. 77 AD) [1]. It is commonly manufactured from the reaction of bauxite ores with sulfuric acid. Evaporation of water in the process results in a dry product, having the approximate formula $Al_2(SO_4)_3 \cdot 14\text{-}18H_2O$, and with an Al content ranging from 7.4 to 9.5%. It is supplied as powder or in liquid form and its applications include water or wastewater treatment.

A product prepared from alum and sulfuric acid is acidified aluminium sulfate (acid alum). It is supplied as a liquid and is used mainly in paper-making industry. Other Al coagulants are the aluminium chloride ($AlCl_3.6H_2O$) and the sodium aluminate ($NaAlO_2$). The first one is supplied as liquid (containing 10.5% Al) and is used widely for sludge conditioning. The latter is usually supplied as a viscous, strongly alkaline and corrosive liquid (containing 13% Al). It is rarely used alone, but generally with alum to obtain some specific results, i.e. in the treatment of highly colored waters [2].

Iron-based coagulants can be used over a pH range from 4 to 11. A common product is ferric sulfate ($Fe_2(SO_4)_3.3H_2O$), which is available in solid (usually 20-21% Fe w/w) or liquid form (usually 11.5% Fe w/w). Ferrous sulfate ($FeSO_4.7H_2O$) is available as water soluble crystals or granules containing 20% Fe. Ferrous sulfate reacts with alkalinity to form soluble ferrous hydroxide ($Fe(OH)_2$), which has to be oxidized to ferric hydroxide ($Fe(OH)_3$) in order the coagulant to work (e.g. with aeration or chlorine addition). Other iron products include chlorinated ferrous sulfate and ferric chloride, which is available in the liquid, crystal or anhydrous forms. The liquid form is the most common, but care should be taken during storage and handle, as the solution is extremely corrosive [2, 3].

The application of simple metal coagulants is widespread, especially due to the relatively low cost and the simpler application route. After the addition of a conventional coagulant in the treated sample, hydrolysis of metal cation occurs, accompanied by polymerization reactions in an unknown extent (chapter 2.2). Figure 6 illustrates the reaction pathways that the aluminium hydrolysis/polymerization species may follow [4]. Pathways A and B represent the reaction of metal hydrolysis products with metal binding sites of the contaminants, with subsequent charge neutralization if the particles are negatively charged. If coagulant dosage increases, $Al(OH)_3$ is formed (pathways C and D), NOM reacts with $Al(OH)_3$ and adsorbs in a certain extent on the micro-crystals. Further increment of coagulant dosage leads to sweep flocculation, whereas pathway E represents the possibility of soluble metal species formation. The action of simple metal salts usually relies on the sweep flocculation mechanism, which requires excessive coagulant dosage and pH adjustment before treatment (pH range of minimum solubility of $M(OH)_3$, pH 6-7 for Al and 7-8 for Fe coagulants respectively). pH adjustment usually is needed after treatment also, as the excessive coagulant dosage results in a significant drop of sample pH. Generally, treatment with conventional coagulants is pH and temperature depended, sensitive to sample characteristics and composition, and often leads to excessive sludge production.

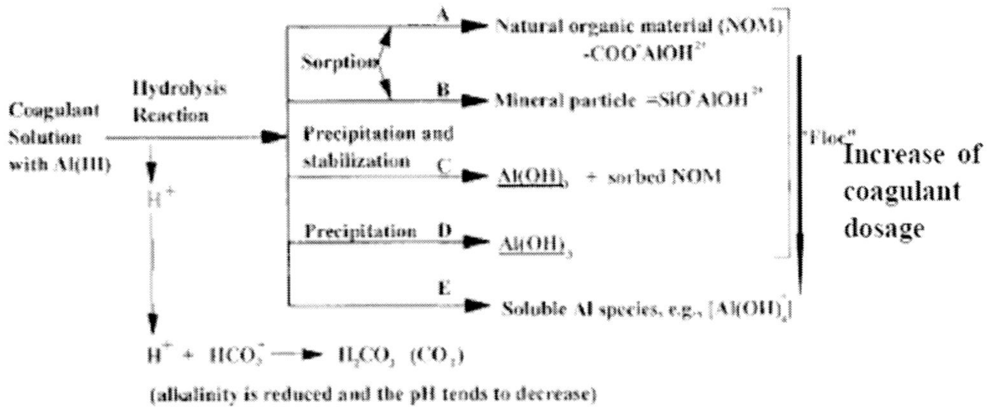

Figure 6. Metal salts coagulation reaction pathways [4].

3.1.2. Pre-Polymerized Coagulants or Inorganic Polymeric Flocculants (IPFS)

The disadvantages of conventional coagulants and the increasing demand for more efficient coagulation treatment were the main reasons for the development of coagulants with improved properties, in comparison with the conventional ones.

After studying the aquatic chemistry and behavior of simple metal salts, the way for improvement seemed to be their (partial) polymerization. The result of these efforts was the production of a range of pre-polymerized solutions, referred as polyaluminium chlorides (PACl), polyaluminium sulfates (PAS), or polyaluminium chloro-sulfates (PACS), in the case of aluminium and poly-feric sulfates (PFS) or poly-feric chlorides (PFC) in the case of iron, with variable degrees of polymerization. These products and especially PACl and PFS are used extensively worldwide since the middle '80s, with an ever increasing demand. Their properties were intensively examined and have proved to be more efficient in lower dosages, in wider pH, temperature and colloids concentration ranges, than the conventional simpler ones, leading to cost and operative more effective treatment [3, 5].

Their superiority is related to the different metal species distribution in the polymerized, as compared with the non-polymerized solutions. Particularly, the aforementioned hydrolysis – polymerization reactions (chapter 2.2) which occur after the dissolution of a metal salt are entirely uncontrolled in the case of conventional coagulants. The coagulation efficiency is based more to $M(OH)_3$ formation (around the pH 6.5-8) and less to charge neutralization, as the hydrolysis reactions are proceeding fast and hence the concentration of positively charged Al/Fe ions is restricted. In the case of IPFs, the pre-polymerization aims to control up to a certain extent these reactions resulting in products containing pre-formed polymerized species. Such species are resistant to further hydrolysis and consequently, they possess superior coagulation properties when used in water treatment and they result in a minor impact of pH value after treatment. Particularly, in the case of aluminium the better coagulation properties are attributed mainly to the presence of a specific polymeric compound, Al_{13} ($AlO_4Al_{12}(OH)_{24}(H_2O)_{12}^{6+,7+}$) with ε-Keggin structure. In the case of iron, no specific compound is distinguished, but the sum of the medium polymeric species is thought to be responsible for the higher efficiency.

Polymerized coagulant agents can be manufactured with various substitutions and polymerization degrees, according to the nature of the sample to be treated. The preparation of PACl generally involves the production of two solutions that contain Al [6-8]. The first is called basic aluminium chloride and is produced through processing an aluminium containing raw material (Al_2O_3, Al containing mineral, e.g. bauxite) under vigorous conditions (e.g. high temperature, pressure, extended contact time) with concentrated acid (e.g. HCl, H_3PO_4 or mixture of them). The second is called sodium aluminate (formulated as $NaAlO_2$, $Na_2O.Al_2O_3$ or $Na_2Al_2O_4$ in solid state) and can be produced by the dissolution of the aluminium raw material in a strong alkaline solution (e.g. NaOH, Na_2CO_3, $NaHCO_3$, KOH, etc.). The production of PACl is accomplished by the introduction of sodium aluminate solution into the basic aluminium chloride solution under stirring, resulting in a product that can be described with the generic formula $Al_n(OH)_mCl_{3n-m}$ (when HCl is used), where $0<m<3n$. The reactions that can describe the preparation of the two solutions are the following (with the use of hydrated alumina, HCl and NaOH):

$Al_2O_3.3H_2O + 2n\ HCl \rightarrow 2\ AlCl_n(OH)_{3-n} + 2n\ H_2O$ (basic aluminium chloride)

$Al_2O_3 + 2\ NaOH \rightarrow 2\ NaAlO_2 + H_2O$ (sodium aluminate)

Various PACl preparation methods exist, based on different starting materials [e.g. 7-12] or on certain modifications during the preparation procedure, e.g. electrochemical dissolution of Al raw material [e.g. 13], without use of alkaline solution [e.g. 9, 14], or use of stabilizing agents [e.g. 15]. More recently, alternative PACl preparation methods appeared in the literature, suggesting relatively simpler and more cost effective PACl preparation procedures [e.g. 16].

A process for commercially preparing polyferric sulfate coagulant is the following [17]: acidic solution of ferrous sulfate is oxidized to form ferric sulfate in an oxidation stage at ambient pressure and relatively high temperature (90° C) and in the absence of oxidation catalyst, using an oxidizing agent, preferably nitric acid. Then, the ferric sulfate is partially neutralized by the addition of sodium bicarbonate solution. The ferric sulfate is kept in contact with the base at the temperature selected (i.e. 50° C) for a time sufficient for polymerization of the hydrolyzed ferric sulfate, notably 3-5 hours. The PFS structure can be described by the simplified formula $[Fe_2(OH)_n(SO_4)_{(3-n/2)}]_m$ where m is a function of n and $n=2y/x$ with $n \leq 2$.

The nature and composition of the final product varies according to the preparation method and the starting materials used. The preparation parameters that affect the distribution of the metal species include the temperature, the base addition rate, the mixing rate, aging time, presence of other anions (e.g. Cl^-, SO_4^-, CO_3^{2-}) and the concentration of Al(III), or Fe(III) ions. The most important factor, however, is claimed to be the molar ratio of bound hydroxide to the concentration of metal ions regarding the total metal hydrolysis products, i.e. [OH]/[M] (also referred as r or m/n) and is used to describe the polymerization degree of coagulants that are prepared mainly in laboratory scale. Usually it takes values from 0.5 to 2.5 and 0.25-1.5 for aluminium and iron products respectively, as for values higher than 2.5 or 1.5 it has been observed that the solutions are unstable and precipitation occurs. The higher the basicity of a coagulant, the lower is the alkalinity consumption, and the lesser is the pH reduction of the treated sample.

In the case of aluminium commercial products, the term relative (%) basicity is more common, representing the degree at which the hydrogen ions produced by hydrolysis of the metal ions are pre-neutralized [3]:

$$\text{Relative basicity (\%)} = ([OH] / [M] * z) * 100$$

Where z corresponds to the charge on metal species (for Al z=3). The influence of all the above mentioned parameters has been extensively examined [18-27].

Aluminium or iron pre-polymerized products are available in solid or liquid form with variable degrees of polymerization and metal content, whereas standards for several of these products have been recently established. In particular, according to the American Water Works Association (AWWA), the main characteristics of a PAC solution should be: aluminium content 2.5-13 %w/w and basicity 10-83 [28].

3.2. Organic Polymers

3.2.1. General

Organic polymers are long-chain (linear or branched) molecules, consisting of repeating small subunits or monomers. If the repeating units are of the same molecular structure, the compound is referred as homo-polymer. If more than one type of subunits are existing, the compound is called copolymer. According to their solubility in water, polymers can be water-insoluble, water-soluble, or water-dispersible. The existence of polar or ionizable functional groups designates whether a polymer is water soluble or not. Water-soluble polymers contain several ionizable groups, such as $-C=O$, $-OH$, $-NH_2$, $-CONH_2$, $-COOH$, $-SO_3H$, and are termed polyelectrolytes. When these groups dissociate, the polymer becomes charged, either positively or negatively, and is called cationic or anionic polyelectrolyte respectively. Polymers that possess both positively and negatively charged sites are referred to as ampholytic. If no ionizable groups are present, the polymer is referred as non-ionic.

The water soluble polymers are of great importance regarding water or wastewater treatment. They have been utilized in coagulation/flocculation process for water purification for more than four decades (synthetic), whereas natural polymers are known as flocculants since ancient times. Their principal uses are as primary coagulants (cationic polyelectrolytes) as well as in the more traditional flocculation step of binding already formed small flocs into larger aggregates (flocculant aids, anionic and non-ionic polyelectrolytes). The main mechanisms of polymers action are the electrostatic patch mechanism and the bridging mechanism. The bridging mechanism enhances the rate of orthokinetic flocculation, by inducing the destabilized particles (small flocs which are formed after the addition of a coagulant) to collect into larger aggregates (large flocs). The whole process of coagulation-flocculation with the supplementary use of a flocculant aid can be described as following:

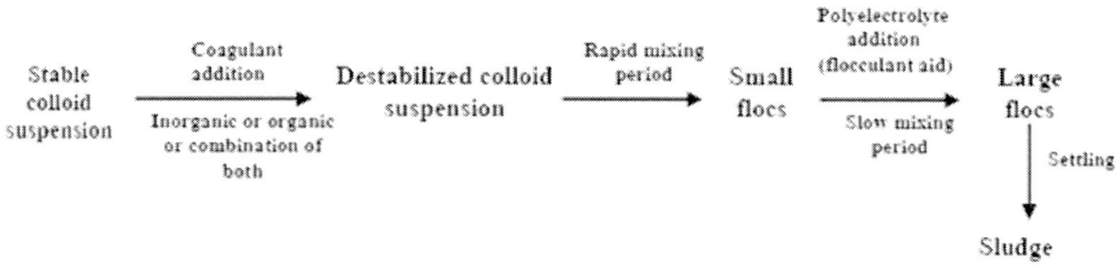

The bridging mechanism involves four stages, the dispersion of polyelectrolyte in the suspension, the adsorption at the solid-liquid interface of the destabilized particles, compression or settling down of the adsorbed polyelectrolyte and collision of adjacent polyelectrolyte-coated particles to form bridges and thereby increasingly larger flocs [2]. Figure 7 represents schematically these stages [29]. The bridging mechanism describes the action of anionic or non-ionic polyelectrolytes as flocculant aids. On the other hand, the electrostatic patch mechanism is more suitable to describe the action of the cationic polyelectrolytes as primary coagulants, and perhaps also the action of anionic polyelectrolytes applied in a destabilized dispersion which may be positively charged. According to this mechanism, the polyelectrolyte after the dispersion will adsorb where there are domains of opposite charge to that of the polyelectrolyte, leading to an excess of charge associated with the polymer. When the polyelectrolyte has a high concentration of charges, it is likely that it will adsorb onto the particles in a rather flat configuration, leading to high-intensity patches or domains with localized excesses of the polymer's charge. Direct electrostatic attraction between the particles is hence encouraged and domains of opposite charge in adjacent particles will combine, resulting in agglomeration [2, 29, 30].

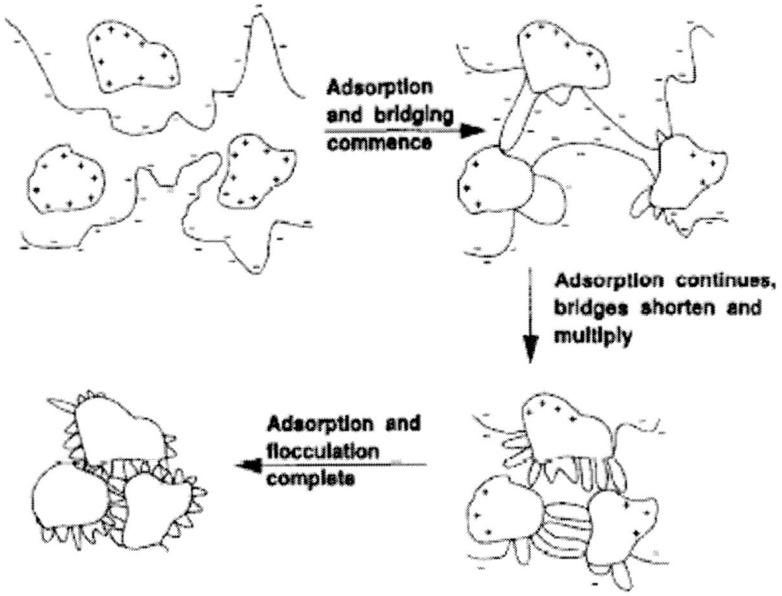

Figure 7. The bridging mechanism [29].

The most important properties of a polymer, which designate at a certain degree the most suitable application and the eventually treatment efficiency, are the charge density (CD) and the molecular weight (MW). Charge density is the number of charged groups per monomer unit of the polymer, and is often presented as a mole fraction (mole %) of the amount of charged monomer moles divided by the total mole of monomers in the macromolecule [31]. Charge density is also cited as C g^{-1} or meq g^{-1}. The molecular weight of a polymer characterizes the extent of polymerization; high molecular weight indicates long chains, whereas low molecular weight indicates short chains. In table 2, a classification of polymers according to their molecular weight and charge density is presented [29, 31].

Generally, low molecular weight ($MW < 10^5$) polymers with high (cationic) charge density (CD 50-100%) are preferred as primary coagulants, whereas high molecular weight ($MW > 10^6$) polymers with weak anionic charge density (CD < 25%) or no charge are used as flocculant aids. Independently of their molecular weight and charge density, the polymers used in water treatment can be divided into two main categories; the natural and the synthetic polymers. These two categories will be described further, as well as several implications regarding the use of polymers that should be taken into consideration.

3.2.2. Natural Polymers

Natural polymers have been used as flocculants since the ancient times. They can be cationic, anionic or non ionic and can be obtained from a number of sources, including potatoes (starches), tapioca, plant seed extracts, crushed kernels. Usually they are applied in conjunction with inorganic metal salts, even if they are used as primary coagulants. Their advantages are that they are toxic free, environmentally biodegradable and the raw products are often locally available, reducing the production costs. However, their effectiveness is questionable and they have to be tested in every case before their possible application. Some examples of natural polymers are given below.

A common cationic natural polymer is chitosan, obtained from chitin shells (crab, lobster, etc.). The commercial product is of medium MW and has a CD which is pH-dependent and can be as high as 80 mol%. Its applications include decolorizing dye-house effluents, treatment of food processing wastes, metal ion removal and sludge conditioning [29]. Appropriately modified starches can be converted to cationic polymers, which usually have quaternary ammonium group substitutions.

Table 2. Classification of polymers according to their molecular weight and charge density

Classification	Molecular weight range (Da)	Charge Density (mol %)
Low	$< 10^5$	~10
Low-medium	$10^5 - 2.5 * 10^5$	
Medium	$2.5 * 10^5 - 10^6$	~25
Medium-high	$10^6 - 5 * 10^6$	
High	$5 * 10^6 - 10^7$	50 - 100
Very high	$> 10^7$	

The product has medium MW and the CD can be low or medium. Finally, lignin based cationic polymers have also been reported, prepared by modifying kraft lignin to give it cationic character.

Natural starches without any modifications have basically anionic character (carboxylic substitutions). They can be processed from various sources, including potato, corn, cassava (manioc), arrowroot, and yams. Starches are branched and nonlinear gluco-pyranose polymers, have extremely low CD, MW of several million Da and are used in concentrations of 1 to 10 mg/L in water purification, preferably together with aluminium salts. A natural anionic modified polysaccharide used is lignin sulfonate, prepared by sulfonating low-MW kraft pine lignin [2, 3, 29].

Non-ionic natural polymers include starches, galactomannans, cellulose derivatives, microbial polysaccharides, gelatins and glues. Tannins have been also used extensively in drinking water, wastewater and industrial effluent treatment applications. The non-ionic natural polymers are relatively unaffected by pH and ionic strength, but they are subject to (enzymatic) degradation reactions during storage [2, 29].

3.2.3. Synthetic Polymers

Although natural polyelectrolytes are biodegradable and have the advantage of being toxic-free, the use of synthetic polyelectrolytes is more widespread. They are in general more effective as flocculants, principally due to the possibility of controlling properties, such as the number and type of charged units and the molecular weight during the manufacturing process. Synthetic polymers with the desired properties indicated for a certain application can be purchased, resulting in better treatment efficiency. Moreover, they are cheaper, than those made from natural sources.

Cationic Polymers

The main application of synthetic cationic polyelectrolytes is as primary coagulants. Their advantages are the lower sludge production, the lower alkalinity consumption, thus having a minor impact on pH of treated sample, than the inorganic metal coagulants. Moreover, the performance of cationic polymers is practically independed of treated sample pH and alkalinity. However, their efficiency is not always the same with the efficiency of metal coagulants, and usually they are used together with inorganic metal coagulants, achieving satisfactory treatment results with reduced metal coagulant dosage (40-80 %).

There is a great variety of molecular structures in the synthetic cationic polymers. Generally, they contain quaternary ammonium groups so that the formal positive charge is present irrespective of the pH level. Weak cationic polyelectrolytes versions are based on polyamines containing primary, secondary or tertiary amino groups, or mixtures of them. The preparative techniques cover polymerizations by chain growth and step growth mechanisms, often in simple aqueous solutions, but also as water-in-oil emulsions, and the modification of existing polymers [30].

Several cationic polymers are derived from acrylamide polymerization. After poly-acrylamide preparation through homo-polymerization of acrylamide, the amide group reacts with formaldehyde and a secondary amine, to produce the so-called Mannich bases [31]. Other examples of cationic poly-acrylamides are the copolymers of quaternized dimethyl-amino-ethyl acrylate or methacrylate, which are extensively used in the water industry (especially for biologically sludge conditioning) [2]. The cationic content in these polymers can vary from 10 to 100 mol% and several products are commercially available, with well studied properties. Another important group of cationic polyelectrolytes are those derived from diallyl-dimethyl-ammonium chloride (DADMAC) polymerization. DADMAC is synthesized from allyl chloride and dimethyl-amine. The homo-polymer derived from DADMAC polymerization is referred as p-DADMAC and can have a MW up to 2-3 10^6 Da. p-DADMAC products are available as viscous aqueous solutions at concentrations up to 40% w/w and are used extensively in water or wastewater treatment [2, 29, 31]. In Figure 8 the synthesis route of p-DADMAC is presented [31].

Anionic Polymers

Anionic synthetic polyelectrolytes are widely used as flocculant aids in water or wastewater treatment. The most common are those containing weakly acidic carboxylic acid groups, with a pH-depending charge density. Polymers with sulfonic acid groups (strongly acidic groups) are also utilized [29].

The predominant anionic polymer types are those derived from acrylamide monomer, the generally termed poly-acrylamides. Acrylamide monomer is produced by the catalytic hydration of acrylonitrile, which is produced from propylene. Acrylamide has a very fast propagation rate and a high exothermic heat of polymerization [31], thus at high concentrations (25-30% w/w) quick homo-polymerization occurs with subsequent formation of a rubbery gel (Figure 9a). By controlling the synthesis conditions (e.g. appropriately modified surfactant system) dry products or water-in-oil emulsions can be produced.

Poly-acrylamide homo-polymer carries no charge, thus it is a non-ionic polymer. The poly-acrylamide anionic polymers are produced mainly according to two procedures. The first one includes partial polymerization of poly-acrylamide and subsequent reaction with, e.g., sodium hydroxide at elevated temperatures, leading to polymers containing acrylamide and acrylate groups. The second is more common nowadays and includes co-polymerization of acrylamide with other vinyl monomers which already have anionic charged groups, such as acrylic acid, or one of its alkali metal or ammonium salts. The products contain carboxylic acid groups; have usually very high molecular weight and low charge density.

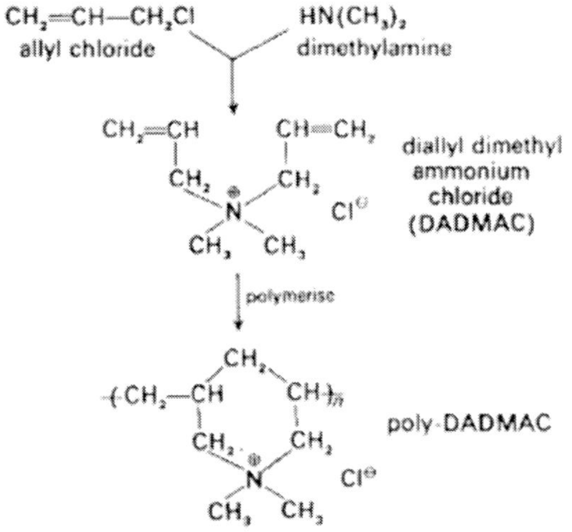

Figure 8. Synthesis route of p-DADMAC [31].

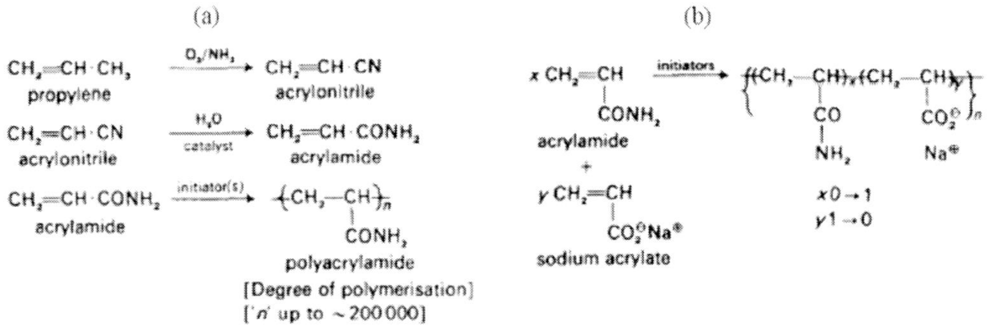

Figure 9. (a) General synthesis pathway of polyacrylamide, (b) Anionic acrylamide copolymers [31].

The majority of anionic poly-acrylamides are produced from acrylamide and sodium acrylate (Figure 9b) [31]. For specific applications (i.e. dispersion of scale-forming precipitates and the inhibition of crystallization of potential deposits), very low MW (< 10000) and high CD polymers can be produced by homo-polymerization of acrylic or methacrylic acids, or by extensive hydrolysis of acid precursor groups.

Anionic acrylate or acrylamide polymers containing sulfonate groups are commercially available as high-MW flocculants or low-MW deposit control agents. These include homo-polymers of sodium 2-methacryoyloxyethyl sulfonate and sodium 2-acrylamido-2-methylpropyl sulfonate, and their copolymers with acrylamide or acrylic acid [29]. Similar useful polymeric flocculants have been made from hydroxy-substituted species such as 3-methacryloyloxy-2-hydroxypropyl sulfonate by starting with the appropriate monomers, or by reacting poly-acrylamide or copolymers of acrylamide and acrylic acid with N-(2-hydroxy-3-sulfopropyl)amide. Sulfated hydroxy polymers are also known, an example being the sulfate ester made from poly(viny1 alcohol) by reaction with chloro-sulfonic acid, but these are hydrolytically unstable at extreme pH levels.

Finally, phosphorus containing polymers are also known, but their use is mainly in deposit control and as sequestering agents because of their chelating properties. Examples are the poly-vinyl-phosphonic acid and the phosphonated poly-ethylene [29].

Non-Ionic Polymers

Non-ionic synthetic polymers include homo- or co-polymers of acrylamide. Because of the hydrolysis of the amide groups, it is possible that non-ionic polymers of that type may contain some anionic groups, usually 1-3%. This content can be minimized by appropriately controlling the monomer concentration, pH, temperature and initiator type. Another type of non-ionic polymers is the poly-ethylene oxide polymers, which can have low to high molecular weight.

3.2.4. Implications Regarding the Use of Organic Polymers

Synthetic organic polyelectrolytes are commercially available in the form of dry powder, granules, beads, aqueous solutions, aqueous gels, and oil-in-water emulsions. Generally, liquid systems are preferred because they require less floor space, reduce labor requirements, and reduce the potential for side reactions because the concentrate can be diluted in the automatic dispensing systems. Their applications include [29, 31, 33]:

- Primary coagulation in drinking water or wastewater treatment
- Flocculant aids in drinking water or wastewater treatment
- Industrial wastewater treatment
- Municipal sewage treatment
- Weighted flocculation
- Sludge thickening
- Paper and board production
- Mining industry applications
- Oil drilling and recovery
- Metal ion removal

The use of organic polymers exhibits several advantages. They can be quite effective in water or wastewater treatment with significantly lower concentrations of inorganic salts (reduction of inorganic salt conc. up to 80% wt.). They do not add to the total solids concentration and the volume of sludge produced can be reduced by 50 to 90%. Polymeric coagulants do not affect pH and moreover their efficiency is independent of the pH value of treated sample, therefore no pH correction is needed after or before treatment [29].

Despite their advantages, there are several implications regarding the application of organic polymers that should be taken into consideration before application. The overall operating costs may be higher, especially when the polymer is supplied at dry form, as a separated dissolution set-up is needed in this case. Specific polymer types are suitable for each type of water to be treated, which can be determined only through extensive experimental procedures, which are becoming more difficult as in many cases the provision of chemical and technical data by manufacturers is inadequate. A very important matter regarding synthetic polymers is the residual toxicity that possibly implies their use. The toxicity is mainly due to remaining unreacted monomers, as the polymers used in water treatment are generally of very low toxicity. Cationic types are more toxic than anionic or non-ionic, and several countries have constricted the use of synthetic polymers in drinking water treatment (e.g. Japan, Switzerland) [29]. Finally, it has been reported that cationic polymers as primary coagulants in wastewater treatment are unable to remove phosphates [34].

3.3. Latest Developments – Modified Pre-Polymerized Metal Salts

3.3.1. General

Coagulation – flocculation is a very important step in water and in wastewater treatment. Regardless of the nature of the treated sample (e.g. various types of water or wastewater) and the overall applied treatment scheme, coagulation-flocculation is usually included, either as pre-, or as post-treatment step. The efficiency of coagulation-flocculation strongly affects the overall treatment performance; hence, the increase of the efficiency of coagulation stage seems to be a key factor for the improvement of the overall treatment efficiency.

The first effort to improve the coagulation-flocculation efficiency was the development of the pre-polymerized coagulation reagents. However, in spite of their improved properties, the pre-polymerized coagulants performance still remains inferior, when compared to the performance of organic polyelectrolytes, e.g. when used in water treatment [35]. Considering the drawbacks of organic polymers, especially when used as primary coagulants (cationic polyelectrolytes), the increasing need for drinking water of high quality, and the stricter discharge criteria of wastewaters, it is evident that more effective inorganic-based reagents are needed.

In the beginning of the 21st century, the relevant research in the coagulation-flocculation field has focused mainly in understanding the behaviour and aquatic chemistry of pre-polymerized coagulants, such as polyaluminium chloride or poly-feric sulfate, and to the further improvement of their properties. The main reason for the higher efficiency of organic polymers is their higher molecular weight (MW), which implies better flocculation properties. Thus, the increase of molecular weight and size of the pre-polymerized coagulants is thought to be the way for further improvement. Two different approaches were taken into consideration; the increase of the effective polymeric species proportion and the addition of other components to produce new composite coagulants.

According to the first one, several efforts have been made to establish an alternative preparation method which will increase the content of the effective metal species. For example, the synthesis of PACl by utilizing a membrane reactor was proposed [36], or even electrochemical processes were also adopted [37], resulting in PACl solutions with higher content of the active specie Al_{13}. In the case of iron, a microwave assisted preparation of PFS was proposed [38]. However, these techniques were applied only in laboratory scale experiments and have not been evaluated using higher (commercial grade) Al or Fe content. In industrial scale, despite the achievement of high basicity values (up to 80%), the greater content of metals inhibits the effective polymeric species formation, as it is more possible that the formed species aggregate to form bigger molecules, with no benefits for coagulation.

The second approach seems to be more promising. The general concept followed is, the introduction of various additives in the structure of a pre-polymerized coagulant, in order to produce a homogenous, stable product with higher MW and improved coagulation-flocculation performance, than the initial reagent. The challenge to confront is the desirable combination of higher efficiency and cost-effectiveness, which are the basic prerequisites for the development of new products. Various additives have been examined, which can be classified into two main categories; inorganic and organic. These compounds and the respective products derived will be described in the following.

3.3.2. Modified Pre-Polymerized Metal Coagulants with Inorganic Additives

A suitable additive for the improvement of flocculation properties is thought to be polymerized silica. Hasegawa et al. [39] noticed that by introducing metal ions into polymerized silicic acid solution, the molecular weight of the product increased and the respective stability and coagulation performance were improved. In this case, the product was rather an inorganic metal-poly-silicate flocculant, where silica was the main component, than a pre-polymerized inorganic coagulant. More recently, research has focused in the incorporation of silica within the pre-polymerized metal solutions, aiming to products with larger molecular weight by the application of two techniques; i.e. either by introducing polymerized silica in the pre-polymerized metal solution, or by introducing polymerized silica in the metal solution, followed by polymerization. The first method is referred as composite polymerization, whereas the latter one as co-polymerization.

Polymerized silica can be obtained when the concentration of mono-silicic acid ($Si(OH)_4$) in aqueous solution exceeds the solubility limit of amorphous silica, or otherwise by bringing the solution temporarily into the pH-concentration range of over-saturation with respect to amorphous silica [40, 41]. Excess monomers in the solution gradually disappear to form polymers. The polymers, or poly-silicic acid, consist of silica tetrahedrons that are linked via silicon-oxygen-silicon bonds [42], e.g.

$$\equiv Si\text{-}OH + HO\text{-}Si\equiv \rightarrow \equiv Si\text{-}O\text{-}Si\equiv + H_2O$$

Various species can be formed, e.g. $Si_2O_3(OH)_4^{2-}$, $Si_3O_5(OH)_5^{3-}$ etc., and exact nature and the extent of their formation is affected by the temperature, the concentration of silica, the presence of other ions, the pH value and the aging time. Soluble, internally hydrated, polymeric silicic acids of colloidal dimensions, or discrete particles of colloidal size, can be present. The most important parameter that affects the kinetics of silica polymerization is the pH value. Particularly, the polymerization is thought to be catalyzed by hydroxyl anions and is very fast at neutral or slightly alkaline pH range. Over pH value 9, the reaction rate decreases. Furthermore, by aging the polymerization can lead to the formation of bigger aggregates and finally, to gel formation (amorphous $(SiO)_n$). In order to avoid gelation and to preserve the poly-silicates at soluble form, the solution has to be brought outside the insolubility range. This can be achieved by acidifying the solution to pH values 2–3. Particularly, at pH around 2 the polymerization rate is retarded enough and the poly-silicates are stable in soluble form for several days [40-43].

Based on these observations and the use of a suitable silicic acid source, i.e. the sodium silicate solution or water glass, several efforts have been made for the introduction of acidified poly-silicates into the structure of pre-polymerized coagulants. The mixed product in the case of aluminium is called polyaluminium silicate chloride and in the case of iron poly-feric silicate sulfate (or chloride), though it should be rather referred as a category of coagulation reagents, because with different combinations of OH/M or M/Si molar ratios, various reagents can be produced.

Through studies examining the chemical species distribution in these new products, it was revealed that the incorporation of silica in the structure of polymerized metal solutions resulted in the decrease of active polymeric species content (Al_{13} in the case of aluminium) [44-47] and the reduction in the charge neutralization capability of coagulants [44, 48, 49], as compared to the respective PACl or PFS solutions. However, the new coagulant agents have shown superior coagulation performance, than the conventional or simple pre-polymerized coagulants in water or wastewater treatment (see chapter 4.2). The better performance can be attributed to the alteration in a certain degree of the coagulants composition. Specifically, it has been proved that the introduction of silica chains into the structure of PACl or PFS increases the molecular weight of the coagulants components with the formation of metal-silicate complexes, thus leading in enhanced aggregating power and bigger-denser flocs formation, and consequently, in more effective coagulation performance [35, 49-52].

3.3.3. Modified Pre-Polymerized Metal Coagulants with Organic Additives

As alternative additives in order to increase the MW of the inorganic coagulants, organic polymers are also under consideration. The combination of an inorganic salt (coagulant) and a suitable poly-electolyte (flocculant) into one reagent is thought to improve, apart from the treatment efficiency of the inorganic salt, the overall cost-effectiveness. No addition of a flocculant aid will be needed any more, whereas the introduction of the polyelectrolyte into the structure of the metal salt will also significantly reduce the eventual residual toxicity.

These suggestions lead to the production of modified inorganic coagulants with organic additives, a field that is under investigation nowadays. All the aforementioned charged or not charged polyelectrolytes (chapter 3.2) can be used for this purpose, with different impact in the nature of the new product. Particularly, by the introduction of a cationic polyelectrolyte into the structure of a pre-polymerized metal coagulant, the charge neutralization capability is mainly expected to be improved, and in a lower extent, the aggregating effect as well. Moreover, the usage of an anionic polyelectrolyte with higher molecular weight (compared to cationic) would further improve the aggregating ability. In this case, however, the anionic polymer should have weaker anionic charge, in order not to deteriorate extensively the charge neutralization strength of the combined coagulant agent. A non-ionic polyelectrolyte would have a minor effect in the coagulants charge density (possible negative), but would increase significantly the flocculation ability. Key factors affecting strongly the properties of the combined, modified coagulant are the charge, MW and content of the polyelectrolyte [53].

Some efforts have been reported, regarding the combination of PACl with various anionic, cationic and non ionic organic polymers [54] and the combination of PACl or PFC with poly-DADMAC [55-57]. Although the research in this field is in an initial stage, promising results were obtained regarding their coagulation performance (see chapter 4.2).

The introduction of the polyelectrolyte can be accomplished by two procedures, as in the case of silica addition, i.e. with the co-polymerization, or the composite polymerization. The content of the polyelectrolyte (PE) in the final product is usually expressed as the Al/PE ratio (mg/L), and together with the aforementioned parameters in the case of pre-polymerized coagulants, designates the properties of the final product.

It should be mentioned that the research in the coagulation-flocculation field is continuing. Optimalization of the preparation conditions is needed, especially in the case of composite coagulants with organic additives, and determination of the specific properties that imply superior coagulation performance. Coagulants with different basicities and/or additives content may have different applications (water/wastewater treatment) (see chapter 4.2). Moreover, suggestions for the preparation of new reagents appear in the literature continually, i.e. different combinations of Al and Fe in one reagent [58] and possible addition in a second step of an inorganic or organic compound (e.g. poly-aluminium-ferric-silicate-chloride) [59]. It can be concluded that the tendency in the coagulation field nowadays is the production of modified composite coagulants, which they are becoming more and more complicated, regarding their composition. The variety of the possible additives and the different possible mixing ratios of them state clearly that the concept of "mixing" to produce superior chemicals is in an initial stage and more research efforts can be devoted to this purpose.

References (For this Part)

[1] Duan, J., Gregory J., Coagulation by hydrolyzing metal salts, *Advances in Colloid and Interface Science,* **100-102** (2003) 475-502.

[2] Bratby, J., *Coagulation and Flocculation in Water and Wastewater Treatment,* second ed., IWA Publishing, London, 2006.

[3] J.C. Crittenden, R.R. Trussel, D.W. Hand, K.J. Howe, G. Tchobanoglous (eds.), Coagulation, mixing and flocculation, in: Water Treatment: Principles and Design, second ed., John Wiley and Sons, New Jersey, 2005, p.p. 643-779.

[4] Letterman, R.D., Amirtharajah, A., O'Melia, C.R., Chapter 6. Coagulation and Flocculation, in Letterman, R.D. (ed.): *Water Quality and Treatment*, 5th Ed., McGraw-Hill, NY, 1999, pp. 6.17-6.30.

[5] S. Sinha, Y. Yoon, G. Amy, J. Yoon, Determining the effectiveness of conventional and alternative coagulants through effective characterization schemes, *Chemosphere* **57** (2004) 1115-1122.

[6] Dulko, J.M., 1999. *Poly-aluminum chlorides and poly-aluminum chloro-sulfates methods and compositions*. United States Patent 5985234.

[7] Kvant, M. and Stendahl, K. 1993. Process for the preparation of poly-aluminum compounds. United States Patent 5182094.

[8] Dufour, P., 1999. *Process for the preparation of basic poly-aluminum chloro-sulphates and applications thereof.* United States Patent 5879651.

[9] Carlsson, O., 1987. *Process for producing a flocculating agent*. United States Patent 4654201.

[10] Gunnarsson, L. K., Nilsson, R., 1986. *Method for producing basic aluminum sulphate (III).* United States Patent 4563342.

[11] Pozzoli, B., 2003. *Aluminium poly-chloro-sulphates, process for their preparation and use thereof.* United States Patent 6548037.

[12] Gancy, A.B., Wamser, C., 1980. *Amorphous poly-aluminum sulfate compositions.* United States Patent 4238347.

[13] Kudermann, G., Blaufuss, K.-H., 1993. *Method for increasing the molar ratio of aluminum to chlorine in poly-aluminum chloride solutions*. United States Patent 5254224.

[14] Haake, G., Geiler, G., Haupt, F., 2001. *Process for preparing sulfate-containing basic solutions of Poly-aluminum chloride.* United States Patent 6241958.

[15] Lindahl, G., 1985. *Stable solutions of basic aluminium sulphate containing poly-nucleate aluminium hydroxide sulphate complexes*. United States Patent 4536384.

[16] A.I. Zouboulis and N. Tzoupanos, Alternative cost-effective preparation method of polyaluminium chloride (PAC) coagulant agent: characterization and comparative application for water/wastewater treatment, *Desalination*, in press (2008).

[17] Graham N., *Preparation and uses of polyferric sulphate*, US Patent 5.785.862, 1998.

[18] Leprince, A., Flessinger, F., Bottero, JY., Polymerised iron Chloride: An improved inorganic coagulant, *Journal of American Water Works Association*, **76** (1984) 93-97.

[19] Bertsch, P.M., Conditions for Al_{13} polymer formation in partially neutralized aluminum solutions, *Soil Science Society America Journal,* **51** (1987) 825-828.

[20] Tsai, P.P., Hsu, P.H., Aging of partially neutralized aluminum solutions of sodium hydroxide/aluminum molar ratio=2.2, *Soil Sci. Soc. Am. J.*, **49** (1985) 1060-1065.

[21] Tang H.X, Tian B.Z., Luan Z.K., Zhang Y., Inorganic polymer flocculant polyferric chloride, its properties, efficiency and production, In Chemical Water and Wastewater treatment, Vol III, eds. R. Klute and In Hahn, Springer–Verlag, Berlin, 1990.

[22] Jiang, J-Q., Graham, N.J.D., Preparation and characterization of an optimal polyferric sulphate (PFS) as a coagulant for water treatment, *Journal of Chemical Technology and Biotechnology*, **73** (1998c) 351-358.

[23] Cheng W.P., Hydrolysis characteristics of polyferric sulfate coagulant and its optimal condition of preparation, *Colloids and Surfaces A*, **182** (2001) 57-63.

[24] Fan M., Brown R.C., Sung S., Huang C.P., Ong S.K., van Leeuwen J.H., Comparison of polymeric and conventional coagulants in arsenic (V) removal, *Water Environmental Research*, **75** (2003) 308-3113.

[25] Wang, D., Sun, W., Xu, Y., Tang, H., Gregory, J., Speciation stability of inorganic polymer flocculant-PACl, *Colloids and Surfaces A*, **243** (2004) 1-10.

[26] Zhang, P., Hahn, H.H., Hoffmann, E., Zeng, G. "Influence of some additives to aluminium species distribution in aluminium coagulants", *Chemosphere*, **57** (2004) 1489-1494.

[27] Zouboulis A.I., Moussas P.A., Vasilakou F., Poly-ferric Sulphate: Preparation, characterization and application in coagulation experiments, submitted in *Journal of Hazardous Materials*, February 2007.

[28] AWWA, AWWA standard for liquid poly-aluminum chloride, American Water Works Association, Denver, Colorado USA, 1999.

[29] Bolto, B.A., Soluble polymers in water treatment, *Progress in Polymer Science*, **20** (1995) 987-1041.

[30] Hoover, M. F., Cationic quaternary polyelectrolytes-A literature review, *Journal of Macromolecular Science-Chemistry,* **A4** (1970) 1327.

[31] Mortimer, D.A. Synthetic polyelectrolytes-A review, *Polymer International*, **25**, 29-41, 1991.

[32] Bache, D.H., Gregory, R. (eds) *Flocs in Water Treatment*, IWA Publishing, London, UK, 2007.

[33] Wang, L.K., Wang, M.H., Kao, J.F., Application and determination of organic polymers, *Water, Air, and Soil Pollution*, **9** (1978) 337-348.

[34] Sorensen, J. and Larsson, S.G. (1992): "Particle separation in wastewater treatment". In Chem. Water Wastewater Treatment II, *Proc. of 5th Gothenburg Symp.*, (R. Klute and H. H. Hahn Ed.), pp. 181-190, Springer Verlag, Berlin.

[35] Gao, B.Y., Hahn, H.H., Hoffmann, E., Evaluation of aluminum-silicate polymer composite as a coagulant for water treatment, *Water Research*, **36** (2002) 3573-3581.

[36] Jia, Z., He, F. and Liu, Z., Synthesis of poly-aluminum chloride with a membrane reactor: operating parameter effects and reaction pathways, *Ind. Chem. Res.* **43** (2004) 12-17.

[37] Liu, H., Qu, J., Hu, C. and Zhang, S., Characteristics of nanosized poly-aluminum chloride coagulant prepared by electrolysis process, *Colloids and Surfaces A*, **216** (2003) 139-147.

[38] Cheng W.P., Yu R.F., Microwave assisted preparation of polyferric sulfate coagulant, *Separation Science and Technology*, **38**, No 1 (2003) 39-55.

[39] Hasegawa, T., Hashimoto, T., Onitsuka, T., Goto, K., Tambo, N., Characteristics of metal poly-silicate coagulants, *Water Science and Technology*, Vol. 23 (1991) 1713-1722.

[40] Chan, S.H., A review on solubility and polymerization of silica, *Geothermics*, **18** (1989) 49-56.

[41] Stumm, W., Huper, H., Champlin, L., Formation of poly-silicates as determined by coagulation effects, *Environmental Science and Technology,* **1** (1967) 221-227.

[42] Dietzel, M., Dissolution of silicates and the stability of poly-silicic acid, *Geochimica et Cosmochimica Acta*, **64** (2000) 3275-3281.

[43] Chan, S.H., A review on solubility and polymerization of silica, *Geothermics*, **18** (1989) 49-56.

[44] Stumm, W., Huper, H., Champlin, L., Formation of poly-silicates as determined by coagulation effects, *Environmental Science and Technology*, **1**, (1967) 221-227.

[45] Zhang, P., Hahn, H.H., Hoffmann, E., Zeng, G., Influence of some additives to aluminium species distribution in aluminium coagulants, *Chemosphere,* **57** (2004) 1489-1494.

[46] Boisvert, J.-P. and Jolicoeur, C., Influences of sulfate and/or silicate present in partially pre-hydrolyzed Al(III) flocculants on Al(III) speciation in diluted solutions, *Colloids and Surfaces* A, **155** (1999) 161-170.

[47] Gao, B.Y., Yue, Q.Y., Wang, B.J., The chemical species distribution and transformation of polyaluminium silicate chloride coagulant, *Chemosphere*, **46** (2002) 809-813.

[48] Gao, B.-Y., Yue, Q.Y., Wang, B.-J., Properties and coagulation performance of coagulant poly-aluminium-ferric-silicate-chloride in water and wastewater treatment, *Journal of Environmental Science and Health - Part A',* **41** (2006) 1281-1292.

[49] Zouboulis A.I. and Tzoupanos, N.D., Polyaluminium Silicate Chloride – a systematic study for the preparation and application of an efficient coagulant for water or wastewater treatment, *Journal of Hazardous Materials*, 2008, doi:10.1016/j. jhazmat.2008.06.019.

[50] Cheng, P.C., Chi, F.G., Yu, R.F., Shi, P.Z., Evaluating the coagulants of polyaluminium silicate chlorides on turbidity removal, *Separation Science and Technology*, **41** (2005) 297-309.

[51] [51] Gao, B.Y., Q.Y. Yue, B.J. Wang, The chemical species distribution and transformation of polyaluminium silicate chloride coagulant, *Chemosphere*, **46** (2002) 809-813.

[52] Song, Y.H., Luan, Z.K., Tang, H.X., Preparation and characterisation of polyaluminium silicate chloride coagulant, *Environmental Technology*, **24** (2003) 319-327.

[53] Zouboulis, A.I., Tzoupanos N.D., Moussas, P.A., *Inorganic Pre-Polymerised Coagulants: Current Status and Future Trends*, 3rd IASME/WSEAS International Conference, S. Nikolaos, Greece, 24-26 July 2007, pp. 292-300.

[54] Tang, H., Shi, B., The characteristics of composite flocculants synthesized with inorganic poly-aluminum and organic polymers, in Hahn, H., Hoffmann, E., Odegaard, H. (eds) Chemical Water and Wastewater Treatment VII, *Proc. of the 10th Gothenburg Symposium 2002*, IWA Publishing, Gothenburg, Sweden, 2002, pp. 17-28.

[55] Leprince, A., Flessinger, F., Bottero, JY., Polymerised iron Chloride: An improved inorganic coagulant, *Journal of American Water Works Association*, **76** (1984) 93-97.

[56] Bertsch, P.M., Conditions for Al_{13} polymer formation in partially neutralized aluminum solutions, *Soil Science Society America Journal*, **51** (1987) 825-828.

[57] Wang, Y., Gao, B.Y., Yue, Q.Y., Wei, J.C., Zhou, W.Z., Novel Composite Flocculent Poly-ferric Chloride- Poly-dimethyl-diallyl-ammonium Chloride (PFC-PDMDAAC): Its Characterization and Flocculation efficiency, *Water Practice and Technology*, 2006, 1 (No 3).

[58] Gao, B., Yue, Q., Miao, J., Evaluation of polyaluminium ferric chloride (PAFC) as a composite coagulant for water and wastewater treatment, *Water Science and Technology*, **47** (2002) 127-132.

[59] Gao, B.-Y., Yue, Q.Y., Wang, B.-J., Properties and coagulation performance of coagulant poly-aluminium-ferric-silicate-chloride in water and wastewater treatment, *Journal of Environmental Science and Health - Part A'*, **41** (2006) 1281-1292.

4. Application of Coagulation Reagents in Water and Wastewater Treatment

4.1. Performance of Pre-Polymerized Coagulation Reagents in Comparison with Conventional Non –Polymerized Coagulation Reagents

Pre-polymerized inorganic coagulants have been used in the full scale treatment of waste water and industrial effluents, as well as of potable water treatment. Other applications of polymeric coagulants include the pre-treatment of wastewaters of various industries, such as dyes, tannery, paper, pulp and the treatment of landfill leachates. Additionally coagulation process is regarded as an important option for the tertiary treatment of wastewater, intended for specific reuse purposes (prevention of eutrophication).

Several studies [1-3] suggested that the polymeric coagulants have a higher treatment efficiency in terms of turbidity, color, metals and COD removal in comparison with the conventional non-polymerized coagulants Their advantages are attributed to a) high concentration of polymeric species, b) wider working pH range, c) lower sensitivity to low water temperature, d) lower dose requirements for achieving equivalent performance with the conventional coagulants, e) lower residual metal - ion concentration and f) lower sludge production.

However, several studies reported that polyferric chloride (PFC) is less efficient than the respective non-polymerized ferric chloride (FC) for the removal of color and NOM from surface water [4, 5, 6]. On the other hand, it is well shown that polyferric sulphate (PFS) exhibits a better performance compared with the conventional coagulation reagents, i.e. ferric sulphate, or ferric chloride for removing turbidity and natural organic matter [7-10]. Additionally, the usage of PFS is expected to reduce the volume of the produced sludge as the required coagulant dosage for an effective treatment was significantly reduced. Jiang et al. [11] also studied the performance of PFS in removing algae and algae-derived organic matter with excellent results.

Zouboulis et al [12] investigated the efficiency of various commercially - available iron based coagulants such as ferrous and ferric salts and pre-polymerized polyferric sulphate (PFS, obtained from a company based on China). The results show that the trivalent iron based coagulants are far more efficient than the divalent ones. However, PFS reagent did not exhibit similarly good performance as described in relevant literature [7-11], which may be attributed to the specific type of PFS used in these applications.

Moreover, Zouboulis and Traskas [13] reported a comparable evaluation of commercially available aluminium-based coagulants for the treatment of surface (from Aliakmon River, Greece) water and for the post-treatment of urban wastewater. As shown in Figure 1, the removal of suspended solids (SS) and NOM is significantly higher by the addition of PAC than that of alum. In addition, the residual aluminium concentration was lower than that of alum and it is highly dependent on the pH value of the sample to be treated. Lower pH values are associated with lower residual concentrations. Furthermore, results from the treatment of biologically pre-treated municipal wastewater showed that PAC is more efficient in SS and NOM removal, while alum was slightly better in terms of phosphate ions removal.

Similarly, Zouboulis et al (2008) comparatively evaluated the efficiency of PAC and alum during full scale experiments in a drinking water treatment plant of N. Greece. The results demonstrated that PAC is more efficient, as lower applied dosages of PAC resulted in a production of water with lower residual turbidity and lower residual aluminium concentration than that produced after coagulated with alum.

Significant research interest has also been devoted to the study of NOM removal by the application of pre-polymerized coagulants [13-17]. Unlike the removal of mineral colloidal particles, which is considered to be proceeded by two distinct coagulation mechanisms, namely the charge neutralization and the sweep flocculation mechanism [18], too much uncertainty lies on the possible mechanisms in removing NOM. It has been suggested that the formation of insoluble complexes between NOM and coagulant species along with the adsorption of organic matter onto the surface of metal hydroxide precipitates could be the determining mechanisms [17-20].

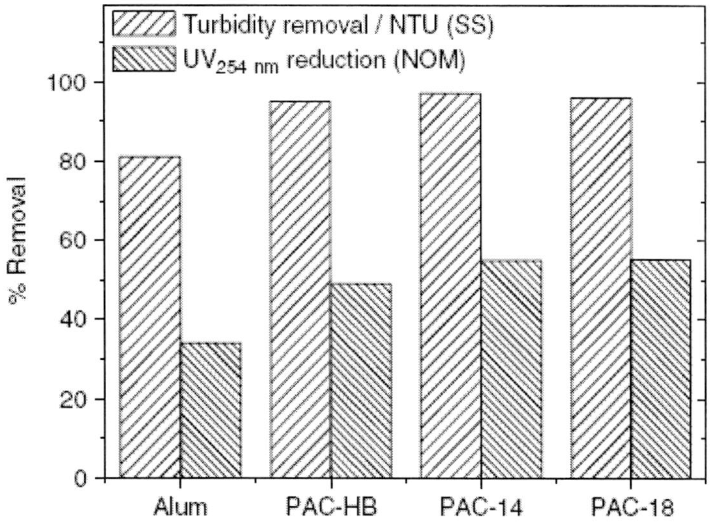

Figure 10. Comparative results of NOM and SS removal for different types of studied coagulants at pH 7 [32].

Duan and Gregory [18] described the two mechanisms in detail; a) Adsorption of coagulant (metal) species to negatively charged surfaces, thus neutralising their charge and consequently reducing their solubility. The produced metal humic acid complexes could precipitate, thus leading to the formation of particles that can be removed by other processes, such as sedimentation and/or filtration. b) Adsorption of humic substances on amorphous metal hydroxide precipitate. At pH values approximately 5-6, the humic substances are negatively charged, while the Al or Fe hydroxides are positively charged, which would give strong adsorption phenomena.

Shi et al. [17] studied the coagulation of humic acids by preformed and non – preformed Al species, suggesting that decreasing pH could significantly reduce the coagulant dosage for complete UV_{254} removal regardless of the Al speciation differences. Additionally, Cheng and Chi [15] investigated the coagulation mechanism of PFS on humic acid through the fluorescence-quenching effect. It is suggested that the complex formation of PFS-HA through charge neutralization was the major coagulation mechanism between PFS and HA at both low and middle pH range. Additionally, Cheng [14] reported that the humic acid removal by PFS is more controllable than that by ferric chloride, while humic acid sludge from PFS coagulation is more stable and easily handled than that from ferric chloride coagulation.

Coagulation flocculation (along with chemical precipitation) has also been used for the phosphate removal [2, 21-23]. Due to the problems arise from the algal blooms and eutrophication of surface waters, the reduction of phosphate concentration entering water bodies is regarded as more than necessary. As it has already been mentioned, coagulation process is regarded as an important option for the tertiary treatment of wastewater, intended for specific reuse purposes (prevention of eutrophication). The mechanism of chemical phosphate removal may involve the two mechanisms separately or in combination and it is highly influenced by the properties of the water to be treated such as the pH and alkalinity. In general, two major mechanisms have been considered to be related to the coagulation of phosphate ions with Fe (or Al) species. The first mechanism is based on the formation of Fe/Al-hydroxo-phosphate complexes which can either adsorb on the surface of positively charged Fe/Al(III) hydrolysis species or act as a center of precipitation for Fe/Al (III) hydrolysis species. The second mechanism is the adsorption of phosphate ions on the surface of the Fe/Al (III) hydrolysis species. However, in the case of using pre-polymerized coagulant such as PFS, the polymeric species are predominant and therefore the phosphate ions may also form a complex with the positively charged polymeric species [2]. However, it still remains unclear the precise mechanism under which the removal of phosphate ions takes place.

Zouboulis et al [24] investigated the efficiency of PFS when treating a biologically pre-treated municipal wastewater aiming at the reduction of phosphate concentration well below the regulation limits. The experiments include a comparison of different PFS reagents (having a different polymerization degree) with the respective non – polymerized reagent, ferric sulphate (FS). The results demonstrated that PFS is highly efficient in removing the phosphate ions from biologically pre-treated municipal wastewater. Moreover, PFS samples exhibit a significantly better coagulation performance, than the respective conventional coagulant, i.e. ferric sulphate, in treating such a wastewater under the applied experimental conditions. In particular, treating the wastewater with PFS removed the phosphate ions successfully (well below 1 mg/L, which is the limit concentration in wastewaters to be discharged in sensitive areas) even by using the lower coagulant dosages (5 mg/L). On the contrary, the ferric sulphate dosage required for achieving a phosphate ions concentration below this limit was higher than 10 mg/L, highlighting the superior performance of the pre-polymerised coagulants. These findings agree with that presented by Odegaard et al, when soft and hard water was treated [25]. However, as Jiang and Graham [22] concluded, there is significant uncertainty about the precise mechanism through which the P removal occurs and this is a gap which needs to be investigated further and in detail.

4.2. Coagulation Performance of Modified Pre-Polymerized Metal Salt Reagents

Several efforts have been made in order to evaluate the coagulation efficiency of the new modified pre-polymerized metal salt reagents and to ascertain if they exhibit advantages compared to simple or pre-polymerized metal coagulants. In the case of metal-poly-silicate coagulants, the research is in a more skilled stage than the respective research in the case of metal coagulants with organic additives.

Gao et al. (2002) [26] examined the coagulation behaviour of polyaluminium silicate coagulants (PASiC) with OH/Al molar ratio 1.5-2 and Al/Si molar ratio 10-20 in treating deionized water contaminated with SiO_2 micro-particles and $NaHCO_3$. The evaluation of coagulation efficiency was based on turbidity measurements, and they concluded that polyaluminium silicate chloride was more efficient than PACl. Song et al. (2003) [27] came to the same conclusion, by examining the coagulation performance of PASiC with OH/Al molar ratio 2 and Al/Si molar ratio 13-40 in synthetic turbid water (100 mg/L kaolin suspension) treatment. Cheng et al. (2004) [28] conducted a more detailed study on turbidity removal from kaolin containing synthetic water sample. They examined several PASiC coagulants derived with the co-polymerization technique, and with OH/Al 0.5-2.3 and Al/Si 5-20 molar ratios (overall 20 different reagents). They observed better treatment performance of PASiC coagulants with certain properties (i.e. base and silica content) than alum. Moreover, it was one of the first attempts to define the characteristics of the more effective new coagulants. According to their results, the optimum values of OH/Al and Al/Si in the silica-enriched coagulants should be 1.5-2 and 5 respectively.

In the aforementioned publications, however, the coagulation performance was evaluated considering only one qualitative parameter, i.e. turbidity. The examination of as many parameters as possible is desirable to get an intergraded approach of coagulation performance, especially regarding the residual aluminium concentration and the removal of natural organic matter. The increased residual Al concentration is a common problem when treating surface waters with conventional coagulants and difficult to keep it under the EU maximum permissible concentration limit of Al in drinking water (200 μg/L) [29]. Organic matter is responsible for taste, odor and appearance (turbidity, color) related problems, is the main precursor to the formation of disinfection byproducts, when chlorine is applied as disinfection agent, and its removal is essential for the overall coagulation performance. Finally, it is clear, that not all composite coagulants are more effective than PACl, but only those with specific properties in each case.

A comparison of the aluminium concentration that remains in a kaolin-contaminated water sample after treatment with PACl and PASiC coagulants was conducted by Gao et al. (2004) [30]. They found lower residual Al conc. for the silica enriched coagulants; even at different pH values (examined pH range 6.13-8.52). They also pointed out that the residual Al conc. decreases with the increase of OH/Al and decrease of Al/Si molar ratios, whereas the composite coagulants prepared with the co-polymerization technique showed superior coagulation performance than those prepared with the co-polymerization technique.

Zouboulis and Tzoupanos (2008) [31] investigated thoroughly the impact of all the aforementioned parameters that designate the nature and the properties of the final composite coagulant, i.e. OH/Al, Al/Si molar ratios and the preparation method, in the coagulation performance. The purpose was to define the optimum values of these parameters and the optimum preparation method in order to produce a superior coagulation reagent than PACl or conventional coagulants. For this purpose, 32 different polyaluminium silicate chloride coagulants were investigated (OH/Al 1-2.5, Al/Si 5-20) in treating a kaolin-humic acids model sample. The respective model sample can be characterized as difficult to tread, as it has low turbidity (16.5 NTU) and relatively high content of organic matter (5 mg/L). According to the respective coagulation performance data, they determined the optimized conditions for a superior coagulation reagent as follows: OH/Al ratio 1.5-2, Al/Si ratio 10-15, whereas as preparation technique the co-polymerization should be preferred. The composite coagulant with these properties exhibits better performance in turbidity and organic matter removal, whereas the residual Al conc. is significantly reduced, compared to conventional coagulants (alum) and PACl coagulants (commercial and laboratory prepared). Moreover, the superior performance of PASiC proved to be more resistant in pH variation, than PACl (PASiC is effective over a pH range of 6.5 to 9, where PACl, is less sufficient above the pH 8.0).

A matter that has not been investigated in detail is the application of polyaluminium silicate chloride coagulants in wastewater treatment. Gao et al. (2007) [32] examined the performance of PASiC in treating wastewater from oil refinery and municipal wastewater, and found PASiC to be more efficient than PACl in organic matter removal (measured by UV absorbance at 254 nm or chemical oxygen demand, COD) and phosphorus removal. They also noticed that the efficiency of PASiC increased with increasing OH/Al molar ratio. However, their results are in contrast with the results obtained by Zouboulis and Tzoupanos [31], who investigated phosphorus removal from biologically pre-treated urban wastewater. In particular, they noticed that the most effective between the composite coagulants are those with relatively low OH/Al = 1 molar ratio (Al/Si = 10). Moreover, they observed that polyaluminium silicate chloride does not exhibit significant advantage against alum or PACl regarding phosphorus removal. The same conclusion made Boisvert et al. (1997) [33], who studied the phosphates adsorption, by using alum and poly-aluminium-silicate-sulphate (PASS) samples. They observed that the presence of silicate at low level (i.e. Al/Si ratio 25) does not present any notable effect on PO_4 adsorption. According to Ratnaweera et al. (1992) [34], the dominant mechanism for phosphates removal when low basicity (or low OH/Al molar ratio) coagulants are applied, is the sweep flocculation (insoluble $Al(OH)_3$ formation). $Al(OH)_3$ formation plays a crucial role for the removal of phosphates, which can also explain the good performance of conventional coagulants (alum), as generally the major mechanism of coagulation by using alum is the sweep flocculation. On the other side, for the case of pre-polymerized coagulants (including PASiC), the existence of polymerized Al species, which are resistant to further hydrolysis, inhibits (at a certain degree) the $Al(OH)_3$ formation, thus no significantly higher removal efficiencies were observed with PASiC.

Polyaluminium silicate chloride was applied also for the post-treatment of biologically pre-treated landfill leachates, a special category of wastewaters, which are difficult to treat further, mainly due to their bio-refractory organic content (humic substances) [35]. The results suggest better performance of PASiC regarding turbidity, organic matter (by means of UV absorbance at 254 nm and COD measurements) and color removal, compared to the performance of alum and PACl. The better performance of PASiC was higher at pH values greater than 6 (i.e. 7-9), the usual pH range of such type of wastewaters. However, again PASiC has not shown any advantages in phosphates removal, and again the most efficient of the different PASiC coagulants were those with OH/Al = 1 (Al/Si 10 or 15).

Generally, rather little information was published, regarding the coagulation performance of the new silica-based coagulants, when applied in wastewater treatment. It seems that the nature and the composition of treated aqueous sample, as well as the main treatment purposes are the major factors, which designate the performance of a coagulant and its ability for a specific application. Composite coagulants with different properties can be most efficient in different cases, and more relevant research is needed.

Regarding iron-based inorganic coagulants with silica addition, Wang and Tang (2001) [36] studied the properties and coagulation behavior of modified polyferric silicate, which resulted after the addition of three different types of silica into ferric chloride and then polymerization. They observed enhanced coagulation performance for the modified coagulant agent compared to poly-feric chloride for the treatment of a kaolin containing model sample (evaluated by means of turbidity removal), despite the deterioration of the charge neutralization ability after silica addition. Zouboulis and Moussas conducted detailed studies [36, 37] about the properties and coagulation behavior of several poly-feric silicate sulphate (PFSiS) coagulants with different OH/Fe and Fe/Si molar ratios in water or wastewater treatment. They also observed that silica addition in poly-feric sulphate (PFS) results in a decrease of the charge neutralization ability of the modified coagulants, as the cationic charge of iron species is partially compromised with the addition of negatively charged silica species. However, a coagulation kinetics study [37] revealed that this disadvantage is compensated by the increased aggregation ability, as the addition of silica results in the formation of large polymeric chains, thereby promoting the formation of big and tenacious flocs via the bridge forming mechanism. As a result, they found that PFSiS exhibits superior coagulation performance compared to ferric sulphate (FS, simple salt) or PFS (pre-polymerized coagulant) in treating kaolin – humic acid suspension (5 mg/L for both), by monitoring turbidity and absorbance at 254 nm reduction. Moreover they found that the silica enriched coagulants work efficiently over a wide pH range (i.e. 5-9). Regarding wastewater treatment, the modified coagulants were applied for the post treatment of biologically pre-treated landfill leachates [37]. The results suggest greater turbidity and phosphate ions removal efficiencies for PFSiS compared to FS. Song and Ren (2008) [38] reported superior coagulation performance of PFSiS (relatively high concentrated, i.e. 0.75 M Fe) compared to PFS in treating kaolin model sample (turbidity measurements). Moreover, they observed that the size of the flocs formed during the jar tests increases with the increase of silica content in the modified coagulants (i.e. decrease of Fe/Si molar ratio).

The initial efforts for the development of composite inorganic-organic coagulant agents appeared in the literature in the beginning of the 21st century. Tang and Shi (2002) [39] investigated the combination of PACl and several cationic, anionic and non-ionic polyelectrolytes into one reagent. They observed higher turbidity reduction of the tested sample for the composite coagulants compared to PACl, and particularly between the composite coagulants those with cationic additives performed better. However, they mention that the performance of the composite coagulants depends in a high extent on the nature of the sample, i.e. if it contains relatively large amount of organic substances (increased negative charge), a composite coagulant with high cationic charge would be more efficient (combination of inorganic coagulant and cationic polyelectrolyte). Gao et al. (2005) [40] studied the properties and coagulation performance of composite coagulants derived from the combination of PACl and p-DADMAC in wastewater treatment. They compared their efficiency with the respective efficiency of PACl and found the composite coagulant to work better, evaluated by the COD reduction. Tzoupanos and Zouboulis [41] applied several PACl-pDADMAC composite coagulants in the treatment of surface water, and compared their efficiency with the efficiency of the combined application of PACl and p-DADMAC but as separated reagents. They found superior coagulation performance of composite coagulants with significantly lower concentration of polyelectrolyte. Moreover, they observed that the most important advantage of the composite coagulants is the significantly lower residual Al concentration that remains in the treated sample. The advantages of the composite coagulants were attributed to the formation of flocs of greater size.

Composite coagulants derived from iron salts and p-DADMAC have also been reported in the literature. In particular, a new coagulant agent produced from the combination of p-DADMAC and poly-feric sulphate (PFC) was tested for the treatment of several types of wastewater, i.e. textile wastewater [42] and paper making intermediate wastewater [43]. In the first case, higher color removal efficiencies were observed for the composite coagulant compared to PFC or the combined application of PFC and p-DADMAC but as separated reagents. The alteration of the composite coagulant properties which affects the respective mechanisms of coagulation (charge neutralization and bridging mechanisms) was given as the possible explanation for the superiority of the composite coagulant. Furthermore, the same composite coagulant was applied in the treatment of surface water (river water) [43], and it was found that it was more efficient than PFC or PFC and p-DADMAC (as separated reagents) in turbidity and organic mater removal (COD and absorbance at 254 nm).

References (For this Part)

[1] Du M., Treatment of dying and printing wastes by the combined process of polyferric sulphate coagulation and chlorination oxidation, *Environmental Engineering*, **9**, (1990), 1-4.

[2] Jiang J-Q., Graham N.J.D., Pre-polymerised inorganic coagulants and phosphorus removal by coagulation – a review, *Water SA*. **24** (3) (1998) 237-244.

[3] Joo D. J., Shin W. S., Choi J-H., Choi S. J., Kim M.-C., Han M.H, Ha T. W., Kim Y-H, Decolorization of reactive dyes using inorganic coagulants and synthetic polymer, *Dyes and Pigments* **73** (2007) 59-64.

[4] Tang H.X., Stumm W., The coagulating behaviors of Fe(III) polymeric species, *Water Research,* **21** (1987) 115-121.

[5] Leprince A., Flessinger F., Bottero JY., Polymerised iron Chloride: An improved inorganic coagulant. *Journal of American Water Works Association*, **76** (1984) 93-97.

[6] Gray KA, Yao C., O'Melia CR, Polymeric inorganic coagulants, *Journal of American Water Works Association,* **87** (1995) 136-146.

[7] Jiang J-Q., Graham N.J.D., Coagulation of upland coloured water with polyferric sulphate compared to conventional coagulants, *J. Water SRT – Aqua*, **45**(3) (1996) 143-154

[8] Jiang J-Q., Graham N.J.D. Enhanced coagulation using Al/Fe (III) coagulants: Effect of coagulant chemistry on the removal of colour-causing NOM, *Environmental Technology*, **17** (1996) 937-950.

[9] Jiang J-Q., Graham N.J.D., Observations of the comparative hydrolysis/ precipitation behavior of poly-ferric sulphate and ferric sulphate, *Water Research*., **32** (3) (1998) 930-935.

[10] Jiang J-Q., Graham N.J.D., Preparation and characterisation of an optimal poly-ferric sulphate (PFS) as a coagulant for water treatment, *Journal of Chemical Technology and Biotechnology*, **73** (1998) 351-358.

[11] Jiang J-Q., Graham N.J.D. Comparison of polyferric sulphate with other coagulants for the removal of algae and algae-derived organic matter. *Water Science and Technology*, **27** (1993) 221-230.

[12] Zouboulis A.I., Traskas G. and Ntolia A., Comparable evaluation of iron-based coagulants for the treatment of surface water, *Separation Science and Technology* **42** (4) (2007) 803-817.

[13] Zouboulis A.I., Traskas G., Comparable evaluation of various commercially available aluminium-based coagulants for the treatment of surface water and for the post-treatment of urban wastewater, *Journal of Chemical Technology and Biotechnology*, **80** (10) (2005) 1136-1147.

[14] Cheng, W.P., Comparison of hydrolysis/coagulation behavior of polymeric and mono-meric iron coagulants in humic acid solution. *Chemosphere*, **47** (2002) 963-969.

[15] Cheng W.P., Chi F.H., A study of coagulation mechanisms of polyferric sulphate reacting with humic acid using a fluorescence-quenching method, *Water Research*, **36** (2002) 4583-4591.

[16] Elfarissi F., Pefferkorn E., Kaolinite/humic acid interaction in the presence of aluminium ion, *Colloids and Surfaces A,* **168** (2000) 1-12.

[17] Shi B., Wei Q., Wang D., Zhu Z., Tang H., Coagulation of humic acid: The performance of preformed and non-preformed Al species, *Colloids and Surfaces A*, **296** (2007) 141-148.

[18] Duan, J., Gregory, J., Coagulation by hydrolysing metal salts, *Advances in Colloid and Interface Science*, **100-102** (2003) 475-502.

[19] Dempsey, B., Ganho, R., O'Melia, C., The coagulation of humic substances by means of aluminium salts. *Journal of American Water Works Association* **76** (1984) 141-149,

[20] Lu X., Chen Z., Yang X., Spectroscopic study of the aluminum speciation in removing humic substances, *Water Research* **33**, (1999), 3271-3280.

[21] Xie W., Wang Q., Ma H., Ohsumi Y., Ogawa H., Study on phosphorus removal using a coagulation system, *Process Biochemistry*, **40** (2005) 2623–2627.

[22] Diamantopoulos E., Vlachos C., Aluminium hydrolysis effects on phosphorus removal from wastewaters, *Journal of Water Pollution Control,* **56** (1984), 1165-1172.

[23] Georgantas D.A., Grigoropoulou H.P., Orthophosphate and metaphosphate ion removal from aqueous solution using alum and aluminum hydroxide, *Journal of Colloid and Interface Science* **315** (2007) 70–79.

[24] Zouboulis A.I., Moussas P.A., Vasilakou F., Polyferric Sulphate: Preparation, characterization and application in coagulation experiments, *Journal of Hazardous Materials* **155** (3) (2008) 459-468.

[25] Odegaard H., Fettig J., Ratnaweera HC., " Coagulation with pre-polymerised metal salts", In Hahn and R Klute (eds.) Chemical Water and Wastewater Treatment. *Springer* – Verlag, Berlin, 1990, 3-27.

[26] B.Y. Gao, H.H Hahn, E. Hoffmann, Evaluation of aluminum-silicate polymer composite as a coagulant for water treatment, *Water Research*, **36** (2002) 3573-3581.

[27] Y.H. Song, Z.K. Luan, H.X. Tang, Preparation and characterisation of polyaluminium silicate chloride coagulant, *Environmental Technology*, **24** (2003) 319-327.

[28] P.C. Cheng, F.G. Chi, R.F. Yu, P.Z. Shi, Evaluating the coagulants of polyaluminium silicate chlorides on turbidity removal, *Separation Science and Technology*, **41** (2005) 297-309.

[29] EU Directive 98/83/EC, *Drinking water quality intended for human consumption*, 1998.

[30] B.Y. Gao, Q.Y. Yue, B.J. Wang, Coagulation efficiency and residual aluminium content of polyaluminium silicate chloride in water treatment, *Acta Hydrochimica et Hydrobiologica*, **32** (2004) 125-130.

[31] Zouboulis A.I. and Tzoupanos, N.D., Polyaluminium Silicate Chloride – a systematic study for the preparation and application of an efficient coagulant for water or wastewater treatment, *Journal of Hazardous Materials*, 2008, doi:10.1016/j.jhazmat.2008.06.019.

[32] Gao, B.Y., Yue, Q.Y., Wang, Y., Coagulation performance of poly-aluminum silicate chloride (PASiC) for water and wastewater treatment, *Separation and Purification Technology*, **56** (2007) 225-230.

[33] J.-P. Boisvert, T.G. To, A. Berrak, C. Jolicoeur, Phosphate adsorption in flocculation processes of aluminium sulphate and poly-aluminium-silicate-sulphate, *Water Res.* **31** (1997) 1939-1946.

[34] Ratnaweera, H., Odegaard, H., Fettig, J., Coagulation with pre-polymerised aluminium salts and their influence on particle and phosphate removal. *Water Science and Technology*, **26** (5-6) (1992) 1229-1237.

[35] Tzoupanos, N.D., Zouboulis A.I., Zhao Y.-C., The application of novel coagulant reagent (polyaluminium silicate chloride) for the post-treatment of landfill leachates. Chemosphere (2008), doi:10.1016/j.*chemosphere.*2008.06.051.

[36] Wang, D., Tang, H., Modified inorganic polymer flocculant-PFSi: its preparation, characterization and coagulation performance, *Water Research*, **35** (2001) 3418-3428.

[37] Zouboulis, A.I., Moussas, P.A., Polyferric silicate sulphate (PFSiS): Preparation, characterization and coagulation behaviour, *Desalination*, **224** (2008) 307–316.

[38] Moussas, P.A., Zouboulis, A.I., A study on the properties and coagulation behaviour of modified inorganic polymeric coagulant—Polyferric silicate sulphate (PFSiS), Separation and Purification Technology (2008), doi:10.1016/j.seppur.2008.06.009.

[39] Song, Z., Ren, N., Properties and coagulation mechanisms of poly-feric silicate sulphate, *Journal of Environmental Sciences*, **20** (2008) 129-134.

[40] Gao, B.Y., Wang, Y., Yue, Q.Y., The chemical species distribution of aluminum in composite flocculants prepared from polyaluminium chloride (PAC) and poly-dimethyl-diallyl-ammonium chloride (PDMDAAC), Acta Hydrochimica and *Hydrobiologica*, **33** (2005) 365-371.

[41] Zouboulis A.I., Tzoupanos, N.D., Novel inorganic-organic composite coagulants based on aluminium, AQUA 2008: 3rd International Conference, *Water Science and Technology*, 16-19 Oct. 2008, Athens, Greece.

[42] Gao, B.Y., Wang, Y., Yue, Q.Y., Wei, J.C., Li, Q., Color removal from simulated dye water and actual textile wastewater using a composite coagulant prepared by polyferric chloride and poly-dimethyl-diallyl-ammonium chloride, *Separation and Purification Technology*, **54** (2007) 157-163.

[43] Wang, Y., Gao, B., Yue, Q., Wie, J., Li, Q., The characterization and flocculation efficiency of composite flocculant iron salts-poly-dimethyl-diallyl-ammonium chloride, *Chemical Engineering Journal*, **142** (2008) 175-181.

In: Industrial Waste...
Editor: John P. Samuelson, pp. 325-348
ISBN: 978-1-60692-720-5

Chapter 12

RECYCLING OF INDUSTRIAL AND AGRICULTURAL WASTES FOR WATER AND SOIL REMEDIATION

Shaobin Wang
Department of Chemical Engineering, Curtin University of Technology, GPO Box U1987, Perth, WA 6845, Australia

Abstract

Red mud and fly ash are important solid wastes from alumina refinery industry and coal-firing power plants. Disposal of red mud and fly ash has been a significant issue in environmental protection. Meanwhile, chitin and chitosan, as the largest wastes from food processing industry, also produce a problem in their disposal. Recycling and utilisation of these wastes will help to reduce solid wastes and save resources. In this work, red mud and fly ash have been tested for soil amendments for humic acid adsorption,. Their characteristics were analysed using N_2 adsorption, XRD, SEM and FT-IR. It is found that red mud exhibits higher adsorption capacity than fly ash due to higher surface area and positive surface charge. The adsorption capacities of red mud and fly ash are 31.0 and 16.6 mg/g, respectively at 30 °C and pH 7. However, treatment of red mud using physical and chemical methods including thermal treatment, acid or base treatment, and H_2O_2 oxidation will decrease the adsorption capacity. For red mud, the presence of NaCl in solution will produce a competitive adsorption and thus reducing humic acid adsorption. Humic acid adsorption on red mud is also decreased with increasing solution pH.

For chitin and chitosan, they were also investigated for humic acid adsorption in aqueous solution. It is found that chitosan exhibits much higher adsorption capacity than chitin. Humic acid adsorption on chitin and chitosan follows the parallel first-order kinetics and the adsorption isotherm can be fitted by the Langmuir and Freundlich isotherms. Further investigations also indicate that humic acid adsorption on chitin was influenced by solution pH and ionic strength. Higher pH will reduce HA adsorption on chitin. The presence of cations and anions in solution will increase HA adsorption. For cation, Mg^{2+} will produce higher adsorption than Na^+ and K^+ while for anions, Cl^- will have stronger effect than NO_3^-. Isothermal calculation indicates that the maximum adsorption of HA on chitin and chitosan is 28 and 202 mg/g, respectively, at pH 7, 30 °C. These investigations indicate that red mud, fly ash, chitin and chitosan can be used as good materials for water and soil remediation to retain humic acid.

1. Introduction

Humic substances are typical naturally occurring compounds that account for 30 – 80% of the dissolved organic matters in natural water. Humic substances are mainly consisted of humic acid, fulvic acid, and humin. The presence of humic acid (HA) in water is not directly toxic but has undesirable effects on the appearance and taste of water and is very harmful in water treatment processes. The most serious health hazard originates from the formation of chloroform during chlorination. Thus removal of humic acid in water treatment has attracted much of environmental and health considerations. Various methods have been studied for humic acid removal from surface water, including adsorption by activated carbon (Newcombe, 1994; Dastgheib et al., 2004), clay (Jiang and Cooper, 2003; Peng et al., 2006), zeolite (Capasso et al., 2005; Wang et al., 2006), coagulation (Exall and Vanloon, 2000; Duan et al., 2002), filtration (Yuan and Zydney, 1999), and oxidation (Murray and Parsons, 2004; Selcuk et al., 2004).

Fly ash and red mud are two important industrial wastes, generated from coal-firing and alumina refinery industries, respectively. Currently, fly ash and red mud are usually disposed in landfills. Some approaches for recycling of them have been developed mainly in civil engineering for cement production, road fillers. In the past a few years, use of fly ash and red mud for wastewater treatment has been widely investigated. It has been found that fly ash and red mud can be employed as low cost adsorbents for adsorption of heavy metals, inorganic anions, and organics from water (Wang et al., 2005; Wang and Wu, 2006; Wang et al., 2008). In addition, fly ash and red mud have been explored as soil amendment to improve plant growth and production (Phillips, 1998; Snars et al., 2004; Jala and Goyal, 2006). However, no research has been reported for humic acid adsorption or retention on fly ash and red mud, while this is important in applying red mud and fly ash for soil amendment. Recently, our preliminary investigations have shown that fly ash can be used for humic acid adsorption and unburned carbon in fly ash plays an important role in adsorption (Wang et al., 2007; Wang and Zhu, 2007).

Chitin is a polysaccharide found in the outer skeleton of insects, crabs, shrimps, and lobsters and in the internal structures of other invertebrates. Chitin is the second most abundant polysaccharide in nature (after cellulose). The structure of chitin is shown in Fig.1. Due to the biodegradable and high adsorption capacities of chitin, it has been utilised as an adsorbent for variety of substrates. Some investigations have been reported on chitin utilisation as an adsorbent for removal of dyes (Juang et al., 1996; Annadurai et al., 1999; Prado et al., 2004; Akkaya et al., 2007), phenol (Dursun and Kalayci, 2005) and heavy metals (da Silva and da Silva, 2004; Kim et al., 2006; Barriada et al., 2007). Chitosan is a N-deacetylated product of chitin (Fig.1). It also has been widely used as an adsorbent for transition metals (Varma et al., 2004; Gerente et al., 2007) and organics (Crini, 2005, 2006).

For humic acid adsorption, some investigators have employed chitosan as an adsorbent (Wu et al., 2002; Zhang and Bai, 2003; Chang and Juang, 2004; Yan and Bai, 2005). However, few investigations have been reported for humic acid adsorption on chitin. Ngah and Musa investigated the adsorption of humic acid onto chitin and chitosan. The uptake of humic acid on chitosan was greater than that on chitin. However, the Langmuir isotherm showed that chitin and chitosan had similar adsorption capacities of 27.3 and 28.8 mg/g, respectively (Ngah and Musa, 1998). Production of chitosan from chitin involves deacetylation by the alkaline hydrolysis of chitin. Depending on the processing conditions, the soluble chitosan form can have a variable degree of acetylation. Thus using chitosan as adsorbent is more costly than chitin.

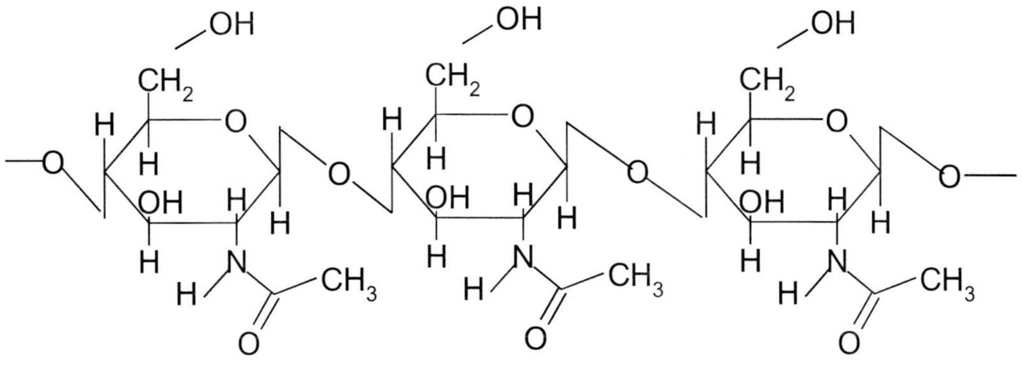

Chitin

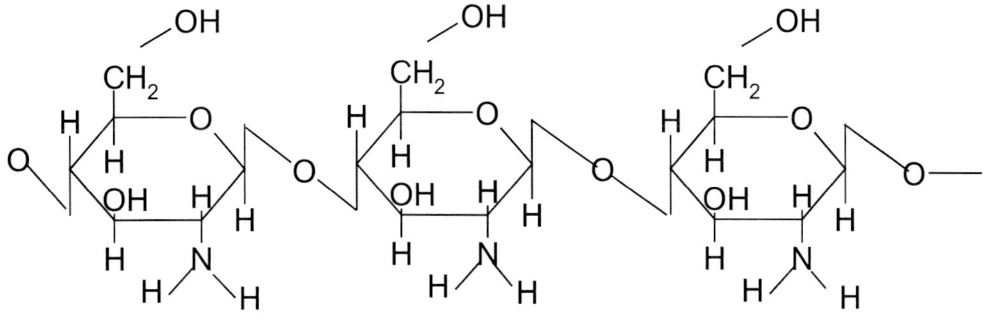

Chitosan

Figure 1. Structures of chitin and chitosan.

In this paper, we report an investigation of HA adsorption on red mud and compare the adsorption behaviour with fly ash. The physicochemical properties of red mud were characterised by N_2 adsorption, XRD, XRF, SEM and FT-IR. The effect of modification of red mud by several physical and chemical methods such as heat treatment, acid or base treatment and H_2O_2 oxidation on HA adsorption were investigated. Moreover, the effects of some important adsorption conditions were also reported. In addition, we also report an investigation of humic acid adsorption on chitin, the kinetics and isotherm and also investigation of the effects of some adsorption conditions on humic acid adsorption on chitin. Furthermore, humic acid adsorption on chitin was compared with chitosan.

2. Experimental

2.1. Adsorbent Materials and Chemicals

Fly ash (FA) was collected using electrostatic precipitators from the Western Power, Australia. A raw red mud sample was collected from a downstream slurry pond from the Worsley Alumina, Australia. The solid material was washed with distilled water several times and dried at 110 °C overnight (RM). For the treatment of RM, one sample was obtained by heat-treatment of the washed RM at 800 °C for overnight (RM-Heat). Other two samples were obtained by acid-treatment in 1 N HNO_3 solution (RM-HNO_3) or base treatment in 1 N NaOH (RM-NaOH) for 24 h at room temperature (10 g solids in 20 mL solution). Another sample was treated with H_2O_2 (RM-H_2O_2) at room temperature for 24 h (10 g in 30 ml 30% H_2O_2). After treatment, the three samples were filtered, washed and dried at 110 °C overnight for adsorption tests.

Chitin and chitosan flake prepared from crab shell were obtained from Aldrich-Sigma. The degree of acetylation of the chitosan is 85%. Humic acid in sodium form was also obtained from Aldrich-Sigma. Other chemicals, NaOH, HNO_3, H_2O_2, NaCl, KCl, $NaNO_3$ and $Mg(NO_3)_2$, are of analytic grade and obtained from Chem-supply.

2.2. Characterisation

The surface area and total pore volume were determined by N_2 adsorption under –196 °C using Autosorb (Quantachrome Corp.). All samples were degassed at 200 °C for 4 h, prior to the adsorption experiments. The BET surface area was obtained by applying the BET equation to the adsorption data. Total pore volume (V_{tot}) was obtained from the adsorption at $P/P_0 = 0.96$.

X-ray diffraction (XRD) patterns of all samples were obtained with an automated Siemens D500 Bragg-Brentano instrument using Cu Kα radiation at 40 kV and 40 mA over the range (2θ) of 5 –70 °.

The pH of samples was measured as follows: 0.5 g of FA and RM were mixed with 10 mL of distilled water and shaken for 24 h at 30 °C. After filtration, the pH of solution was determined by a pH meter (Radiometer PHM250 Ion Analyser).

Microstructure and surface morphology of the FA and RM samples were characterised by a JEOL 6400 field emission scanning electron microscope (SEM) with an accelerating voltage 12.0 kV at different magnifications.

Fourier transfer infrared (FT-IR) spectra were collected on a Nicolet 6700 with a resolution of 4 cm^{-1} by using attenuated total reflectance (ATR) techniques. The spectrum was scanned from 400 - 4000 cm^{-1}.

2.3. Adsorption Studies

The tests of HA adsorption onto various fly ash and red mud samples were conducted in batch mode. In these experiments, 50 mg adsorbents, along with 100 ml HA solutions of concentrations 10 – 100 ppm, were put into 250 ml bottles. The temperature of all solutions was controlled at 30 °C and pH 7, otherwise indicated, was maintained using 0.1 M HNO_3 and NaOH solution. The tests were conducted in an orbital-shaker at 100 rpm. The samples were taken out for centrifugation separation and then analysed. A UV-vis spectrophotometer (Metertech UV-Vis SP-8001) was used for the determination of HA concentration at 258 nm.

The pH effect on HA adsorption was studied similarly as described above. In batch experiments, 50 mg adsorbents mixed with 100 ml HA solution at 50 ppm with different initial pHs in 250 ml bottles were set at 30 °C in the orbital-shaker at 100 rpm. The samples were taken out at different time interval for analysis.

The effect of ionic strength on HA adsorption on red mud samples was also investigated. Solutions of sodium chloride at a concentration of 0.010, 0.050 and 0.10 M were used. For each run, 50 mg solid with 100 ml of HA solution at 50 ppm at solution pH 7 were put in the orbital-shaker with set shaken rate of 100 rpm and 30 °C. The samples were then taken out for analysis at varying time interval using the spectrophotometer.

The tests of HA adsorption onto chitin and chitosan were conducted also in batch mode. In these experiments, 50 mg adsorbents, along with 100 ml HA solutions of concentrations 10 – 100 ppm, were put into 250 ml bottles. The temperature of all solutions was controlled at 30 °C, otherwise indicated, and pH 7 was maintained using 0.1 M HNO_3 and NaOH solution. The tests were conducted in an orbital-shaker at 100 rpm. The samples were taken out for centrifugation separation and then analysed. A UV-vis spectrophotometer (Metertech UV-Vis SP-8001) was used for the determination of HA concentration at 258 nm.

The pH effect on HA adsorption was studied similarly as described above. In batch experiments, 50 mg adsorbents mixed with 100 ml HA solution at 50 ppm with different initial pH values was set at 30 °C in the orbital-shaker at 100 rpm. The samples were taken out at different time interval for analysis.

The effect of different salts on HA adsorption on chitin was also investigated. Sodium chloride, potassium chloride, potassium nitrate and magnesium nitrate at varying concentrations were used. For each run, 50 mg solid with 100 ml of HA solution at 50 ppm at solution pH 7 was put in the orbital-shaker with a set shaking rate of 100 rpm and 30 °C. The samples were then taken out for analysis at varying time interval using the spectrophotometer.

3. Results and Discussion

3.1. Characteristics of FA and RM

Textural properties of RM and FA were determined by N_2 adsorption. BET surface areas of RM and FA are 22 and 15 m^2/g, respectively. XRD analyses reveal that the major chemical phases of FA are quartz, mullite, with minor hematite and magnetite while the RM contains hematite, goethite, quartz and calcite. XRF analyses shows the major chemical compositions of FA as SiO_2 (55%), Al_2O_3(29%) and Fe_2O_3(8.8%), whereas the chemical compositions of RM are Fe_2O_3(60%), Al_2O_3(15%), Na_2O (16%) and SiO_2 (5%). Solution pH of FA and RM shows that FA presents acidity (pH = 5) while RM has strong basicity (pH = 12).

Fig.2 presents SEM photos of RM and FA. As seen, the FA is mainly constituted by compact or hollowed spheres of different size. On the sphere surface the existence of solid deposits or small crystals can be observed. Some other vitreous unshaped fragments can also be seen in the FA sample. For RM, most of particles are in irregular shape. Some crystals can also be found in the RM sample. The RM presents strong surface porosity.

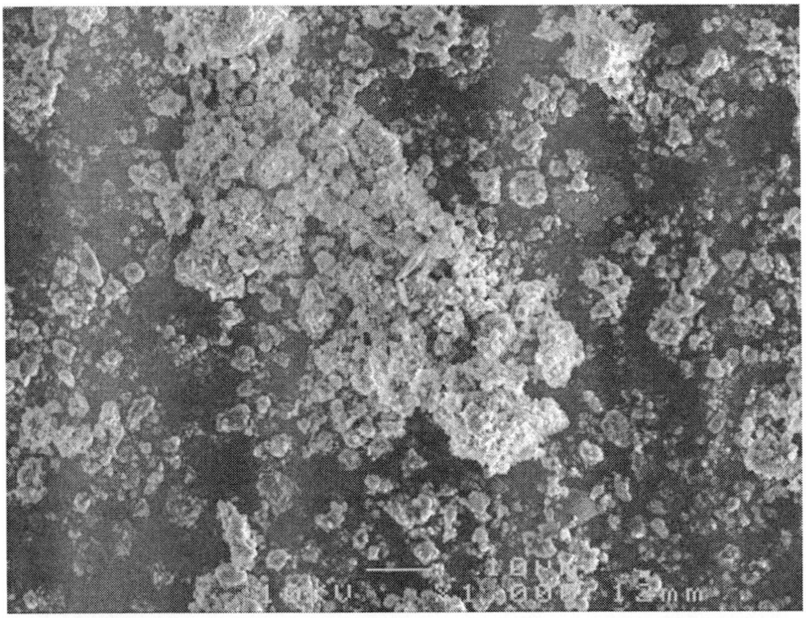

(A)

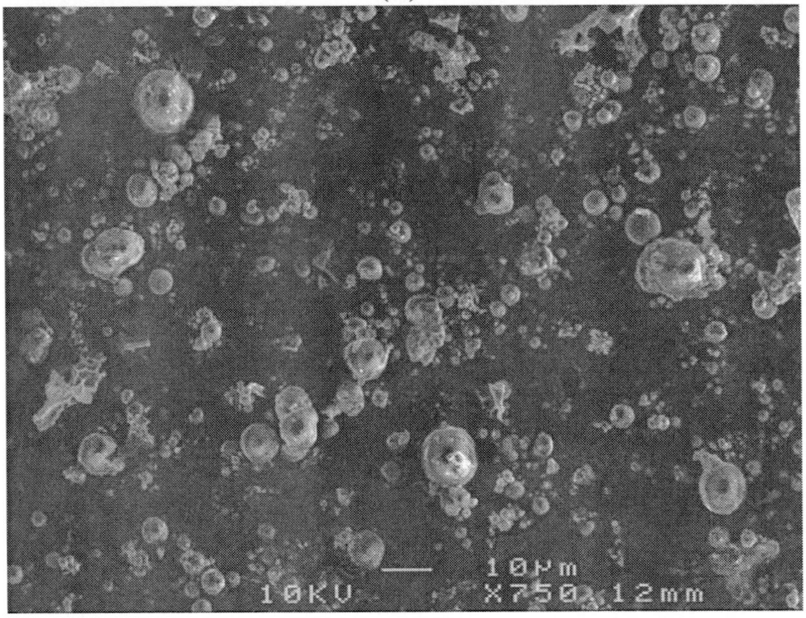

(B)

Figure 2. SEM photos of RM and FA. (a) RM (b) FA.

Fig.3 shows the FT-IR spectra of RM and FA. The FTIR spectrum of FA shows one broad component at 1050 cm^{-1}, ascribed to T–O stretching vibrations (T = Si or Al) and another located at 427 cm^{-1} ascribed to ν_4(O–Si–O) bending modes of SiO_4 tetrahedra. The band appearing at 780–790 cm^{-1}, corresponds to the quartz present in FA and the band at

534 cm^{-1} corresponds to the mullite present in the FA (Fernandez-Jimenez and Palomo, 2005). These observations confirm the XRD and XRF results.

For RM, the broad peak centred at 3100 cm^{-1} corresponding to OH vibrations appears. Two intense hydroxyl deformation bands at 874 and 802–806 cm^{-1} are also seen. The 1630 cm^{-1} is for the stretching vibrations of Fe^{3+}-O^{2-} bond. The peaks at 1450 cm^{-1} represent the existence of $CaCO_3$ and the peak at 960 cm^{-1} is for the SiO_2 (Alp and Goral, 2003).

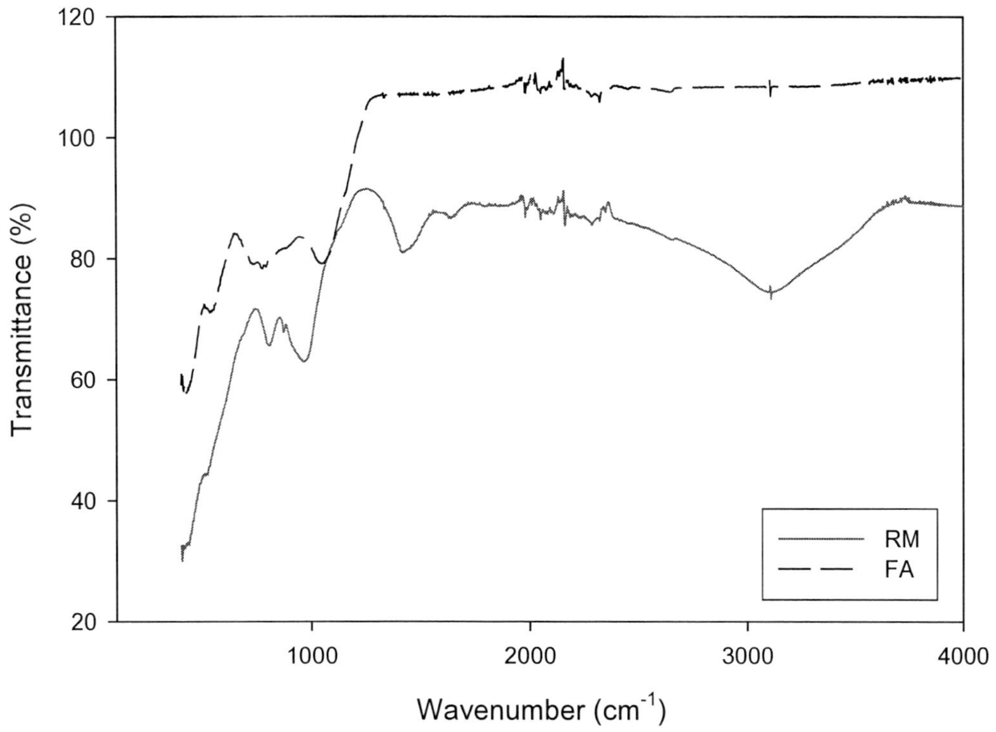

Figure 3. FT-IR spectra of RM and FA.

3.2. Dynamic Adsorption of Humic Acid and Mechanism

Fig.4 displays a comparison of dynamic adsorption of HA on FA and RM. As shown, HA adsorption on the FA and RM is approaching to equilibrium around 60 h. RM is exhibiting higher adsorption than FA. At equilibrium, HA adsorption on RM and FA is about 25 and 10 mg/g, respectively.

Investigation of adsorption kinetics can provide adsorption mechanism. Two commonly used kinetic models, pseudo first-order and pseudo second-order kinetics (Ho and McKay, 1999; Ho and Chiang, 2001), have been employed to describe the dynamic adsorption of various adsorbate on adsorbents and will be used for humic acid adsorption here.

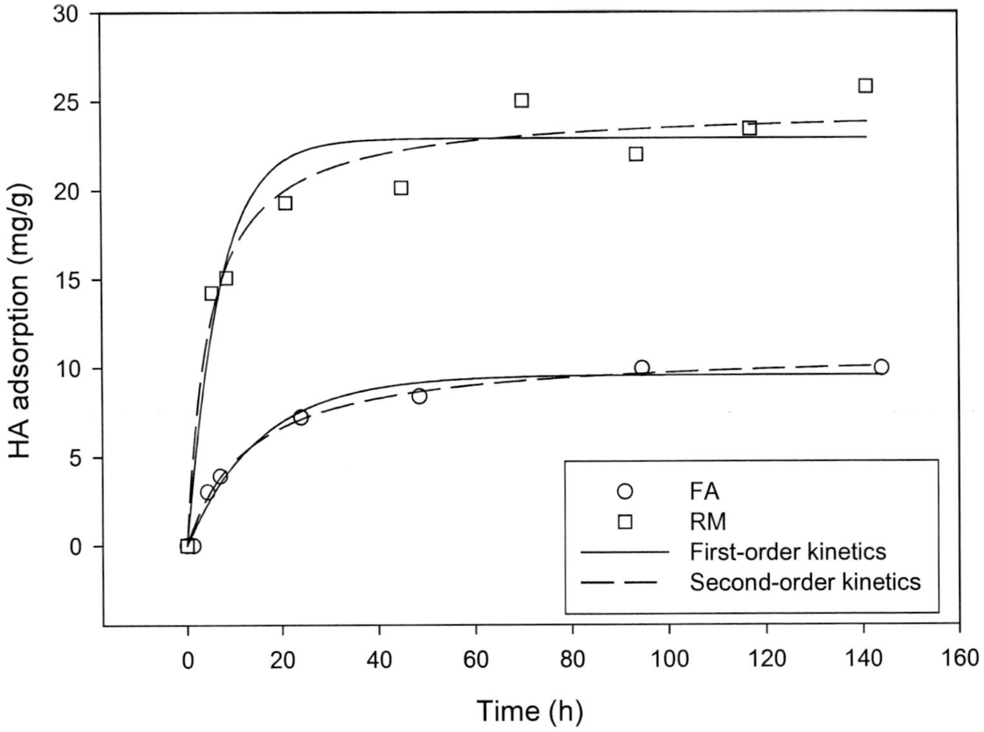

Figure 4. Comparison of HA adsorption on FA and RM.

The Lagergren first-order model is given by the equation

$$\log(q_e - q_t) = \log q_e - \frac{k_1}{2.303}t \tag{1}$$

where, q and q_e represent the amount of adsorbate adsorbed (mg g^{-1}) at any time t and at equilibrium time, respectively, and k_1 represents the adsorption rate constant (h^{-1}). After transformation, the nonlinear form for curve fitting can be obtained as follows:

$$q_t = q_e(1 - e^{-k_1 t}) \tag{2}$$

The pseudo-second-order kinetics in linear form is:

$$\frac{t}{q_t} = \frac{1}{k_2 q_e^2} + \frac{1}{q_e}t \tag{3}$$

where k_2 is the pseudo-second order rate constant of sorption (g mg^{-1} h^{-1}), q_e is the amount of adsorbate sorbed (mg g^{-1}) at equilibrium and q_t is the amount of adsorbate on the surface of the sorbent (mg g^{-1}) at any time t. The nonlinear form is

$$q_t = \frac{q_e^2 k_2 t}{(1 + q_e k_2 t)} \tag{4}$$

Another kinetic model, two-site kinetic model was also employed, which gives the non-linear form as

$$q_t = q_{e1}(1 - e^{-k_1 t}) + q_{e_2}(1 - e^{-k_2 t}) \tag{5}$$

where, q_{e1} and q_{e2} are the amount of adsorbate sorbed (mg g^{-1}) at two active sites at equilibrium, respectively. k_1 and k_2 are the rate constants of each active site, respectively.

Fig.4 presents the curves of the first-order and second-order kinetics and the kinetic parameters are given in Table 1. One can see that the first-order and second-order kinetics can fit the experimental data very well. This suggests that the adsorption of HA on FA and RM involves physisorption and chemisorption. Based on the regression coefficients and equilibrium adsorption, it is seen that the second-order kinetics presents slightly better results.

Table 1. Kinetic parameters of HA adsorption on FA and RM.

Sample	**First order kinetics**			**Second-order kinetics**		
	q_e (mg/g)	**k_1(h^{-1})**	**R^2**	**q_e (mg/g)**	**k_2(g/mg h)**	**R^2**
FA	9.57	0.0649	0.981	11.0	6.94×10^{-3}	0.990
RM	22.9	0.142	0.936	24.6	8.39×10^{-3}	0.966

SEM photos have shown that FA and RM present some porous surface. The BET surface areas of FA and RM also indicate that the RM has a slightly higher value of surface area than FA, which may contribute to the higher adsorption of HA on RM. XRD, XRF and FT-IR results also demonstrate that FA and RM have different chemical composition and phases. FA is mainly composed of silica and alumina while the major components of RM are FeOOH and Fe_2O_3. In addition, the solution pHs of FA and RM indicate that FA and RM present different slurry pH. Previous research has indicated that the slurry pH could be similar to the point of zero charge of particle (Wang and Lu, 1998; Bourikas et al., 2003). Thus, at pH 7, FA has negative surface charge but RM has positive surface charge. It is known that HA contains functional groups of phenolic (-OH) and carboxylic groups (-COOH) and will produce negatively charged ions at pH 7. Due to the electrostatic attraction, HA will be favourably adsorbed on RM with positive charge, resulting in higher adsorption than that on FA.

According to the results of FT-IR, the surface chemical groups of FA are Si-OH and Al-OH while the surface functional groups on RM would be Fe-OH, the adsorption mechanism for HA on FA and RM can be proposed as

$$S^*Fe\text{-}OH + HA \rightarrow FeA + H_2O \tag{6}$$

$$S^*Si\text{-}OH + HA \rightarrow SiA + H_2O \tag{7}$$

$$S^*Al\text{-}OH + HA \rightarrow AlA + H_2O \tag{8}$$

S* represents particle surface.

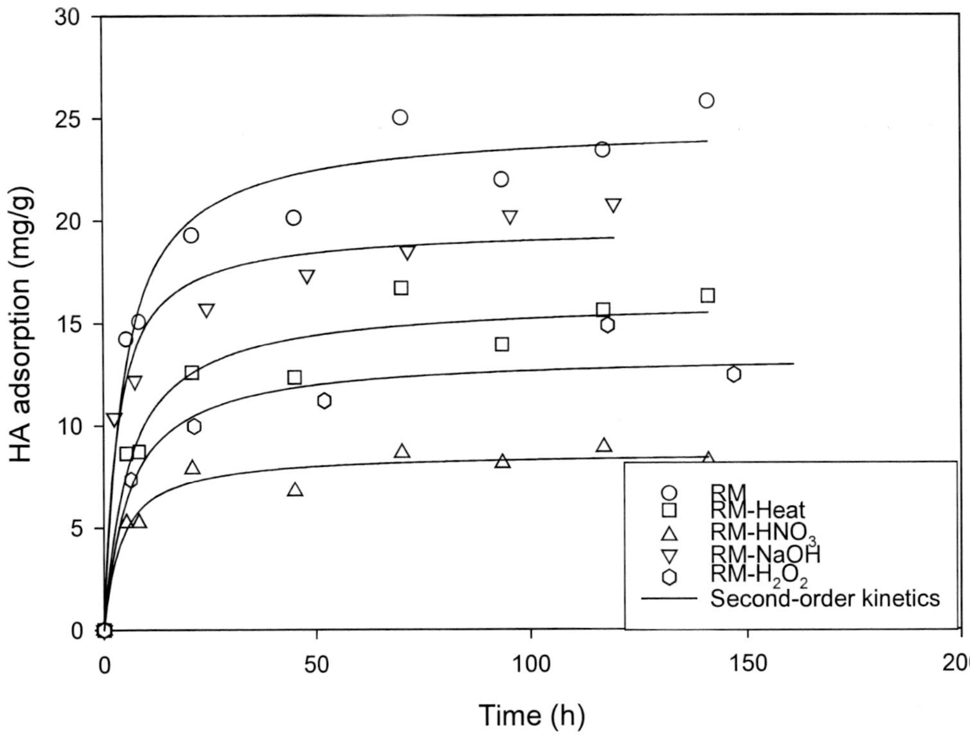

Figure 5. HA adsorption on various RMs.

Fig.5 shows the dynamic adsorption of HA on various RM samples treated by physical and chemical methods. As shown, all treated RMs exhibit lower HA adsorption than the washed RM. The adsorption equilibrium can be reached after 50 h. The equilibrium adsorption of RM-NaOH is about 20 mg/g while RM-HNO_3 only has 8 mg/g at equilibrium. The second-order kinetics was also used to describe the dynamic adsorption and the results are presented in Table 2. It is seen that the capacity follows an order of RM > RM-NaOH > RM-Heat > RM-H_2O_2 > RM-HNO_3.

It has been reported that acid or heat treatment of RM will change the composition and adsorption capacity. Our previous investigation of RM using acid or heat treatment for methylene blue adsorption has shown that both acid and heat treatment would reduce the adsorption capacity (Wang et al., 2005). Gupta et al. have used H_2O_2 oxidation combining with heat treatment to modify RM and tested for adsorption of dyes (Gupta et al., 2004b) and chlorophenols (Gupta et al., 2004a). The results showed that the RM exhibited high removal efficiency, which is different from this investigation. In this investigation, all the treatments result in a decrease of HA adsorption on RM. It has been found that heat treatment will bring the decomposition of hydroxyl on RM and reduce the surface area. Although acid treatment can increase the surface area but it will change the surface charge by removal hydroxyl compounds, resulting in the change of point of zero charge to low pH. H_2O_2 oxidation can also change the surface functional groups, resulting in low adsorption.

Table 2. The parameters of second-order kinetics for HA adsorption on various RMs.

Sample	Second-order kinetics		
	q_e (mg/g)	k_2(g/mg h)	R^2
RM-NaOH	19.6	0.0156	0.958
RM-Heat	16.1	0.0104	0.951
RM-H_2O_2	13.4	0.0124	0.961
RM-HNO_3	8.6	0.0282	0.956

3.3. Comparison of Humic Acid Adsorption on Chitin and Chitosan

Fig.6 shows the dynamic adsorption of HA on chitin and chitosan. It is seen that humic acid adsorption on chitin and chitosan is quite different. The adsorption rate on chitosan is much higher than that on chitin. The adsorption equilibrium on chitin is approaching to 25 mg/g while the adsorption value is close to 65 mg/g. The observation is similar to the finding of Ngah and Musa (1998), who found that the dynamic uptake of humic acid on chitosan was greater than that on chitin.

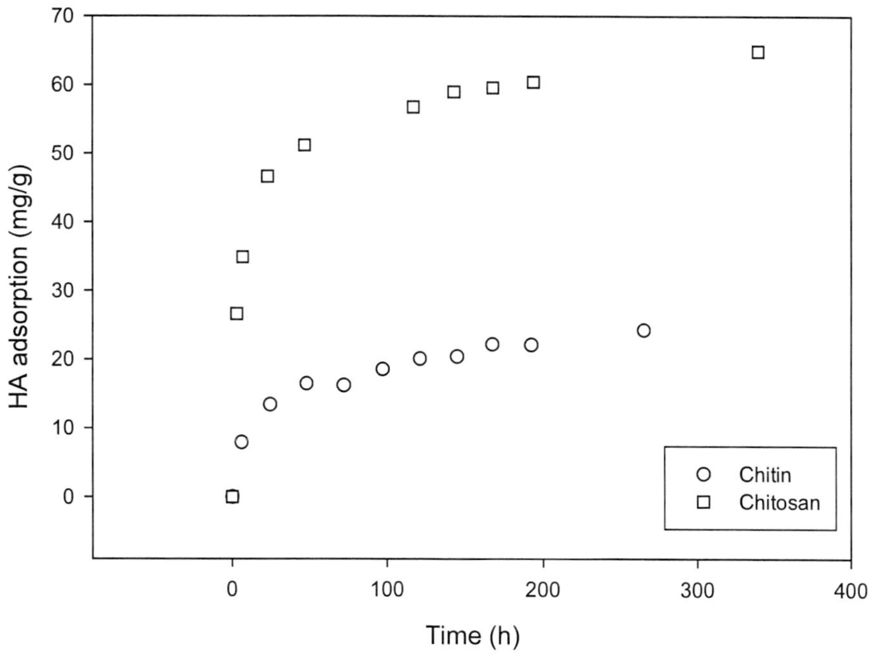

Figure 6. Comparison of dynamic HA adsorption on chitin and chitosan.

Table 3. Parameters of kinetic models using curve fitting to experimental results.

Adsorbent	First-order kinetics			Second-order kinetics			Two-site kinetics				
	k_1 (h^{-1})	q_e (mg/g)	R^2	k_2 (g/mg h)	q_e (mg/g)	R^2	k_1 (h^{-1})	q_1 (mg/g)	k_2 (h^{-1})	q_2 (mg/g)	R^2
Chitin	0.0347	21.2	0.914	2.11x10^{-3}	23.8	0.962	0.159	11.9	5.39x10^{-3}	16.3	0.995
Chitosan	0.146	57.4	0.937	3.27x10^{-3}	60.9	0.981	0.286	41.7	8.97 x10^{-3}	23.6	0.996

Fig.7 presents the comparison of the three different kinetic models for HA adsorption on chitin and chitosan and the kinetic parameters obtained from the models are given in Table 3. The results show that the first-order and second-order kinetics can fit the experimental data but the two-site kinetics model will give the best fitting in terms of regression coefficients. This suggests that there are two active sites or mechanisms responsible for HA adsorption on chitin and chitosan. The first active site usually has a higher adsorption rate constant. Also, chitosan has higher adsorption rate constant adsorption capacity than chitin.

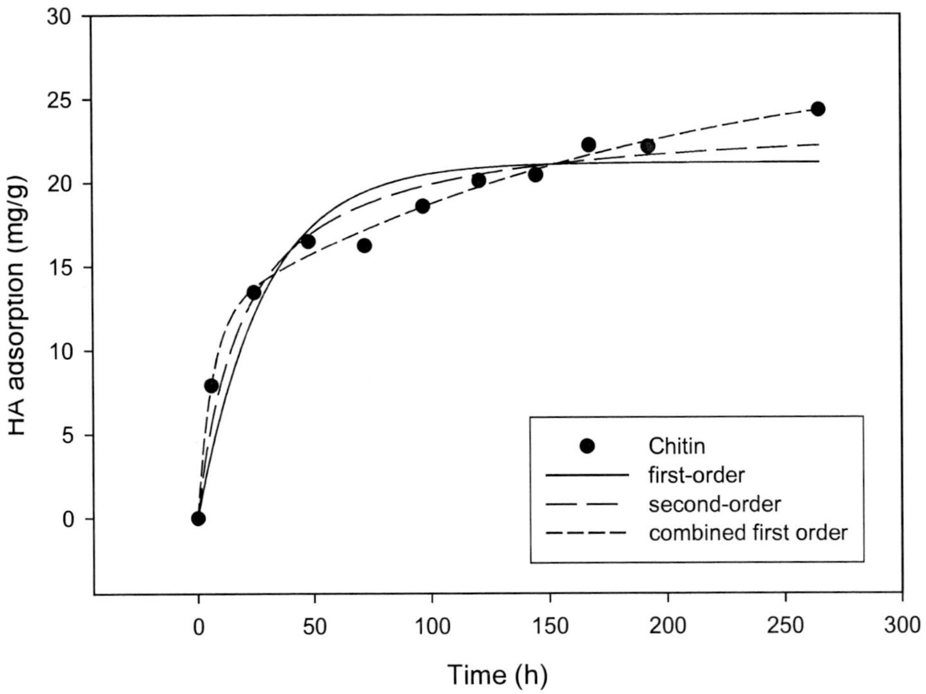

Figure 7. Continued on next page.

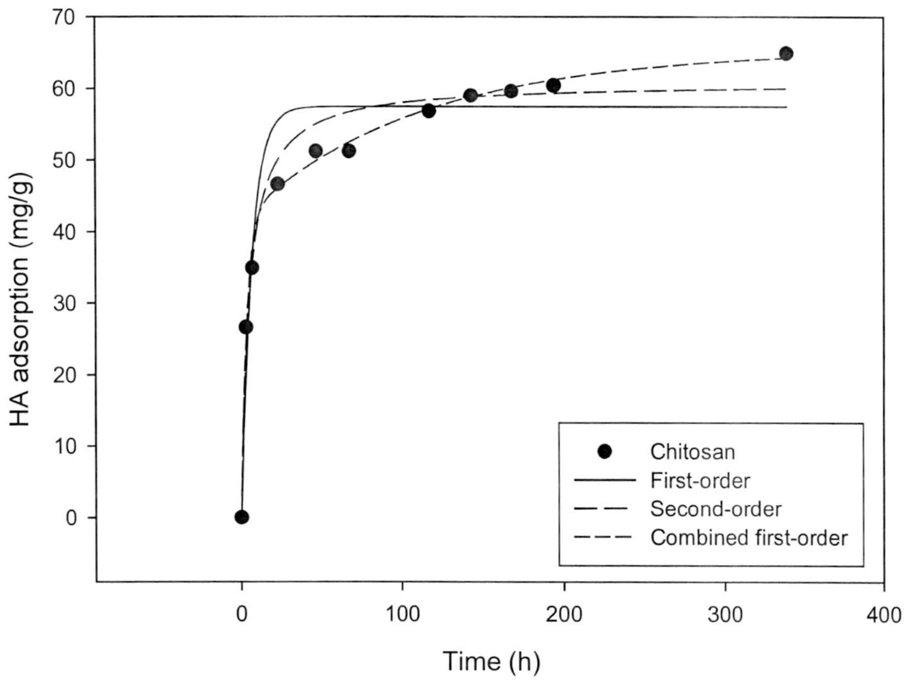

Figure 7. Comparison of kinetic models of HA adsorption on chitin and chitosan.

(A)

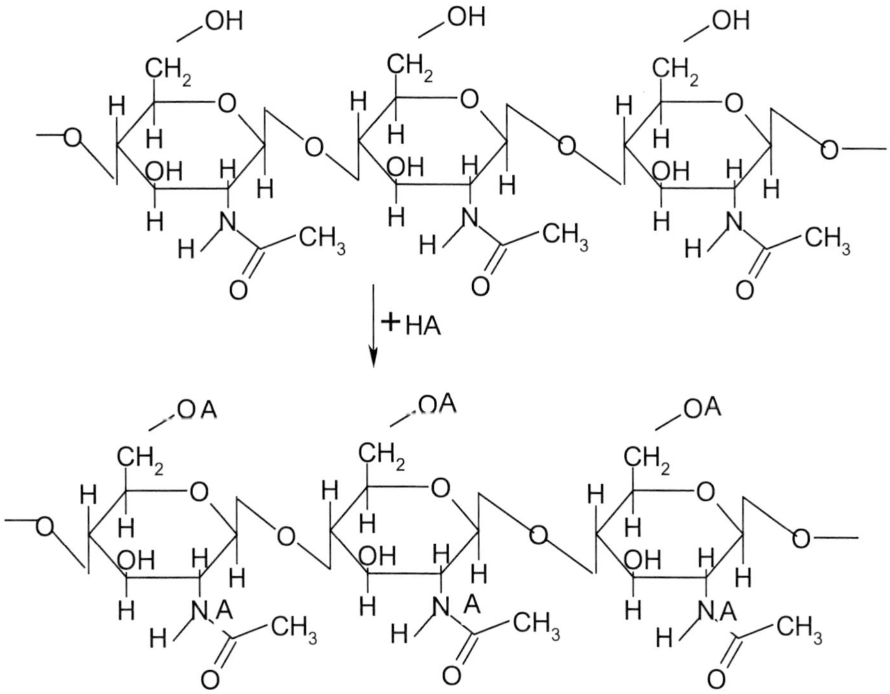

Figure 8. Continued on next page.

(B)

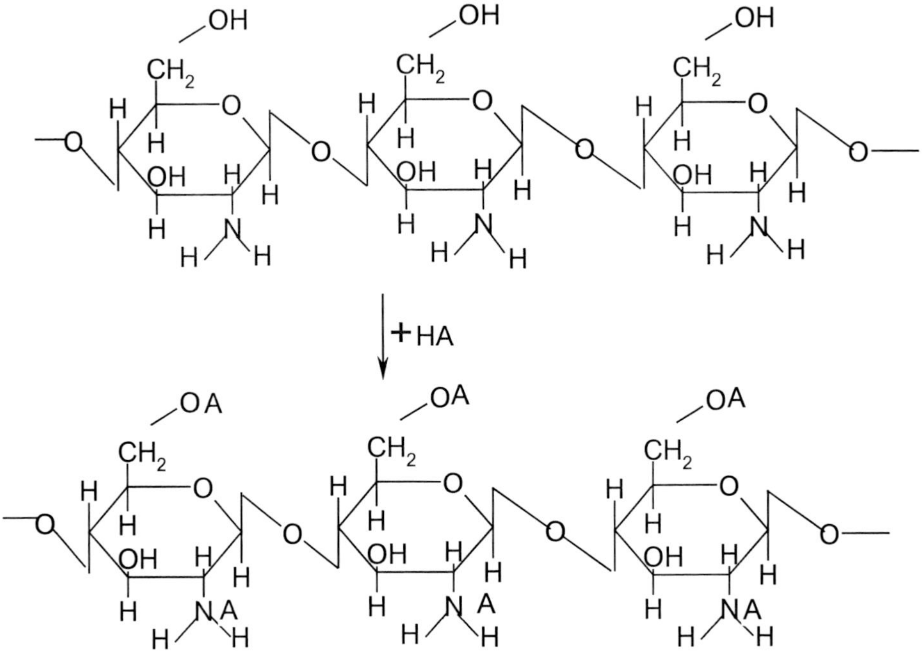

Figure 8. HA adsorption mechanism on chitin and chitosan. (a) Chitin (b) Chitosan.

HA is known to have several functional groups such as phenolic (-OH), carboxyl (-COOH) and amino (-NH_2). For chitin and chitosan, the hydroxyl (-OH) and amino (-NH_2) can serve as the coordination and reaction sites. It is found that chitosan has much higher HA adsorption than chitin, suggesting that deacetylation has a great contribution to HA adsorption by providing strong affinity of the active sites to HA. Thus, it is deduced that amino (-NH_2) groups will play significant role for adsorption. Also the kinetic studies indicate that there are two active sites for HA adsorption and the rate constants are significant different. Therefore, the adsorption mechanism of HA on chitin and chitosan is proposed as shown in Fig.8.

3.3. Effect of Cation and Anion on HA Adsorption on Chitin and Red Mud

It has been known that ionic strength is an important factor influencing HA adsorption on some adsorbents. Fig.9 presents the HA adsorption on chitin at varying concentration of NaCl and KCl. As shown that the presence of NaCl or KCl will increase HA adsorption and higher concentration of NaCl will produce high HA adsorption. At 0.013 N NaCl, HA adsorption is around 30 mg/g while at 0.1 N NaCl, HA adsorption will be increased to 40 mg/g. For different salts, KCl and NaCl, HA adsorption shows little difference when their concentration is the same, which suggests that the positive charge ions, Na^+ and K^+, have the same effect on HA adsorption on chitin.

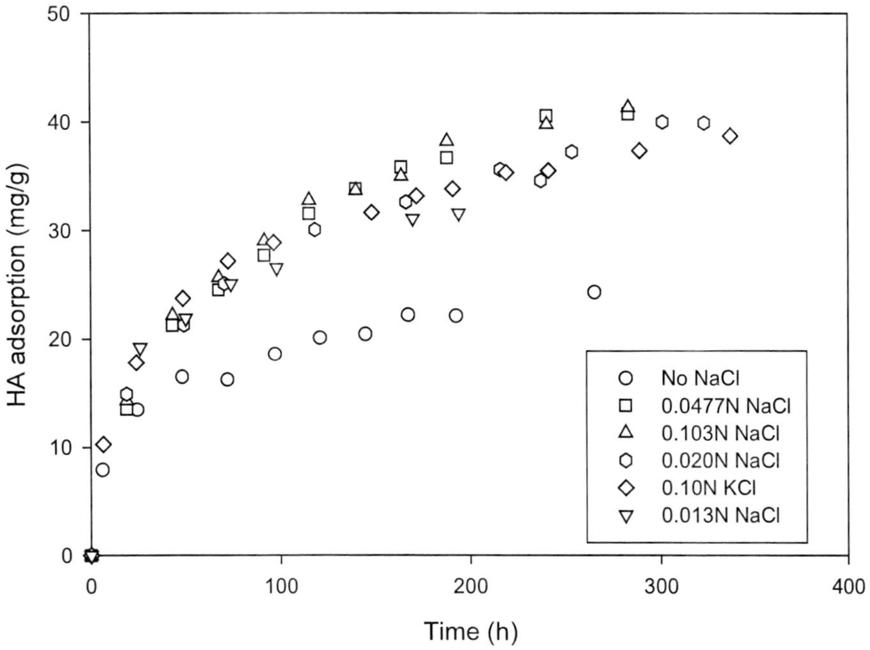

Figure 9. Effect of ion strength on HA adsorption on chitin.

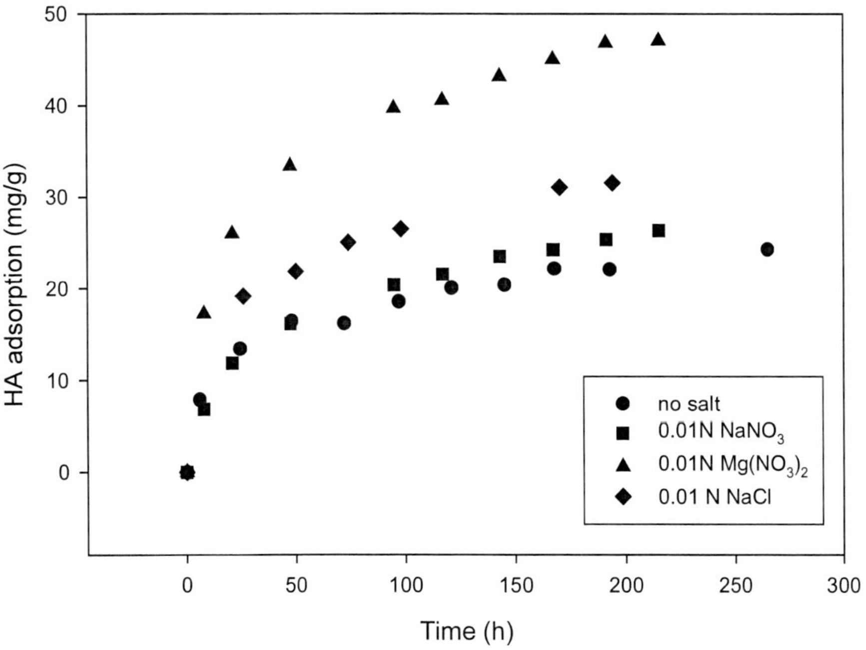

Figure 10. Effects of cation and anion of salts on HA adsorption on chitin.

Fig.10 compares the adsorption of HA on chitin with the presence of different salts, It is seen that HA adsorption presents different amounts at the presence of $Mg(NO_3)_2$, $Na(NO_3)$

and NaCl at the same concentration. For cations, Mg^{2+} produces a stronger effect on HA adsorption than on Na^+, while for anions, Cl^- will also have a better effect on HA adsorption than NO_3^-. In the presence of 0.01 N $NaNO_3$, HA adsorption will be slightly enhanced. While at 0.01 N NaCl and $Mg(NO_3)_2$, HA adsorption is about 30 and 50 mg/g, respectively.

Daifullah et al. (2004) investigated the effect of certain cationic species in solution on humic acid adsorption on activated carbon and found that cationic species, Ca^{2+}, Mg^{2+}, Na^+ and Al^{3+}, would enhance humic acid adsorption in an order of $Ca^{2+} > Mg^{2+} > Na^+ > Al^{3+}$, which is similar to the result of this investigation, $Mg^{2+} > Na^+ \sim K^+$. They also found that anion nitrate would reduce humic acid adsorption as the concentration of nitrate increased. Our previous investigation of humic acid adsorption on unburned carbon at the presence of various anionic species, nitrate, chloride, and sulphate, showed that these anionic species presented a similar effect on humic acid adsorption (Wang and Zhu, 2007).

Fig.11 shows the dynamic adsorption of HA on RM at varying NaCl concentrations. It is seen that the presence of NaCl produces a negative effect on HA adsorption. At 0.01 N NaCl, HA adsorption shows a slight reduction and the equilibrium adsorption is around 20 mg/g. At 0.05 N NaCl, HA adsorption is significantly reduced to 18 mg/g, but the adsorption is quite similar when NaCl concentration is further increased to 0.1 N.

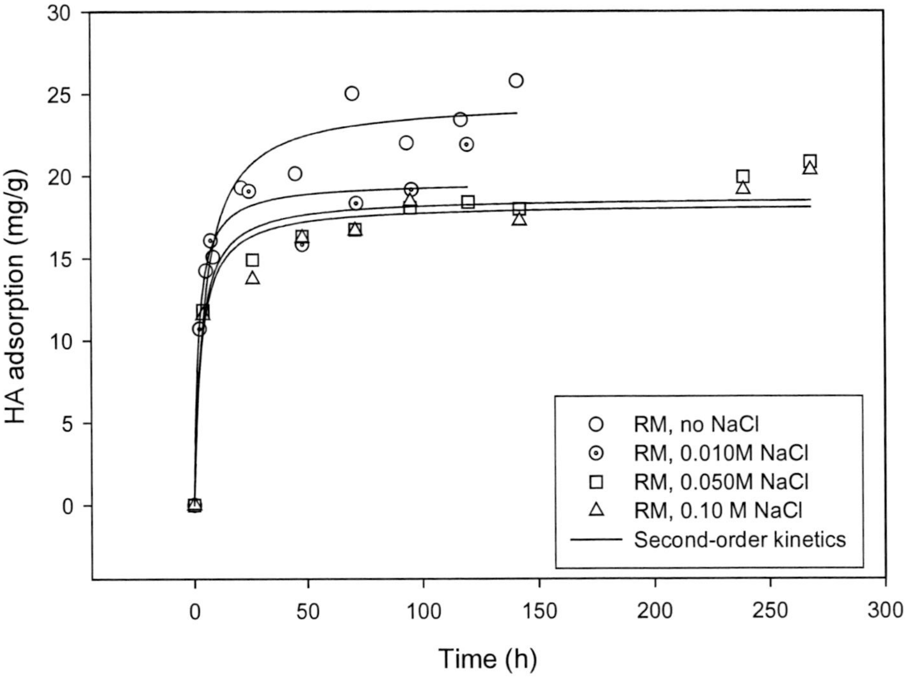

Figure 11. Effect of NaCl concentration on HA adsorption on RM.

The presence of salts in aqueous solution of HA with adsorbent can produce two effects on adsorption. One is to change the interaction between salts and HA resulting in the neutralisation of its negatively charged functional groups, rendering the molecules both smaller and more hydrophobic, which is known as the Fuoss effect (Stevenson, 1977). This effect will bring in an enhancement in HA adsorption. The other effect is the competitive adsorption of between anionic ions with HA. Many previous investigations have found that the increase of ionic strength would increase HA adsorption on some adsorbents and the extent depended on metal charge (Daifullah et al., 2004; Capasso et al., 2005; Peng et al., 2006). However, in this investigation, the negative effect of NaCl on HA adsorption suggests that there is strong competitive adsorption between Cl^- and HA.

3.4. Effect of pH on HA Adsorption on Red Mud and Chitin

Fig.12 further displays the effect of solution pH on HA adsorption on chitin. As shown, high pH will result in low adsorption. At pH 4.7, the adsorption is approaching 50 mg/g. When the pH is increased to 7.5, HA adsorption is only 20 mg/g. Several investigations of HA adsorption on chitosan have shown the similar observations (Ngah and Musa, 1998; Zhang and Bai, 2003). Humic acid contains both hydrophobic and hydrophilic moieties with several functional groups of carboxylic, phenolic, carbonyl and hydroxyl. At low pH, more of these groups are in an uncharged state, hence more adsorbable on hydrophobic surface of chitin (Daifullah et al., 2004). When pH is increased, carboxylic and phenolic functional groups on humic acid will compete with OH^- to be adsorbed on chitin, resulting in low adsorption.

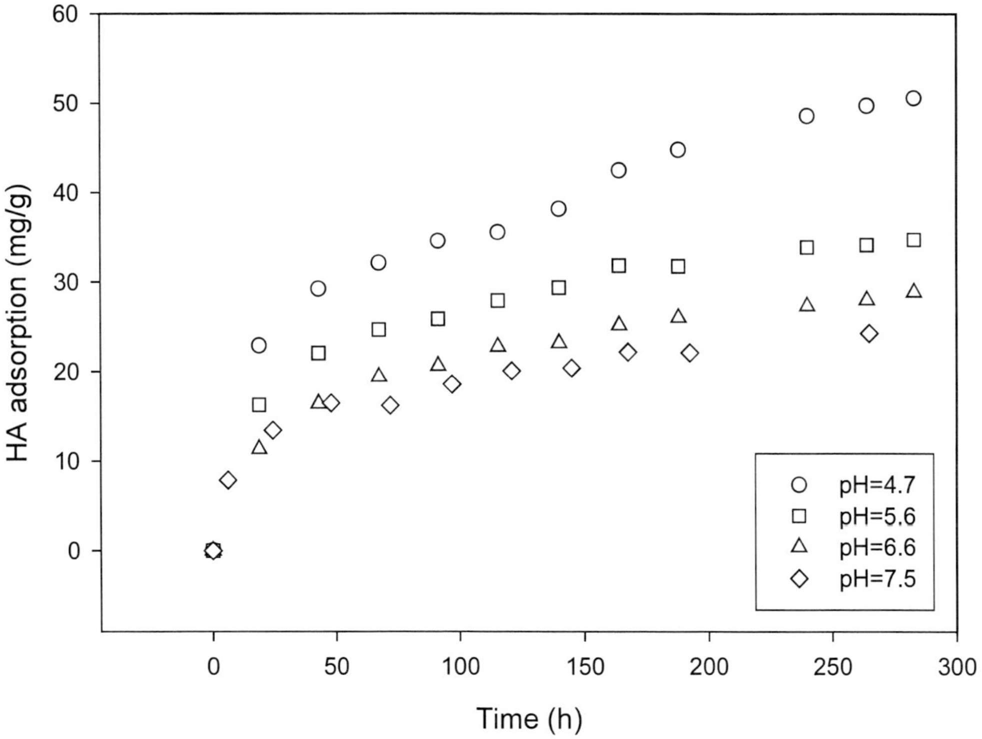

Figure 12. Effect of pH on HA adsorption on chitin.

The effect of solution pH on HA adsorption on RM is presented in Fig.13. It is seen that higher pH results in lower HA adsorption on RM. At pH 4.7, HA adsorption is much higher at 80 mg/g while at pH 10.6, the adsorption is only 11 mg/g, only one-eighth of the value at pH 4.7. Many investigations have reported the similar effect of pH on HA adsorption on adsorbents (Tombacz et al., 2004; Peng et al., 2006). The effect of pH is also related with particle surface charge. As known, humic acids in aqueous solutions can develop dominantly negative charges due to the dissociation of (mainly) acidic functional groups bound chemically to the more or less cross-linked carbon network. At low pH, red mud possess positive charge which favours the adsorption of HA while at high pH, the surface charge of red mud is changed to negative, which will resist the adsorption of HA. In addition, as the pH is lowered, carboxylate groups are protonated, minimising the intramolecular electrostatic repulsion and contracting the HA molecules. Decrease of the molecular volume can facilitate the adsorption process due to greater availability of pores where HA molecules can penetrate.

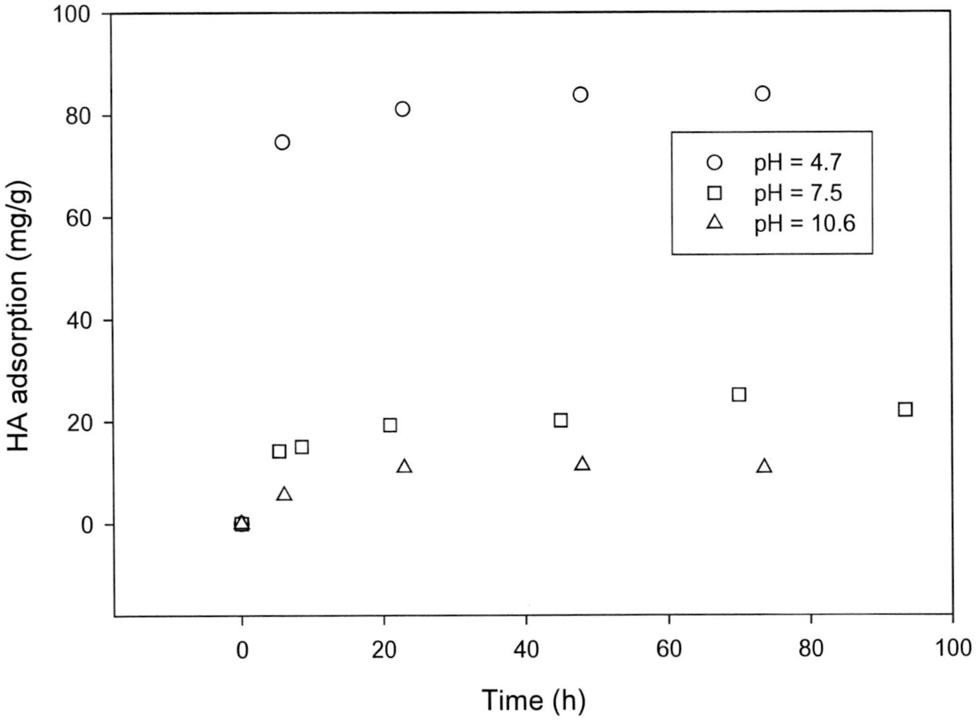

Figure 13. Effect of solution pH on HA adsorption on RM.

3.5. Adsorption Isotherms

Fig.14 displays the adsorption isotherm of HA on chitin and chitosan. As seen that HA adsorption increases with the increasing HA concentration and chitosan exhibits higher adsorption than chitin. As HA concentration is higher than 30 ppm, the adsorption approaches equilibrium on chitin. However, the adsorption of HA on chitosan is far from the equilibrium.

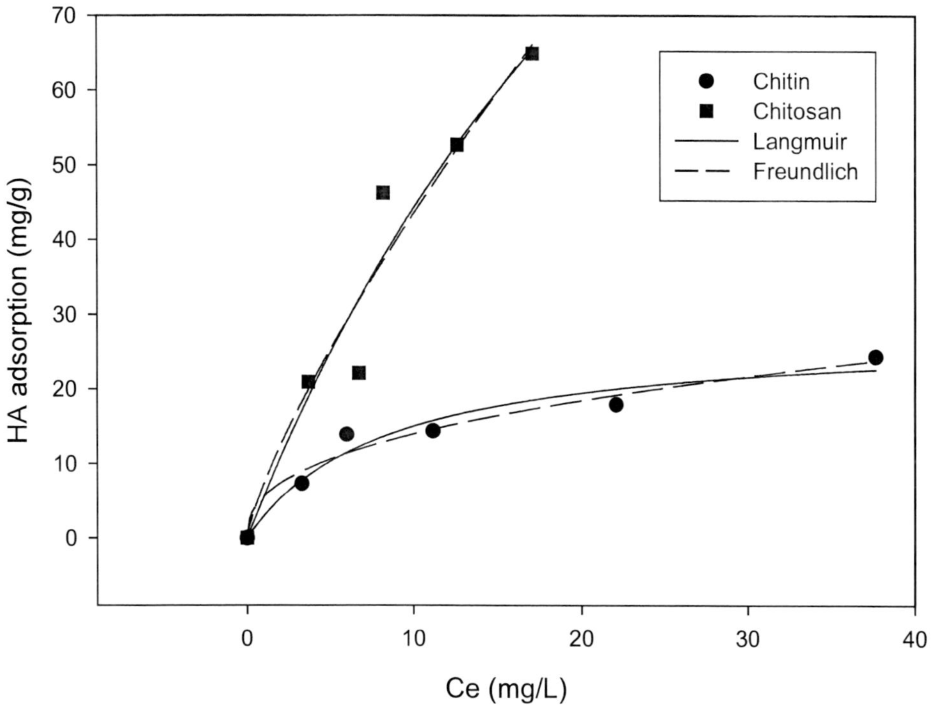

Figure 14. Adsorption isotherm of HA on chitin and chitosan.

Two adsorption isotherm models, Langmuir and Freundlich isotherms, were employed to fit the data. The Langmuir isotherm theory assumes monolayer coverage of adsorbate over a homogenous adsorbent surface. Once a HA molecule occupies a site, no further adsorption can take place at that site. The generalised Langmuir isotherm can be written in the form

$$Q_e = \frac{Q_{max} K_b C_e}{1 + K_b C_e} \tag{9}$$

Q_e is the adsorbed amount of the HA, C_e is the equilibrium concentration of the HA in solution, Q_{max} is the monolayer adsorption capacity and K_b is the constant related to the free energy of adsorption.

The Freundlich expression (Eq.10) is an exponential equation and therefore, assumes that as the adsorbate concentration increases so too does the concentration of adsorbate on the adsorbent surface. In this equation K and n are the Freundlich constants. This expression is characterised by the extent of the adsorption (K) and the heterogeneity factor (n), and so the Freundlich isotherm may be used to describe heterogeneous systems.

$$Q_e = K C_e^{1/n} \tag{10}$$

Fig.14 also shows the curve fitting of the above two isotherms. The parameters obtained from the curve fittings are given in Table 4. It is seen that the data can be fitted with both the

Langmuir and Freundlich isotherms. The regression coefficients are similar for two models. Based on the Langmuir isotherm, the maximum HA adsorption on chitin and chitosan is 28 and 202 mg/g, respectively.

Table 4. Parameters of adsorption isotherm of humic acid on chitin.

Adsorbent	Langmuir isotherm			Freundlich isotherm		
	Q_{max} (mg/g)	K_b (L/mg)	R^2	K	1/n	R^2
Chitin	27.6	0.119	0.957	5.55	0.400	0.969
Chitosan	202.7	0.0279	0.940	7.16	0.783	0.938

Ngah and Musa (1998) reported that the adsorption was correlated well with the Langmuir isotherm equation, as compared to the Freundlich isotherm equation. The adsorption capacities of humic acid on chitin and chitosan were 27.3 and 28.9 mg/g, respectively. Yan and Bai (2005) investigated humic acid adsorption on chitosan hydrogel beads (85% deacetylation) at different pHs and found that the adsorption was around 0.6 mg/g. Zhang and Bai (2003) also reported humic acid adsorption on chitosan (85% deacetylation) coated polyethyleneterephthalate granules as 0.407 mg/g. However, Chang and Juang found that dried chitosan-fiber composite beads with 97.2% deacetylation were quite effective for humic acid adsorption at capacity of 118 mg/g. Thus it is seen that chitin can be a good adsorbent for HA adsorption though its capacity is lower than chitosan.

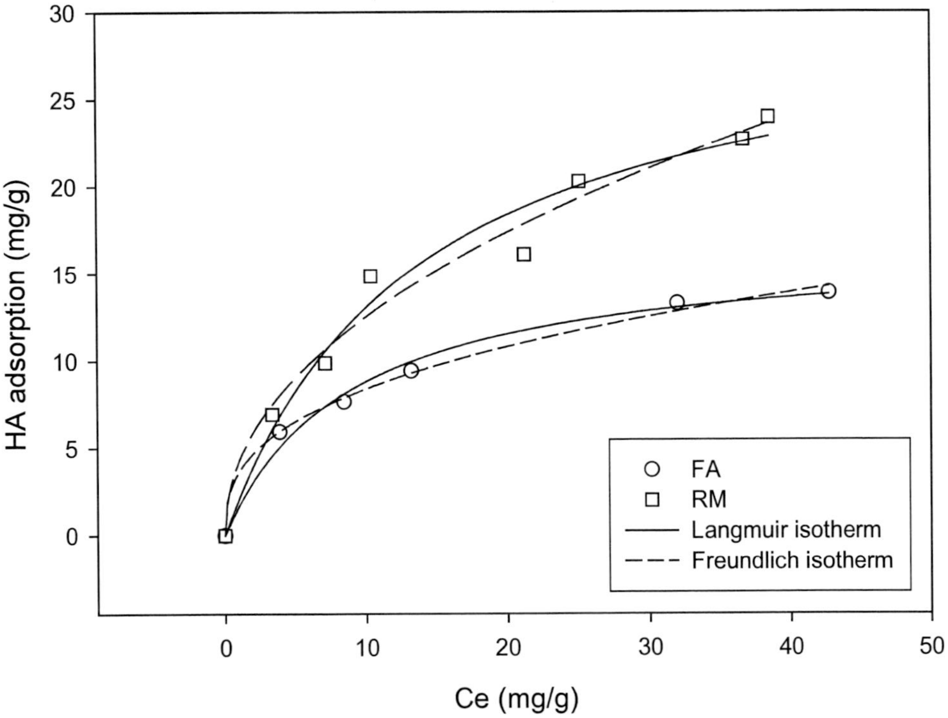

Figure 15. Adsorption isotherms of HA on FA and RM.

Fig.15 illustrates the HA adsorption isotherms on FA and RM at 0.01 N NaCl. It is seen that HA adsorption increases with HA equilibrium concentration. For FA, the adsorption will level off when HA concentration is higher than 30 mg/L. RM will present higher adsorption capacity than FA and the adsorption still shows an increasing trend after 30 mg/L.

The fitted adsorption isotherms of the Langmuir and Freundlich models are also shown in Fig.15. The parameters from curve fitting are produced in Table 5. As shown, two isotherms are fitting quite well with the experimental data. The regression coefficients from two models are similar and greater than 0.98. From Table 5, the HA adsorption capacities of FA and RM are derived as 17 and 31 mg/g, respectively.

Table 5. The parameters of HA adsorption isotherms on FA and RM.

Sample	Langmuir isotherm			Freundlich isotherm		
	Q_m (mg/g)	k (L/mg)	R^2	K (mg/g)	1/n	R^2
FA	16.6	0.112	0.990	3.62	0.364	0.997
RM	31.0	0.0724	0.975	4.22	0.472	0.980

Some investigations on HA adsorption on other clay-based adsorbents have been reported. Koyama (1995) found that the saturated adsorption amount of the HA on Ca-montmorillonite was 20-38 mg/g at pH 6-7 and in a NaCl concentration range of 0.004-0.028 N. Wibulswas et al. (1998) have reported a maximum adsorption capacity of 23.4 mg/g of HA on alumina-pillared clays. Tombacz et al. (2004) investigated the interaction of humic acid on clay mineral and iron oxide particles. They found that humic acid can be bound to the most reactive surface sites of clay and oxide particles, i.e. to Al-OH mainly at the edges of clay lamellae, and to Fe-OH on iron oxides. The adsorption of humic acid is much higher on hematite and magnetite than on montmorillonite and kaolinite. This investigation is similar to the above finding and suggests that FA and RM exhibit similar adsorption as clays and can be used as adsorbent and soil amendment.

4. Conclusion

RM has been treated by various physicochemical methods and tested for HA adsorption. Their adsorption behaviour was compared with FA. The adsorption results indicate that the RM shows higher adsorption capacity than FA due to the higher surface area and different surface charge. However, treatment of RM will result in low adsorption capacity. Several adsorption conditions will also influence the HA adsorption on RM. The presence of NaCl induced a competitive adsorption and thus producing a negative effect of HA adsorption. Higher concentration of NaCl than 0.1 N will not make significant change in adsorption. Solution pH will also affect the adsorption and HA adsorption will decrease with increasing pH.

HA presents strong adsorption on chitin and chitosan. The adsorption studies indicate that amino functional groups will play significant role for HA adsorption. Chitosan exhibits much higher adsorption capacity than chitin. For HA adsorption on chitin, higher ionic strength will result in higher adsorption and the extents depend on cation positive charge and anion properties. The effect of Mg^{2+}, Na^+ and K^+ ions show an order of $Mg^{2+} > Na^+ \sim K^+$ while chloride in solution will produce higher HA adsorption than nitrate. Solution pH also affects HA adsorption on chitin and higher pH will reduce HA adsorption. Kinetic studies show that humic acid adsorption on chitin and chitosan follows a two-site kinetics. The maximum adsorption of humic acid on chitin and chitosan based on the Langmuir isotherm is 28 and 202 mg/g, respectively.

References

Akkaya, G., Uzun, I., Guzel, F., 2007. Kinetics of the adsorption of reactive dyes by chitin. *Dyes and Pigments* **73**, 168-177.

Alp, A., Goral, M.S., 2003. The influence of soda additive on the thermal properties of red mud. *Journal of Thermal Analysis and Calorimetry* **73**, 201-207.

Annadurai, G., Chellapandian, M., Krishnan, M.R.V., 1999. Adsorption of reactive dye on chitin. *Environmental Monitoring and Assessment* **59**, 111-119.

Barriada, J.L., Herrero, R., Prada-Rodriguez, D., de Vicente, M.E.S., 2007. Waste spider crab shell and derived chitin as low-cost materials for cadmium and lead removal. *Journal of Chemical Technology and Biotechnology* **82**, 39-46.

Bourikas, K., Vakros, J., Kordulis, C., Lycourghiotis, A., 2003. Potentiometric mass titrations: experimental and theoretical establishment of a new technique for determining the point of zero charge (PZC) of metal (hydr)oxides. *J. Phys. Chem. B* **107**, 9441-9451.

Capasso, S., Salvestrini, S., Coppola, E., Buondonno, A., Colella, C., 2005. Sorption of humic acid on zeolitic tuff: a preliminary investigation. *Applied Clay Science* **28**, 159-165.

Chang, M.Y., Juang, R.S., 2004. Adsorption of tannic acid, humic acid, and dyes from water using the composite of chitosan and activated clay. *Journal of Colloid and Interface Science* **278**, 18-25.

Crini, G., 2005. Recent developments in polysaccharide-based materials used as adsorbents in wastewater treatment. *Progress in Polymer Science* **30**, 38-70.

Crini, G., 2006. Non-conventional low-cost adsorbents for dye removal: A review. *Bioresource Technology* **97**, 1061-1085.

Daifullah, A.A.M., Girgis, B.S., Gad, H.M.H., 2004. A study of the factors affecting the removal of humic acid by activated carbon prepared from biomass material. *Colloids and Surfaces A: Physicochemical and Engineering Aspects* **235**, 1-10.

Dastgheib, S.A., Karanfil, T., Cheng, W., 2004. Tailoring activated carbons for enhanced removal of natural organic matter from natural waters. *Carbon* **42**, 547-557.

Duan, J.M., Wang, J.H., Graham, N., Wilson, F., 2002. Coagulation of humic acid by aluminium sulphate in saline water conditions. *Desalination* **150**, 1-14.

Dursun, A.Y., Kalayci, C.S., 2005. Equilibrium, kinetic and thermodynamic studies on the adsorption of phenol onto chitin. *Journal of Hazardous Materials* **123**, 151-157.

Exall, K.N., Vanloon, G.W., 2000. Using coagulants to remove organic matter. *Journal American Water Works Association* **92**, 93-102.

Fernandez-Jimenez, A., Palomo, A., 2005. Mid-infrared spectroscopic studies of alkali-activated fly ash structure. *Microporous and Mesoporous Materials* **86**, 207-214.

Gerente, C., Lee, V.K.C., Le Cloirec, P., McKay, G., 2007. Application of chitosan for the removal of metals from wastewaters by adsorption - Mechanisms and models review. *Critical Reviews in Environmental Science and Technology* **37**, 41-127.

Gupta, V.K., Ali, I., Saini, V.K., 2004a. Removal of chlorophenols from wastewater using red mud: An aluminum industry waste. *Environmental Science & Technology* **38**, 4012-4018.

Gupta, V.K., Suhas, Ali, I., Saini, V.K., 2004b. Removal of rhodamine B, fast green, and methylene blue from wastewater using red mud, an aluminum industry waste. *Industrial & Engineering Chemistry Research* **43**, 1740-1747.

Ho, Y.S., Chiang, C.C., 2001. Sorption studies of acid dye by mixed sorbents. *Adsorption* **7**, 139-147.

Ho, Y.S., McKay, G., 1999. Pseudo-second order model for sorption processes. *Process Biochem* **34**, 451-465.

Jala, S., Goyal, D., 2006. Fly ash as a soil ameliorant for improving crop production - a review. *Bioresource Technology* **97**, 1136-1147.

Jiang, J.Q., Cooper, C., 2003. Preparation of modified clay adsorbents for the removal of humic acid. *Environmental Engineering Science* **20**, 581-586.

Juang, R.S., Tseng, R.L., Wu, F.C., Lin, S.J., 1996. Use of chitin and chitosan in lobster shell wastes for color removal from aqueous solutions. *Journal of Environmental Science and Health Part A-Environmental Science and Engineering & Toxic and Hazardous Substance Control* **31**, 325-338.

Kim, S.H., Song, H., Nisola, G.M., Ahn, J., Galera, M.M., Lee, C.H., Chung, W.J., 2006. Adsorption of lead(II) ions using surface-modified chitins. *Journal of Industrial and Engineering Chemistry* **12**, 469-475.

Koyama, M., 1995. Adsorption of Humic-Acid on Ca-Montmorillonite. *Soil Science and Plant Nutrition* **41**, 215-223.

Murray, C.A., Parsons, S.A., 2004. Comparison of AOPs for the removal of natural organic matter: performance and economic assessment. *Water Science and Technology* **49**, 267-272.

Newcombe, G., 1994. Activated carbon and soluble humic substances - adsorption, desorption, and surface-charge effects. *Journal of Colloid and Interface Science* **164**, 452-462.

Ngah, W.S.W., Musa, A., 1998. Adsorption of humic acid onto chitin and chitosan. *Journal of Applied Polymer Science* **69**, 2305-2310.

Peng, X.J., Luan, Z.K., Zhang, H.M., 2006. Montmorillonite-Cu(II)/Fe(III) oxides magnetic material as adsorbent for removal of humic acid and its thermal regeneration. *Chemosphere* **63**, 300-306.

Phillips, I.R., 1998. Use of soil amendments to reduce nitrogen, phosphorus and heavy metal availability. *Journal of Soil Contamination* **7**, 191-212.

Prado, A.G.S., Torres, J.D., Faria, E.A., Dias, S.C.L., 2004. Comparative adsorption studies of indigo carmine dye on chitin and chitosan. *Journal of Colloid and Interface Science* **277**, 43-47.

Selcuk, H., Sene, J.J., Sarikaya, H.Z., Bekbolet, M., Anderson, M.A., 2004. An innovative photocatalytic technology in the treatment of river water containing humic substances. *Water Science and Technology* **49**, 153-158.

Da Silva, K.M.P., da Silva, M.I.P., 2004. Copper sorption from diesel oil on chitin and chitosan polymers. *Colloids and Surfaces a-Physicochemical and Engineering Aspects* **237**, 15-21.

Snars, K., Hughes, J.C., Gilkes, R.J., 2004. The effects of addition of bauxite red mud to soil on P uptake by plants. *Australian Journal of Agricultural Research* **55**, 25-31.

Stevenson, F.J., 1977. Nature of divalent transition-metal complexes of humic acids as revealed by a modified potentiometric titration method. *Soil Science* **123**, 10-17.

Tombacz, E., Libor, Z., Illes, E., Majzik, A., Klumpp, E., 2004. The role of reactive surface sites and complexation by humic acids in the interaction of clay mineral and iron oxide particles. *Organic Geochemistry* **35**, 257-267.

Varma, A.J., Deshpande, S.V., Kennedy, J.F., 2004. Metal complexation by chitosan and its derivatives: a review. *Carbohydrate Polymers* **55**, 77-93.

Wang, S., Ang, H.M., Tadé, M.O., 2008. Novel applications of red mud as coagulant, adsorbent and catalyst for environmentally benign processes, *Chemosphere*, **72**, 1621-1635.

Wang, S.B., Boyjoo, Y., Choueib, A., Zhu, Z.H., 2005. Removal of dyes from aqueous solution using fly ash and red mud. *Water Research* **39**, 129-138.

Wang, S.B., Lu, G.Q., 1998. Effects of acidic treatments on the pore and surface properties of Ni catalyst supported on activated carbon. *Carbon* **36**, 283-292.

Wang, S., Terdkiatburana, T., Tadé, M.O., 2007. Single and co-adsorption of heavy metals and humic acid on fly ash. *Separation and Purification Technology*, **58**, 353-358.

Wang, S.B., Wu, H.W., 2006. Environmental-benign utilisation of fly ash as low-cost adsorbents. *Journal of Hazardous Materials* **136**, 482-501.

Wang, S., Zhu, Z.H., 2007. Humic acid adsorption on fly ash and its derived unburned carbon. *Journal of Colloid and Interface Science* **315**, 41-46.

Wang, S.G., Gong, W.X., Liu, X.W., Gao, B.Y., Yue, Q.Y., 2006. Removal of fulvic acids using the surfactant modified zeolite in a fixed-bed reactor. *Separation and Purification Technology* **51**, 367-373.

Wibulswas, R., White, D.A., Rautiu, R., 1998. Removal of humic substances from water by alumina-based pillared clays. *Environmental Technology* **19**, 627-632.

Wu, F.C., Tseng, R.L., Juang, R.S., 2002. Adsorption of dyes and humic acid from water using chitosan-encapsulated activated carbon. *Journal of Chemical Technology and Biotechnology* **77**, 1269-1279.

Yan, W.L., Bai, R.B., 2005. Adsorption of lead and humic acid on chitosan hydrogel beads. *Water Research* **39**, 688-698.

Yuan, W., Zydney, A.L., 1999. Humic acid fouling during microfiltration. *Journal of Membrane Science* **157**, 1-12.

Zhang, X., Bai, R.B., 2003. Mechanisms and kinetics of humic acid adsorption onto chitosan-coated granules. *Journal of Colloid and Interface Science* **264**, 30-38.

In: Industrial Waste...
Editor: John P. Samuelson, pp. 349-367

ISBN: 978-1-60692-720-5

Chapter 13

USE OF MUNICIPAL AND INDUSTRIAL SLUDGE IN AGRICULTURAL SOIL: RESPONSES OF SOIL MICROBIALS AND SOIL ENZYMES

***Ademir Sérgio Ferreira de Araújo*[1,a], *Wanderley José de Melo*[2,b], *Joseany Andrade Santos*[3] and *Valéria Peruca de Melo*[2]**

[1] Universidade Federal do Piauí – Centro de Ciências Agrárias, Campus da Socopo, 64000-000, Teresina, PI, Brazil

[2] Universidade Estadual Paulista – Faculdade de Ciências Agrárias e Veterinárias, CEP 14884-900, Jaboticabal, SP, Brazil

[3] Programa de Pós-Graduação em Agronomia - Universidade Federal do Piauí, Centro de Ciências Agrárias, Campus da Socopo, 64000-000, Teresina, PI, Brazil

Abstract

Industrialization has contributed to the increase in the generation of potentially toxic solid residues. Textile and tannery industries produce residues, called sludge, with high organic and inorganic content, including heavy metals. The most widely-used method for disposal of tannery, textile and sewage sludge is sanitary sites, presenting a high risk for environmental contamination. Concerns about environmental quality have led to the introduction of alternative disposal methods, such as the use of sludge as a nutrient source for plants and as soil conditioners. However, the use of sludge in agricultural soils needs defined action, in order to avoid damage to the environment, mainly to the soil. The knowledge of short- and long-term effects of soil amendments becomes essential when considering, for example, the conservation of soil quality. In recent years, soil microbiological parameters, as such microbial activity and biomass, along with soil enzyme activity, have been seen to be early and sensitive indicators of soil stress and can be used to predict long-term trends in soil quality. Several studies have been done aiming to evaluate the effects of use of these sludges in agricultural soil. In this way, this chapter describes the state of the art about the effect of textile, tannery and sewage sludge on soil microorganisms and soil enzyme activity.

Key words: Soil microbial biomass, soil quality, soil enzymes, sewage sludge, tannery sludge

[a] E-mail address: asfaruaj@yahoo.com.br
[b] E-mail address: wjmelo@fcav.unesp.br

1. Municipal and Industrial Wastes

Sludge is a solid waste produced after wastewater treatment that has been affected in quality by anthropogenic activities. It comprises wastes discharged by domestic residences, commercial properties, industries, and agriculture. Usually, it refers to the municipal and industrial sludge that contains a great number of contaminants resulting from the mixture of wastewaters from different sources. Municipal and industrial sludge, in an ample sense, includes domestic, municipal, or industrial liquid waste. Increasing urbanization and industrialization have resulted in a dramatic increase in the volume of sludge produced around the world. Generally, these sludges are released into the environment. For example, American industrial activity generates and disposes of approximately 7.6 billion tons of industrial solid waste each year (EPA, 2007).

Table 1. Macronutrients in sewage, textile and tannery sludge and other organic wastes used in agriculture (Melo, 2007)

Waste	C	N	P	K	Ca	Mg	S
	-------------------- g kg^{-1} (dry basis)--------------------						
Bovine waste	486	27	18	32	30	9	3
Chicken waste	311	31	18	16	51	11	4
Swine waste	273	32	9	23	55	14	na
MSW compost	278	10	3	5	19	2	3
Sewage sludge	340	32	16	4	32	12	4
Textile sludge[1]	222	52	16	6	21	17	na
Textile sludge compost[1]	365	10	87	34	139	43	na
Tannery sludge[2]	407	23	2	6	46	27	na
Vinasse	200	12	2	60	20	8	10

MSW – Municipal solid waste; na= not available.
[1] Araújo and Monteiro, 2006; [2] Teixeira et al., 2006

Table 2. Micronutrients in sewage, textile and tannery sludge and other wastes used in agriculture (Melo et al., 2002b)

Waste	B	Cu	Fe	Mn	Mo	Zn
	------------------ mg kg^{-1} (dry basis) ------------------					
Bovine manure	na	160	7336	552	16	128
Hen litter	na	na	na	240	na	210
MSW compost	1	229	23325	304	22	340
Chicken litter	na	na	na	360	na	280
Sewage sludge	118	98	42224	242	9	1868
Textile sludge[1]	na	203	na	na	36	543
Textile sludge compost[1]	na	110	na	na	< 4	397
Tannery sludge[2]	na	na	na	na	na	Na

MSW – Municipal solid waste; na= not available.
[1] Araújo and Monteiro, 2006; [2] Teixeira et al., 2006

Table 3. Heavy metals in sewage, textile and tannery sludge and other organic wastes used in agriculture (Melo, 2007)

Waste	Pb	Cd	Ni	Cr	Hg
	------------------- mg kg^{-1} (dry basis) -------------------				
Bovine manure	1.52	0	3.0	na	na
Hen manure	38	4.4	4.4	na	na
Swine manure	13	na	8.3	na	na
MSW compost	1.3-2240	0.01-100	0.9-279	1.8-410	0.09-2.1
Sewage sludge	2-7000	0-3410	6-5300	8-40600	1-260
Textile sludge[1]	71	5.6	104	111	na
Textile sludge compost[1]	33	< 0.3	30	73	na
Tannery sludge[2]	na	na	na	43	na

MSW – Municipal solid waste; na= not available.
[1] Araújo and Monteiro, 2006; [2] Teixeira et al., 2006

The main concern with environmental quality has caused an increase in sludge collection and treatment. It is urgent to find a way to ecologically dispose of this residue without risk to the environment and to human health, recycling the elements it contains if possible. An alternative is the use of sludge as a source for plant nutrients and soil conditioners, mainly owing to the high content of organic matter, macronutrients and micronutrients when compared to the others organic residues (Tables 1 and 2). However, there is in these sludges a relative quantity of heavy metals (Table 3) that can have detrimental effects on soil quality and plant growth.

1.1. Sewage Sludge

Sewage sludge originates from wastewater treatment stations (WTS), and for many years this residue was called biosolids, but more recently there has been pressure to use the term *sewage sludge*, a term adopted in Brazilian legislation (CONAMA, 2006). Sewage sludge composition varies widely depending on its origin and the process used in the WTS (Abreu Júnior et al., 2005). It is rich in organic matter, important for soils in the tropical regions, which are at the end of the weathering scale, with the minerals presenting low cation exchange capacity (CEC). This residue also contains all of the macro and micronutrients. Sewage sludge can also contribute to improving soil physical properties such as density, permeability, capacity of water retention, and water infiltration (Melo et al., 2004; Souza et al., 2005). Consequently, it has been considered for use in agricultural soil in order to improve the physical, chemical and biological properties of soil, and to supply nutrients to plants.

But sewage sludge also contains heavy metals and other pollutants, so its application to soil represents a risk to the environment, to plant growth, and to animal and human health (Figure 1). Additionally, sewage sludge obtained from the treatment of domestic wastewater may contain a higher concentration of pathogenic agents such as helminth eggs.

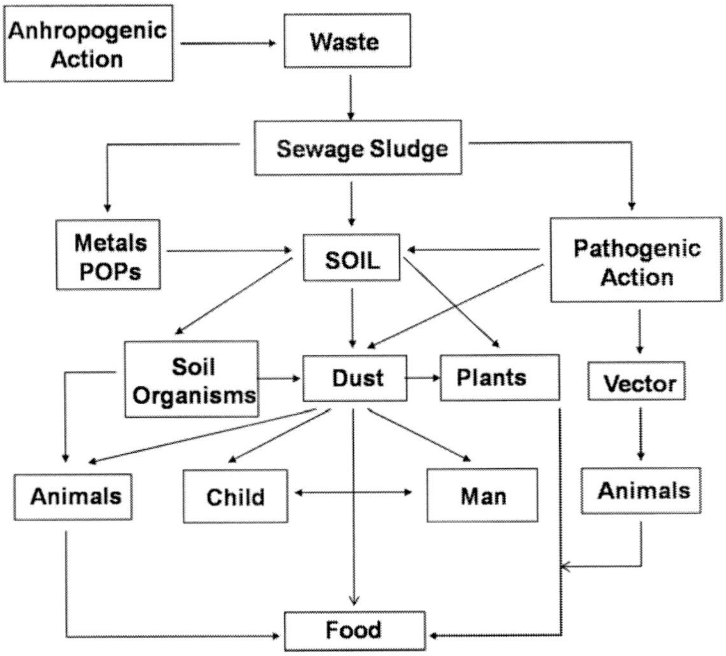

Figure 1. Routes of human contamination by heavy metals and other contaminants present in sewage sludge when the residue is applied to soil.

The use of sewage sludge in agricultural soils must be very judicious, subordinated to a very rigorous legislation based on data obtained during long-term field experiments and with a mechanism of annual evaluation that are able to detect any loss in soil quality. Some authors have developed researches on the use of sewage sludge as a mechanism for the reclamation of degraded areas which lost their organic matter or were contaminated by heavy metals. In studies to evaluate the use of sewage sludge in agriculture, soil enzymatic activity may play an important participation, since it responses quickly to environmental alterations.

1.2. Textile and Tannery Sludge

Textile and tannery industries are large industrial consumers of waters as well as producers of wastewater. The wastewater treatment step concentrates the various pollutants in the wastewater into sludge, normally containing between 1 and 2% by weight dry solids (Priestly, 1991). Because of the dramatic increase in volume of wastewater treated large volume of sludge need to be disposed of in an environmentally safe manner. There is therefore the need to effect proper treatment to sludge before disposal or reuse. The principal disposal method for these wastes is in sanitary sites that can cause environmental impact. Concerns about environmental quality have led to the introduction of alternative disposal methods such as the use as nutrient source for plants and as soil conditioners.

Textile and tannery sludge have a variable composition and normally contain high organic matter, N, P, K and micronutrients contents. Additionally, heavy metals and pathogenic microorganisms may be present in these wastes. Thus, the addition of textile and tannery sludge to agricultural soil improves chemical and physical characteristics and promotes plant growth. Canellas et al. (2001) reported that the use of industrial organic matter in agricultural lands can be justified by the need of finding an appropriate destination for waste recycling. However, there is concern over the use of these sludges, mainly due to presence of potential toxic elements, as heavy metals. In this way, the use of industrial sludge in agricultural soils needs of defined action, in order to not cause damage to environment, mainly to the soil.

The knowledge of effects of these wastes on soil environment is essential when considering, for example, the conservation of soil quality. In recent years, soil microbiological parameters, as microbial activity and biomass, N_2 fixers and soil enzymes have been seen to be early and sensitive indicators of soil stress and can be used to predict long-term trends in the soil quality (Saviozzi et al., 2002).

2. Soil Microorganisms and Soil Enzymes

Soil is a complex environment, where microorganisms play a crucial role in nutrient cycling and the degradation of different pollutants (for example, pesticides and industrial wastes) contributing in this way to the maintenance of soil quality (Haney et al., 2002; Watanabe et al., 2003; Araújo et al., 2003; Araújo and Monteiro, 2006). Additionally, the soil microorganisms performs others important functions, as to form symbiotic associations with roots, to act as antagonist of pathogens, to influence the weathering and solubilization of minerals and to contribute to soil structure and aggregation (Sparling, 1997). The role of the microbial fractions in mediating soil process, and their relatively high rate in turnover, suggest that the microbial fraction could be sensitive indicator and early predictor of the changing soil organic matter processes (Powlson et al., 1987).

Additionally, all soil inhabitants produce different enzymes that take place in the different metabolic pathways. Some enzymes are extracellular in order to break down great molecules as starch, cellulose, proteins in small molecules as glucose that can be absorbed. Others are intracellular enzymes. A bacterial cell contains about 1000 enzymes (Paul and Clark, 1989). When the soil organisms dead and the cell membrane disrupt, or when changes occur in the cell membrane permeability, the intracellular enzymes are put into the soil matrix and may form complexes with soil colloids (Dick et al., 1996). The complexes enzyme-colloid protects the enzymes against the effects of pH, temperature, desiccation, and attack by other enzymes and microorganisms. Consequently, they stay active for a period of time. They are called abiotic enzymes. Soil aggregates and their constituent clays influence the interaction of enzymes with their substrates. The clay particle with its large external and internal surface areas is capable of adsorbing enzymes as urease and protease. Enzymes adsorbed or intertwined with humic constituents are protected from the hydrolysis by other enzymes. Adsorption also makes the catalytic sites less available. A small molecule as urea can diffuse readily to a urease site and there undergo decomposition. A large molecule, such as a protein, would not diffuse as readily to a protease site and consequently would be broken down at a much lower rate than would urea (Paul and Clark, 1989). In the soil matrix, adsorption of enzymes to the soil humates moves their optimum pH and temperature to higher values (Longo and Melo, 2005).

Enzymes are mainly proteins and consequently their structure is affected by factors as temperature, pH, salt concentration, heavy metals, and other contaminants. Considering that sewage sludge contains in its composition heavy metals in different concentrations, it may affect the soil organisms growth and soil enzyme activity.

As enzymes are involved in cell metabolism or may be active even as abiotic enzyme, they are the first soil property that show alteration when the system is disturbed. Soil enzyme activity has been considered an indicator of soil quality since a long time because it controls the supplying of nutrients to plants and the microbial growth (Burn, 1978). Some individual or group of soil enzymes have correlated to plant growth in different soils and soil management.

2.1. Response of Soil Microbial Biomass to Application of Textile and Tannery Sludge

The soil microbial biomass comprises all soil organisms with a volume of less than about 5 x 10^3 um^3, other than living plant tissue, and can thus be considered as the living part of soil organic matter (Brookes, 2001). The microbial biomass comprises about 1 to 4% of soil organic matter, being a important labile reservoir of essential plant nutrients, e.g. nitrogen (N), phosphate (P) and sulphate (S). Because it is living, the microbial biomass responds much more quickly to changing soil conditions, particularly decreases or increases in plant or animal residues, than does soil organic matter as a whole. Typical methods of determination are by fumigation-incubation, fumigation-extraction or by substrate-induced respiration methods (Sparling and Ross, 1993). Recently, others methods can be used to estimates microbial biomass, as Biolog™ system (Biolog, Inc., California), phospholipid ester-linked fatty acid analysis (PLFA) (Zelles et al., 1992) and denaturing gradient gel electrophoresis (DGGE) method (Muller et al., 2002).

Table 4. Changes in microbial biomass C (C_{mic}), bacteria and fungi number in samples of unamended soil (control) and in samples after application of untreated and composted sludge (Araújo & Monteiro, 2006)

	Time, day after sludge application	
	15 days	60 days
C_{mic} (mg kg^{-1})		
Control soil	272 c	301 a
Soil plus untreated sludge	302 b	190 c
Soil plus composted sludge	396 a	218 b
Bacteria (10^6) (colony forming units g^{-1})		
Control soil	2.3 b	3.0 b
Soil plus untreated sludge	2.0 b	3.2 b
Soil plus composted sludge	8.1 a	9.0 a
Fungi (10^4) (colony forming units g^{-1})		
Control soil	1.7 a	2.8 a
Soil plus untreated sludge	1.1 a	1.6 a
Soil plus composted sludge	2.1 a	3.6 a

Values followed by the same letter within each column are not significantly different at 5% level by Duncan's test. The rates of untreated and composted sludge used were 6.4 and 19 t ha^{-1}, respectively.

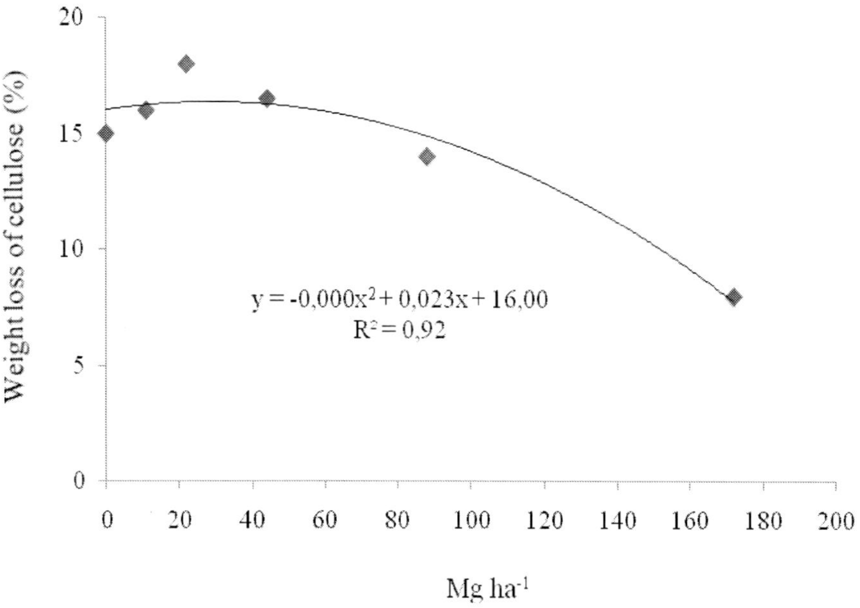

Figure 2. Effect of different rates of tannery sludge on cellulose decomposition estimated by the loss of cellulose filter disks buried in a soil (Araújo et al., 2005).

Several studies were conduced to evaluate the effect of textile and tannery sludge on soil microbial biomass (Araújo et al., 2005; Araújo and Monteiro, 2006). In the laboratory study, Araújo and Monteiro (2006) examined the effect of application of untreated and composted textile sludge on microbial biomass in a Brazilian soil. The soil was amended with untreated and composted sludge at rates equivalent of 6.4 Mg ha^{-1} and 19 Mg ha^{-1}, respectively, and were incubated for 60 days. The application of composted sludge increased significantly the microbial biomass and bacteria number of soil (Table 4). There were not differences in the microbial activity and bacteria number among the control and untreated sludge amended soils. In conclusion, after 2 months of incubation, the effects of the two amendments on soil microorganisms were: microbial biomass, soil respiration and bacteria number were increased only in composted sludge treated soil.

Previously, Araújo et al. (2005) verified the effect of tannery sludge on cellulose decomposition in soil. The amounts of tannery sludge were 0, 11, 22, 44, 88 and 172 Mg ha^{-1}. The authors observed that the waste applied in high rates inhibited the cellulose decomposition (Figure 2). The decrease in cellulose decomposition indicates that the waste reduced the fungi population, conform related by Anderson et al. (1981).

2.2. Responses of N_2 Fixers to Application of Textile and Tannery Sludge

Biological Nitrogen Fixation (BNF) is the process by which atmospheric nitrogen gas (N_2) is converted into ammonia (NH_3) and its is subsequently available for plants. In agricultural settings, perhaps 80% of this biologically fixed N_2 comes from symbioses involving leguminous plants and bacteria of the family *Rhizobiaceae*. The family *Rhizobiaceae* currently includes six genera: *Rhizobium, Sinorhizobium, Mesorhizobium, Allorhizobium, Azorhizobium,* and *Bradyrhizobium,* which are collectively referred to as rhizobia (Vance, 1998). They have the ability to infect the roots of legumes and to produce nodules. The differentiated forms of rhizobia in the nodule, called bacteroids, fix atmospheric nitrogen into ammonia and export the fixed nitrogen to the host plant (Long, 1989). Symbiotic nitrogen fixation is of great importance not only in the production of leguminous crops but also in the global nitrogen cycle. Nitrogen-fixing nodules are formed as a consequence of a series of interactions between rhizobia and leguminous host plants.

The rhizobial-legume symbiosis is characterized by high host specificity and the BNF and numbers of nodules has been proposed as an indicator of soil pollution based on the sensitivity of rhizobial to pesticides, urban and industrial wastes and heavy metals (Chaudri et al., 1993). Studies suggest that leguminous crops dependent on symbiotic N fixation may be sensitive to the toxic effects of heavy metals present in sludge (Heckman et al., 1986; Reddy et al., 1983, Obbard, 2001; Chaudhary et al., 2004). Biological nitrogen fixation has been used to evaluate the effect of sewage sludge on soil (Reddy et al., 1983; Kinkle et al., 1987; Angle et al., 1992; Abd-Alla et al., 1999; Vieira, 2001; Vieira et al., 2004). The results are conflicting on the effects of sewage sludge on soil population of rhizobia, N fixation and yield of legumes. Reddy et al. (1983) studied the survival of *Brdyrhizobium* in soil after application of sewage sludge and found a decline in bacterial population. They conclude that this decline was the result of heavy metal toxicity. In the clover-*Rhizobium leguminosarum* association, reduced nitrogen fixation and clover yields have been associated with poorer survival of the rhizobial in contaminated soil with heavy metal (McGrath, 1994). Angle et al. (1992) attributed a rapid decline in population of *Bradyrhizobium* as to toxic concentration in soluble salts in soil after addition of sewage sludge. On the other hands, studies on the effects of sewage sludge on symbiotic nitrogen fixation not demonstrated negative effects on nodulation (Kinkle et al., 1987; Angle et al. 1992; Ibekwe, 1997; Vieira, 2001; Vieira et al., 2004). In a study to evaluate the effect of sewage sludge application on nitrogen fixation in soybean, Vieira (2001) observed that the waste promoted a beneficial effect on nodulation, however, according to the same author, attention must be paid to heavy applications of sewage sludge to soils, which can negatively affect *Bradyrhizobium* and indirectly inhibit nodulation. Previously, Kinkle et al. (1987) observed an increase in the number of rhizobial in soil, which received high rates of sludge application. They concluded that the application of heavy metal-containing sewage sludge did not have a long-term detrimental effect on soil rhizobial numbers.

Others studies evaluated the effects of textile (Araújo et al., 2007) and tannery (Teixeira et al., 2006; Araújo et al., 2006) sludge on nodulation and biological nitrogen fixation (BNF) in legumes. Some studies were conduced for evaluate the effect of tannery sludge on growth and nodulation in legumes in Leucaena and Prosopis (Araújo et al., 2006) and cowpea (Teixeira et al., 2006). The tannery sludge was applied in four rates (0, 11.5, 23 and 46 Mg ha^{-1}) and the results showed that the tannery sludge did not have a detrimental effect on nodulation in Leucaena and Prosopis. However, for cowpea, the tannery sludge, in high rates, caused an inhibition in the nodulation. The authors attributed this behavior to high content of salts presents in the waste (Table 5).

Table 5. Nodule number of Leucaena, Prosopis and cowpea in soil plus tannery sludge and inoculation (Teixeira et al., 2006; Araújo et al., 2006)

Rate of tannery sludge	Nodule number (nodule per plant)		
	Leucaena	Prosopis	Cowpea
0 Mg ha^{-1}	44.7 b	39.6 a	53.0 a
11.5 Mg ha^{-1}	97.0 a	36.3 a	43.0 a
23 Mg ha^{-1}	75.3 a	63.3 a	39.3 ab
46 Mg ha^{-1}	39.0 c	44.3 a	22.6 b

In each column the means followed by the same letter do not differ statistically ($P<0.05$) from each other, according to Duncan's Test.

Table 6. Effect of different rates of composted textile sludge on nodule glutamine synthetase and leghemoglobin content of soybean and cowpea (Araújo et al., 2007)

Compost rates	Soybean	Cowpea	Soybean	Cowpea
	GS activity (µmoles of glutamyl h^{-1} g^{-1} nodules)		Leghemoglobin content (mg g^{-1} nodules)	
0x	101.3 a	176.8 a	63.4 b	37.5 b
0.5x	126.9 a	160.7 a	91.6 a	45.5 a
1x	125.7 a	169.8 a	78.5 ab	40.9 ab
2x	109.2 a	165.7 a	74.3 ab	38.2 ab

In each column the means followed by the same letter do not differ statistically ($P<0.05$) from each other, according to Duncan's Test.

The same authors proposed the composting as an alternative method for detoxification of textile sludge. Previously, Araújo et al. (2001) and Araújo and Monteiro (2005) observed that the composting decreased or eliminated the toxicity of textile sludge. Additionally, Kaushik and Garg (2003) evaluated the composting of textile sludge using worms, and concluded that the waste could be converted into a stabilized product for agricultural use. Thus, Araújo et al. (2007) evaluated the effect of composted textile sludge on growth, nodulation and nitrogen fixation of soybean and cowpea. The compost was applied in the soil at four rates (0, 9.5, 19 and 38 Mg ha^{-1}) and were evaluated nodulation, shoot nitrogen accumulation and nodulins activity. The results shows that nodulation and nodule glutamine synthetase (GS) activity and leghemoglobin content of the soybean and cowpea were not affected by the application of compost (Table 6).

2.3. Responses of Soil Enzyme Activity to Application of Sewage Sludge

Soil enzyme activity generally increases with the application of sewage sludge, since the residue contains organic matter, microorganisms, nutrients and also microorganisms with their own enzymes and probably free enzymes. Garcia et al. (1993) have evaluated the activity of urease, phosphatase and protease in sewage sludge submitted or not to composting. They have observed that organic matter extracted from non-composted sewage sludge had a higher enzyme activity than that from the composed sewage sludge. The organic fraction with the highest enzyme activity was that of intermediate molecular weight (ranging from 1,000 to

10,000), except for urease, which exhibited the highest activity in the 210-day compost and in the lowest molecular weight fraction (<1,000). Composting influenced urease and phosphatase activity in the sludge.

If the presence of organic carbon, nutrients and microorganisms cells in sewage sludge causes an expectative that the sewage sludge added to the soil will cause an increase in its enzyme activity, the presence of heavy metals and other enzyme inhibitors can promote a decrease in soil enzyme activity. This antagonic effect of sewage sludge, associated to other environment factors that affect soil enzyme activity has caused difficulty to establish indexes for soil quality based on soil enzyme activity. For example, Dar (2006) found decrease in soil enzyme activity caused by high Cd content resulted by the application of sewage sludge.

The evaluation of a single enzyme must be avoided, considering that enzymes catalyze specific reactions making use of specific substrates (Jimenez et al., 2002). A pool of enzymes must be selected which are involved in the cycles of C (amylases, dehydrogenases), N (proteases, urease, fluorescein diacetate (FDA) hydrolysis), P (phosphatases), S (arylsulphatase) and other plant nutrients, since these enzymes reflect changes in the soil metabolic processes (Karaca et al., 2002). In this way it is possible to select those enzymes that are the best indicators of soil quality and that were affected by anthropogenic actions as the addition of sewage sludge or tannery sludge.

Several authors have studied the application of sewage sludge and other residues as tannery sludge to soil and the effect on soil enzyme activity (Melo et al., 1997, Moreno et al., 1999; Pascual et al., 1999; Albiach et al., 2000; Madejón et al., 2001; Karaca et al., 2002; Zaman et al., 2004; Fernandes et al., 2005, Revoredo and Melo, 2007).

2.3.1. Amylase (3.2.1.-)

The name amylase is applied to a group of enzymes classified as hydrolases, which catalyze the hydrolysis of starch. In this group are enzymes as α-amylase and β-amylase. The starch hydrolysis produces small molecules as glucose and maltose, which are absorbed by soil microorganisms and used as C and energy source.

Table 7. Amylase activity in a Typic Haplorthox treated with increasing rates of sewage sludge and cropped with sorghum at different sampling times (Melo et al., 2002a)

Sewage Sludge	Sampling Time (days)					
t ha^{-1}	0	30	60	90	120	170
0	3.3 e	2.5 ab	2.6 ab	2.5	2.0	3.9 b
2	5.1 c	3.1 ab	2.5 ab	3.0	2.7	2.8 b
4	5.9 d	2.9 ab	2.7 ab	2.7	2.9	3.7 b
8	7.4 b	3.2 a	2.2 b	2.6	2.7	4.0 b
16	8.6 a	2.2 b	3.3 a	2.9	2.7	6.0 a

Means followed by the same letters are not different by the Tukeys's test ($P<0.05$).

According to Stroo and Jenckes (1985), soil amylase activity may be used as indicator of its biological activity.In a pot experiment, Melo et al (2002a) found that amylase activity increased soon after sewage sludge was applied to soil, decreased at 30 days and again increased at 170 days (Table 7). The authors attributed the last increase to the plant root system effect.

Fernandes et al. (2005) observed increase in amylase activity when applied sewage sludge obtained at "Companhia de Saneamento Básico do Estado de São Paulo" (SABESP), which treats the wastewater from the Great São Paulo, SP, Brazil, to a clayish Oxisol for 4 consecutive years. The rates of sewage sludge used were 1, 2, 4 and 8 times the required to supply N to maize. The amylase active ranged from 3.52 to 6.30 g glucose g^{-1} h^{-1} in the layer 0-0.10 m. In a similar experiment, but with a sandy Oxisol, the enzyme activity was in the interval 2.16 - 5.35 g glucose g^{-1} h^{-1}.

2.3.2. Cellulase (3.2.1.4 β-glucan 4-glucanohydrolase)

Cellulase is an enzymatic complex that hydrolysis cellulose initially to cellobiose and then to glucose. Cellulase activity were not correlated with total organic matter content in the soil but were correlated with the type of organic matter added to the soil during a plant succession (Panckoly and Rice, 1973).

Fortes Neto (2000) observed correlation between cellulase activity and the decomposition of sewage sludge added to a soil cropped with eucalyptus. In a pot experiment where sewage sludge was added to a Typic Haplorthox at the rates 0, 2, 4, 8 and 16 Mg ha^{-1} and sorghum was cultivated, cellulase activity increased until 90 days after sewage sludge application and then decreased. Sorghum roots did not affect cellulose activity (Melo et al., 202a).

2.3.3. Phosphatases

Phosphatases are important enzymes related to the phosphorus cycle in soil. They transform organic species of phosphorus to inorganic, which are available to plants and soil organisms absorption. Various organic orthophosphoric mono and di-esters are hydrolyzed when incubated with soil, indicating phosphatase activity. This is an important mechanism of organic-P mineralization when wastes are added to soil.

Phosphatase activity in a sandy soil incubated (1:1 w/w) with different mixtures of sewage sludge anaerobically digested and coal fly ash was the highest at 5% coal fly ash-sludge mixture amended soil and no general trend was observed with time (Lai et al., 1999).

In a soil treated with sterilized and non-sterilized, aerobically or anaerobically digested sewage sludges, phosphatase activity was inhibited, fact the authors attributed to the presence of heavy metals (Bonnati et al., 1985). Revoredo and Melo (2007) added a sewage sludge (10 Mg ha^{-1}, dry basis) previously contaminated with nickel chloride (329, 502, 746 and 119 mg kg^{-1}) and cropped sorghum in a pot experiment. At 128 days after sowing, alkaline phosphatase negatively correlated with soil Ni determined by a mixture of acids (hot HNO_3, H_2O_2 and HCl) in open vessel (USEPA, 1986), while acid phosphatase negatively correlated with soil-Ni determinate by the USEPA's method and with the Ni extracted by the Mehlich 1 extractant (Defelipo and Ribeiro, 1981).

Moreno et al. (2003) studied the effect of sewage sludge addition to soil on the ED 50 for Ni and Cd. Ecological dose (ED) values of Cd and Ni were calculated from three mathematical models which described the inhibition of the enzymatic activities with increasing concentrations of heavy metal in the soil. For phosphatase activity, the ED values for Cd and Ni increased after application of sewage sludge to soil, indicating a decrease in Cd and Ni toxicity.

Balik et al. (2007) evaluated the effect of a long-term sewage sludge and manure application on phosphates activity in the rhizosphere of a Cambisol. The long-term application of sewage sludge and manure increased phosphatase activity and acidic phosphatase activity was higher in rape compared to wheat. The treatment with manure revealed an increased of both the acidic and alkaline phosphatases compared to sewage sludge.

2.3.4. Proteases

When proteins arrive at the soil, they are hydrolyzed by proteases supplying amino acids to the soil environment, a reaction that is catalyzed by proteases. The amino acids may be deaminated and the ammonium-N may be absorbed by plants and microorganisms (Karaca, 2002) or may be oxidized to nitrate-N (Burns, 1978), that also may be used by plants and microorganisms or may be leached to the underground water.

Moreno et al. (2003) studied the effect of sewage sludge addition to soil on the ED 50 for Ni and Cd. Protease activity was not sensitive to Cd or Ni contamination and it was difficult to determine whether the addition of sewage sludge had affected the inhibition of the enzyme by the tested heavy metals.

Several authors have observed that protease activity is higher in the upper layer of the soil in studies on the effect of applying sewage sludge to soil (Zaman et al., 2004). Soil protease activity of a gray-brown podzolic soil was significantly stimulated by the addition of sewage sludge from a diary plant enriched or not with lignite ash and the enzyme activity decreased with time (Jezierska-Tys and Frac, 2006).

2.3.5. Urease (3.5.1.5. urea amidohydrolase)

Urease is an enzyme that catalysis the hydrolysis of urea, the most important mineral fertilizer used actually. Urea hydrolysis produces ammonium-N, which can be absorbed by plants and microorganisms or oxidized to nitrate-N, which also can be absorbed or leached from the soil. Urease activity has been considered an indicator for soil quality. A contaminant concentration causing a decrease of urease activity to 50% of its initial value has been recently proposed as an ecological dose (ED50), and can be used as an indicator of toxicity of the contaminant.

The enzyme activity is influenced by many factors as pH, clay content, temperature, moisture, soil depth, vegetation (Longo and Melo, 2005b). Increase in pH and high clay content promote increase in urease activity, while heavy metals promote decrease in the enzyme activity (Moreira and Siqueira, 2002) Urease activity decreases with the time of incubation (Kizilkaya and Bayrakli, 2005).

Many authors have related increase in soil urease activity caused by the application of organic residues, fact that has been attributed to the effect of organic matter on soil microbial biomass (Moreno et al., 1999; Albiach et al., 2000; Madejón et al., 2001. Karaca et al., 2002; Zaman et al., 2004; Fernandes et al., 2005).

A sandy soil receiving 10% coal fly ash-sludge mixture and incubate showed increase in urease activity and, in general, enzyme activity decreased with increasing incubation time (Lai et al., 1999). In a soil treated with sterilized and non-sterilized, aerobically or anaerobically digested sewage sludges, urease activity was inhibited, fact the authors attributed to the presence of heavy metals (Bonnati et al., 1985).

In an experiment where three soils were treated with increasing loading rates of a sewage sludge containing high concentrations of heavy metals and incubated for 0, 3, 7, 14 and 30 days, urease activity was often inhibited in the presence of low rates of the residue and enhanced in the higher rates. The authors attributed the inhibition to the presence of heavy metals and the enhancing as an effect of the organic matter and the nutrients, which stimulated microbial activity and subsequent urease synthesis (Frenkenberger et al., 1983).

Moreno et al. (2003) studied the effect of sewage sludge addition to soil on the urease ED 50 for Ni and Cd. The ED values for Cd and Ni increased after application of sewage sludge to soil, indicating a decrease in Cd and Ni toxicity. The addition of sewage sludge at 60 g kg^{-1} to a sandy soil poor in organic matter and N promoted increase in urease activity (Bustamante-Munõz et al., 2002).

Urease activity was determined in a citrate buffer containing pure enzyme, a suspension containing soil never irrigated with wastewater but amended with pure enzyme and a suspension of the same soil which was irrigated with wastewater amended with the same enzyme, all the treatments containing different concentrations (0.1–1 mg kg^{-1}) of chromium (VI) and (III) (Samborska et al., 2004). For Cr (VI), the ED50 values were 0.4 mg Cr kg^{-1}, 0.5 mg Cr kg^{-1} and 0.5 mg Cr kg^{-1}, respectively. For Cr(III), the ED50 0.2 mg Cr kg^{-1}, 0.4 mg Cr kg^{-1} and 0.6 mg Cr^{+3} kg^{-1}. As can be seen, sewage sludge attenuated the effect of Cr (III) on urease activity.

Revoredo and Melo (2007) added 10 t ha^{-1} sewage sludge (dry basis) previously contaminated with increasing rates of Ni (329, 502, 746 and 119 mg kg^{-1} as nickel chloride) to a Typic Haplustox and cropped sorghum in a pot experiment. Urease activity was higher in soil samples soon after sewage sludge was applied and decreased with time and with Ni concentration in the residue. In samples taken at the end of sorghum cycle urease activity increased with Ni concentration in sewage sludge.

2.3.6. Arylsulphatase (3.1.6.1 Aryl-sulphate sulphohydrolase)

Arylsulphatase is an important enzyme of the sulphur cycle. It hydrolyses organic sulphate esters, resulting available sulphur to plant absorption. Kizilkaya and Bayrakli (2005) have evaluated the effect of sewage sludge rates (0, 100, 200, 300 t ha^{-1}) with different C to N ratio (9 to 1, 6 to 1 and 3 to 1) on the arylsulphatase activity of a clay loam soil. The addition of different doses and C to N ratios of the sludge caused a rapid and significant increase in the enzymatic activity, and the highest values were obtained with high doses of sewage sludge and high C to N ratio. The activity reached a peak value at 30 days incubation and then gradually decreased up to 90 days.

2.3.7. Dehydrogenases

Dehydrogenases are a group of enzymes included in the class of the oxidoredutase. Their activity is evaluated by incubating soil sample with 2, 3, 5 triphenyltetrazolium chloride and measuring the production of triphenylformazan (Casida et al., 1964). Toluol is not used in the method since it destroys the enzyme activity. Thus, the contribution to the observed activity by proliferating microorganisms is unknown but can be substantial in those assays where soils are amended with organic-C substrates (as sewage sludge and tannery sludge) and incubated for 24 h.

Active dehydrogenases are considered to exist in soils as integral parts of intact cells and dehydrogenase activities are thought to reflect the total range of oxidative activities of the soil microflora (Burns, 1978). In an experiment performed to evaluate the effect of the mixture alkaline coal fly ash-anaerobically digested sewage sludge applied to a sandy soil, dehydrogenases activity was higher in the treatment 5% ash-sludge than in those receiving 0, 10, 35 and 50%, w/w (Lai et al., 1999).

A Eutric Histosol flood-irrigated 10 times per yeas with 60-75 mm and 120-150 mm with municipal wastewater treated mechanically and biologically, dehydrogenase activity increased 44 and 24% in the treatments with high and low irrigation doses, respectively (Brzezinska et al., 2001). Plant cover significantly affected dehydrogenase activity.

Soil dehydrogenase activity of a gray-brown podzolic soil was not significantly stimulated by the addition of sewage sludge from a diary plant enriched or not with lignite ash (Jezierska-Tys and Frac, 2006). Enzyme activity quickly decreased in the 7 first days of soil incubation and then stayed practically constant until 240 days.

In soils amended with Cr^{3+}, Cr^{6+}, tannery sludge, tannery sludge (Cr^{3+}) and tannery sludge (Cr^{6+}), dehydrogenase activity correlated with the fractions bound to iron oxides and to organic matter (Barajas-Aceves et al., 2007). Cr^{6+} reduced dehydrogenase activity in the following order: outside the canopy, cultivated and under the canopy soils.

2.3.8. Fluorescein Diacetate (FDA) Hydrolysis

Fluorescein diacetate hydrolysis is catalyzed by a group of enzymes including lipases, proteases and esterases produced by bacteria and fungi, considered primary organic matter decomposers. Because of this FDA hydrolysis is considered a good indicator for estimating soil organic matter decomposers and a good indicator for the heterotrophic activity in soil (Moreira and Siqueira, 2002). In an extend scale, FDA hydrolysis is considered a good index for evaluating soil biologic activity (Vepsäläinen et al., 2001). Swisher and Carroll (1980) demonstrated that FDA hydrolysis is directly proportional to the microbial growth. But in some experiments it was not observed correlation between FDA hydrolysis and soil microbial activity.

When added sewage sludge to a soil cropped with eucalyptus, Fortes Neto (2000) did not observe a correlation between the FDA hydrolysis and the microbial activity during the residue decomposition.

3. Concluding Remarks

Soil microbials are affected by a great number of factors such as pH, temperature, climate, presence of activators and inhibitors and other soil properties, sampling and storage, and methodology for activity estimation. Soil microorganisms and soil enzyme activity are considered a good indicator of soil quality, since they respond quickly to soil alterations caused by anthropogenic actions. Thus, the response of these microbial indicators can be evidence of benefits or damage from applications of municipal and industrial sludge. In recent years, a great number of research studies evaluating the effect of application of sewage sludge and other organic residues to soil have been carried out throughout the world. However, no definitive conclusion was generically accepted as an index for using soil microorganisms and enzyme activity as a control for the application of sewage or textile or tannery sludge to agricultural soils. More research studies, mainly carried out under field conditions and long-term, must be realized in order to gather data that make soil microorganism and soil enzyme activity useful tools for controlling sludge application to soil.

References

Abd-Alla, HM, Yan, F, Schubert, S. Effects of sewage sludge application on nodulation, nitrogen fixation and plant growth of faba bean, soybean and lupin. *Journal of Applied Botany*, 1999, 73, 69–75.

Albiach, R, Canet, R, Pomares, F, Ingelmo, F. Microbial biomass content and enzymatic activities after the application of organic amendments to a horticultural soil. *Bioresource Technology*, 2000, 75, 43-48.

Angle, JS, Madariaga, GM, Heger, EA. Sewage sludge effects on growth and nitrogen fixation of soybean. *Agriculture, Ecosystem & Environment*, 1992, 41, 231–239.

Araújo, ASF, Carvalho, AJS, Santos, FJS, Carvalho, EMS, Santos, VB. Growth and nodulation of Leucaena and Prosopis seedlings in soil plus tannery sludge. *Caatinga*, 2006, 19, 20-24.

Araújo, ASF, Monteiro, RTR, Carvalho, EMS. Effect of composted textile sludge on growth, nodulation and nitrogen fixation of soybean and cowpea. *Bioresource Technlogy*, 2007, 97, 1028-1032.

Araújo, ASF, Monteiro, RTR. Microbial biomass and activity in a Brazilian soil plus untreated and composted textile sludge. *Chemosphere*, 2006, 64, 1043-1046.

Araujo, ASF, Monteiro, RTR. Plant bioassays to assess toxicity of textile sludge compost. *Scientia Agricola*, 2005, 62, 286–290.

Araujo, ASF, Sahyoun, FK, Monteiro, RTR. Evaluation of toxicity of textile sludge compost on seed germination and root elongation of soybean and wheat. *Ecossistema*, 2001, 26, 117–120.

Araújo, ASF, Araújo, RS. Nodulação e sobrevivência do *Rhizobium tropici* em sementes de feijão tratadas com fungicidas. *Ciência Rural*, 2006, 36, 973-976.

Araújo, ASF, Monteiro, RTR, Abrakeli, RB. Effect of glyphosate on the microbial activity of two Brazilian soils. *Chemosphere*, 2003, 52, 799-804.

Balik, J, Pavlikova, D, Vaněk, V, Kulhánek, M, Kotková, B. The influence of long-term sewage sludge application on the activity of phosphatases in the rhizosphere of plants. *Plant Soil Environment*, 2007, 53, 375-381.

Barajas Aceves, M, Ocampo Velásquez, R, Rodriguez Vásquez, R. Effects of Cr^{3+}, Cr^{6+} and tannery sludge on C and N mineralization and microbial activity in semi-arid soils. *Journal of Hazardous Materials*, 2007, 143, 522-531.

Bonnati, M, Pujola, M, Sana, J, Soliva, M, Felipo, MT, Garau, M, Ceccanti, B, Nannipieri, P. Chemical properties, populations of nitrige oxidizers, urease and phosphatase activities in sewage sludge-amended soils. *Plant and Soil*, 1985, 84, 79-91.

Brookes, P. The soil microbial biomass: concept, measurements and applications in soil ecosystem research. *Microbes and Environment*, 2001, 16, 131-140.

Brzezinska, M, Stepniewiska, Z, Stepniewski, W, Dehydrogenase and catalase activity of soil irrigated with municipal wastewater. *Polish Journal of Environmental Studies*, 2001, 10, 307-311.

Burns, RG. Soil enzymes. New York: Academic Press, 1978. 380p.

Bustamante-Munõz, MA, Gómes-Licas, I, Casado-Vela, J, Navarro-Pedreño, J, Mataix-Solera, J, Garcia, F. Influence of sewage sludge compost on biological activity in soils.

Sustainable use and management of soils in arid and semirarid regions. Cartagena, Espanha, 2002. pp. 114-115.

Casida, LE, Klein, DA, Santoro, T. Soil dehydrogenase activity. *Soil Science*, 1964, 98, 371-376.

Chaudhary, P, Dudeja, SS, Kapoor, KK. Effectivity of host-*Rhizobium leguminosarum* symbiosis in soils receiving sewage water containing heavy metals. *Microbiological Research*, 2004, 159, 121–127.

Chaudri, AM, McGrath, SP, Giller, KE, Rietz, E, Sauerbeck, DR. Enumeration of indigenous *Rhizobium leguminosarum* biovar *trifolii* in soils previously treated with metal-contaminated sewage sludge. *Soil Biology & Biochemistry*, 1993, 25, 301–309.

Dar, GH. Effects of cadmium and sewage-sludge on soil microbial biomass and enzyme activities. *Bioresource Technology*, 1996, 56, 141-145.

Defelipo, BV, Ribeiro, AC. Análise química de solo: metodologia. Viçosa: UFV, 1981, 17p. (Boletim de Extensão 29).

Dick, RP, Breakwill, D, Turco, R. Soil enzyme activities and biodiversity measurements as integrating biological indicators. In: Doran, JW, Jones, AJ. Editors. *Handbook of methods for assessment of soil quality*. Madison: Soil Science Society American, 1996, 247-272.

EPA, *Guide for Industrial Waste Management*. 2007.

Fernandes, SAP, Bettiol, W, Cerri, CC. Effect of sewage sludge on microbial biomass, basal respiration, metabolic quotient and soil enzymatic activity. *Applied Soil Ecology*, 2005, 30, 65-77.

Fortes Neto, P. Degradação de biossólido incorporado ao solo avaliada através de medidas microbiológicas. Piracicaba, ESALQ/USP, 2000. 113p. (PhD Thesis).

Fox, JE, Gulledge, J, Engelhaupt, E, Burow, ME, McLachlan, JA. Pesticides reduce symbiotic efficiency of nitrogen-fixing rhizobia and host plants. *Proceeding National Academic of Science,* 2007, 104, 10282-10287.

Frenkenberger, WT, Johanson, JB, Nelson, CO. Urease activity in sewage sludge-amended soils. *Soil Biology and Biochemistry*, 1983, 15, 543-549.

Garcia, C, Hernandez, T, Costs, F, Ceccanti, B, Ganni, A. Hydrolases in the Organic Matter Fractions of Sewage Sludge: Changes with Composting *Bioresource Technology*, 1993, 45, 47-52.

Heckman, JR, Angle, JC, Chane, RL. Soybean nodulation and nitrogen fixation on soil previously amended with sewage sludge. *Biology & Fertility of Soils*, 1986, 2, 181-185.

Ibekwe, AM, Angle, JS, Chaney, RL, Van Berkum, P. Enumeration and N_2 fixation potential of *Rhizobium leguminosarum* biovar *trifolii* grown in soil with varying pH values and heavy metal concentrations. *Agriculture, Ecosystems and Environment,* 1997, 61, 103-111.

Jezierska-Tys, S, Frac, M. Enzymatic activity in gray-brown podzolic soil enriched with sewage sludge from a dairy plant. *Polish Journal of Soil Science*, 2006, 39, 33-42.

Jimenez, MP, Horra, AM, Pruzzo, L, Palma, RM. Soil quality: a new index based on microbiological and biochemical parameters. *Biology and Fertility of Soils*, 2002, 35, 302-306.

Karaca, A, Naseby, DC, Lynch, JM. Effect of cadmium contamination with sewage sludge and phosphate fertiliser amendments on soil enzyme activities, microbial structure and available cadmium. *Biology and Fertility of Soils*, 2002, 35, 428-434.

Kaushik, P, Garg, VK. Vermicomposting of mixed solid textile mill sludge and cow dung with the epigeic earthworm Eisenia foetida. *Bioresource Technology*, 2003, 90, 311-316.

Kinkle, BK, Angle, JS, Keyser, HH. Long-Term Effects of Metal-Rich Sewage Sludge Application on Soil Populations of *Bradyrhizobium japonicum*. *Applied Environmental Microbiology*, 1987, 53, 315–319.

Kizilkaya, R, Bayrakli, B. Effects of N-enriched sewage sludge on soil soil enzyme activities. *Applied Soil Ecology*, 2005, 30, 192-202.

Lai, KM, Ye, DY, Wong, WC. Enzyme activity in a sandy soil amended with sewage sludge and coal fly ash. *Water, Air and Soil Pollution*, 1999, 113, 261-272.

Long SR. *Rhizobium*–legume nodulation: life together in the underground. *Cell*, 1989, 56, 203–214.

Longo, RM, Melo, WJ. Hidrólise da uréia em latossolos: efeito da concentração da uréia, temperatura, pH, armazenamento e tempo de incubação. *Revista Brasileira de Ciência do Solo*, 2005a, 29, 651-657.

Longo, RM, Melo, WJ. Atividade da urease em latossolos sob influência da cobertura vegetal e da época de amostragem . *Revista Brasileira de Ciência do Solo*, 2005b, 29, 645-650,.

Madejón, E, Burgos, P, López, R, Cabrera, F. Soil enzymatic response to addition of heavy metals with organic residues. *Biology and Fertility of Soils*, 2001, 34, 144-150.

Melo, VP, Beutler, NA, Souza, ZM, Centurion, JF, Melo, WJ. Atributos físicos de latossolos adubados durante cinco anos com biossólido. *Pesquisa Agropecuária Brasileira*, 2004, 39, 67-72.

Melo, WJ. Uso de Resíduos na Agricultura e Qualidade Ambiental. In: Silveira, APD, Freitas, SS. editores. Microbiota do solo e qualidade ambiental. Parte III: Atuação da microbiota do solo em situações de estresse. Campinas: IAC, 2007. 312p. pp. 275-297.

Melo, WJ, Aguiar, PS, Melo, GMP, Melo, VP. Nickel in a tropical soil treated with sewage sludge and cropped with maize in a long-term field study. *Soil Biology & Biochemistry*, 2007, 39, 1341-1347.

Melo, WJ, André, EM, Melo, GMP, Marques, MO, Melo, VP, Chelli, RA, Leite, SAS. Effects of fertilization with sewage sludge on soil urease activity and nitrogen supply to sorghum. Proceedings of the 8th Meeting of the International Humic Substances Society. 1997, 967-972.

Melo, WJ, Marques, MO, Ferreira, ME, Melo, GMP, Melo, VP. Chemical properties and enzyme activity in a sewage sludge-treated soil. *Communications in Soil Science and Plant Analysis*, 2002a, 33, 1643-1659.

Melo, WJ, Marques, MO, Melo, VP. O uso agrícola do biossólido e as propriedades do solo. In: Tsutyia, MT. editor. Biossólido na Agricultura. São Paulo:ABES, 2002b, 289-363.

Moreira, FMS, Siqueira, JO. Microbiologia e bioquímica do solo. 1st edition. Lavras: UFLA, 2002. 626p.

Moreno, JL, Garcia, C, Hernández, T. Toxic effect of cadmium and nickel on soil enzymes and the influence of adding sewage sludge. *European Journal of Soil Science*, 2003, 54, 377-386.

Moreno, JL, Hernández, T, Garcia, C. Effects of a cadmium-contaminated sewage sludge compost on dynamics of organic matter and microbial activity in an arid soil. *Biology and Fertility of Soils*, 1999, 28, 230-237.

Obbard, JP. Ecotoxicological assessment of heavy metals in sewage sludge amended soils. *Applied Geochemistry* , 2001, 16, 1405–14 11

Panckoly, SK, Rice, RL. Soil enzymes in relation to old field succession? amylase, cellulose, invertase, dehydrogenase, and urease. *Soil Science Society American Proceeding*, 1973, 37, 47-50.

Pascual, JA, García, C, Hernandez, T. Lasting microbiological and biochemical effects of the addition of municipal solid waste to an arid soil. *Biology and Fertility of Soils*, 1999, 30, 1-6.

Paul, EA, Clark, FE. Soil Microbiology and Biochemistry. 2nd edition. Academic Press, London, 1989. 340p.

Priestly AT. Report on Sewage Sludge Treatment and Disposal-Environmental Programs and Research Needs from an Australian Perspective. CSIRO, Division of chemicals and Polymers. 1991. pp. 1-44.

Reddy, GB, Cheng, CN, Dunn, SJ. Survival of *Rhizobium japonicum* in soil-sludge environment. *Soil Biology & Biochemistry*, 1983, 15, 343– 345.

Revoredo, MD, Melo, WJ. Enzyme activity and microbial biomass in an Oxisol amended with sewage sludge contaminated with nickel. *Scientia Agricola*, 2007, 64, 61-67.

Samborska, A, Stepniewska, Z, Stepniewski, W. Influence of different oxidation states of chromium (VI and II) on soil urease activity. *Geoderma*, 2004, 122, 317-322.

Saviozzi, A, Bufalino, P, Levi-Minzi, R, Riffald, R. Biochemical activities in a degraded soil restored by two amendments: a laboratory study. *Biology & Fertility of Soils*, 2002, 35, 96-101.

Souza, ZM, Beutler, NA, Melo, VP, Melo, WJ. Estabilidade de agregados e resistência à penetração em latossolos adubados por cinco anos com biossólido. *Revista Brasileira de Ciência do Solo*, 2005, 29, 117-123..

Stroo, HF, Jenckes, EM. Effects of sewage sludge in microbial activity in and old abandoned mine soil. *Journal of Environmental Quality*, 1985, 14, 301-304.

Swisher, R, Carroll, GC. Fluorescein diacetate hydrolysis as an estimator of microbial biomass on coniferous needle surfaces. *Microbial Ecology*, 1980, 6, 217-226.

Teixeira, KRG, Gonçalves Filho, LAR, Carvalho, SEM, Araújo, ASF, Santos, VB, Efeito da adição de lodo de curtume na fertilidade do solo, nodulação e rendimento de matéria seca do caupi. *Ciência e Agrotecnologia*, 2006, 30, 1071-1076

Tsutiya, MT. Alternativas de disposição final de biossólidos. In: Tsutiya, MT, Comparini, JB, Sobrinho, PA, Hespanhol, I, Carvalho, PCT, Melfi, AJ, Melo, WJ, Marques, MO. Editores. Biossólidos na Agricultura. São Paulo: ABES, 2001, 133-180.

USEPA. Standards for the use and disposal of sewage sludge. Washington, EPA, 1986. (Code of Federal Regulations 40 CRF, part 503).

Vance, CP. Legume symbiotic nitrogen fixation: agronomic aspects. In: Spaink, HP. editor. *The Rhizobiaceae*. Dordrecht: Kluwer Academic Publishers, 1998, pp. 509-530. Vieira, RF. Sewage sludge effects on soybean growth and nitrogen fixation. *Biology & Fertility of Soils*, 2001, 34, 196–200.

Vieira, RF, Tsai, SM, Teixeira, MA. Efeito do lodo de esgoto no crescimento e fixação simbiotica de N_2 em feijoeiro (*Phaseolus vulgaris* L.). Jaguariuna: Embrapa Meio Ambiente, Boletim de Pesquisa e Desenvolvimento, 2004, 18p.

Vepsäläinen, M, Kukkonen, S, Vestberg, M, Sirviö, H, Niemi, R.M. Application of soil enzyme activity test kit in a field experiment. *Soil Biology & Biochemistry*, 2001, 33, 1665-1672.

Zaman, M, Matsushima, M, Chang, SX, Inubushi, K, Nguyen, L, Goto, S, Kaneko, F, Yoneyama, T. Nitrogen mineralization, N_2O production and soil microbiological properties as affected by long-term applications of sewage sludge composts. *Biology and Fertility of Soils*, 2004, 40, 101-109.

Short Communications

In: Industrial Waste...
Editor: John P. Samuelson, pp. 371-386

ISBN: 978-1-60692-720-5

INTERFERENCES OF PB(II) AND CR(III) ON CU(II) BIOSORPTION BY ALGAE *GELIDIUM*, ALGAL WASTE AND COMPOSITE MATERIAL

***Vítor J.P. Vilar*[a], *Cidália M.S. Botelho*[b] *and Rui A.R. Boaventura*[c]**

LSRE-Laboratory of Separation and Reaction Engineering
Departamento de Engenharia Química, Faculdade de Engenharia da Universidade do Porto, Rua Dr. Roberto Frias, 4200-465 Porto, Portugal

Abstract

Biosorption of Cr(III), Cu(II) and Pb(II) from binary metal solutions onto *Gelidium* algae, algal waste and a composite material was investigated in batch systems, at pH 4 for Cu(II)/Cr(III) and pH 5.3 for Cu(II)/Pb(II). Three biosorbents have been tested, namely, algae *Gelidium*, algal waste from agar extraction industry and a composite material. The maximum uptake capacity followed the order: Pb > Cu > Cr. The interference of lead on copper uptake is stronger than on chromium. The experimental equilibrium data were well described by a discrete and a continuous equilibrium model. Three dimensional sorption surfaces were generated, describing the metal uptake as a function of the equilibrium concentration of both metals. From the discrete model, it has been possible to obtain the Langmuir affinity constant. The presence of co-cations decreases the apparent affinity of copper ions for the biosorbent binding sites.

Keywords: *Gelidium* algae, algal waste, composite material, binary metal mixtures, biosorption

[a] E-mail address: vilar@fe.up.pt
[b] E-mail address: cbotelho@fe.up.pt
[c] E-mail address: bventura@fe.up.pt. Tel.: +351225081683; Fax.: +351225081674 (Corresponding author)

1. Introduction

Heavy metal pollution is a threat to human health, animals, plants, and the planet itself, and is mainly caused by industrialization and its consequences. *Gelidium* algae and an algal waste from the agar extraction industry have shown a good capacity to remove several metal ions (Cu(II), Pb(II), Cd(II), Zn(II) and Cr(III)) at different pH, temperature and ionic strength conditions [1-5]. In the metal biosorption process from real wastewaters, the effect of different metal ions in solution must be considered. In this work, the interferences of Pb(II) and Cr(III) on Cu(II) uptake by *Gelidium* algae and algal waste (in granular and powdered forms) has been studied. It has been concluded from previous works that lead ions have the highest affinity to these biosorbents, followed by copper, cadmium and zinc ions. The affinity sequence is probably related to the increase of the hydration effects and the decrease of the covalent character of the binding [6, 7].

Biosorption of Cr(III), Cu(II) and Cd(II) from binary metal solutions onto peat were studied by Ma and Tobin [8]. These authors found uptake capacities in the order: Cr(III) > Cu(II) > Cd(II). Co-ion competition resulted in an up to 70 percent decrease of the primary metal adsorption. Two extended Langmuir-type models were found to exhibit good fit to the experimental data. Puranik and Paknikar (1999) studied the influence of co-cations on Cu(II) and Pb(II) biosorption onto *Streptoverticillium cinnamoneum* and *Penicillium chrysogenum* biomass and found that Pb(II) was preferentially adsorbed.

The multi-component equilibrium of different binary systems Pb(II)/Cd(II), Pb(II)/Zn(II) and Cd(II)/Zn(II) has been described by a Non Ideal Competitive Adsorption (NICA) model [6, 7]. This model considers a continuous distribution of the carboxylic affinity constants given by the Sips distribution.

2. Materials and Methods

2.1. Biosorbents

An algal waste from the agar extraction industry, the same material granulated with polyacrylonitrile (PAN), and the algae *Gelidium* (the raw material for agar extraction) were the biosorbents selected for this study. The characterization of all materials and their preparation procedure were presented in previous works [2, 9].

2.2. Metal Solutions

Pb(II), Cu(II) and Cr(III) solutions were prepared by dissolving weighed amounts of anhydrous lead (II) chloride (Sigma-Aldrich, 98%), copper (II) chloride dehydrate (Riedel-de Haën, 99%) and chromium (III) nitrate nonahydrate (Carlo Erba, 98%) in distilled water. The pH was controlled by adding 0.01 M HCl or 0.01 M NaOH solutions.

2.3. Sorption Equilibrium Studies

All experiments were conducted in duplicate, using 100 ml Erlenmeyer flasks, at constant pH = 5.3 for Cu(II)/Pb(II) and pH = 4.0 for Cu(II)/Cr(II) and temperature of 20°C. Metal solutions of Cu(II) + Pb(II) and Cu(II) + Cr(II), in the concentration range of 0 to 300 mg l^{-1}, of each metal, were used. A known weight of material was added to 100 ml of the binary metal solution and stirred at 100 rpm. The pH was controlled by addition of NaOH and HCl diluted solutions. The temperature was kept constant by using a HOTTECOLD thermostatic refrigerator. Once equilibrium was reached, one hour later, samples were taken and centrifuged (Eppendorf Centrifuge 5410) and the supernatant was analysed for the metals in solution.

2.4. Analytical Procedure

The concentration of residual Pb(II), Cu(II) and Cr(III) in the supernatants was determined by atomic absorption spectrometry (GBC 932 Plus Atomic Absorption Spectrometer). The amount of each metal adsorbed per gram of biosorbent was calculated as follows:

$$q = \frac{V\left(C_i - C_f\right)}{W} \tag{1}$$

where q is the metal uptake (mmol metal g^{-1} of the biosorbent), C_i and C_f the initial and final metal concentrations in the solution (mmol metal l^{-1}), V the volume of solution (l), and W the dry weight of the added biosorbent (g).

3. Modelling of Metal Biosorption

3.1. Discrete Model

The discrete model assumes that the biosorbent surface has one kind of predominant active sites (carboxylic groups), which are responsible for metal biosorption at pH < 7.0, and that competition exists between metal ions and protons. The model defines apparent equilibrium binding constants, K_H and K_{M_i} for H^+ ions and metal cations, M_i, respectively, sorbing onto binding sites in the biomass surface, according to the following equation [6, 7]:

$$q_{M_1} = \frac{Q_{max}\, K_{M_1}\, C_{M_1}}{1 + K_H\, C_H + K_{M_1}\, C_{M_1} + K_{M_2}\, C_{M_2}} \tag{1}$$

Dividing Eq. (2) by $1 + K_H C_H + K_{M_2} C_{M_2}$, the Langmuir equation is obtained:

$$q_{M_1} = \frac{Q_{max}\, K_L''[M_1]C_{M_1}}{1+K_L''\,[M_1]C_{M_1}},\; K_L''[M_1] = \frac{K_{M_1}}{1+K_H\, C_H + K_{M_2}\, C_{M_{21}}} \quad (2)$$

where $K_L''[M_1]$ is the Langmuir constant, which depends on the pH value and on the second metal concentration. A similar equation can be obtained for M_2. The total metal uptake (M_1 + M_2) is then given by the following expression:

$$q_T = q_{M_1} + q_{M_2} = \frac{Q_{max}\left(K_{M_1}C_{M_1} + K_{M_2}C_{M_2}\right)}{1+K_HC_H + K_{M_1}C_{M_1} + K_{M_2}C_{M_2}} \quad (3)$$

3.2. Continuous Model

For heterogeneous surface active sites, with a continuous distribution of affinities given by the Sips distribution [10], considering the competition between metal ions and protons, and assuming that metal ions adsorb only on carboxylic sites (pH < 7.0), the continuous model (NICA model) is defined as [11]:

$$\theta_{T,X} = \frac{\left(K_x'C_x\right)^{n_X}}{\sum_X \left(K_X'C_X\right)^{n_X}} \times \frac{\left[\sum_X \left(K_X'C_X\right)^{n_X}\right]^p}{1+\left[\sum_X \left(K_X'C_X\right)^{n_X}\right]^p} \quad (1)$$

where K_X' is the average value of the affinity constant distribution for species X and n_X reflects the overall non-ideality, which can be due to lateral interactions and/or stoichiometry effects . The effect of the intrinsic heterogeneity of the carboxylic group (p) can be isolated from the non-ideal behaviour of species X (n_X): $m_X = n_X \times p$. The total bound amount, q_X, is given by:

$$q_X = Q_{max}\left(\theta_{T,X}\right)\left(\frac{n_X}{n_H}\right) \quad (2)$$

where the factor n_X / n_H is added to give thermodynamic consistency and has some other advantages [12]. Assuming that metal ion biosorbs only on carboxylic groups (pH < 7.0), the total metal uptake ($M_1 + M_2$) model can be expressed as follows:

$$q_T = \frac{Q_{max}}{n_H}\left\{n_{M_1}\left(K'_{M_1}C_{M_1}\right)^{n_{M_1}} + n_{M_2}\left(K'_{M_2}C_{M_2}\right)^{n_{M_2}}\right\}\frac{\left\{\left(K'_H C_H\right)^{n_H} + \left(K'_{M_1}C_{M_1}\right)^{n_{M_1}} + \left(K'_{M_2}C_{M_2}\right)^{n_{M_2}}\right\}^{p-1}}{1+\left\{\left(K'_H C_H\right)^{n_H} + \left(K'_{M_1}C_{M_1}\right)^{n_{M_1}} + \left(K'_{M_2}C_{M_2}\right)^{n_{M_2}}\right\}^{p}} \quad (3)$$

3.3. Parameter Estimates

The experimental data obtained from equilibrium studies, with three independents variables (C_H, C_{M_1} and C_{M_2}), were fitted to mathematical models using the Solver application (Quasi-Newton algoritm). The objective function, F_{obj}, for the optimal regression has been defined as:

$$\text{Minimizing} \quad F_{obj} = \sum_{i=1}^{n}\left(q_{i,M_1}^{exp} - q_{i,M_1}^{teo}\right)^2 + \sum_{i=1}^{n}\left(q_{i,M_2}^{exp} - q_{i,M_2}^{teo}\right)^2 + \sum_{i=1}^{n}\left(q_{i,T}^{exp} - q_{i,T}^{teo}\right)^2 \quad (4)$$

where q_{i,M_1}^{exp}, q_{i,M_2}^{exp} and $q_{i,T}^{exp}$ are the experimental metal biosorption uptakes (M_1, M_2, and total), q_{i,M_1}^{teo}, q_{i,M_2}^{teo} and $q_{i,T}^{teo}$ are the equilibrium metal uptakes calculated by the model, *i* is the experimental sample number and *n* is the number of samples.

In order to determine the parameters deviation the inverse matrix approach was used [13]. The algorithm has been described in a previous work [6].

4. Results and Discussion

4.1. Equilibrium Models

The Pb(II) and Cr(III) interferences on Cu(II) uptake by algae *Gelidium*, algal waste and composite material were studied in a batch system, considering the sorption isotherms Cu(II)/Pb(II) at pH = 5.3 and Cu(II)/Cr(III) at pH = 4.0, for the same initial concentration of both metal ions (mg l^{-1}). In order to quantify the co-cation effect, the experimental data were fitted to a discrete (Eq. 2) and a continuous model (Eq. 7).

The discrete model parameters were also determined using the single-metal experimental data at different pH [3-5]. The obtained parameters are presented in Table 1. To fit the continuous model to the experimental data, the parameters Q_{max}, pK'_H and m_H were fixed at those values obtained from potentiometric titrations [14], and *p* was obtained from the single metal adsorption experiment [3]. The model gives the pK'_M and n_M values for each metal ion in each binary system (Table 2).

Table 1. Adjustable parameters for the multi-component discrete equilibrium model (Eq. 2) (value ± standard error)

Biosorbent	System	Metal	Discrete model: Q_{max} (mmol g^{-1})	Discrete model: pK_M	Discrete model: pK_H	R^2	S_R^2 (mmol g^{-1})^2
Gelidium	Pb^{2+} / Cu^{2+}	Pb^{2+}	0.40 ± 0.01	3.6 ± 0.1	4.2 ± 0.1	0.912	1.1×10^{-3}
		Cu^{2+}		3.2 ± 0.1			
	Cu^{2+} / Cr^{3+}	Cu^{2+}	0.41 ± 0.02	3.1 ± 0.1	4.2 ± 0.1	0.917	1.0×10^{-4}
		Cr^{3+}		3.1 ± 0.1			
Algal waste	Pb^{2+} / Cu^{2+}	Pb^{2+}	0.24 ± 0.01	3.8 ± 0.1	4.6 ± 0.1	0.884	6.5×10^{-3}
		Cu^{2+}		3.4 ± 0.1			
	Cu^{2+} / Cr^{3+}	Cu^{2+}	0.24 ± 0.01	3.5 ± 0.1	4.9 ± 0.1	0.939	2.5×10^{-4}
		Cr^{3+}		3.5 ± 0.1			
Composite material	Pb^{2+} / Cu^{2+}	Pb^{2+}	0.15 ± 0.01	4.2 ± 0.1	4.9 ± 0.1	0.771	7.5×10^{-4}
		Cu^{2+}		3.9 ± 0.1			
	Cu^{2+} / Cr^{3+}	Cu^{2+}	0.16 ± 0.004	4.3 ± 0.1	5.6 ± 0.1	0.945	1.4×10^{-4}
		Cr^{3+}		4.1 ± 0.1			

Table 2. Adjustable parameters for the multi-component continuous equilibrium model (Eq. 7), using only the experimental data points from binary systems (value ± standard error)

Biosorbent	System	Metal	Continuous model: pK'_M	Continuous model: n_M	R^2	S_R^2 (mmol g^{-1})^2
Gelidium	Pb^{2+} / Cu^{2+}	Pb^{2+}	3.8 ± 0.1	0.64 ± 0.03	0.940	7.5×10^{-4}
		Cu^{2+}	3.1 ± 0.1	0.79 ± 0.03		
	Cu^{2+} / Cr^{3+}	Cu^{2+}	3.1 ± 0.1	0.83 ± 0.04	0.938	7.9×10^{-4}
		Cr^{3+}	3.3 ± 0.1	0.67 ± 0.04		
Algal waste	Pb^{2+} / Cu^{2+}	Pb^{2+}	3.9 ± 0.1	0.72 ± 0.04	0.919	4.1×10^{-4}
		Cu^{2+}	3.5 ± 0.1	0.73 ± 0.04		
	Cu^{2+} / Cr^{3+}	Cu^{2+}	3.3 ± 0.1	0.76 ± 0.05	0.934	3.3×10^{-4}
		Cr^{3+}	3.3 ± 0.1	0.74 ± 0.05		
Composite material	Pb^{2+} / Cu^{2+}	Pb^{2+}	4.0 ± 0.1	0.79 ± 0.08	0.869	3.3×10^{-4}
		Cu^{2+}	3.3 ± 0.1	1.0 ± 0.09		
	Cu^{2+} / Cr^{3+}	Cu^{2+}	3.2 ± 0.1	1.0 ± 0.1	0.834	4.5×10^{-4}
		Cr^{3+}	3.1 ± 0.1	1.0 ± 0.1		

The same parameters were also determined using all the experimental data (single-metal and two-metal experiments) for each biosorbent. In this case, the model gives only a value of $pK_M^{'}$ and n_M for each metal and biosorbent (Table 3).

Table 3. Adjustable parameters for the multi-component continuous equilibrium model (Eq. 7), using all the experimental data points for all metal ions (value ± standard error)

Biosorbent	Metal	Continuous model $pK_M^{'}$	Continuous model n_M	$\frac{n_M}{n_H}$	$\frac{n_H}{n_M}$	m_M	R^2	S_R^2 $(\mathrm{mmol\ g^{-1}})^2$
Gelidium	Cu^{2+}	3.1 ± 0.1	0.78 ± 0.03	1.7	0.60	0.71	0.944	5.4×10^{-4}
	Pb^{2+}	3.7 ± 0.1	0.64 ± 0.03	1.4	0.74	0.57		
	Cd^{2+}	2.9 ± 0.1	0.53 ± 0.05	1.1	0.90	0.48		
	Zn^{2+}	2.7 ± 0.3	0.53 ± 0.09	1.1	0.89	0.48		
	Cr^{3+}	3.2 ± 0.3	0.66 ± 0.09	1.4	0.72	0.59		
Algal waste	Cu^{2+}	3.5 ± 0.1	0.74 ± 0.04	1.3	0.76	0.49	0.928	2.9×10^{-4}
	Pb^{2+}	4.2 ± 0.1	0.66 ± 0.04	1.2	0.84	0.44		
	Cd^{2+}	3.2 ± 0.1	0.63 ± 0.05	1.1	0.89	0.42		
	Zn^{2+}	2.9 ± 0.3	0.70 ± 0.09	1.3	0.80	0.46		
	Cr^{3+}	3.5 ± 0.3	0.74 ± 0.09	1.3	0.75	0.49		
Composite material	Cu^{2+}	3.3 ± 0.1	1.0 ± 0.1	1.3	0.75	0.44	0.885	2.0×10^{-4}
	Pb^{2+}	4.0 ± 0.1	0.75 ± 0.08	1.0	1.0	0.33		
	Cd^{2+}	2.9 ± 0.1	0.7 ± 0.1	0.9	1.1	0.30		
	Zn^{2+}	2.8 ± 0.3	0.8 ± 0.1	1.1	0.94	0.35		
	Cr^{3+}	3.2 ± 0.3	1.0 ± 0.1	1.3	0.75	0.44		

The statistical parameters in Table 1 and 2, show that the experimental data can be described by both models. Comparing the two models with the F-test (95% confidence level), it was concluded that, although for some two-metal systems one model is better than the other, no significant statistical difference exists between the two models when considering all the binary systems. So, as it is easier to fit the discrete model, it can be used in the mathematical mass transfer modelling, both in batch and continuous systems. Moreover, as the continuous model considers a heterogeneous distribution of binding sites, obtained by titration experiments, a better interpretation of the experimental results can be obtained with this model.

Fitting all experimental data to the continuous model, a set of parameters for each metal ion in all binary systems (8 parameters) is obtained, instead of a set of parameters for each binary system, which gives four parameters for each two-metal system. Similar average pK_M values were obtained from discrete (Table 1) and continuous (Table 2) models. $pK_M^{'}$ and n_M values obtained from the two regression procedures (Table 2 and 3) are similar, suggesting that metal ions affinity for biosorbents is independent of the co-cation. n_M / n_H values in Table 3 are always higher than unity suggesting that cations M_i are more adsorbed than the proton. This could reflect some degree of cooperativity [11].

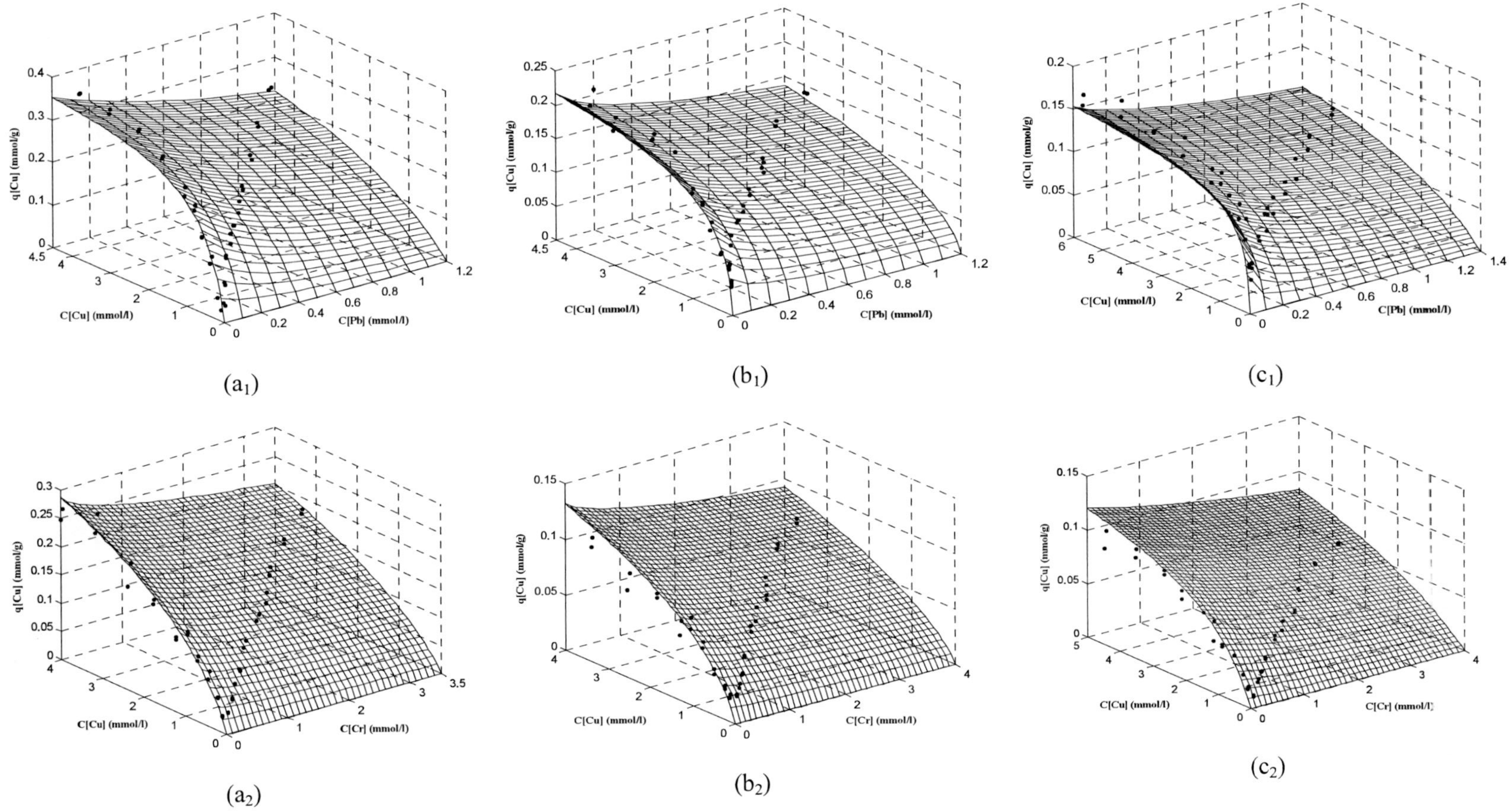

Figure 1. Copper uptake as a function of the equilibrium concentration of: Pb/Cu, pH = 5.3; Cu/Cr, pH = 4.0, and T = 20ºC. (a_i) Algae *Gelidium*, (b_i) algal waste and (c_i) composite material: i = 1 – Pb/Cu; i = 2 – Cu/Cr.

The ratio n_H / n_M has a particularly strong influence on the exchange ratio (r_{ex}) since it is linearly related to it. r_{ex} can change over the range $0 < r_{ex} < n_H / n_M$, the greatest value being found for small p (high heterogeneity), low metal concentration and low pH. This is true when most of the sites are occupied by protons or other ions as potassium, sodium, calcium and magnesium and the binding reaction approaches a "normal" ion exchange reaction. Under these conditions, the ratio n_H / n_M represents some kind of reaction stoichiometry [11]. n_H / n_M values close to one (Table 3) were obtained, suggesting a 1:1 stoichiometry exchange.

The most homogeneous binding constants distribution was obtained for algae *Gelidium*, followed by algal waste and composite material, as given by the highest m_M value (narrowest distribution).

4.2. Co-cations Interferences

The most appropriate and correct way of representing the biosorption equilibrium of a two-metal system is by a three-dimensional (3-D) sorption isotherm plot, whereby the metal uptake is plotted as a function of the other metal equilibrium concentrations at constant pH. Figs. 1 (a_1 and a_2), (b_1 and b_2) and (c_1 and c_2) represent copper uptake as a function of lead and chromium equilibrium concentrations. These surfaces were generated by MATLAB using the continuous model (Table 2) and experimental data points for single and binary systems. It can be concluded that, due to competition to the binding sites, copper adsorption decreases in the presence of Pb(II) and Cr(III).

The 3-D isotherm surfaces are the summary of the two-metal equilibrium results. The selected cuts through the 3-D diagrams presented in Fig. 2 for Cu/Cr, represent the quantitative trends observed in the two-metal sorption systems. These cuts are called as "iso-concentrations", representing the total metal uptake vs. the equilibrium concentration of one metal, for a constant equilibrium concentration of the other metal. For a given C_{Cu}, as C_{Cr} increases, the uptake capacity of copper by the three biosorbents also increases (Fig. 2 (a), (b) and (c)). However, this increase becomes gradually smaller at higher Cr concentrations,

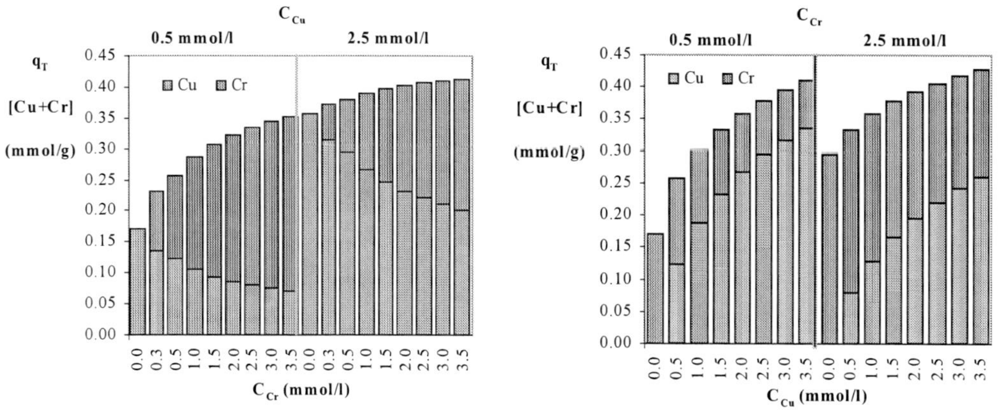

(a)

Figure 2. Continued on next page.

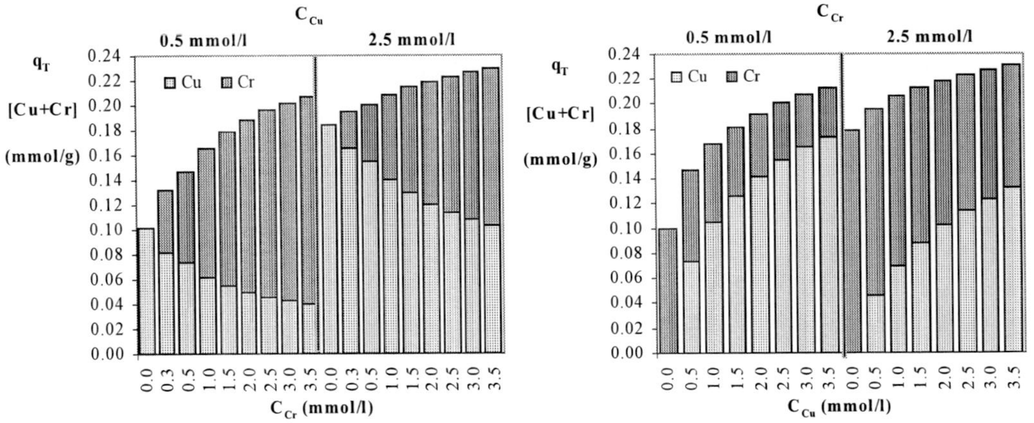

(b)

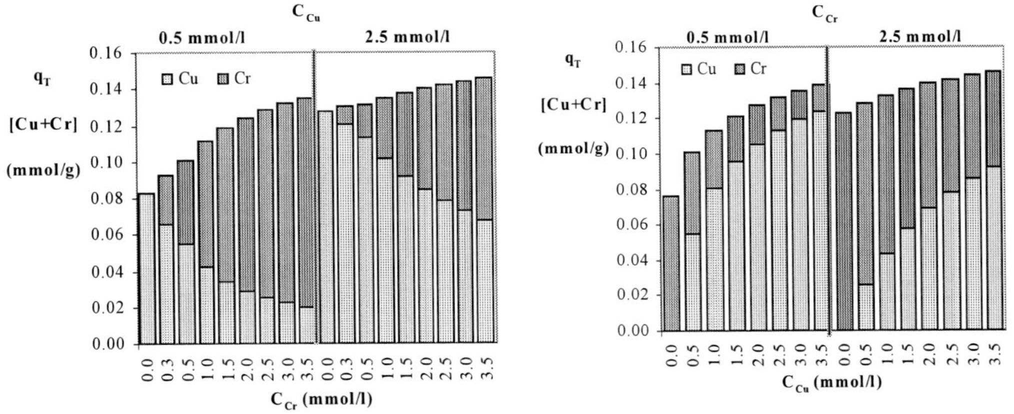

(c)

Figure 2. Iso-concentration cuts of Pb/Cu and Cu/Cr sorption isotherm surfaces: total sorption uptake vs. the equilibrium concentration of one metal, for a constant concentration of the other metal, pH = 4 and 20°C. (a_i) Algae *Gelidium*, (b_i) algal waste and (c_i) composite material: i = 1 – Pb/Cu; i = 2 – Cu/Cr.

probably due to saturation of the active sites. The total metal uptake increases with C_M of one metal, when C_M of the other metal was held constant. For the same Cu(II) and Cr(III) equilibrium concentrations more than 50% of the total metal uptake is due to copper biosorption. The "iso-concentrations" plots for the Pb/Cu binary system were presented in a previous work [6]. For the same equilibrium concentrations of both metal ions, the copper uptake is around 35% of the total binding. The same uptake capacity for both metals was achieved for a copper equilibrium concentration 2.5 times higher than that of lead.

Figure 3 represents the uptake percentage of Cu(II) as a function of the co-cation concentration, at pH 5.3. The curves resulting from the cuts, are well described by the Langmuir model. The Langmuir parameter K_L represents the apparent affinity of the metal ion towards the binding sites and depends on the co-cation and proton equilibrium concentrations and equilibrium affinity constants and also of the primary metal (M_1) equilibrium affinity constant (Eq. (3)). Figure 4 represents the primary metal (copper) K_L values as a function of

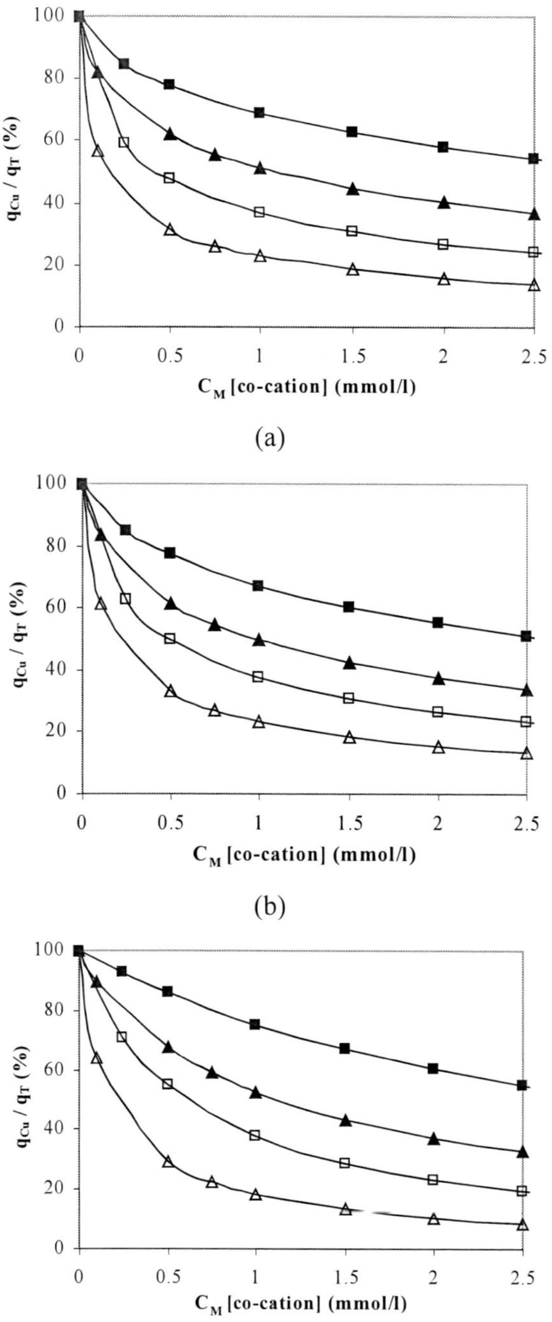

(c)

Figure 3. Effects of co-cation (M_2) on the uptake of the primary metal (M_1):(ρ) C_{M_1} = 0.5 mmol/l, M_2: Pb; (π) C_{M_1} = 2.5 mmol/l, M_2: Pb; (■) C_{M_1} = 0.5 mmol/l, M_2: Cr; (□) C_{M_1} = 2.5 mmol/l, M_2: Cr. (a) Algae *Gelidium*, (b) algal waste and (c) composite material.

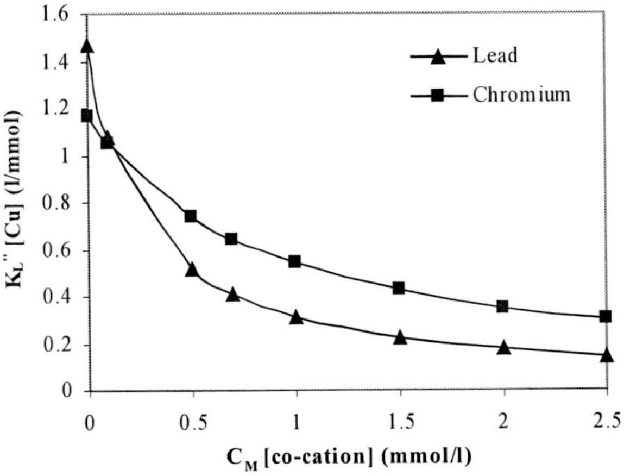

(a)

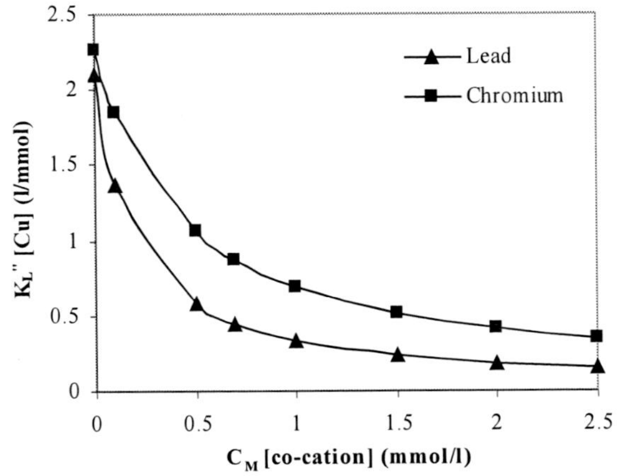

(b)

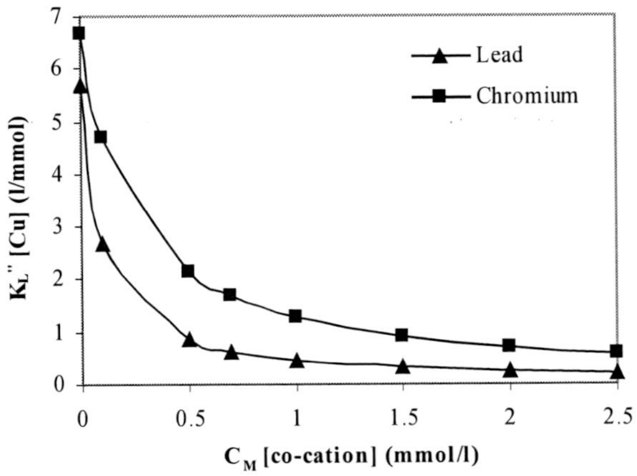

(c)

Figure 4. Effects of co-cation on the apparent Langmuir affinity value, K_L'', for the primary metal given by Eq. (3). (a) Algae *Gelidium*, (b) algal waste and (c) composite material.

the co-cation equilibrium concentration in solution. For this discussion two values were arbitrary chosen to represent "low" (0.5 mmol/l) and "high" (2.5 mmol/l) concentrations. The obtained results suggest that the binding of copper is more sensitive to the presence of lead than that of chromium. Langmuir K_L, values for chromium decrease 50-90% in the presence of 1.0 mmol/l of copper. For high Cu(II) concentrations the adsorbed Cr decreases around 35%. From the results it can be concluded that the biosorbents exhibited a metal preference in the order: Pb > Cu ≈ Cr > Cd ≈ Zn. This order is in agreement with the affinity order (Table 3).

Similar sorption characteristics have been reported by Puranik and Paknikar [15] using *Streptoverticillium cinnamoneum* and *Penicillium chrysogenum* biomass in the biosorption of metal binary systems. In the systems Pb/Cd, Pb/Zn, Pb/Cu, Pb/Ni, Pb/Co, Cu/Zn, Cd/Zn, Ni/Zn and Co/Zn, *Streptoverticillium cinnamoneum* and *Penicillium chrysogenum* exhibited preferential sorption orders Pb > Zn ≈ Cu > Cd > Ni > Co and Pb > Cu > Zn > Ni > Co, respectively.

Silva et. al. [16] studied the adsorption of two metal ions, Cr(III) and Cu(II), as single-component and binary systems, by *Sargassum sp.* The maximum uptake capacities obtained were 1.30 mmol Cr(III) g^{-1} and 1.08 mmol Cu(II) g^{-1} at pH 3.5 and T = 30°C. In the binary system the uptake capacity of Cu(II) showed higher sensibility to the presence of Cr(III) than the opposite, suggesting that Cr(III) has a higher affinity to the biosorbent than Cu(II).

Ma and Tobin (2003) studied Cr(III), Cu(II) and Cd(II) biosorption onto peat from binary metal solutions (Cr/Cd, Cr/Cu and Cu/Cd at pH = 4.0), in batch systems, and concluded that the uptake capacity decreased in the following order: Cr ≥ Cu > Cd. The maximum values were 0.4 mmol g^{-1} for Cr(III) and Cu(II) and 0.2 mmol g^{-1} for Cd(II), giving a total capacity of 0.8 mmol g^{-1}. Co-ion competition resulted in up to 70 percent decrease on primary metal uptake [8].

Kim (2003) used crab shell to remove heavy metals. In single heavy metal ion systems, the removals of Cr(III) and Pb(II) were much higher than that of Cd(II). In binary systems, Cd(II) did not affect Pb(II) removal while Cr(III) produced a severe inhibition effect on Pb(II) removal. Cd(II) uptake was slightly affected by the presence of Pb(II); however, it was severely affected by the presence of Cr(III). The inhibitory effect of Cd(II) on Cr(III) uptake was relatively lower than that of Pb(II) [17].

The three biosorbents used in this work are characterized essentially by the presence of carboxylic and hydroxyl groups [14]. As the pH of the equilibrium experiments is lower than 7, only the carboxylic groups (donor – O) participate in the metal binding.

Some characteristics of the three ions are listed in Table 4. It can be seen that, according to the (z^2 / r_{hyb}) criterion, the strength of ionic bonds is in the order: Cr > Pb > Cu > Cd > Zn, with small differences among metals, except for chromium, which presents an higher value due the high charge. Using the ΔX_m or the $1-exp\left(-\Delta X_m^2/4\right)$ criteria, the binding of Zn is more ionic (electrostatic) than the binding of Cd, Cr, Cu or Pb. The binding of Pb is highly covalent. The parameter $X_m^2\left(r_{cryst}+0.85\right)$ is a measure of the strength of covalent bonding, where 0.85 stands for the contribution of N or O donors to the bond distance [18]. "Hard" ions are characterized by $X_m^2\left(r_{cryst}+0.85\right)$ < ≈ 4.2, and "soft" ions are characterized by $X_m^2\left(r_{cryst}+0.85\right)$ > 7. The $X_m^2\left(r_{cryst}+0.85\right)$ criterion confirms that the relative contribution of covalent bonding is Pb > Cu > Cd > Zn > Cr. All criteria indicate that Zn and Cr should be the cations with the lowest tendency to covalent binding. The sequence of binding strength found for the studied metal ions (Pb > Cu > Cr > Cd > Zn) is probably related with the decrease on the covalent character of the binding and the increase on the hydration strength (z^2 / r_{cris}).

Table 4. Parameters characterizing the metal binding strength

	H	Na	K	Ca	Mg	Zn	Cd	Cr	Cu	Pb
z (charge)	1	1	1	2	2	2	2	3	2	2
r_{cryst} [a] (Å)	---	1.02	1.51	1.00	0.72	0.74	0.95	0.62	0.73	1.19
r_{hyb} [b] (Å)	2.82	3.58	3.31	4.12	4.28	4.30	4.26	4.61	4.19	4.01
X_m [c]	2.20	0.93	0.82	1.00	1.31	1.65	1.69	1.66	1.90	1.8
z^2 / r_{cryst} [d1] (1/Å)	---	0.98	0.66	4.00	5.56	5.41	4.21	14.1	5.48	3.36
z^2 / r_{hyb} [d2] (1/Å)	0.35	0.28	0.30	0.97	0.93	0.93	0.94	1.95	0.95	1.00
$X_m^2\left(r_{cryst}+0.85\right)$ [e] (Å)	---	1.51	1.51	1.85	2.26	4.07	5.20	4.05	6.32	6.61
ΔX_m [f]	1.3	2.6	2.7	2.5	2.3	1.9	1.8	1.8	1.5	1.7
$1-exp\left(-\Delta X_m^2/4\right)$ [g]	0.34	0.82	0.84	0.79	0.73	0.59	0.56	0.56	0.43	0.51
ξ [h] $= z^2/r_{hyb}/1-exp\left(-\Delta X_m^2/4\right)$ (1/Å)	1.03	0.34	0.36	1.23	1.27	1.56	1.69	3.49	2.22	1.94

[a] Shannon crystal radii (r_{cryst}) [19]
[b] Nightingale hydrated ion radii (r_{hyb}) [20]
[c] Pauling electronegativity (X_m) [21]
[d] Parameter for hydration ([1]) or ionic bonding ([2]) strength [22]
[e] Parameter for covalent bond character (0.85 is an appropriate constant assumed to reflect the radius of O and N donor atoms) [18]
[f] Parameter for ionic bond character (Electronegativity of the metal relatively to the oxygen) [21]
[g] Fraction of ionic bond character [23]
[h] Parameter for total binding strength

5. Conclusions

From the binary systems Pb(II)/Cu(II) and Cu(II)/Cr(III), it can be concluded that the three biosorbents bind Pb preferentially to Cu and Cr. The effect of Pb on Cu sorption is strong, relative to the effect of Cr. These inhibitory effects on metal adsorption can be attributed to the characteristic class behaviour of each cation as suggested by Pearson. According to the ΔX_m or the $1-exp\left(-\Delta X_m^2/4\right)$ criteria, the binding of Cr is more ionic (electrostatic) than the binding of Cu or Pb. The use of the Nieboer and Richardson criteria confirm that the relative contribution of covalent bonding follows the order Pb > Cu > Cr.

Acknowledgements

Financial support for this work was in part provided by the national research grant FCT/POCTI/AMB/57616/2004 and by LSRE financing from FEDER/POCI/2010, for which the authors are thankful. V. Vilar acknowledges his Ph.D. scholarship by FCT (SFRH/BD/7054/2001).

References

[1] Vilar VJP, Botelho CMS, Boaventura RAR. Influence of pH, ionic strength and temperature on lead biosorption by *Gelidium* and agar extraction algal waste. *Process Biochemistry* **40**(10): 3267-3275, 2005.

[2] Vilar VJP, Botelho CMS, Boaventura RAR. Equilibrium and kinetic modelling of Cd(II) biosorption by algae *Gelidium* and agar extraction algal waste. *Water Research* **40**(2): 291-302, 2006.

[3] Vilar VJP, Botelho CMS, Boaventura RAR. Kinetics and equilibrium modelling of lead uptake by algae *Gelidium* and algal waste from agar extraction industry. *Journal of Hazardous Materials* **143**(1-2): 396-408, 2007.

[4] Vilar VJP, Botelho CMS, Boaventura RAR. Chromium and Zinc Uptake by Algae Gelidium and Agar Extraction Algal Waste: Kinetics and Equilibrium. *Journal of Hazardous Materials* **149**(3): 643-649, 2007.

[5] Vilar VJP, Botelho CMS, Boaventura RAR. Copper removal by algae *Gelidium*, agar extraction algal waste and granulated algal waste: kinetics and equilibrium. *Bioresource Technology* **99**(4): 750-762, 2008.

[6] Vilar VJP, Botelho CMS, Boaventura RAR. Effect of Cu(II), Cd(II) and Zn(II) on Pb(II) biosorption by algae *Gelidium*-derived materials. *Journal of Hazardous Materials* **154** (711-720, 2008.

[7] Vilar VJP, Botelho CMS, Boaventura RAR. Interferences of Zn^{2+} and Pb^{2+} on Cd^{2+} biosorption onto algae *Gelidium*, algal waste and a composite material. In: Brown SE, Welton WC, eds. Heavy Metal Pollution,vol New York, Nova Science Publishers, 2008.

[8] Ma W, Tobin JM. Development of Multimetal Binding Model and Application to Binary Metal Biosorption Onto Peat Biomass. *Water Research* **37**(3967-3977, 2003.

[9] Vilar VJP, Sebesta F, Botelho CMS, Boaventura RAR. Equilibrium and kinetic modelling of Pb^{2+} biosorption by granulated agar extraction algal waste. *Process Biochemistry* **40**(10): 3276-3284, 2005.

[10] Sips R. On the Structure of a Catalyst Surface. *Journal of Chemical Physics* **16**(490-495, 1948.

[11] Kinniburgh DG, Riemsdijk WHv, Koopal LK, Borkovec M, Benedetti MF, Avena MJ. Ion Binding to Natural Organic Matter: Competition, Heterogeneity, Stoichiometry and Thermodynamic Consistency. *Colloids and Surfaces A: Physicochemical and Engineering Aspects* **151**(147-166, 1999.

[12] Koopal LK, Saito T, Pinheiro JP, Riemsdijk WHv. Ion binding to natural organic matter: General considerations and the NICA-Donnan model. *Colloids and Surfaces A: Physicochemical and Engineering Aspects* **265**(40-54, 2005.

[13] Chapra SC, Canale RP. Numerical Methods for Engineers. McGraw-Hill, 1998.

[14] Vilar VJP, Botelho CMS, Boaventura RAR. Modelling Equilibrium and Kinetics of Metal Uptake by Algal Biomass in Continuous Stirred and Packed Bed Adsorbers *Adsorption* **13**(5-6): 587-601, 2007.

[15] Puranik PR, Paknikar KM. Influence of Co-Cations on Biosorption of Lead and Zinc-a Comparative Evaluation in Binary and Multimetal Systems. *Bioresource Technology* **70**(269-276, 1999.

[16] Silva EA, Cossich ES, Tavares CG, Filho LC, Guirardello R. Biosorption of Binary Mixtures of Cr(III) and Cu(II) Ions by *Sargassum sp. Brazilian Journal Chemical Engineering* **20**(3): 1-22, 2003.

[17] Kim DS. The Removal by Crab Shell of Mixed Heavy Metal ions in Aqueous Solution. *Bioresource Technology* **87**(355-357, 2003.

[18] Nieboer E, McBryde WAE. Free-Energy Relationship in Coordination Chemistry. III. A Comprehensive Index to Complex Stability. *Canadian Journal Chemistry* **51**(2512-2524, 1973.

[19] Evans HT. *Ionic Radii in Crystals*. Boca Raton, In: Lide, D.R. (Ed.), CRC Handbook of Chemistry and Physics, CRC press, 1993.

[20] Marcus Y, Kertes AS. *Ion Echange and Solvent Extraction of Metal Complexes.* London, Wiley Interscience, John Wiley & Sons, 1969.

[21] Dean JA. Lange´s *Handbook of Chemistry*. New York, McGraw-Hill, 1999.

[22] Phillips CSG, Williams RJP. Inorganic Chemistry. *Principles and Non-Metals.* New York, Oxford University Press, 1965.

[23] Pauling L. *Nature of the Chemical Bond.* Ithaca, NY, Cornell University Press, 1967.

In: Industrial Waste...
Editor: John P. Samuelson, pp. 387-391
ISBN: 978-1-60692-720-5

NOVEL APPROACHES IN PHOSPHATE-FERTILIZER PRODUCTION BASED ON WASTES DERIVED FROM ROCK PHOSPHATE MINING AND THE FOOD PROCESSING INDUSTRY

Nikolay Vassilev[1,2,*], *Edward Someus*[3], *Mercedes Serrano, Vicente Bravo, Miguel Garcia Roman, Antonia Reyes and Maria Vassileva*

[1] Department of Chemical Engineering, Faculty of Sciences, University of Granada, c/Fuentenueva s/n, Granada-18071, Spain
[2] Institute of Biotechnology, University of Granada, Spain
[3] Terra Humana Ltd, Hungary-Sweden

Abstract

Phosphorus is a vital nutrient for all life forms, but is a finite non-renewable element with a progressively declining availability and non-sustainable exploitation rate. The Commentary deals with novel trends in exploring alternative phosphate-bearing sources and corresponding treatment techniques aimed at substituting for the existing resources and environmentally damaging chemical processing. Approaches are presented that include biological treatment of low-grade (waste) phosphate rocks deriving from the mining industry, as well as biosolubilization of industrial ashes of animal bone meal. The novel biotechnology-based processing may employ a wide range of agro-industrial residues as substrates in fermentation operations.

The significant growth in worldwide raw materials consumption may lead to supply bottlenecks in the future, at least in the case of certain raw materials, such as rock phosphate

* Corresponding author: Prof. Nikolay Vasslev, Department of Chemical Engineering, Faculty of Sciences, University of Granada, c/Fuentenueva s/n, Granada-18071, Spain

(RP) which is used in fertilizer industry. What is the actual situation regarding RP processing, P fertilizer production, application, and soil P agricultural importance?

Although phosphorus (P) is quite abundant in many soils, it is one of the major nutrients limiting plant growth. P is added to soil in the form of phosphate fertilizers, but the overall P use efficiency is low because although a fraction of soluble P is utilized by plants, the rest rapidly forms insoluble complexes with soil constituents. Therefore, frequent application of soluble forms of inorganic P is needed, well above what would be really necessary under ideal conditions. Even under adequate P fertilization, only 20% or less of that applied is removed by the first year's growth. As the capacity of soil to bind P is limited, this means that many soils receive P in excess of crop requirements, which often results in its leaching into the groundwater. The runoff from P-loaded soil is accepted as the main factor in eutrophication of natural water reservoirs (Del Campillo et al. 1999). Traditional P fertilizer production is based on chemical processing of insoluble mineral phosphate high-grade ore, which includes an energy-intensive treatment with sulfuric acid at high temperature. This process is environmentally undesirable, not least because of the release of contaminants into the main product, gas streams and by-products. It should be noted that a lower concentration of phosphate in RP and lower quality deposit generate more waste materials and, moreover, more energy and chemicals are required per tonne of useful phosphate produced.

The major parts of the phosphate minerals in RPs are, in fact, species of apatite, which is calcium phosphate in varying combinations with quartz, calcite, dolomite, clay and iron oxide. Phosphate rock often contains carbonates and alkalic compounds. Phosphate content in mined RPs can range from over 40% to below 5%. The rock phosphate is further processed to remove the bulk of the contained impurities, thus improving the rock phosphate quality. Consequently, the resulting concentrate contains improved apatite content and characteristics. The content of P_2O_5 after beneficiation of phosphate rock ranges from 26% to about 34% and up to as much as 42%. Low-grade RPs are normally considered waste. High-grade ores with high P_2O_5 concentration and ores of good quality, i.e., containing few contaminants, are being progressively depleted (Vance, 2001).

Three facts should be taken into account when assessing the current situation in this field. One of the main reasons for concern in both the RP-mining and P-fertilizer industries is that, by some calculations, inexpensive, high-grade rock phosphate reserves could be exhausted in as little as 60 to 80 years, as phosphate-bearing ore is a finite non-renewable resource (Runge-Metzger, 1995). P fertilizer use increased four- to five-fold between 1960 and 2000 and is projected to increase in the first three decades of the twenty-first century by 20 Tg per year. Second, there are no substitutes for phosphorus in agriculture but, on the other hand, phosphorus can be recycled mainly by use of animal manure and sewage sludge (although there are serious concerns about these alternatives). Finally, it is important to note that the cultivation soil available per capita on a global level is lowering while the world population is increasing by 250,000 people every day (approximately 80 million per year). This situation will result in enhanced need for food and, consequently, phosphate demand will increase. Therefore, there is an urgent need to find novel phosphate sources and/or novel techniques for ensuring phosphate availability. Here, we would like to mention some additional approaches aimed at improving the sustainable use of phosphate-bearing sources.

One of the most studied approaches is the biological treatment of low-grade rock phosphates and particularly microbial solubilization of insoluble phosphates. In recent years, various techniques for RP solubilization have been proposed, with increasing emphasis on application of P-solubilizing microorganisms (Rodriguez and Fraga 1999; Whitelaw 2000; Vassilev et al. 2001). It is well known that microorganisms play a central role in the natural P cycle (Gadd 1999). A number of in vitro studies have shown that bacteria, fungi and actinomycetes are able to liberate phosphate ions from sparingly soluble inorganic P-bearing compounds. The P-solubilizing activity is determined by the microbial biochemical's ability to produce and release metabolites such as organic acids that, through their hydroxyl and carboxyl groups, chelate the cations (mainly calcium) bound to phosphate, the latter being converted into soluble forms (Kpomblekou and Tabatabai, 1994). However, metabolizable C compounds must be applied as energy source to the microbial solubilizers to ensure their growth, organic acid production, and, simultaneously, RP solubilization. In a number of studies, we have shown that wastes from the food industry (production of sugar, olive oil, wine, fruit juices, etc.) can be efficiently used as substrates in solubilization of low-grade (12-13% phosphate) phosphate rocks (Vassilev et al., 2003, 2006). Further introduction of the resulting biotechnological products into soil-plant systems resulted in significantly higher plant growth and enhanced soil properties and biological (including biocontrol) activity. These processes were carried out employing various technologies, including solid-state fermentations, composting, and submerged fermentations (in the case of liquid food-industry wastes). Goldstein (2000) recently proposed application of gluconic-acid-producing fungi in heap bioreactors, spraying a cheap substrate to support microbial growth and P solubilization. It is also recognized that agro-industrial wastes, due to their low price, are the most important route towards realization of cost-effective biotechnological solubilization of RP (Vassilev and Vassileva, 2003).

Another very attractive approach includes utilization of residues of combusted animal bones. Until recently, residual materials of meat production have been efficiently used as high-protein components of animal feed. However, as a result of the bovine spongiform encephalopathy (BSE) crisis in the European meat industry, the use of these animal wastes is strictly controlled. The main disposal option is incineration, producing huge amounts of ashes, the valorization of which becomes a major concern. Industrial ashes of meat and bone meal are characterized by high calcium and phosphorus content (up to 47%), as the minimal part of bone of a higher animal is composed of calcium and phosphate in a fixed ratio. Therefore, apart from the efforts to employ bone meal for its energy content (approximately 17 000 kJ per kg), valorization of these wastes can be realized by application in the fertilizer industry. So far, soil enrichment with bone meal ashes has shown very efficient results in remediation of heavy metal contaminated sites (Hodson et al., 2000). Similar results were achieved in treatment of metal-contaminated aqueous effluents (Deydier et al., 2003). However, we do hope that in the near future studies will be focused on the use of industrial bone meal ashes and bone char as novel and efficient sources of phosphate. One of the most important advantages of heat-treated bones is their low metal content. While rock phosphates contain cadmium, lead, cooper, arsenic, and other heavy metals, the latter cannot be detected in bone char and bone ashes (Deydier et al., 2005). Therefore, the use of these novel P-sources in agricultural soils will avoid contamination with metals and particularly with cadmium. It is a question of time whether a radical change will occur in the P-fertilizer industry that will be based on the use of non-traditional and inexpensive sources of phosphorus, mainly industrial wastes.

Acknowledgements

This work was supported by Project MEC CTM2005-06955, R&C Programme, and the EU "Waste-to-Products" project.

References

Del Campillo, S.E., Van der Zee, S.E.A.T.M. & Torrent, J. (1999) Modelling long-term phosphorus leaching and changes in phosphorus fertility in excessively fertilized acid sandy soils. *European Journal of Soil Science,* **50**, 391–399

Deydier, E., Guilet, R. & Scharrock, P. (2003) Beneficial use of meat and bone meal combustion residue: an efficient low cost material to remove lead from aqueous effluent. *Journal of Hazardous Materials B,* **101**, 55–64.

Deydier, E., Guilet, R., Sarda, S. & Sharrock, P. (2005) Physical and chemical characterisation of crude meat and bone meal combustion residue: "waste or raw material?" *Journal of Hazardous Materials* **B121**, 141–148

Hodson, M., Valsami-Jones, E. & Cotter-Howells, J.D. (2000) Bone meal additions as a remediation treatment for metal contaminated soil. *Environmental Science and Technology,* **34**, 3501–3507.

Gadd, G. (1999) Fungal production of citric and oxalic acid: Importance of metal specification, physiology and biogeochemical processes. *Advances in Microbial Physiology,* **44**, 47–92

Goldstein, A.H. (2000) Bioprocessing of rock phosphate ore: essential technical considerations for the development of a successful commercial technology. IFA Technical Conference, New Orleans, La., pp 1–21, http://goldsteinlab.alfred.edu/publications.html

Kpomblekou, K. & Tabatabai, M.A. (1994) Effect of organic acids on release of phosphorus from phosphate rocks. *Soil Science,* **158**, 442–453

Rodriguez, H. & Fraga, R. (1999) Phosphate solubilizing bacteria and their role in plant growth promotion. *Biotechnology Advances,* **17**, 319–339

Runge-Metzger, A. (1995) Closing the cycle: obstacles to efficient P management for improved global security. *In* H Tiessen, ed, Phosphorus in the Global Environment. John Wiley and Sons Ltd, Chichester, UK, pp 27–42

Vance, C. P. (2001) Symbiotic nitrogen fixation and phosphorus acquisition. Plant nutrition in a world of declining renewable resources. *Plant Physiology,* **127**, 390–397

Vassilev, N., Vassileva, M., Fenice, M. & Federici, F. (2001) Immobilized cell technology applied in solubilization of insoluble inorganic (rock) phosphates and P plant acquisition. *Bioresource Technology,* **79**, 263–271

Vassilev, N. & Vassileva, M. (2003) Biotechnological solubilization of mineral phosphates on media containing agro-industrial wastes. *Applied Microbiology and Biotechnology,* **61**, 435-440.

Vassilev N, Vassileva M & Nikolaeva I (2006) Simultaneous P-solubilizing and biocontrol activity of microorganisms: Potentials and future trends. *Applied Microbiology and Biotechnology,* **71**, 137-144

Whitelaw, M.A. (2000) Growth promotion of plants inoculated with phosphate-solubilizing fungi. *Advances in Agronomy,* **69**, 99–151.

The following publications could be useful for the reader:

Garcia R. A., Rosentrater K. A. (2008) Concentration of key elements in North American meat & bone meal. *Biomass and Bioenergy*, doi:10.1016/j.biombioe.2007.12.011

Haarr A. (2004) The reuse of Phosphorus. Eureau position paper EU2-04-SL09 European Economic and Social Committee (2006*) Risks and problems associated with the supply of raw materials to European industry*. CCMI/028 - CESE 964/2006

INDEX

A

B

C

D

E

F

G

H

I

J

K

L

M

N

O

P

Q

R

S

T

U

V

W

X

Y

Z